Theoretical and Mathematical Physics

The series founded in 1975 and formerly (until 2005) entitled *Texts and Monographs in Physics* (TMP) publishes high-level monographs in theoretical and mathematical physics. The change of title to *Theoretical and Mathematical Physics* (TMP) signals that the series is a suitable publication platform for both the mathematical and the theoretical physicist. The wider scope of the series is reflected by the composition of the editorial board, comprising both physicists and mathematicians.

The books, written in a didactic style and containing a certain amount of elementary background material, bridge the gap between advanced textbooks and research monographs. They can thus serve as basis for advanced studies, not only for lectures and seminars at graduate level, but also for scientists entering a field of research.

Akihito Hora   Nobuaki Obata

Quantum Probability and Spectral Analysis of Graphs

With a Foreword by Professor Luigi Accardi

With 48 Figures

Springer

Professor Dr. Akihito Hora
Graduate School of Mathematics
Nagoya Universtiy
Nagoya 464-8602, Japan

Professor Dr. Nobuaki Obata
Graduate School of Information Sciences
Tohoku University
Sendai 980-8579, Japan

Akihito Hora and Nobuaki Obata, *Quantum Probability and Spectral Analysis of Graphs*, Theoretical and Mathematical Physics (Springer, Berlin Heidelberg 2007)
DOI 10.1007/b11501497

Library of Congress Control Number: 2006940905

ISSN 0172-5998
ISBN-13 978-3-540-48862-0 Springer Berlin Heidelberg New York

Springer is a part of Springer Science+Business Media
springer.com

Typesetting: by the authors and techbooks using a Springer LaTeX macro package
Cover design: eStudio Calamar, Girona/Spain

Printed on acid-free paper SPIN: 11501497 55/techbooks 5 4 3 2 1 0

Foreword

It is a great pleasure for me that the new Springer Quantum Probability Programme is opened by the present monograph of Akihito Hora and Nobuaki Obata.

In fact this book epitomizes several distinctive features of contemporary quantum probability: First of all the use of specific quantum probabilistic techniques to bring original and quite non-trivial contributions to problems with an old history and on which a huge literature exists, both independent of quantum probability. Second, but not less important, the ability to create several bridges among different branches of mathematics apparently far from one another such as the theory of orthogonal polynomials and graph theory, Nevanlinna's theory and the theory of representations of the symmetric group.

Moreover, the main topic of the present monograph, the asymptotic behaviour of large graphs, is acquiring a growing importance in a multiplicity of applications to several different fields, from solid state physics to complex networks, from biology to telecommunications and operation research, to combinatorial optimization. This creates a potential audience for the present book which goes far beyond the mathematicians and includes physicists, engineers of several different branches, as well as biologists and economists.

From the mathematical point of view, the use of sophisticated analytical tools to draw conclusions on discrete structures, such as, graphs, is particularly appealing. The use of analysis, the science of the continuum, to discover non-trivial properties of discrete structures has an established tradition in number theory, but in graph theory it constitutes a relatively recent trend and there are few doubts that this trend will expand to an extent comparable to what we find in the theory of numbers.

Two main ideas of quantum probability form the unifying framework of the present book:

1. The quantum decomposition of a classical random variable.
2. The existence of a multiplicity of notions of quantum stochastic independence.

The authors establish original and fruitful connections between these ideas and graph theory by considering the adjacency matrix of a graph as a classical random variable and then by decomposing it in two different ways:

(i) either using its quantum decomposition;
(ii) or decomposing it into a sum of independent quantum random variables (for some notion of quantum independence).

The former method has a universal applicability but depends on the choice of a *stratification* of the given graph. The latter is applicable only to special types of graphs (those which can be obtained from other graphs by applying some notion of product) but does not depend on special choices.

In both cases these decompositions allow to reduce many problems related to the asymptotics of large graphs to traditional probabilistic problems such as quantum laws of large numbers, quantum central limit theorems, etc. Given the central role of these two decompositions in the present volume, it is maybe useful for the reader to add some intuitive and qualitative information about them.

The quantum decomposition of a classical random variable, like many other important mathematical ideas, has a long history. Its first examples, the representation of the Gaussian and Poisson measures on $\mathbb{R}^d$ in terms of creation and annihilation operators, were routinely used in various fields of quantum theory, in particular quantum optics. Its continuous extension, obtained by the usual second quantization functor, played a fundamental role in Hudson–Parthasarathy quantum stochastic calculus and a few additional examples, going beyond the Gaussian and Poisson family appeared in the early 1990s in papers by Bożejko and Speicher.

However, the realization that the quantum decomposition of a classical random variable is a universal phenomenon in the category of random variables with moments of all orders came up only in connection with the development of the theory of interacting Fock spaces. This theory provided the natural conceptual framework to interpret the famous Jacobi relation for orthogonal polynomials in terms of a new class of creation, annihilation and preservation operators generalizing in a natural way the corresponding objects in quantum mechanics.

Most of the present monograph deals with the quantum decomposition of a single real valued random variable for which the quantum decomposition is just a re-interpretation of the Jacobi relation. The situation radically changes for $\mathbb{R}^d$-valued random variables with $d \geq 2$ for which a natural (i.e. intrinsic) extension of the Jacobi relation could only be formulated in terms of interacting Fock space.

An interesting discovery of the authors of the present book is that examples of this more complex situation also arise in connections with graph theory. This will be surely a direction of further developments for the theory developed in the present monograph.

The intimately related notions of quantum decomposition of a classical random variable and of interacting Fock space have been up to now two of the most fruitful and far reaching new ideas introduced by quantum probability. The authors of the present monograph have developed in the past years a new approach to a traditional problem of mathematics, the asymptotics of large graphs, which puts to use in an original and creative way both the above-mentioned notions.

The results of their efforts enjoy the typical merits of inspiring mathematics: elegance and depth. In fact a vast multiplicity of results, previously obtained at the cost of lengthy and ad hoc calculations or complicated combinatorial arguments, are now obtained through a unified method based on the common intuition that the quantum decomposition of the adjacency matrix of the limit graph should be the limit of the quantum decompositions of the adjacency matrices of the approximating graphs. This limit procedure involves central limit theorems which, in the previous approaches to the asymptotics of large graphs, were proved within the context of classical probability. In the present monograph they are proved in their full quantum form and not just in their reduced classical (or semiclassical) form. This produces the usual advantage of quantum central limit theorems with respect to classical ones namely that, by considering various types of self-adjoint linear combinations of the quantum random variables, one obtains the corresponding central limit theorem for the resulting classical random variable.

Thus in some sense a quantum central limit theorem is equivalent to infinitely many classical central limit theorems. This additional degree of freedom was little appreciated in the early quantum central limit theorems, concerning Boson, Fermion, q-deformed, free random variables, because, before the discovery of the universality of the quantum decomposition of classical random variables, a change in the coefficients of the linear combination, could imply a radical change (i.e., not limited to a simple change of parameters within the same family) in the limit classical distribution, only at some critical values of the parameters (e.g., if a^+, a^- are Boson Fock random variables, then independently of z the Boson Fock vacuum distribution of $za^+ + \bar{z}a^- + \lambda a^+a^-$ is Gaussian for $\lambda = 0$ and Poisson for $\lambda \neq 0$).

The emergence of the interacting Fock space produced the first examples (due to Lu) in which a continuous interpolation between radically different measures could occur by continuous variations of the coefficients of the linear combinations of a^+ and a^-. This bring us to the second deep and totally unexpected connection between quantum probability and graphs, which is investigated in the present monograph starting from Chap. 8. To explain this idea let us recall that one of the basic tenets of quantum probability since its development in the early 1970s has been the multiplicity of notions of independence. The first examples beyond classical independence (Bose and Fermi independence) where motivated by physics and the first notions of independence going beyond these physically motivated ones were introduced by von

Waldenfels in the early 1970s. However, it is only with the birth of free probability, in the late 1980s, that the notions of stochastic independence begin to proliferate and to motivate theoretical investigations trying to unify them within some common framework.

An important step in this direction, because of its constructive and not merely descriptive nature, was Lenczewski's tensor representation of the Boolean m-free and free independence, extended to the monotone case by Franz and Muraki (this extension was also implicitly used in an earlier paper by Liebscher). This tensor representation turned out to be absolutely crucial in the connection between notions of independence and graphs, which can be described by the following general abstract ansatz: 'there exist many different notions of products among graphs and, if π is such a notion, the adjacency matrix of a π-product of two graphs can be decomposed as a non-trivial sum of $\mathcal{I}_\pi$-independent quantum random variables where $\mathcal{I}_\pi$ denotes a notion of independence determined by the product π and by a vector in the l^2-space of the graph'. It is then natural to call this decomposition the π-decomposition of the adjacency matrix of the product graph.

Comparing this with a folklore ansatz of quantum probability, namely: 'to every notion of π-product among algebras, one can associate a notion $\mathcal{I}_\pi$ of stochastic independence' one understands that the analogy between the two statements is a natural fact because, by exploiting the equivalence (of categories) between sets and complex valued functions on them, one can always translate a notion of product of graphs into a notion of product of algebras and conversely.

Historically, the first example which motivated the above-mentioned ansatz was the discovery that the adjacency matrix of a comb product of a graph with a rooted graph can be decomposed as the sum of two monotone independent random variables (with respect to a natural product vector). In other words: the above ansatz is true if π is the comb product among graphs and $\mathcal{I}_\pi$ the notion of monotone independence. In addition the π-decomposition of the adjacency matrix is nothing but a particular realization of the tensor representation of two monotone independent random variables.

As expected, if π is the usual cartesian product the corresponding independence notion $\mathcal{I}_\pi$ is the usual tensor (or classical) independence. The fact that, if π is the star-product of rooted graphs, then the associated notion of independence $\mathcal{I}_\pi$ is Boolean independence was realized in a short time by a number of people. Strangely enough the fourth notion of independence in Schürman's axiomatization, i.e. free independence, was the hardest one to relate to a product of graphs in the sense of the above ansatz. This is strange because the free product of graphs was introduced by Znoĭko about 30 years ago and then studied by many authors, in particular Gutkin and Quenell, thus it would have been natural to conjecture that the free product of graphs should be related to free independence.

That this is true has been realized only recently, but the relation is not as simple as in the case of the previous three independences. In fact, in the formerly known cases, the adjacency matrix of the π-product of two graphs was decomposed into a sum of two $\mathcal{I}_\pi$-independent quantum random variables, but in the free case the π-decomposition involves infinitely many free independent random variables. Another special feature of the free product is that it can be expressed by 'combining together' (in some technical sense) the comb (monotone) and the star (Boolean) products.

These arguments are not dealt with in the present book because fortunately the authors realized that, if one decides to include all the important latest developments in a field evolving at the pace of quantum probability, then the present monograph would have become a Godot.

Another important quality of the present volume is the authors' ability to condensate a remarkably large amount of information in a clear and self–contained way. In the structure of this book one can clearly distinguish three parts, approximatively of the same length (about 100 pages). The first part introduces all the basic notions of quantum probability, analysis and graph theory used in the following. The second part (from Chaps. 4 to 8) deals with different types of graphs and the last part (from Chaps. 9 to 12) includes an introduction to Kerov's theory of the asymptotics of the representations of the permutation group $S(N)$, for large N, and the extensions of this theory in various directions, due to various authors themselves and other researchers.

The clarity of exposition, the ability to keep the route firmly aimed towards the essential issues, without digressions on inessential details, the wealth of information and the abundance of new results make the present monograph a precious reference as well as an intriguing source of inspiration for all those who are interested in the asymptotics of large graphs as well as in any of the multiple applications of this theory.

Roma
December, 2006

Luigi Accardi

Preface

Quantum probability theory provides a framework of extending the measure-theoretical (Kolmogorovian) probability theory. The idea traces back to von Neumann [219], who, aiming at the mathematical foundation for the statistical questions in quantum mechanics, initiated a parallel theory by making a self-adjoint operator and a trace play the roles of a random variable and a probability measure, respectively. During the recent development, quantum probability theory has been related to various fields of mathematical sciences beyond the original purposes. We focus in this book on the spectral analysis of a large graph (or of a growing graph) and show how the quantum probabilistic techniques are applied, especially, for the study of asymptotics of spectral distributions in terms of quantum central limit theorem.

Let us explain our basic idea with the simplest example. The coin-toss is modelled by a Bernoulli random variable X specified by

$$P(X = +1) = P(X = -1) = \frac{1}{2}, \tag{0.1}$$

or more essentially by its distribution, i.e., the probability measure μ defined by

$$\mu = \frac{1}{2}\,\delta_{-1} + \frac{1}{2}\,\delta_{+1}. \tag{0.2}$$

The moment sequence is one of the most fundamental characteristics of a probability measure. For μ in (0.2) the moment sequence is calculated with no difficulty as

$$M_m(\mu) = \int_{-\infty}^{+\infty} x^m \mu(dx) = \begin{cases} 1, & \text{if } m \text{ is even}, \\ 0, & \text{otherwise}. \end{cases} \tag{0.3}$$

When we wish to recover a probability measure from the moment sequence, we meet in general a delicate problem called *determinate moment problem*. For the coin-toss there is no such an obstacle and we can recover the Bernoulli distribution from the moment sequence.

Now we discuss, somehow abruptly, elementary linear algebra. We set

$$A = \begin{pmatrix} 0 & 1 \\ 1 & 0 \end{pmatrix}, \qquad e_0 = \begin{pmatrix} 0 \\ 1 \end{pmatrix}, \qquad e_1 = \begin{pmatrix} 1 \\ 0 \end{pmatrix}. \tag{0.4}$$

Then $\{e_0, e_1\}$ is an orthonormal basis of the two-dimensional Hilbert space $\mathbb{C}^2$ and A is a self-adjoint operator acting on it. It is straightforward to see that

$$\langle e_0, A^m e_0 \rangle = \begin{cases} 1, & \text{if } m \text{ is even}, \\ 0, & \text{otherwise}, \end{cases} \tag{0.5}$$

which coincides with (0.3). In other words, the coin-toss is also modelled by using the two-dimensional Hilbert space $\mathbb{C}^2$ and the matrix A. In our terminology, letting $\mathcal{A}$ be the $*$-algebra generated by A, the coin-toss is modelled by an algebraic random variable A in an algebraic probability space $(\mathcal{A}, e_0)$. We call A an *algebraic realization* of the random variable X.

Once we come to an algebraic realization of a classical random variable, we are naturally led to the non-commutative paradigm. Let us consider the decomposition

$$A = A^+ + A^- = \begin{pmatrix} 0 & 1 \\ 0 & 0 \end{pmatrix} + \begin{pmatrix} 0 & 0 \\ 1 & 0 \end{pmatrix}, \tag{0.6}$$

which yields a simple proof of (0.5). In fact, note first that

$$\langle e_0, A^m e_0 \rangle = \langle e_0, (A^+ + A^-)^m e_0 \rangle = \sum_{\epsilon_1, \dots, \epsilon_m \in \{\pm\}} \langle e_0, A^{\epsilon_m} \cdots A^{\epsilon_1} e_0 \rangle. \tag{0.7}$$

Let $\mathcal{G}$ be a connected graph consisting of two vertices e_0, e_1. Observing the obvious fact that (0.7) coincides with the number of m-step walks starting at and terminating at e_0 (see the figure below), we obtain (0.5).

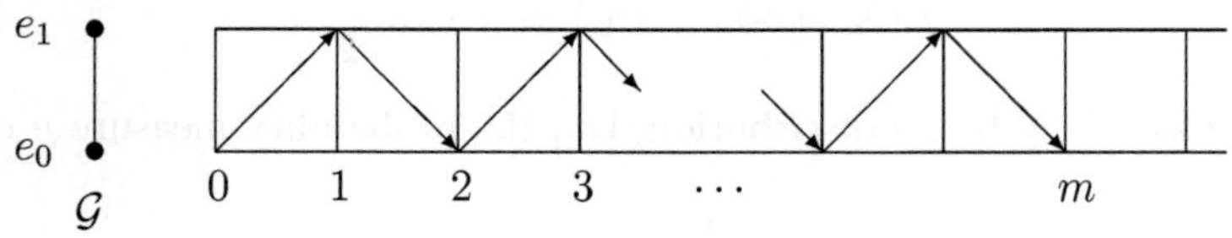

Thus, the computation of the mth moment of A is reduced to counting the number of certain walks in a graph through (0.6). This decomposition is in some sense canonical and is called the *quantum decomposition* of A.

We now note that A in (0.4) is the adjacency matrix of the graph $\mathcal{G}$. Having established the identity

$$\langle e_0, A^m e_0 \rangle = \int_{-\infty}^{+\infty} x^m \mu(dx), \qquad m = 1, 2, \dots, \tag{0.8}$$

we say that μ is the spectral distribution of A in the state e_0. In other words, we obtain an integral expression for the number of returning walks in the graph by means of such a spectral distribution. A key role in deriving (0.8) is again played by the quantum decomposition.

The *method of quantum decomposition* is the central topic of this book. Given a classical random variable, or a probability distribution, we consider the associated orthogonal polynomials. We then introduce the quantum decomposition through the famous three-term recurrence relation and come to the fundamental link with an interacting Fock probability space, which is one of the most basic algebraic probability space. On this basis we shall develop spectral analysis of a graph by regarding the adjacency matrix as an algebraic random variable and illustrate with many concrete examples usefulness of the method of quantum decomposition. Our method is effective especially for the asymptotic spectral analysis and the results are formulated in terms of *quantum central limit theorems*, where our target is not a single graph but a growing graph. Making a sharp contrast with the so-called harmonic analysis on discrete structures, our approach shares a common spirit with the asymptotic combinatorics proposed by Vershik and is expected to contribute also the interdisciplinary study of evolution of networks. Spectral analysis of large graphs is an interesting field in itself, which has a wide range of communications with other disciplines. At the same time it enables us to see pleasant aspects in which quantum probability essentially meets profound classical analysis.

This book is organized as follows: Chapter 1 is devoted to assembling basic notions and notations in quantum probability theory. A special emphasis is placed on the interplay between interacting Fock probability spaces and orthogonal polynomials. The Stieltjes transform and its continued fraction expansion is concisely and self-containedly reviewed.

Chapter 2 gives a short introduction to graph theory and explains our main questions. The idea of quantum decomposition is applied to the adjacency matrix of a graph.

Chapter 3 deals with distance-regular graphs which possess a significant property from the viewpoint of quantum decomposition. We shall establish general framework for asymptotic spectral distributions of the adjacency matrix and derive the limit distributions in terms of intersection numbers.

Chapter 4 analyses homogeneous trees as the first concrete example of growing distance-regular graphs. We shall derive the Wigner semicircle law from the vacuum state and the free Poisson distribution from the deformed vacuum state. The former is a reproduction of the free central limit theorem.

Chapter 5 studies the Hamming graphs which form a growing distance-regular graph. Both Gaussian and Poisson distributions emerge as the central limit distributions.

Chapter 6 discusses the Johnson graphs and odd graphs as further examples of growing distance-regular graphs. As the central limit distributions, we shall obtain the exponential distribution and the geometric distribution from the Johnson graphs, and the two-sided Rayleigh distribution from the odd graphs.

Chapter 7 focuses on growing regular graphs. We shall prove the central limit theorem under some natural conditions, which cover many concrete examples.

Chapter 8 surveys four basic notions of independence in quantum probability theory. The adjacency matrix of an integer lattice is decomposed into a sum of commutative independent random variables, which is also observed through Fourier transform. While, the adjacency matrix of a homogeneous tree is decomposed into a sum of free independent random variables, which provide a prototype of free central limit theorem of Voiculescu. For the rest notions of independence, i.e., the Boolean independence and the monotone independence, we assign a particular graph structure called *star product* and *comb product* and study asymptotic spectral distributions as an application of the associated central limit theorems.

Chapter 9 is devoted to assembling basic notions and tools in representation theory of the symmetric groups. The analytic description of Young diagrams, which is essential for the study of asymptotic behaviour of a representation of $S(n)$ as $n \to \infty$, is also concisely overviewed.

Chapter 10 attempts to derive the celebrated limit shape of Young diagrams, which opens the gateway to the asymptotic representation theory of the symmetric groups. Our approach is based on the moment method developed in previous chapters and serves as a new accessible introduction to asymptotic representation theory.

Chapter 11 answers to the natural question about the fluctuation in a small neighbourhood of the limit shape of Young diagrams with respect to the Plancherel measure. The nature of Gaussian fluctuation is described from several points of view, especially as central limit theorem for quantum components of adjacency matrices associated with conjugacy classes.

Finally Chap. 12 studies a one-parameter deformation (called α-deformation) related to the Jack measure on Young diagrams and the Metropolis algorithm on the symmetric group. The associated central limit theorem follows from the quantum central limit theorem (Theorem 11.13), which shows again usefulness of quantum decomposition.

The notes section at the end of each chapter contains supplementary information of references but is not aimed at documentation. Accordingly, the bibliography contains mainly references that we have actually used while writing this book, and therefore, is far from being complete.

We are indebted to many people whose books, papers and lectures inspired our approach and improved our knowledge, especially, K. Aomoto, M. Bożejko, F. Hiai and D. Petz. Special thanks are due to L. Accardi for stimulating discussion, constant encouragement and kind invitation of writing this book.

Okayama and Sendai
January, 2006

Akihito Hora
Nobuaki Obata

Contents

1

Quantum Probability and Orthogonal Polynomials

This chapter is devoted to most basic notions and results in quantum probability theory, especially concerning the interplay of (one-mode) interacting Fock spaces and orthogonal polynomials.

1.1 Algebraic Probability Spaces

Throughout this book by an *algebra* we mean an algebra over the complex number field $\mathbb{C}$ with the identity. Namely, an *algebra* $\mathcal{A}$ is a vector space over $\mathbb{C}$, in which a map $\mathcal{A} \times \mathcal{A} \ni (a, b) \mapsto ab \in \mathcal{A}$, called multiplication, is defined. The multiplication satisfies the bilinearity:

$$(a + b)c = ac + bc, \quad a(b + c) = ab + ac, \quad (\lambda a)b = a(\lambda b) = \lambda(ab),$$

the first two of which are also referred to as the distributive law, and the associative law:

$$(ab)c = a(bc),$$

where $a, b, c \in \mathcal{A}$ and $\lambda \in \mathbb{C}$. Moreover, there exists an element $1_\mathcal{A} \in \mathcal{A}$ such that

$$a1_\mathcal{A} = 1_\mathcal{A}a = a, \quad a \in \mathcal{A}.$$

Such an element is obviously unique and is called the *identity*. The above definition is slightly unconventional though in many literatures an algebra is defined over an arbitrary field and does not necessarily possess the identity. An algebra $\mathcal{A}$ is called *commutative* if its multiplication is commutative, i.e., $ab = ba$ for all $a, b \in \mathcal{A}$. Otherwise the algebra is called *non-commutative*.

A map $a \mapsto a^*$ defined on $\mathcal{A}$ is called an *involution* if

$$(a + b)^* = a^* + b^*, \quad (\lambda a)^* = \bar{\lambda}a^*, \quad (ab)^* = b^*a^*, \quad (a^*)^* = a,$$

hold for $a, b \in \mathcal{A}$ and $\lambda \in \mathbb{C}$. An algebra equipped with an involution is called a $*$-*algebra*. A linear function φ defined on a $*$-algebra $\mathcal{A}$ with values in $\mathbb{C}$ is

A. Hora and N. Obata: Quantum Probability and Orthogonal Polynomials. In: A. Hora and N. Obata, Quantum Probability and Spectral Analysis of Graphs, Theoretical and Mathematical Physics, 1–63 (2007)
DOI 10.1007/3-540-48863-4_1

called (i) *positive* if $\varphi(a^*a) \ge 0$ for all $a \in \mathcal{A}$; (ii) *normalized* if $\varphi(1_{\mathcal{A}}) = 1$; and (iii) a *state* if φ is positive and normalized.

With these terminologies we give the following:

Definition 1.1. An *algebraic probability space* is a pair $(\mathcal{A}, \varphi)$ of a $*$-algebra $\mathcal{A}$ and a state φ on it. If $\mathcal{A}$ is commutative, the algebraic probability space $(\mathcal{A}, \varphi)$ is called *classical*.

A subset $\mathcal{B}$ of a $*$-algebra $\mathcal{A}$ is called a $*$-*subalgebra* if (i) it is a subalgebra of $\mathcal{A}$, i.e., closed under the algebraic operations; (ii) it is closed under the involution, i.e., $a \in \mathcal{B}$ implies $a^* \in \mathcal{B}$; and (iii) $1_{\mathcal{A}} \in \mathcal{B}$. Here again somewhat unconventional condition (iii) is required. If $(\mathcal{A}, \varphi)$ is an algebraic probability space, for any $*$-subalgebra $\mathcal{B} \subset \mathcal{A}$, the restriction $\varphi\lceil_{\mathcal{B}}$ is a state on $\mathcal{B}$ and, hence, $(\mathcal{B}, \varphi\lceil_{\mathcal{B}})$ becomes an algebraic probability space. We denote it by $(\mathcal{B}, \varphi)$ for simplicity.

Let $\mathcal{A}, \mathcal{B}$ be two $*$-algebras. A map $f : \mathcal{B} \to \mathcal{A}$ is called a $*$-*homomorphism* if the following three conditions are satisfied: (i) f is an algebra homomorphism, i.e., for $a, b \in \mathcal{B}$ and $\lambda \in \mathbb{C}$,

$$f(a+b) = f(a) + f(b), \qquad f(\lambda a) = \lambda f(a), \qquad f(ab) = f(a)f(b);$$

(ii) f is a $*$-map, i.e., for $a \in \mathcal{B}$,

$$f(a^*) = f(a)^*;$$

and (iii) f preserves the identity, i.e.,

$$f(1_{\mathcal{B}}) = 1_{\mathcal{A}}.$$

If $f : \mathcal{B} \to \mathcal{A}$ is a $*$-homomorphism, the image $f(\mathcal{B})$ is a $*$-subalgebra of $\mathcal{A}$. Let $(\mathcal{A}, \varphi)$ be an algebraic probability space, $\mathcal{B}$ a $*$-algebra, and $f : \mathcal{B} \to \mathcal{A}$ a $*$-homomorphism. Then $(\mathcal{B}, \varphi \circ f)$ is an algebraic probability space.

A $*$-homomorphism is called a $*$-*isomorphism* if it is bijective. The inverse map of a $*$-isomorphism is also a $*$-isomorphism. If there exists a $*$-isomorphism between two $*$-algebras $\mathcal{A}$ and $\mathcal{B}$, we say that $\mathcal{A}$ and $\mathcal{B}$ are $*$-*isomorphic*. Two algebraic probability spaces $(\mathcal{A}, \varphi)$ and $(\mathcal{B}, \psi)$ are said to be *isomorphic* if there exists a $*$-isomorphism $f : \mathcal{B} \to \mathcal{A}$ such that $\psi = \varphi \circ f$. However, this concept is too strong from the probabilistic viewpoint (see Definition 1.10 and Proposition 1.11).

If $\mathcal{B}$ is a $*$-subalgebra of a $*$-algebra $\mathcal{A}$, the natural inclusion map $\mathcal{B} \to \mathcal{A}$ is an injective $*$-homomorphism. The complex number field $\mathbb{C}$ itself is a $*$-algebra with involution $\lambda^* = \bar{\lambda}$ for $\lambda \in \mathbb{C}$. Then, given a non-zero $*$-algebra $\mathcal{A}$, an injective $*$-homomorphism $f : \mathbb{C} \to \mathcal{A}$ is defined by $f(\lambda) = \lambda 1_{\mathcal{A}}$. The image of f, denoted by $\mathbb{C}1_{\mathcal{A}}$, is a $*$-subalgebra of $\mathcal{A}$ and is $*$-isomorphic to $\mathbb{C}$. We always identify $\mathbb{C}$ with the $*$-subalgebra $\mathbb{C}1_{\mathcal{A}} \subset \mathcal{A}$ and write $1_{\mathcal{A}} = 1$ for simplicity.

We mention basic examples.

Example 1.2. Let X be a compact Hausdorff space and $C(X)$ the space of $\mathbb{C}$-valued continuous functions on X. Equipped with the usual pointwise addition and multiplication, $C(X)$ becomes a commutative algebra. Moreover, equipped with the involution defined by

$$f^*(x) = \overline{f(x)}, \qquad x \in X, \quad f \in C(X), \tag{1.1}$$

$C(X)$ becomes a $*$-algebra. A Borel probability measure μ on X gives rise to a state φ_μ on $C(X)$ defined by

$$\varphi_\mu(f) = \int_X f(x)\mu(dx), \qquad f \in C(X).$$

We denote by $(C(X), \mu)$ the obtained algebraic probability space. It is noted that every state on $C(X)$ is of the form above, which is a consequence of the following celebrated representation theorem.

Theorem 1.3 (Riesz–Markov). *Let X be a compact Hausdorff space. For any state φ on $C(X)$ there exists a unique regular Borel probability measure on X such that*

$$\varphi(f) = \int_X f(x)\mu(dx), \qquad f \in C(X).$$

In this manner, the states on $C(X)$ and the regular Borel probability measures on X are in one-to-one correspondence. In particular, if X is metrizable, the states on $C(X)$ and the Borel probability measures on X are in one-to-one correspondence.

Example 1.4. Let $(\Omega, \mathcal{F}, P)$ be a classical probability space, i.e., Ω is a non-empty set, $\mathcal{F}$ a σ-field over Ω, and P a probability measure defined on $\mathcal{F}$. The mean value of a random variable X is defined by

$$\mathbb{E}(X) = \int_\Omega X(\omega)P(d\omega),$$

whenever the integral exists. Let $L^\infty(\Omega) = L^\infty(\Omega, \mathcal{F}, P)$ be the set of equivalence classes of essentially bounded $\mathbb{C}$-valued random variables. Then, $L^\infty(\Omega)$ becomes a commutative $*$-algebra equipped with similar operations as in Example 1.2, and $\mathbb{E}$ is a state on $L^\infty(\Omega)$. Thus $(L^\infty(\Omega), \mathbb{E})$ becomes an algebraic probability space. Similarly, let $L^{\infty-}(\Omega)$ denote the set of equivalence classes of $\mathbb{C}$-valued random variables having moments of all orders. Then $(L^{\infty-}(\Omega), \mathbb{E})$ is also an algebraic probability space. These algebraic probability spaces contain the statistical information possessed by $(\Omega, \mathcal{F}, P)$.

The above two examples are classical. A typical non-classical example is given below.

Example 1.5. Let $M(n, \mathbb{C})$ be the set of $n \times n$ complex matrices. Equipped with the usual addition, multiplication and involution (defined by complex conjugation and transposition), $M(n, \mathbb{C})$ becomes a $*$-algebra. It is non-commutative if $n \geq 2$. The *normalized trace*

$$\varphi_{\mathrm{tr}}(a) = \frac{1}{n} \operatorname{tr} a = \frac{1}{n} \sum_{i=1}^{n} a_{ii}, \qquad a = (a_{ij}) \in M(n, \mathbb{C}),$$

is a state on $M(n, \mathbb{C})$. We preserve the symbol $\operatorname{tr} a$ for the usual trace.

Example 1.6. A matrix $\rho \in M(n, \mathbb{C})$ is called a *density matrix* if (i) $\rho = \rho^*$; (ii) all eigenvalues of ρ are non-negative; and (iii) $\operatorname{tr} \rho = 1$. A density matrix ρ gives rise to a state φ_ρ on $M(n, \mathbb{C})$ defined by

$$\varphi_\rho(a) = \operatorname{tr}(a\rho), \qquad a \in M(n, \mathbb{C}).$$

Conversely, any state on $M(n, \mathbb{C})$ is of this form. Moreover, there is a one-to-one correspondence between the set of states and the set of density matrices. This algebraic probability space is denoted by $(M(n, \mathbb{C}), \rho)$.

Example 1.7. Let $\mathbb{C}^n$ be equipped with the inner product defined by

$$\langle \xi, \eta \rangle = \sum_{i=1}^{n} \bar{\xi}_i \eta_i, \qquad \xi = \begin{pmatrix} \xi_1 \\ \vdots \\ \xi_n \end{pmatrix}, \quad \eta = \begin{pmatrix} \eta_1 \\ \vdots \\ \eta_n \end{pmatrix}.$$

A matrix $a \in M(n, \mathbb{C})$ acts on $\mathbb{C}^n$ from the left in a usual manner. Choose a unit vector $\omega \in \mathbb{C}^n$ and set

$$\varphi_\omega(a) = \langle \omega, a\omega \rangle, \qquad a \in M(n, \mathbb{C}).$$

Then, φ_ω becomes a state, which is called a *vector state* associated with a *state vector* $\omega \in \mathbb{C}^n$. Thus the obtained algebraic probability space is denoted by $(M(n, \mathbb{C}), \omega)$. The density matrix corresponding to φ_ω is the projection onto the one-dimensional subspace spanned by ω.

We may generalize the situation in Example 1.7 to an infinite-dimensional case. Let $\mathcal{D}$ be a pre-Hilbert space with inner product $\langle \cdot, \cdot \rangle$. If two linear operators a, b from $\mathcal{D}$ into itself are related as

$$\langle \xi, a\eta \rangle = \langle b\xi, \eta \rangle, \qquad \xi, \eta \in \mathcal{D},$$

we say that a and b are mutually adjoint. The adjoint operator is uniquely determined and we write $b = a^*$. Let $L(\mathcal{D})$ be the set of linear operators from $\mathcal{D}$ into itself which admit adjoints. Then $L(\mathcal{D})$ becomes a $*$-algebra. For a unit vector $\omega \in \mathcal{D}$,

$$\varphi_\omega(a) = \langle \omega, a\omega \rangle, \qquad a \in L(\mathcal{D}), \tag{1.2}$$

defines a state on $L(\mathcal{D})$, which is called a *vector state* associated with a *state vector* $\omega \in \mathcal{D}$. Thus the obtained algebraic probability space is denoted by $(L(\mathcal{D}), \omega)$. The right-hand side of (1.2) is denoted often by $\langle a \rangle_\omega$ or more simply by $\langle a \rangle$ when the state vector ω is understood from the context.

Definition 1.8. Let $(\mathcal{A}, \varphi)$ be an algebraic probability space. An element $a \in \mathcal{A}$ is called an *algebraic random variable* or a *random variable* for short. A random variable $a \in \mathcal{A}$ is called *real* if $a = a^*$.

Definition 1.9. Let a be a random variable of an algebraic probability space $(\mathcal{A}, \varphi)$. Then a quantity of the form

$$\varphi(a^{\epsilon_1} a^{\epsilon_2} \cdots a^{\epsilon_m}), \quad \epsilon_1, \dots, \epsilon_m \in \{1, *\}, \quad m = 1, 2, \dots,$$

is called a *mixed moment* of a. For a real random variable a the mixed moments are reduced to the *moment sequence*

$$\varphi(a^m), \quad m = 1, 2, \dots.$$

Statistical properties of an algebraic random variable are determined by the mixed moments so that the following definition is adequate.

Definition 1.10. Let $(\mathcal{A}, \varphi)$ and $(\mathcal{B}, \psi)$ be two algebraic probability spaces. Algebraic random variables $a \in \mathcal{A}$ and $b \in \mathcal{B}$ are called *stochastically equivalent*, denoted as

$$a \stackrel{\mathrm{s}}{=} b$$

if their mixed moments coincide, i.e., if

$$\varphi(a^{\epsilon_1} a^{\epsilon_2} \cdots a^{\epsilon_m}) = \psi(b^{\epsilon_1} b^{\epsilon_2} \cdots b^{\epsilon_m}) \tag{1.3}$$

for any choice of $m = 1, 2, \dots$ and $\epsilon_1, \dots, \epsilon_m \in \{1, *\}$.

The concept of stochastic equivalence is rather weak.

Proposition 1.11. *Let $(\mathcal{A}, \varphi)$ be an algebraic probability space, $\mathcal{B}$ a $*$-algebra, and $f : \mathcal{B} \to \mathcal{A}$ a $*$-homomorphism. Then for any $a \in \mathcal{B}$ we have*

$$a \stackrel{\mathrm{s}}{=} f(a),$$

where the left-hand side is a random variable in the algebraic probability space $(\mathcal{B}, \varphi \circ f)$ and so is the right-hand side in $(\mathcal{A}, \varphi)$.

Proof. Let $\epsilon_1, \dots, \epsilon_m \in \{1, *\}$. Since f is a $*$-homomorphism, we obtain

$$(\varphi \circ f)(a^{\epsilon_1} \cdots a^{\epsilon_m}) = \varphi(f(a^{\epsilon_1} \cdots a^{\epsilon_m})) = \varphi(f(a)^{\epsilon_1} \cdots f(a)^{\epsilon_m}),$$

which proves the assertion. □

The concept of stochastic equivalence of random variables is applied to convergence of random variables.

Definition 1.12. Let $(\mathcal{A}_n, \varphi_n)$ be a sequence of algebraic probability spaces and $\{a_n\}$ a sequence of random variables such that $a_n \in \mathcal{A}_n$. Let b be a random variable in another algebraic probability space $(\mathcal{B}, \psi)$. We say that $\{a_n\}$ *converges stochastically* to b and write

$$a_n \xrightarrow{\mathrm{s}} b$$

if

$$\lim_{n\to\infty} \varphi_n(a_n^{\epsilon_1} a_n^{\epsilon_2} \cdots a_n^{\epsilon_m}) = \psi(b^{\epsilon_1} b^{\epsilon_2} \cdots b^{\epsilon_m})$$

for any choice of $m = 1, 2, \dots$ and $\epsilon_1, \dots, \epsilon_m \in \{1, *\}$.

1.2 Representations

Definition 1.13. A triple $(\pi, \mathcal{D}, \omega)$ is called a *representation* of an algebraic probability space $(\mathcal{A}, \varphi)$ if $\mathcal{D}$ is a pre-Hilbert space, $\omega \in \mathcal{D}$ a unit vector, and $\pi : \mathcal{A} \to L(\mathcal{D})$ a $*$-homomorphism satisfying

$$\varphi(a) = \langle \omega, \pi(a)\omega \rangle, \qquad a \in \mathcal{A},$$

i.e., $\varphi = \omega \circ \pi$.

As a simple consequence of Proposition 1.11 we obtain the following:

Proposition 1.14. *Let $(\mathcal{A}, \varphi)$ be an algebraic probability space and $(\pi, \mathcal{D}, \omega)$ its representation. Then, for any $a \in \mathcal{A}$ we have*

$$a \stackrel{\mathrm{s}}{=} \pi(a),$$

where $\pi(a)$ is a random variable in $(L(\mathcal{D}), \omega)$.

We shall construct a particular representation.

Lemma 1.15. *Let $(\mathcal{A}, \varphi)$ be an algebraic probability space. Then φ is a $*$-map, i.e.,*

$$\varphi(a^*) = \overline{\varphi(a)}, \qquad a \in \mathcal{A}. \tag{1.4}$$

Proof. Since $\varphi((a+\lambda)^*(a+\lambda)) \geq 0$ for all $\lambda \in \mathbb{C}$, we have

$$\varphi(a^*a) + \bar{\lambda}\varphi(a) + \lambda\varphi(a^*) + |\lambda|^2 \geq 0.$$

In particular, $\bar{\lambda}\varphi(a) + \lambda\varphi(a^*) \in \mathbb{R}$. Hence

$$\bar{\lambda}\varphi(a) + \lambda\varphi(a^*) = \lambda\overline{\varphi(a)} + \overline{\lambda\varphi(a^*)}, \qquad \lambda \in \mathbb{C}.$$

Multiplying λ, we obtain

$$\varphi(a) - \overline{\varphi(a^*)} = \frac{\lambda^2}{|\lambda|^2}(\overline{\varphi(a)} - \varphi(a^*)), \qquad \lambda \in \mathbb{C}, \quad \lambda \neq 0.$$

The left-hand side being independent of λ, we obtain (1.4). □

Lemma 1.16 (Schwarz inequality). *Let* $(\mathcal{A}, \varphi)$ *be an algebraic probability space. Then*

$$|\varphi(a^*b)|^2 \le \varphi(a^*a)\varphi(b^*b), \qquad a, b \in \mathcal{A}. \tag{1.5}$$

Proof. Since $\varphi((a + \lambda b)^*(a + \lambda b)) \ge 0$ for all $\lambda \in \mathbb{C}$, we have

$$\begin{aligned} 0 &\le \varphi(a^*a) + \bar{\lambda}\varphi(b^*a) + \lambda\varphi(a^*b) + |\lambda|^2\varphi(b^*b) \\ &= \varphi(a^*a) + \overline{\lambda\varphi(a^*b)} + \lambda\varphi(a^*b) + |\lambda|^2\varphi(b^*b), \end{aligned} \tag{1.6}$$

where Lemma 1.15 is taken into account. It is sufficient to prove (1.5) by assuming that $\varphi(a^*b) \ne 0$. Consider the polar form $\varphi(a^*b) = |\varphi(a^*b)|e^{i\theta}$. Letting $\lambda = te^{-i\theta}$ in (1.6), we see that

$$\varphi(a^*a) + 2t|\varphi(a^*b)| + t^2\varphi(b^*b) \ge 0 \quad \text{for all } t \in \mathbb{R}. \tag{1.7}$$

Note that $\varphi(b^*b) \ne 0$, otherwise (1.7) does not hold. Then, applying to (1.7) the elementary knowledge on a quadratic inequality, we obtain (1.5) with no difficulty. □

Corollary 1.17. $|\varphi(a)|^2 \le \varphi(a^*a)$ *for* $a \in \mathcal{A}$.

Lemma 1.18. *Let* $(\mathcal{A}, \varphi)$ *be an algebraic probability space. Then,* $\mathcal{N} = \{x \in \mathcal{A}\,;\, \varphi(x^*x) = 0\}$ *is a left ideal of* $\mathcal{A}$.

Proof. Let $x, y \in \mathcal{N}$. By the Schwarz inequality (Lemma 1.16) we have

$$|\varphi(x^*y)|^2 \le \varphi(x^*x)\varphi(y^*y) = 0, \qquad |\varphi(y^*x)|^2 \le \varphi(y^*y)\varphi(x^*x) = 0.$$

Hence $\varphi(x^*y) = \varphi(y^*x) = 0$. Therefore

$$\varphi((x + y)^*(x + y)) = \varphi(x^*x) + \varphi(x^*y) + \varphi(y^*x) + \varphi(y^*y) = 0,$$

which shows that $x + y \in \mathcal{N}$. It is obvious that $x \in \mathcal{N}$, $\lambda \in \mathbb{C} \Rightarrow \lambda x \in \mathcal{N}$. Finally, let $a \in \mathcal{A}$ and $x \in \mathcal{N}$. Then, by the Schwarz inequality,

$$|\varphi((ax)^*(ax))|^2 = |\varphi(x^*(a^*ax))|^2 \le \varphi(x^*x)\varphi((a^*ax)^*(a^*ax)) = 0,$$

which implies that $ax \in \mathcal{N}$. □

Theorem 1.19. *Every algebraic probability space* $(\mathcal{A}, \varphi)$ *admits a representation* $(\pi, \mathcal{D}, \omega)$ *such that* $\pi(\mathcal{A})\omega = \mathcal{D}$.

Proof. Let $\mathcal{N}$ be the left ideal of $\mathcal{A}$ defined in Lemma 1.18. Consider the quotient vector space $\mathcal{D}$ and the canonical projection:

$$p : \mathcal{A} \to \mathcal{D} = \mathcal{A}/\mathcal{N}.$$

Since $\mathcal{N}$ is a left ideal, $\varphi(x^*y)$ is a function of $p(x)$ and $p(y)$. Moreover, one can easily check that

$$\langle p(x), p(y) \rangle = \varphi(x^* y), \qquad x, y \in \mathcal{A},$$

becomes an inner product on $\mathcal{D}$. We next define an action of $\mathcal{A}$ on $\mathcal{D}$ by

$$\pi(a)p(x) = p(ax), \qquad a \in \mathcal{A}, \quad p(x) \in \mathcal{D}.$$

It is straightforward to see that this definition is well defined and $\pi : \mathcal{A} \to L(\mathcal{D})$ is a $*$-homomorphism. We set $\omega = p(1_{\mathcal{A}})$, which is a unit vector of $\mathcal{D}$ since $\langle \omega, \omega \rangle = \varphi(1_{\mathcal{A}}^* 1_{\mathcal{A}}) = 1$. That $\langle \omega, \pi(a)\omega \rangle = \varphi(a)$ follows from the simple observation:

$$\begin{aligned} \langle \omega, \pi(a)\omega \rangle &= \langle p(1_{\mathcal{A}}), \pi(a)p(1_{\mathcal{A}}) \rangle \\ &= \langle p(1_{\mathcal{A}}), p(a1_{\mathcal{A}}) \rangle = \varphi(1_{\mathcal{A}}^* a 1_{\mathcal{A}}) = \varphi(a). \end{aligned}$$

Finally, $\pi(\mathcal{A})\omega = \mathcal{D}$ follows from

$$\pi(a)\omega = \pi(a)p(1_{\mathcal{A}}) = p(a), \qquad a \in \mathcal{A}.$$

This completes the proof. □

The argument in the above proof is called the *GNS-construction* and the obtained representation $(\pi, \mathcal{D}, \omega)$ the *GNS-representation* of an algebraic probability space $(\mathcal{A}, \varphi)$. As a result, any state on a $*$-algebra $\mathcal{A}$ is realized as a vector state through GNS-representation. So the bracket symbol $\langle \cdot \rangle$ is reasonable for a general state too.

For uniqueness of the GNS-representation we prove the following:

Proposition 1.20. *For $i = 1, 2$ let $(\pi_i, \mathcal{D}_i, \omega_i)$ be representations of an algebraic probability space $(\mathcal{A}, \varphi)$. If $\pi_i(\mathcal{A})\omega_i = \mathcal{D}_i$, there exists a linear isomorphism $U : \mathcal{D}_1 \to \mathcal{D}_2$ satisfying the following conditions:*

(i) *U preserves the inner products;*
(ii) *$U\pi_1(a) = \pi_2(a)U$ for all $a \in \mathcal{A}$;*
(iii) *$U\omega_1 = \omega_2$.*

Proof. Define a linear map $U : \mathcal{D}_1 \to \mathcal{D}_2$ by $U(\pi_1(a)\omega_1) = \pi_2(a)\omega_2$. To see the well-definedness we suppose that $\pi_1(a)\omega_1 = 0$. Noting that

$$\begin{aligned} \langle \pi_2(a)\omega_2, \pi_2(a)\omega_2 \rangle &= \langle \omega_2, \pi_2(a^*a)\omega_2 \rangle = \varphi(a^*a) \\ &= \langle \omega_1, \pi_1(a^*a)\omega_1 \rangle = \langle \pi_1(a)\omega_1, \pi_1(a)\omega_1 \rangle = 0, \end{aligned} \tag{1.8}$$

we obtain $\pi_2(a)\omega_2 = 0$, which means that U is a well-defined linear map. That U is surjective is apparent. Moreover, (1.8) implies that U preserves the inner product and hence is injective. Condition (ii) follows from

$$\begin{aligned} U\pi_1(a)(\pi_1(b)\omega_1) &= U(\pi_1(ab)\omega_1) = \pi_2(ab)\omega_2 \\ &= \pi_2(a)\pi_2(b)\omega_2 = \pi_2(a)U(\pi_1(b)\omega_1), \qquad a, b \in \mathcal{A}. \end{aligned}$$

Finally (iii) is obvious by definition. □

Hereafter a representation $(\pi, \mathcal{D}, \omega)$ of an algebraic probability space $(\mathcal{A}, \varphi)$ is called a *GNS-representation* if $\pi(\mathcal{A})\omega = \mathcal{D}$.

By a similar argument one may prove without difficulty the following:

Proposition 1.21. *Let a and b be random variables in algebraic probability spaces $(\mathcal{A}, \varphi)$ and $(\mathcal{B}, \psi)$, respectively. Let $\mathcal{A}_0 \subset \mathcal{A}$ and $\mathcal{B}_0 \subset \mathcal{B}$ be $*$-subalgebras generated by a and b, respectively. Let $(\pi_1, \mathcal{D}_1, \omega_1)$ and $(\pi_2, \mathcal{D}_2, \omega_2)$ be GNS-representations of $(\mathcal{A}_0, \varphi)$ and $(\mathcal{B}_0, \psi)$, respectively. If a and b are stochastically equivalent, there exists a linear isomorphism $U : \mathcal{D}_1 \to \mathcal{D}_2$ preserving the inner products such that*

$$U\pi_1(a^{\epsilon_1} \cdots a^{\epsilon_m}) = \pi_2(b^{\epsilon_1} \cdots b^{\epsilon_m})U,$$

for any choice of $m = 1, 2, \dots$ and $\epsilon_i \in \{1, *\}$.

We mention a simple application of GNS-representation.

Lemma 1.22. *Let $(\mathcal{A}, \varphi)$ be an algebraic probability space and $(\pi, \mathcal{D}, \omega)$ its GNS-representation. For $a \in \mathcal{A}$ satisfying $\varphi(a^*a) = |\varphi(a)|^2$ (Schwarz equality) we have $\pi(a)\omega = \varphi(a)\omega$.*

Proof. In fact,

$$\begin{aligned}
\|\pi(a)\omega - \varphi(a)\omega\|^2 &= \|\pi(a)\omega\|^2 + \|\varphi(a)\omega\|^2 - 2\mathrm{Re}\,\langle \pi(a)\omega, \varphi(a)\omega\rangle \\
&= \langle \omega, \pi(a^*a)\omega\rangle + |\varphi(a)|^2 - 2\mathrm{Re}\,\varphi(a)\overline{\langle \omega, \pi(a)\omega\rangle} \\
&= \varphi(a^*a) - |\varphi(a)|^2 = 0,
\end{aligned}$$

as desired. □

Proposition 1.23. *Let $(\mathcal{A}, \varphi)$ be an algebraic probability space. If $a \in \mathcal{A}$ satisfies $\varphi(a^*a) = \varphi(aa^*) = |\varphi(a)|^2$, we have $a \stackrel{\mathrm{s}}{=} \varphi(a)1_{\mathcal{A}}$.*

Proof. Let $(\pi, \mathcal{D}, \omega)$ be a GNS-representation of $(\mathcal{A}, \varphi)$. By Lemma 1.22 we have

$$\pi(a)\omega = \varphi(a)\omega, \qquad \pi(a^*)\omega = \varphi(a^*)\omega = \overline{\varphi(a)}\,\omega.$$

Then, for any $\epsilon_1, \dots, \epsilon_m \in \{1, *\}$,

$$\pi(a^{\epsilon_m} \cdots a^{\epsilon_1})\omega = \pi(a^{\epsilon_m}) \cdots \pi(a^{\epsilon_1})\omega = \varphi(a)^{\epsilon_m} \cdots \varphi(a)^{\epsilon_1}\omega,$$

where $\varphi(a)^* = \overline{\varphi(a)}$. Hence

$$\begin{aligned}
\varphi(a^{\epsilon_m} \cdots a^{\epsilon_1}) &= \langle \omega, \pi(a^{\epsilon_m} \cdots a^{\epsilon_1})\omega\rangle \\
&= \langle \omega, \varphi(a)^{\epsilon_m} \cdots \varphi(a)^{\epsilon_1}\omega\rangle \\
&= \varphi((\varphi(a)1_{\mathcal{A}})^{\epsilon_m} \cdots (\varphi(a)1_{\mathcal{A}})^{\epsilon_1}),
\end{aligned}$$

which proves the assertion. □

For later use, we introduce the group $*$-algebra of a discrete group G. Let $C_0(G)$ denote the set of $\mathbb{C}$-valued functions on G with finite supports. We equip $C_0(G)$ with the pointwise addition and the scalar multiplication. Moreover, the *convolution product* is defined by

$$\begin{aligned}(a * b)(g) &= \sum_{h\in G} a(gh^{-1})b(h) \\ &= \sum_{h\in G} a(h)b(h^{-1}g), \qquad a,b \in C_0(G), \quad g \in G.\end{aligned}$$

The identity is given by

$$\delta_e(g) = \begin{cases} 1, & g = e, \\ 0, & \text{otherwise.} \end{cases}$$

Furthermore, the involution is given by

$$a^*(g) = \overline{a(g^{-1})}, \qquad a \in C_0(G), \quad g \in G.$$

Equipped with these operations, $C_0(G)$ becomes a $*$-algebra. The most important state on $C_0(G)$ is the *vacuum state* defined by

$$\varphi_e(a) = a(e), \qquad a \in C_0(G). \tag{1.9}$$

Let us check the positivity. We define an inner product on $C_0(G)$ by

$$\langle a, b\rangle = \sum_{g\in G} \overline{a(g)}\, b(g), \qquad a,b \in C_0(G). \tag{1.10}$$

We readily have

$$\varphi_e(a) = \langle \delta_e, a\rangle, \qquad a \in C_0(G),$$

and

$$\varphi_e(a^* * b) = (a^* * b)(e) = \langle a, b\rangle, \qquad a,b \in C_0(G),$$

from which $\varphi_e(a^* * a) \ge 0$ follows immediately. Thus φ_e is a state so that $(C_0(G), \varphi_e)$ becomes an algebraic probability space. We often write $\varphi_e = \delta_e$ for simplicity of notation.

Let us consider the GNS-representation $(\pi, \mathcal{D}, \omega)$ of $(C_0(G), \varphi_e)$. Set $\mathcal{D} = C_0(G)$ and $\omega = \delta_e$. Then $\mathcal{D}$ becomes a pre-Hilbert space equipped with the inner product defined by (1.10) and ω a unit vector. With each $a \in C_0(G)$ we associate an operator $\pi(a) \in L(\mathcal{D})$ by

$$\pi(a)b = a * b, \qquad a \in C_0(G), \quad b \in \mathcal{D}.$$

It is noted that

$$\langle \omega, \pi(a)\omega\rangle = \langle \delta_e, a * \delta_e\rangle = \varphi_e(a), \qquad a \in C_0(G).$$

For convenience we use another notation. Let $\mathbb{C}[G]$ denote the set of formal sums of the form

$$\sum_{g\in G} a(g)g, \qquad a \in C_0(G).$$

The addition and the scalar multiplication are naturally defined, where G is regarded as a linear basis of $\mathbb{C}[G]$. The multiplication is defined by

$$\Big(\sum_{g\in G} a(g)g\Big)\Big(\sum_{h\in G} b(h)h\Big) = \sum_{g,h\in G} a(g)b(h)gh, \qquad a,b \in C_0(G),$$

and the involution by

$$\Big(\sum_{g\in G} a(g)g\Big)^* = \sum_{g\in G} \overline{a(g)}\, g^{-1}, \qquad a \in C_0(G).$$

It is easily seen that $\mathbb{C}[G]$ becomes a $*$-algebra. Moreover, we see from

$$\Big(\sum_{g\in G} a(g)g\Big)\Big(\sum_{h\in G} b(h)h\Big) = \sum_{g\in G} a * b(g)\, g,$$

$$\Big(\sum_{g\in G} a(g)g\Big)^* = \sum_{g\in G} a^*(g)\, g$$

that $\mathbb{C}[G]$ and $C_0(G)$ are $*$-isomorphic. The $*$-algebra $C_0(G)$ or $\mathbb{C}[G]$ is called the *group $*$-algebra* of G.

1.3 Interacting Fock Probability Spaces

We shall define a family of algebraic probability spaces, which are concrete and play a central role in spectral analysis of graphs.

Definition 1.24. A sequence $\{\omega_n\,;\, n = 1, 2, \dots\}$ is called a *Jacobi sequence* if one of the following two conditions is satisfied:

(i) [infinite type] $\omega_n > 0$ for all n;
(ii) [finite type] there exists a number $m_0 \geq 1$ such that $\omega_n = 0$ for all $n \geq m_0$ and $\omega_n > 0$ for all $n < m_0$.

By definition $\{0, 0, 0, \dots\}$ is a Jacobi sequence. We identify a finite sequence of positive numbers with a Jacobi sequence of finite type by concatenating an infinite sequence consisting of only zero.

Consider an infinite-dimensional separable Hilbert space $\mathcal{H}$, in which a complete orthonormal basis $\{\Phi_n;\, n = 0, 1, 2, \dots\}$ is chosen. Let $\mathcal{H}_0 \subset \mathcal{H}$ denote the dense subspace spanned by the complete orthonormal basis $\{\Phi_n\}$.

Given a Jacobi sequence $\{\omega_n\}$ we associate linear operators $B^\pm \in L(\mathcal{H}_0)$ defined by

$$B^+\Phi_n = \sqrt{\omega_{n+1}}\,\Phi_{n+1}, \quad n \ge 0, \tag{1.11}$$

$$B^-\Phi_0 = 0, \qquad B^-\Phi_n = \sqrt{\omega_n}\,\Phi_{n-1}, \quad n \ge 1. \tag{1.12}$$

Lemma 1.25. *B^+ and B^- are mutually adjoint, i.e.,*

$$\langle B^+\Psi, \Psi'\rangle = \langle \Psi, B^-\Psi'\rangle, \qquad \Psi, \Psi' \in \mathcal{H}_0. \tag{1.13}$$

Proof. By (1.11) we have

$$\langle B^+\Phi_m, \Phi_n\rangle = \sqrt{\omega_{m+1}}\,\langle \Phi_{m+1}, \Phi_n\rangle = \sqrt{\omega_{m+1}}\,\delta_{m+1,n}.$$

In a similar manner, by (1.12) we have

$$\langle \Phi_m, B^-\Phi_n\rangle = \sqrt{\omega_n}\,\langle \Phi_m, \Phi_{n-1}\rangle = \sqrt{\omega_n}\,\delta_{m,n-1}, \quad n \ge 1.$$

Since $\sqrt{\omega_{m+1}}\,\delta_{m+1,n} = \sqrt{\omega_n}\,\delta_{m+1,n} = \sqrt{\omega_n}\,\delta_{m,n-1}$, we obtain

$$\langle B^+\Phi_m, \Phi_n\rangle = \langle \Phi_m, B^-\Phi_n\rangle, \quad m \ge 0, \quad n \ge 1. \tag{1.14}$$

Moreover, it is easily seen that (1.14) holds also for $n = 0$. Consequently, (1.13) follows. □

Lemma 1.26. *Let $\Gamma \subset \mathcal{H}_0$ be the linear subspace spanned by $\{(B^+)^n\Phi_0\,;\, n = 0, 1, 2, \dots\}$. Then, Γ is invariant under the actions of $B^\pm$.*

Proof. Obviously, Γ is invariant under B^+. It is noted that

$$(B^+)^n\Phi_0 = \sqrt{\omega_n \cdots \omega_1}\,\Phi_n, \qquad n = 1, 2, \dots,$$

which follows immediately from the definition of B^+. If $\{\omega_n\}$ is of infinite type, Γ coincides with $\mathcal{H}_0$ and is invariant under the actions of B^- too. Suppose that $\{\omega_n\}$ is of finite type and take the smallest number $m_0 \ge 1$ such that $\omega_{m_0} = 0$. Then Γ is the m_0-dimensional vector space spanned by $\{\Phi_n\,;\, n = 0, 1, \dots, m_0 - 1\}$ and is obviously invariant under B^-. □

We thus regard $B^\pm$ as linear operators on the pre-Hilbert space Γ, in which an orthonormal basis $\{\Phi_n\,;\, n = 0, 1, 2, \dots\}$ or $\{\Phi_n\,;\, n = 0, 1, 2, \dots, m_0 - 1\}$ is chosen depending on whether $\{\omega_n\}$ is of infinite type or of finite type. In any case, $B^\pm \in L(\Gamma)$ are again mutually adjoint, as is seen from Lemma 1.25.

Definition 1.27. The quadruple $\Gamma_{\{\omega_n\}} = (\Gamma, \{\Phi_n\}, B^+, B^-)$ is called an *interacting Fock space* associated with a Jacobi sequence $\{\omega_n\}$. We call Φ_n the *nth number vector* and, in particular, Φ_0 the *vacuum vector*. We call B^+ and B^- the *creation operator* and the *annihilation operator*, respectively. The inner product of Γ is denoted by $\langle\cdot,\cdot\rangle_\Gamma$ or by $\langle\cdot,\cdot\rangle$ for brevity.

Let $\Gamma_{\{\omega_n\}} = (\Gamma, \{\Phi_n\}, B^+, B^-)$ be an interacting Fock space. A linear operator $N \in L(\Gamma)$ defined by

$$N\Phi_n = n\Phi_n, \qquad n = 0, 1, 2, \dots,$$

is called the *number operator*. More generally, with an arbitrary sequence $\{\alpha_n\,;\, n = 1, 2, \dots\}$ we associate a diagonal operator $\alpha_{N+1} \in L(\Gamma)$ by

$$\alpha_{N+1}\Phi_n = \alpha_{n+1}\Phi_n, \qquad n = 0, 1, 2, \dots.$$

With this notation we claim the following:

Proposition 1.28. *Let $\Gamma_{\{\omega_n\}} = (\Gamma, \{\Phi_n\}, B^+, B^-)$ be an interacting Fock space. Then*

$$B^+B^- = \omega_N, \qquad B^-B^+ = \omega_{N+1},$$

where we tacitly understand $\omega_0 = 0$ in the first equality.

Proof. By definition we see that

$$B^+B^-\Phi_0 = 0, \quad B^+B^-\Phi_n = \omega_n\,\Phi_n\,, \quad n = 1, 2, \dots,$$

and hence $B^+B^- = \omega_N$ with understanding that $\omega_0 = 0$. Similarly, we have

$$B^-B^+\Phi_n = \omega_{n+1}\,\Phi_n\,, \quad n = 0, 1, 2, \dots,$$

from which $B^-B^+ = \omega_{N+1}$ follows. □

A diagonal operator on an interacting Fock space will often appear.

Proposition 1.29. *Let $\Gamma_{\{\omega_n\}} = (\Gamma, \{\Phi_n\}, B^+, B^-)$ be an interacting Fock space and $B^\circ \in L(\Gamma)$ a diagonal operator. If $\{\omega_n\}$ is of infinite type, there exists a unique sequence of real numbers $\{\alpha_n\,;\, n = 1, 2, \dots\}$ such that $B^\circ = \alpha_{N+1}$. If $\{\omega_n\}$ is of finite type, there exists a unique sequence of real numbers $\{\alpha_n\,;\, n = 1, 2, \dots, m_0\}$ such that $B^\circ = \alpha_{N+1}$.*

We are now in a good position to define an important algebraic probability space.

Definition 1.30. Let $\Gamma_{\{\omega_n\}} = (\Gamma, \{\Phi_n\}, B^+, B^-)$ be an interacting Fock space associated with a Jacobi sequence $\{\omega_n\}$. The algebraic probability space $(L(\Gamma), \Phi_0)$, where Φ_0 is the vacuum state, is called the *interacting Fock probability space associated with a Jacobi sequence* $\{\omega_n\}$.

Particularly interesting random variables of the algebraic probability space $(L(\Gamma), \Phi_0)$ are

$$B^+ + B^-, \qquad (B^+ + \lambda)(B^- + \lambda), \qquad B^+ + B^- + B^\circ,$$

and so forth. For these random variables we only need to consider the $*$-subalgebras of $L(\Gamma)$ generated by $\{B^+, B^-\}$ and $\{B^+, B^-, B^\circ\}$. Such a $*$-subalgebra equipped with the vacuum state Φ_0 is also called an *interacting Fock probability space*. Later on we shall discuss other states as well as the vacuum state Φ_0.

Most basic examples are the following:

Example 1.31. The interacting Fock space associated with a Jacobi sequence given by $\{\omega_n = n\}$ is called the *Boson Fock space* and is denoted by Γ_{Boson}. It follows immediately from Proposition 1.28 that $B^{\pm}$ satisfies

$$B^- B^+ - B^+ B^- = 1,$$

which is referred to as the *canonical commutation relation (CCR).*

Example 1.32. The interacting Fock space associated with a Jacobi sequence given by $\{\omega_n \equiv 1\}$ is called the *free Fock space* and is denoted by Γ_{free}. For the annihilation and creation operators the *free commutation relation* holds:

$$B^- B^+ = 1.$$

Example 1.33. The interacting Fock space associated with a Jacobi sequence given by $\{\omega_1 = 1, \omega_2 = \omega_3 = \cdots = 0\}$ is called the *Fermion Fock space* and is denoted by $\Gamma_{\mathrm{Fermion}}$. Note that

$$B^- B^+ + B^+ B^- = 1,$$

which is referred to as the *canonical anticommutation relation (CAR).*

Remark 1.34. The above three Fock spaces are special cases of the so-called q-Fock space $(-1 \le q \le 1)$, which is defined by a Jacobi sequence:

$$\omega_n = [n]_q = 1 + q + q^2 + \cdots + q^{n-1}, \qquad n \ge 1,$$

which is known as the q-numbers of Gauss. In this case we obtain

$$B^- B^+ - q B^+ B^- = 1,$$

which is referred to as the *q-deformed commutation relation.*

1.4 The Moment Problem and Orthogonal Polynomials

Let $\mathfrak{P}(\mathbb{R})$ be the space of probability measures defined on the Borel σ-field over $\mathbb{R}$. We say that a probability measure $\mu \in \mathfrak{P}(\mathbb{R})$ has a finite moment of order m if

$$\int_{-\infty}^{+\infty} |x|^m \mu(dx) < \infty.$$

In that case the *mth moment* of μ is defined by

$$M_m(\mu) = \int_{-\infty}^{+\infty} x^m \mu(dx). \tag{1.15}$$

Let $\mathfrak{P}_{\mathrm{fm}}(\mathbb{R})$ be the set of probability measures on $\mathbb{R}$ having finite moments of all orders. With each $\mu \in \mathfrak{P}_{\mathrm{fm}}(\mathbb{R})$ we associate the *moment sequence* $\{M_0(\mu) =$

$1, M_1(\mu), M_2(\mu), \dots\}$ defined by (1.15). The classical moment problem asks conditions for a given real sequence $\{M_m\}$ to be a moment sequence of a certain $\mu \in \mathfrak{P}_{\mathrm{fm}}(\mathbb{R})$.

For an infinite sequence of real numbers $\{M_0 = 1, M_1, M_2, \dots\}$ we define the *Hankel determinants* by

$$\Delta_m = \det \begin{pmatrix} M_0 & M_1 & \dots & M_m \\ M_1 & M_2 & \dots & M_{m+1} \\ \vdots & \vdots & & \vdots \\ M_m & M_{m+1} & \dots & M_{2m} \end{pmatrix}, \qquad m = 0, 1, 2, \dots. \tag{1.16}$$

Let $\mathfrak{M}$ be the set of infinite sequences of real numbers $\{M_0 = 1, M_1, M_2, \dots\}$ satisfying one of the following two conditions:

(i) [infinite type] $\Delta_m > 0$ for all $m = 0, 1, 2, \dots$;
(ii) [finite type] there exists $m_0 \geq 1$ such that $\Delta_0 > 0, \Delta_1 > 0, \dots, \Delta_{m_0 - 1} > 0$ and $\Delta_{m_0} = \Delta_{m_0+1} = \dots = 0$.

Theorem 1.35 (Hamburger). *Let $\{M_0 = 1, M_1, M_2, \dots\}$ be an infinite sequence of real numbers. There exists a probability measure $\mu \in \mathfrak{P}_{\mathrm{fm}}(\mathbb{R})$ such that*

$$M_m = \int_{-\infty}^{+\infty} x^m \mu(dx), \qquad m = 0, 1, 2, \dots,$$

if and only if $\{M_m\} \in \mathfrak{M}$. In that case, depending on whether the above condition (i) or (ii) is fulfilled, $\operatorname{supp}\mu$ *consists of infinitely many points or exactly m_0 points.*

Recall that the *support* of $\mu \in \mathfrak{P}(\mathbb{R})$ is a closed subset of $\mathbb{R}$ defined by

$$\operatorname{supp}\mu = \mathbb{R} \setminus \cup\{U \subset \mathbb{R};\ \text{open set such that } \mu(U) = 0\}.$$

A *δ-measure* at $a \in \mathbb{R}$ is defined by

$$\delta_a(E) = \begin{cases} 1 & \text{if } a \in E, \\ 0 & \text{otherwise}, \end{cases} \qquad E \subset \mathbb{R}\text{: Borel set.}$$

Obviously, $\operatorname{supp}\delta_a = \{a\}$. Note that $\operatorname{supp}\mu$ consists of finitely many points if and only if μ is a finite sum of δ-measures.

Theorem 1.35 says that the map $\mathfrak{P}_{\mathrm{fm}}(\mathbb{R}) \to \mathfrak{M}$ is defined and becomes surjective. This map is, however, not injective. We say that $\mu \in \mathfrak{P}_{\mathrm{fm}}(\mathbb{R})$ *is the solution of a determinate moment problem* if the counter image of $\{M_m(\mu)\}$ consists of a single element μ, i.e., if μ is determined uniquely by its moment sequence. In this connection we mention here the following:

Theorem 1.36 (Carleman's moment test). *If $\{M_m\} \in \mathfrak{M}$ satisfies the condition*

$$\sum_{m=1}^{\infty} M_{2m}^{-\frac{1}{2m}} = +\infty, \tag{1.17}$$

there exists a unique $\mu \in \mathfrak{P}_{\text{fm}}(\mathbb{R})$ whose moment sequence is $\{M_m\}$.

Remark 1.37. When $M_{2m} = 0$ occurs for some m, we understand that condition (1.17) is automatically satisfied. In that case, the probability measure is unique and given by δ_0. See also Theorem 1.66 for a determinate moment problem.

Corollary 1.38. *A probability measure $\mu \in \mathfrak{P}_{\text{fm}}(\mathbb{R})$ having a compact support is the solution of a determinate moment problem.*

Let $\mu \in \mathfrak{P}_{\text{fm}}(\mathbb{R})$ and consider the Hilbert space $L^2(\mathbb{R}, \mu)$, the inner product of which is denoted by

$$\langle f, g \rangle = \langle f, g \rangle_\mu = \int_{-\infty}^{+\infty} \overline{f(x)}\, g(x)\, \mu(dx), \qquad f, g \in L^2(\mathbb{R}, \mu).$$

Apparently a polynomial is considered as a function in $L^2(\mathbb{R}, \mu)$; however, we need a care because $L^2(\mathbb{R}, \mu)$ is the space of equivalence classes of functions. To distinguish the double role of a polynomial we introduce some notion and notation.

Let $\mathbb{C}[X]$ denote the set of polynomials in X with complex coefficients. A typical element of $\mathbb{C}[X]$ is of the form

$$F(X) = c_0 + c_1 X + c_2 X^2 + \cdots + c_n X^n, \qquad c_0, c_1, \ldots, c_n \in \mathbb{C}.$$

Equipped with the usual addition and scalar product, $\mathbb{C}[X]$ is a vector space. Note that X is an indeterminate. On the other hand, each polynomial $F \in \mathbb{C}[X]$ gives rise to a $\mathbb{C}$-valued function defined on $\mathbb{R}$ as soon as X is regarded as a variable running over $\mathbb{R}$. When we need to discriminate between a polynomial in $\mathbb{C}[X]$ and a polynomial as a function on $\mathbb{R}$, we call the latter a *polynomial function* and denote it often by $F(x)$ with a lowercase letter x.

Let $\mu \in \mathfrak{P}_{\text{fm}}(\mathbb{R})$. Since every polynomial function belongs to $L^2(\mathbb{R}, \mu)$, we have a linear map

$$\iota : \mathbb{C}[X] \to L^2(\mathbb{R}, \mu).$$

We denote by $\mathcal{P}(\mathbb{R}, \mu)$ the image of ι, namely, the subspace of $L^2(\mathbb{R}, \mu)$ consisting of polynomial functions. We shall characterize the kernel of ι.

Lemma 1.39. *Let $\mu \in \mathfrak{P}_{\text{fm}}(\mathbb{R})$. For a polynomial $g \in \mathbb{C}[X]$ the following conditions are equivalent:*

(i) *$g(x)$ is zero as a function in $L^2(\mathbb{R}, \mu)$, i.e., $g(x) = 0$ for μ-a.e. $x \in \mathbb{R}$;*
(ii) *$\mu(\{x \in \mathbb{R}\,;\, g(x) = 0\}) = 1$;*
(iii) *$\operatorname{supp} \mu \subset \{x \in \mathbb{R}\,;\, g(x) = 0\}$.*

Proof. Condition (i) is equivalent to

$$\mu(\{x \in \mathbb{R}\,;\, g(x) \neq 0\}) = 0. \tag{1.18}$$

Since μ is a probability measure, (1.18) is equivalent to (ii), which proves that (i) $\Leftrightarrow$ (ii). Assume that (1.18) is satisfied. Since $\{x \in \mathbb{R}\,;\, g(x) \neq 0\}$ is an open subset of $\mathbb{R}$, by definition we have

$$\operatorname{supp}\mu \subset \mathbb{R} - \{x \in \mathbb{R}\,;\, g(x) \neq 0\} = \{x \in \mathbb{R}\,;\, g(x) = 0\}.$$

Hence (i) $\Rightarrow$ (iii). Finally, (iii) $\Rightarrow$ (ii) follows from the fact that $\mu(\operatorname{supp}\mu) = 1$. □

Using the above lemma we shall prove the following:

Lemma 1.40. *Let $\mu \in \mathfrak{P}_{\mathrm{fm}}(\mathbb{R})$.*

(1) *If $|\operatorname{supp}\mu| = \infty$, the monomials $\{1, x, x^2, \dots\}$ are linearly independent as a subset of $L^2(\mathbb{R}, \mu)$.*
(2) *If $|\operatorname{supp}\mu| = m_0 < \infty$, the monomials $\{1, x, x^2, \dots, x^{m_0-1}\}$ are linearly independent in $L^2(\mathbb{R}, \mu)$. Moreover, it is a maximal linearly independent subset of $\{1, x, x^2, \dots\}$.*

Proof. (1) Suppose that $\{1, x, x^2, \dots\}$ is not linearly independent in $L^2(\mathbb{R}, \mu)$. Then we may choose $n \geq 1$ and $(c_0, c_1, \dots, c_n) \neq (0, 0, \dots, 0)$ such that

$$\sum_{k=0}^{n} c_k x^k = 0 \quad \text{for } \mu\text{-a.e. } x \in \mathbb{R}. \tag{1.19}$$

Since the left-hand side of (1.19) is a non-zero polynomial of degree at most n, it has at most n zeros in $\mathbb{R}$. It then follows from Lemma 1.39 that $|\operatorname{supp}\mu| \leq n$, which is contradiction.

(2) Set $\operatorname{supp}\mu = \{a_1, \dots, a_{m_0}\}$. We see from Lemma 1.39 that a polynomial function $g(x)$ is zero as a function in $L^2(\mathbb{R}, \mu)$ if and only if it is expressed in the form

$$g(x) = h(x)(x - a_1) \cdots (x - a_{m_0}),$$

where $h(x)$ is an arbitrary polynomial. Since a non-trivial linear combination of $\{1, x, x^2, \dots, x^{m_0-1}\}$ is of degree less than m_0, it is a non-zero function in $L^2(\mathbb{R}, \mu)$, that is $\{1, x, x^2, \dots, x^{m_0-1}\}$ is linearly independent in $L^2(\mathbb{R}, \mu)$.

For maximality it is sufficient to prove that $\{1, x, x^2, \dots, x^{m_0-1}\} \cup \{x^n\}$, where $n \geq m_0$, is not linearly independent. Choose two polynomials $h(x)$, $f(x)$, $\deg f < m_0$, such that

$$x^n = h(x)(x - a_1) \cdots (x - a_{m_0}) + f(x).$$

Then the first term on the right-hand side is zero in $L^2(\mathbb{R}, \mu)$, and hence $x^n = f(x)$ as functions in $L^2(\mathbb{R}, \mu)$. Therefore $\{1, x, x^2, \dots, x^{m_0-1}\} \cup \{x^n\}$ is not linearly independent in $L^2(\mathbb{R}, \mu)$. □

Summing up the above argument,

Proposition 1.41. *Let* $\mu \in \mathfrak{P}_{\mathrm{fm}}(\mathbb{R})$.

(1) *If* $|\mathrm{supp}\,\mu| = \infty$, *then* $\mathrm{Ker}\,\iota = \{0\}$ *and* $\mathbb{C}[X]$ *is linearly isomorphic to* $\mathcal{P}(\mathbb{R},\mu)$. *Hence* $\{1, x, x^2, \dots\}$ *is a linear basis of* $\mathcal{P}(\mathbb{R},\mu)$.
(2) *Assume that* $|\mathrm{supp}\,\mu| = m_0 < \infty$ *and set* $\mathrm{supp}\,\mu = \{a_1, \dots, a_{m_0}\}$. *Then* $\mathrm{Ker}\,\iota = \mathbb{C}[X](x-a_1)\cdots(x-a_{m_0})$ *and* $\{1, x, x^2, \dots, x^{m_0-1}\}$ *is a linear basis of* $\mathcal{P}(\mathbb{R},\mu)$.

Remark 1.42. If $|\mathrm{supp}\,\mu| = m_0 < \infty$, we have $\mathcal{P}(\mathbb{R},\mu) = L^2(\mathbb{R},\mu)$, which are of dimension m_0. If $|\mathrm{supp}\,\mu| = \infty$, then $\mathcal{P}(\mathbb{R},\mu)$ is a proper subspace of $L^2(\mathbb{R},\mu)$. Moreover, if μ is the solution of a determinate moment problem, $\mathcal{P}(\mathbb{R},\mu)$ is a dense subspace of $L^2(\mathbb{R},\mu)$ (the converse is not valid).

We apply the Gram–Schmidt orthogonalization procedure to the sequence of monomials $\{1, x, x^2, \dots\} \subset L^2(\mathbb{R},\mu)$. If $|\mathrm{supp}\,\mu| = \infty$, we obtain an infinite sequence of polynomials

$$P_0(x) = 1, P_1(x), P_2(x), \dots,$$

and, if $|\mathrm{supp}\,\mu| = m_0 < \infty$, the Gram–Schmidt orthogonalization procedure terminates in m_0 steps and we obtain a finite sequence of polynomials:

$$P_0(x) = 1, P_1(x), P_2(x), \dots, P_{m_0-1}(x).$$

Here $P_n(x)$ is a polynomial of degree n and orthogonal to each other, i.e.,

$$\langle P_m, P_n \rangle = \int_{-\infty}^{+\infty} P_m(x) P_n(x)\,\mu(dx) = 0, \quad m \neq n. \tag{1.20}$$

The property of being orthogonal does not change by a constant factor so that, in this book, we adopt the following normalization:

$$P_n(x) = x^n + \cdots, \tag{1.21}$$

i.e., $P_n(x)$ is a monic polynomial. The polynomials $\{P_n(x)\}$ obtained in this way are called the *orthogonal polynomials* associated with μ. The following result is easily verified by induction.

Proposition 1.43. *Let* $\mu \in \mathfrak{P}_{\mathrm{fm}}(\mathbb{R})$.

(1) *If* $|\mathrm{supp}\,\mu| = \infty$, *the polynomials* $\{P_n(x)\,;\, n = 0, 1, 2, \dots\}$ *satisfying* (1.20) *and* (1.21) *are the orthogonal polynomials associated with* μ.
(2) *If* $|\mathrm{supp}\,\mu| = m_0 < \infty$, *a similar assertion holds for a finite sequence of polynomials* $\{P_n(x)\,;\, n = 0, 1, 2, \dots, m_0 - 1\}$.

We are now in a position to prove one of the most important characteristics of orthogonal polynomials.

Theorem 1.44 (Three-term recurrence relation). *Let $\{P_n(x)\}$ be the orthogonal polynomials associated with $\mu \in \mathfrak{P}_{\rm fm}(\mathbb{R})$. Then there exists a pair of sequences $\alpha_1, \alpha_2, \ldots \in \mathbb{R}$ and $\omega_1, \omega_2, \ldots > 0$ uniquely determined by*

$$\begin{aligned} P_0(x) &= 1, \\ P_1(x) &= x - \alpha_1 , \\ xP_n(x) &= P_{n+1}(x) + \alpha_{n+1}P_n(x) + \omega_n P_{n-1}(x), \quad n \geq 1. \end{aligned} \tag{1.22}$$

Here, if $|\text{supp}\,\mu| = \infty$, both $\{\omega_n\}$ and $\{\alpha_n\}$ are infinite sequences, and if $|\text{supp}\,\mu| = m_0 < \infty$ they are finite sequences: $\{\omega_n\} = \{\omega_1, \ldots, \omega_{m_0-1}\}$ and $\{\alpha_n\} = \{\alpha_1, \ldots, \alpha_{m_0}\}$, where the last numbers are determined by (1.22) with $P_{m_0} = 0$.

Proof. Suppose that $|\text{supp}\,\mu| = \infty$. As is seen above, the orthogonal polynomials $\{P_n(x)\}$ form an infinite sequence. By definition $P_0(x) = 1$. Since $P_1(x) = x + \cdots$ and $\langle P_1, P_0\rangle = 0$, we see that

$$\alpha_1 = \int_{-\infty}^{+\infty} x\,\mu(dx). \tag{1.23}$$

Let $n \geq 1$ and consider $xP_n(x)$. Since $xP_n(x)$ is a polynomial of degree $n+1$ of the form $xP_n(x) = x^{n+1} + \cdots$, it is a unique linear combination of $P_0(x), P_1(x), \ldots, P_{n+1}(x)$, say,

$$xP_n(x) = P_{n+1}(x) + \sum_{k=0}^{n} c_{n,k}P_k(x). \tag{1.24}$$

Taking an inner product with P_j with $0 \leq j \leq n-2$ and noticing that xP_j is a linear combination of $P_0(x), P_1(x), \ldots, P_{n-1}(x)$, we obtain

$$c_{n,j}\langle P_j, P_j\rangle = \langle P_j, xP_n\rangle = \langle xP_j, P_n\rangle = 0.$$

Since $\langle P_j, P_j\rangle \neq 0$ we have $c_{n,j} = 0$ for $0 \leq j \leq n-2$ and (1.24) becomes

$$xP_n(x) = P_{n+1}(x) + c_{n,n}P_n(x) + c_{n,n-1}P_{n-1}(x), \qquad n \geq 1.$$

Thus (1.22) is proved with $\alpha_{n+1} = c_{n,n}$ and $\omega_n = c_{n,n-1}$. For the assertion it is sufficient to prove that $\omega_n > 0$ for all n. Integrating (1.22) with $n = 1$ yields

$$\begin{aligned} \omega_1 &= \int_{-\infty}^{+\infty} xP_1(x)\,\mu(dx) \\ &= \int_{-\infty}^{+\infty} (x-\alpha_1)P_1(x)\,\mu(dx) = \int_{-\infty}^{+\infty} P_1(x)^2\,\mu(dx) > 0. \end{aligned}$$

Let $n \geq 2$. We see from (1.22) that

$$\omega_n \langle P_{n-1}, P_{n-1} \rangle = \langle P_{n-1}, xP_n \rangle = \langle xP_{n-1}, P_n \rangle$$
$$= \langle P_n + \alpha_n P_{n-1} + \omega_{n-1} P_{n-2}, P_n \rangle = \langle P_n, P_n \rangle.$$

Hence

$$\omega_n = \frac{\langle P_n, P_n \rangle}{\langle P_{n-1}, P_{n-1} \rangle} > 0, \qquad n \geq 2.$$

This completes the proof for the case of $|\mathrm{supp}\,\mu| = \infty$. The case of $|\mathrm{supp}\,\mu| < \infty$ is proved with small modification. □

Definition 1.45. The pair of sequences $(\{\omega_n\}, \{\alpha_n\})$ determined in Theorem 1.44 is called the *Jacobi coefficient* of the orthogonal polynomials $\{P_n(x)\}$ or of the probability measure $\mu \in \mathfrak{P}_{\mathrm{fm}}(\mathbb{R})$.

During the proof of Theorem 1.44, we have established the following:

Corollary 1.46. *Let $\{P_n(x)\}$ be the orthogonal polynomials associated with $\mu \in \mathfrak{P}_{\mathrm{fm}}(\mathbb{R})$. Then the Jacobi coefficient $(\{\omega_n\}, \{\alpha_n\})$ is determined by*

$$\omega_n \cdots \omega_2 \omega_1 = \int_{-\infty}^{+\infty} P_n(x)^2 \mu(dx), \quad n = 1, 2, \dots,$$
$$\alpha_1 = \int_{-\infty}^{+\infty} x\, \mu(dx),$$
$$\alpha_n \omega_{n-1} \cdots \omega_1 = \int_{-\infty}^{+\infty} x P_{n-1}(x)^2 \mu(dx), \quad n = 2, 3, \dots.$$

In particular, α_1 is the mean of μ and ω_1 is its variance, i.e.,

$$\omega_1 = \int_{-\infty}^{+\infty} (x - \alpha_1)^2\, \mu(dx).$$

Recall that a Jacobi coefficient does not determine a probability measure uniquely because of the uniqueness of a moment problem. Nevertheless, the Jacobi coefficient reflects some properties of a probability measure.

Proposition 1.47. *Let $(\{\omega_n\}, \{\alpha_n\})$ be the Jacobi coefficient of $\mu \in \mathfrak{P}_{\mathrm{fm}}(\mathbb{R})$. If μ is symmetric, i.e., $\mu(-dx) = \mu(dx)$, then $\alpha_n = 0$ for all $n = 1, 2, \dots$.*

Proof. Let $\{P_n(x)\}$ be the orthogonal polynomials associated with μ. Define $Q_n(x) = (-1)^n P_n(-x)$. Then, for $m \neq n$ we have

$$\int_{-\infty}^{+\infty} Q_m(x) Q_n(x) \mu(dx) = (-1)^{m+n} \int_{-\infty}^{+\infty} P_m(-x) P_n(-x) \mu(dx)$$
$$= (-1)^{m+n} \int_{-\infty}^{+\infty} P_m(x) P_n(x) \mu(dx) = 0,$$

where the assumption of μ being symmetric is taken into account. It is obvious that $Q_n(x) = x^n + \cdots$. We then see from Proposition 1.43 that $P_n(x) = Q_n(x)$, that is,

$$P_n(-x) = (-1)^n P_n(x), \qquad n = 0, 1, 2, \dots. \tag{1.25}$$

On the other hand, by replacing x by $-x$ in (1.22) we obtain

$$\begin{aligned} P_1(-x) &= -x - \alpha_1, \\ (-x)P_n(-x) &= P_{n+1}(-x) + \alpha_{n+1}P_n(-x) + \omega_n P_{n-1}(-x). \end{aligned}$$

Then, inserting (1.25) we obtain

$$\begin{aligned} P_1(x) &= x + \alpha_1, \\ xP_n(x) &= P_{n+1}(x) - \alpha_{n+1}P_n(x) + \omega_n P_{n-1}(x). \end{aligned}$$

Comparing with the original recurrence relation (1.22), we see that $\alpha_n = 0$ for all $n = 1, 2, \dots$. □

The converse assertion to Proposition 1.47 being not true in general, we prove the following assertion.

Proposition 1.48. *Let $\mu \in \mathfrak{P}_{\mathrm{fm}}(\mathbb{R})$ and $(\{\omega_n\}, \{\alpha_n\})$ its Jacobi coefficient. If μ is the solution of a determinate moment problem, that $\alpha_n \equiv 0$ implies that μ is symmetric.*

Proof. Define a probability measure $\nu \in \mathfrak{P}_{\mathrm{fm}}(\mathbb{R})$ by

$$\nu(-E) = \mu(E), \qquad E \subset \mathbb{R}\text{: Borel set.}$$

Using Proposition 1.43, one can easily check that $\{Q_n(x) = (-1)^n P_n(-x)\}$ is the orthogonal polynomials associated with ν. On the other hand, since $\alpha_n = 0$ for all n, we see that the three-term recurrence relation for $\{Q_n(x)\}$ coincides with that of $\{P_n(x)\}$. Therefore $P_n(x) = Q_n(x)$. By construction of orthogonal polynomials, for each $m = 1, 2, \dots$ there exist $c_{m,0}, c_{m,1}, \dots, c_{m,m-1} \in \mathbb{R}$ such that

$$x^m = P_m(x) - \sum_{k=0}^{m-1} c_{m,k}P_k(x) = Q_m(x) - \sum_{k=0}^{m-1} c_{m,k}Q_k(x).$$

By orthogonality we obtain

$$M_m(\mu) = \int_{-\infty}^{+\infty} x^m \mu(dx) = c_{m,0} = \int_{-\infty}^{+\infty} x^m \nu(dx) = M_m(\nu),$$

for all $m = 1, 2, \dots$. Since μ is the solution of a determinate moment problem by assumption, we conclude that $\mu = \nu$. □

We next study affine transformations of a probability measure $\mu \in \mathfrak{P}_{\rm fm}(\mathbb{R})$ in terms of the Jacobi coefficient. For $s \in \mathbb{R}$ we define the translation by

$$T_s^*\mu(E) = \mu(E - s), \qquad E \subset \mathbb{R}\colon \text{Borel set}.$$

For $\lambda \in \mathbb{R}$, $\lambda \neq 0$, we define the dilation by

$$S_\lambda^*\mu(E) = \mu(\lambda^{-1}E), \qquad E \subset \mathbb{R}\colon \text{Borel set}.$$

For convention we set $S_0^*\mu = \delta_0$.

Proposition 1.49. *Let $(\{\omega_n\}, \{\alpha_n\})$ be the Jacobi coefficient of $\mu \in \mathfrak{P}_{\rm fm}(\mathbb{R})$. Then the Jacobi coefficients of $T_s^*\mu$ and $S_\lambda^*\mu$ are given by $(\{\omega_n\}, \{\alpha_n + s\})$ and $(\{\lambda^2\omega_n\}, \{\lambda\alpha_n\})$, respectively.*

Proof. The proofs being similar, we prove the assertion only for $S_\lambda^*\mu$, $\lambda \neq 0$. Set

$$Q_n(x) = \lambda^n P_n(\lambda^{-1}x) = x^n + \cdots .$$

Then, for $m \neq n$ we have

$$\begin{aligned}\int_{-\infty}^{+\infty} Q_m(x)Q_n(x)\,(S_\lambda^*\mu)(dx) &= \int_{-\infty}^{+\infty} Q_m(\lambda x)Q_n(\lambda x)\,\mu(dx) \\ &= \lambda^{m+n}\int_{-\infty}^{+\infty} P_m(x)P_m(x)\,\mu(dx) = 0.\end{aligned}$$

Hence $\{Q_m(x)\}$ is the orthogonal polynomials associated with $S_\lambda^*\mu$. The recurrence formula (1.22) for $\{P_n(x)\}$ yields the one for $\{Q_n(x)\}$ by replacing x by $\lambda^{-1}x$, from which the Jacobi coefficient of $S_\lambda^*\mu$ is obtained as desired. □

We shall clarify the relationship among $\mathfrak{P}_{\rm fm}(\mathbb{R})$, $\mathfrak{M}$ and the Jacobi coefficients. Let $\mathfrak{J}$ be the set of pairs of sequences $(\{\omega_n\}, \{\alpha_n\})$ satisfying one of the following conditions:

(i) [infinite type] $\{\omega_n\}$ is a Jacobi sequence of infinite type and $\{\alpha_n\}$ is an infinite sequence of real numbers;
(ii) [finite type] $\{\omega_n\}$ is a Jacobi sequence of finite type and $\{\alpha_n\}$ is a finite real sequence $\{\alpha_1, \dots, \alpha_{m_0}\}$, where $m_0 \geq 1$ is the smallest number such that $\omega_{m_0} = 0$.

Given $\mu \in \mathfrak{P}_{\rm fm}(\mathbb{R})$, constructing the orthogonal polynomials associated with μ we obtain the Jacobi coefficient $(\{\omega_n\}, \{\alpha_n\}) \in \mathfrak{J}$ from the three-term recurrence relation. We thus have a map $\mathfrak{P}_{\rm fm}(\mathbb{R}) \to \mathfrak{J}$.

Since the Gram–Schmidt orthogonalization requires only moments of μ, the Jacobi coefficient of μ is determined by its moment sequence (see also

Corollary 1.46). Therefore, there is a map $\mathfrak{M} \to \mathfrak{J}$ satisfying the following commutative diagram:

$$\begin{array}{ccc} & \mathfrak{P}_{\mathrm{fm}}(\mathbb{R}) & \\ \swarrow & & \searrow \\ \mathfrak{M} & \longrightarrow & \mathfrak{J} \end{array} \tag{1.26}$$

In the next sections we shall prove that the map $\mathfrak{M} \to \mathfrak{J}$ is bijective and obtain an explicit form of the inverse map (Theorem 1.64).

1.5 Quantum Decomposition

We shall obtain a fundamental result which links orthogonal polynomials and interacting Fock spaces, and which brings non-commutative fine structures into a classical random variable.

We introduced in the previous section the vector space $\mathbb{C}[X]$ of polynomials. As is well known, $\mathbb{C}[X]$ is also equipped with a natural multiplication, i.e., a bilinear map $\mathbb{C}[X] \times \mathbb{C}[X] \to \mathbb{C}[X]$ uniquely determined by $X^m X^n = X^{m+n}$. More concretely, for two polynomials

$$F(X) = \sum_{j=0}^{\infty} c_j X^j, \qquad G(X) = \sum_{k=0}^{\infty} d_k X^k,$$

which are in fact finite sums, we define

$$(FG)(X) = \sum_{j=0}^{\infty} \sum_{k=0}^{\infty} c_j d_k X^{j+k} = \sum_{l=0}^{\infty} \Big(\sum_{j+k=l} c_j d_k \Big) X^l.$$

Furthermore, the involution for $F(X)$ given as above is defined by

$$F^*(X) = \sum_{j=0}^{\infty} \bar{c}_j X^j.$$

Equipped with these structures, $\mathbb{C}[X]$ becomes a commutative $*$-algebra called the *polynomial $*$-algebra.* Actually the polynomial $*$-algebra can be defined without using the indeterminate X; however, for algebraic operations the dummy variable X is useful.

Let $\mu \in \mathfrak{P}_{\mathrm{fm}}(\mathbb{R})$. For $F \in \mathbb{C}[X]$ we define

$$\mu(F) = \int_{-\infty}^{+\infty} F(x)\, \mu(dx).$$

Then $(\mathbb{C}[X], \mu)$ becomes an algebraic probability space. Recall that $\mathcal{P}(\mathbb{R}, \mu) \subset L^2(\mathbb{R}, \mu)$ denotes the space of polynomial functions. Set $P_0(x) \equiv 1 \in \mathcal{P}(\mathbb{R}, \mu)$, which is a unit vector.

Proposition 1.50. *Let* $\mu \in \mathfrak{P}_{\mathrm{fm}}(\mathbb{R})$*. For each* $F \in \mathbb{C}[X]$ *we define* $M_F \in L(\mathcal{P}(\mathbb{R}, \mu))$ *by*

$$(M_F f)(x) = F(x) f(x), \qquad f \in \mathcal{P}(\mathbb{R}, \mu), \quad x \in \mathbb{R}. \tag{1.27}$$

Then $(M, \mathcal{P}(\mathbb{R}, \mu), P_0)$ *is a GNS-representation of* $(\mathbb{C}[X], \mu)$*. In particular,*

$$F \overset{\mathrm{s}}{=} M_F, \qquad F \in \mathbb{C}[X].$$

Proof. As is easily verified, $M : \mathbb{C}[X] \to L(\mathcal{P}(\mathbb{R}, \mu))$ is a $*$-homomorphism. Since $P_0(x) = 1$ we have

$$\begin{aligned} \mu(F) &= \int_{-\infty}^{+\infty} F(x)\, \mu(dx) = \int_{-\infty}^{+\infty} \overline{P_0(x)}\, F(x) P_0(x)\, \mu(dx) \\ &= \int_{-\infty}^{+\infty} \overline{P_0(x)}\, (M_F P_0)(x)\, \mu(dx) = \langle P_0, M_F P_0 \rangle, \qquad F \in \mathbb{C}[X], \end{aligned}$$

so that $(M, \mathcal{P}(\mathbb{R}, \mu), P_0)$ is a representation of $(\mathbb{C}[X], \mu)$. The assertion follows from the obvious fact that $\{M_F P_0\,;\, F \in \mathbb{C}[X]\} = \mathcal{P}(\mathbb{R}, \mu)$. □

The linear operator M_F defined in (1.27) is called a *multiplication operator* by $F(x)$. A particularly interesting one is the multiplication operator by x, which is denoted by M_X. Thus,

$$(M_X f)(x) = x f(x), \qquad f \in \mathcal{P}(\mathbb{R}, \mu), \quad x \in \mathbb{R}.$$

We note from Proposition 1.50 that

$$X \overset{\mathrm{s}}{=} M_X \tag{1.28}$$

and

$$\int_{-\infty}^{+\infty} x^m\, \mu(dx) = \mu(X^m) = \langle P_0, M_X^m P_0 \rangle, \qquad m = 1, 2, \dots. \tag{1.29}$$

If $(\{\omega_n\}, \{\alpha_n\})$ is a Jacobi coefficient, then $\{\omega_n\}$ is a Jacobi sequence so that an interacting Fock space $\Gamma_{\{\omega_n\}} = (\Gamma, \{\Phi_n\}, B^+, B^-)$ is defined. We now prove the fundamental link between orthogonal polynomials and interacting Fock spaces.

Theorem 1.51. *Let* $\mu \in \mathfrak{P}_{\mathrm{fm}}(\mathbb{R})$ *and* $(\{\omega_n\}, \{\alpha_n\})$ *its Jacobi coefficient. Let* $\Gamma_{\{\omega_n\}} = (\Gamma, \{\Phi_n\}, B^+, B^-)$ *be an interacting Fock space associated with* $\{\omega_n\}$*. Define a linear map by*

$$U : \Phi_0 \mapsto P_0, \quad \sqrt{\omega_n \cdots \omega_1}\, \Phi_n \mapsto P_n, \quad n = 1, 2, \dots,$$

where $\{P_n\}$ *are the orthogonal polynomials associated with* μ*. Then* $U : \Gamma \to \mathcal{P}(\mathbb{R}, \mu)$ *becomes a linear isomorphism which preserves the inner products. Moreover, it holds that*

$$M_X = U(B^+ + B^- + B^\circ)U^*, \tag{1.30}$$

where B° *is a diagonal operator defined by* $B^\circ = \alpha_{N+1}$.

Proof. It is obvious that U is well defined and becomes a linear isomorphism from Γ onto $\mathcal{P}(\mathbb{R},\mu)$. We see from Corollary 1.46 that

$$\omega_n \cdots \omega_1 = \langle P_n, P_n \rangle, \qquad n = 1, 2, \dots,$$

which means that U is isometric, i.e., preserves the inner product. Moreover, $U^{-1} = U^* : \mathcal{P}(\mathbb{R},\mu) \to \Gamma$ is easily checked. To show (1.30) we rewrite the three-term recurrence relation (Theorem 1.44) as

$$\begin{aligned} M_X P_0 &= P_1 + \alpha_1 P_0, \\ M_X P_n &= P_{n+1} + \alpha_{n+1} P_n + \omega_n P_{n-1}, \quad n = 1, 2, \dots. \end{aligned}$$

Since $P_n = \sqrt{\omega_n \cdots \omega_1}\, U\Phi_n$ by definition, we obtain

$$M_X U\Phi_0 = \sqrt{\omega_1} U\Phi_1 + \alpha_1 U\Phi_0 = U(B^+ + B^- + B^\circ)\Phi_0. \tag{1.31}$$

Similarly,

$$\begin{aligned} M_X U\Phi_n &= \sqrt{\omega_{n+1}}\, U\Phi_{n+1} + \alpha_{n+1} U\Phi_n + \sqrt{\omega_n}\, U\Phi_{n-1} \\ &= U(B^+ + B^- + B^\circ)\Phi_n. \end{aligned} \tag{1.32}$$

It then follows from (1.31) and (1.32) that

$$M_X U = U(B^+ + B^- + B^\circ),$$

which is equivalent to (1.30). □

Now we come to the following important result.

Theorem 1.52. *Let $\mu \in \mathfrak{P}_{\rm fm}(\mathbb{R})$ and $(\{\omega_n\}, \{\alpha_n\})$ its Jacobi coefficient. Consider the interacting Fock space $\Gamma_{\{\omega_n\}} = (\Gamma, \{\Phi_n\}, B^+, B^-)$ and the diagonal operator defined by $B^\circ = \alpha_{N+1}$. Then,*

$$\langle \Phi_0, (B^+ + B^- + B^\circ)^m \Phi_0 \rangle = \int_{-\infty}^{+\infty} x^m \mu(dx), \quad m = 1, 2, \dots. \tag{1.33}$$

Proof. Let $U : \Gamma \to \mathcal{P}(\mathbb{R},\mu)$ be an isometric linear isomorphism defined in Theorem 1.51. Then, in view of (1.30) we have

$$\begin{aligned} \langle P_0, M_X^m P_0 \rangle_\mu &= \langle P_0, U(B^+ + B^- + B^\circ)^m U^* P_0 \rangle_\mu \\ &= \langle \Phi_0, (B^+ + B^- + B^\circ)^m \Phi_0 \rangle_\Gamma, \qquad m = 1, 2, \dots. \end{aligned} \tag{1.34}$$

Then (1.33) follows by combining (1.29) and (1.34). □

Corollary 1.53. *Let $\mu \in \mathfrak{P}_{\rm fm}(\mathbb{R})$ and $(\{\omega_n\}, \{\alpha_n\})$ its Jacobi coefficient. Then,*

$$X \stackrel{\rm s}{=} M_X \stackrel{\rm s}{=} B^+ + B^- + B^\circ, \tag{1.35}$$

where X is a random variable in $(\mathbb{C}[X], \mu)$, M_X in $(L(\mathcal{P}(\mathbb{R},\mu)), P_0)$, and $B^+ + B^- + B^\circ$ in $(L(\Gamma), \Phi_0)$ with $\Gamma_{\{\omega_n\}} = (\Gamma, \{\Phi_n\}, B^+, B^-)$ being the interacting Fock space associated with $\{\omega_n\}$ and $B^\circ = \alpha_{N+1}$.

Proof. That $X \stackrel{\mathrm{s}}{=} M_X$ was already shown in (1.28). That $M_X \stackrel{\mathrm{s}}{=} B^+ + B^- + B^\circ$ follows directly from (1.34). □

Through the stochastic equivalence (1.35) the real algebraic random variables X and M_X are decomposed into a sum of *quantum components* B^+, B^-, B°. This is called the *quantum decomposition* of X or of M_X. In fact, we shall prove that any real random variable in a general algebraic probability space admits a quantum decomposition. We start with the following:

Theorem 1.54. *Let $(\mathcal{A}, \varphi)$ be an algebraic probability space. For a real random variable $a = a^* \in \mathcal{A}$ there exists a probability measure $\mu \in \mathfrak{P}_{\mathrm{fm}}(\mathbb{R})$ satisfying*

$$\varphi(a^m) = \int_{-\infty}^{+\infty} x^m \mu(dx), \qquad m = 1, 2, \dots . \tag{1.36}$$

Proof. For $m = 0, 1, 2, \dots$ we set $M_m = \varphi(a^m)$. We first show that $\Delta_m \geq 0$ for all $m = 0, 1, \dots,$ where Δ_m is the Hankel determinant defined in (1.16). Let $m \geq 1$ be fixed and consider

$$x = c_0 + c_1 a + c_2 a^2 + \dots + c_m a^m, \qquad c_0, c_1, \dots, c_m \in \mathbb{C}.$$

Then $x \in \mathcal{A}$ and

$$0 \leq \varphi(x^* x) = \sum_{i,j=0}^{m} \bar{c}_i c_j \varphi(a^{i+j}) = \sum_{i,j=0}^{m} \bar{c}_i c_j M_{i+j}.$$

Since this holds for any choice of $c_0, c_1, \dots, c_m \in \mathbb{C}$, the $(m+1) \times (m+1)$ matrix (M_{i+j}) is positive definite so that $\Delta_m \geq 0$.

Our assertion is a direct consequence of Theorem 1.35, provided $\{M_m\}$ is verified to satisfy either condition (i) or (ii) therein. For that purpose we suppose $\Delta_m = 0$ happens for some $m \geq 1$. Then, there exists a choice $(c_0, c_1, \dots, c_m) \neq (0, 0, \dots, 0)$ such that

$$\sum_{i,j=0}^{m} \bar{c}_i c_j M_{i+j} = 0.$$

Then, setting $c_{m+1} = 0$, we obtain

$$\sum_{i,j=0}^{m+1} \bar{c}_i c_j M_{i+j} = \sum_{i,j=0}^{m} \bar{c}_i c_j M_{i+j} = 0.$$

Since $(c_0, c_1, \dots, c_m, 0) \neq (0, 0, \dots, 0, 0)$, we have $\Delta_{m+1} = 0$ as desired. □

The probability measure μ mentioned in Theorem 1.54 is referred to as the *distribution* of a real random variable $a \in \mathcal{A}$ (in the state φ) though it is not uniquely specified in general.

Corollary 1.55. *Every real random variable a in an algebraic probability space $(\mathcal{A}, \varphi)$ admits a quantum decomposition*

$$a \overset{\mathrm{s}}{=} B^+ + B^- + B^\circ, \tag{1.37}$$

where $B^\pm$ is the annihilation and creation operators in an interacting Fock space $\Gamma_{\{\omega_n\}}$ and B° is a diagonal operator.

Proof. Take a probability measure $\mu \in \mathfrak{P}_{\mathrm{fm}}(\mathbb{R})$ as in Theorem 1.54 and consider an algebraic probability space $(\mathbb{C}[X], \mu)$. Then (1.36) means that X and a are stochastically equivalent. Since X admits a quantum decomposition by Corollary 1.53, so does a. □

Uniqueness of the quantum decomposition (1.37) will be proved in Corollary 1.65, for which we prepare some results.

Proposition 1.56. *The map $\mathfrak{P}_{\mathrm{fm}}(\mathbb{R}) \to \mathfrak{J}$ defined in (1.26) is surjective, hence so is the map $\mathfrak{M} \to \mathfrak{J}$.*

Proof. Given $(\{\omega_n\}, \{\alpha_n\}) \in \mathfrak{J}$, let $\Gamma_{\{\omega_n\}} = (\Gamma, \{\Phi_n\}, B^+, B^-)$ be the interacting Fock space associated with $\{\omega_n\}$ and $B^\circ = \alpha_{N+1}$ the diagonal operator defined by $\{\alpha_n\}$. Since $B^+ + B^- + B^\circ$ is a real random variable, we see from Theorem 1.54 that there exists a probability measure $\mu \in \mathfrak{P}_{\mathrm{fm}}(\mathbb{R})$ such that

$$\langle \Phi_0, (B^+ + B^- + B^\circ)^m \Phi_0 \rangle = \int_{-\infty}^{+\infty} x^m \mu(dx), \qquad m = 1, 2, \dots. \tag{1.38}$$

It is sufficient to prove that the Jacobi coefficient of μ coincides with the given $(\{\omega_n\}, \{\alpha_n\}) \in \mathfrak{J}$.

We consider the algebraic probability space $(\mathbb{C}[X], \mu)$. Let $\pi : \mathbb{C}[X] \to L(\Gamma)$ be a $*$-homomorphism determined by $\pi(X) = B^+ + B^- + B^\circ$. Without difficulty we see from (1.38) that

$$\langle \Phi_0, \pi(F) \Phi_0 \rangle = \int_{-\infty}^{+\infty} F(x) \mu(dx) = \mu(F), \qquad F \in \mathbb{C}[X].$$

Thus, (π, Γ, Φ_0) is a representation of $(\mathbb{C}[X], \mu)$. We shall prove by induction that

$$\{\Phi_0, \Phi_1, \dots\} \subset \pi(\mathbb{C}[X])\Phi_0. \tag{1.39}$$

Obviously, $\Phi_0 \in \pi(\mathbb{C}[X])\Phi_0$. Suppose that $n \geq 1$ and $\Phi_0, \Phi_1, \dots, \Phi_{n-1} \in \pi(\mathbb{C}[X])\Phi_0$. Note first that

$$(B^+ + B^- + B^\circ)^n \Phi_0 = \sqrt{\omega_n \cdots \omega_1}\, \Phi_n + R(\Phi_0, \Phi_1, \dots, \Phi_{n-1}), \tag{1.40}$$

where the second term stands for a linear combination of $\Phi_0, \Phi_1, \dots, \Phi_{n-1}$. Since $\pi(X^n)\Phi_0 = (B^+ + B^- + B^\circ)^n \Phi_0$, we see that $\Phi_n \in \pi(\mathbb{C}[X])\Phi_0$. Since

(1.39) implies that $\pi(\mathbb{C}[X])\Phi_0 = \Gamma$, so (π, Γ, Φ_0) is a GNS-representation of $(\mathbb{C}[X], \mu)$.

On the other hand, Proposition 1.50 says that $(\pi, \mathcal{P}(\mathbb{R}, \mu), P_0 = 1)$ is another GNS-representation of $(\mathbb{C}[X], \mu)$. Then by uniqueness of a GNS-representation (Proposition 1.20) there exists a linear isomorphism $U : \Gamma \to \mathcal{P}(\mathbb{R}, \mu)$ preserving the inner product such that

$$U\Phi_0 = P_0, \qquad U(B^+ + B^- + B^\circ)^n = U\pi(X^n) = M_X^n U. \tag{1.41}$$

Define

$$P_n = U(B^+)^n \Phi_0 = \sqrt{\omega_n \cdots \omega_1}\, U\Phi_n, \qquad n = 1, 2, \dots.$$

We shall prove by induction that $P_n(x) = x^n + \cdots$ is a monic polynomial for all $n = 0, 1, 2, \dots$. This is true for $n = 0$ by definition. Suppose the assertion is true up to $n-1$. Applying U to both sides of (1.40), we obtain

$$\begin{aligned} U(B^+ + B^- + B^\circ)^n \Phi_0 &= \sqrt{\omega_n \cdots \omega_1}\, U\Phi_n + R(U\Phi_0, U\Phi_1, \dots, U\Phi_{n-1}) \\ &= P_n + R(P_0, P_1, \dots, P_{n-1}). \end{aligned} \tag{1.42}$$

On the other hand, by (1.41) we have

$$U(B^+ + B^- + B^\circ)^n \Phi_0 = M_X^n U\Phi_0 = M_X^n P_0.$$

Since $M_X^n P_0(x) = x^n$, (1.42) becomes

$$x^n = P_n(x) + R(P_0(x), P_1(x), \dots, P_{n-1}(x)),$$

which means that $P_n(x) = x^n + \cdots$ as desired. We next notice that $\langle P_j, P_k \rangle = 0$ for $j \neq k$, which follows from the fact that U preserves the inner product. Hence $\{P_n\}$ is the orthogonal polynomials associated with μ. Finally, since

$$\begin{aligned} M_X P_n &= \sqrt{\omega_n \cdots \omega_1}\, M_X U\Phi_n \\ &= \sqrt{\omega_n \cdots \omega_1}\, U(B^+ + B^- + B^\circ)\Phi_n \\ &= \sqrt{\omega_n \cdots \omega_1}\, U(\sqrt{\omega_{n+1}}\, \Phi_{n+1} + \sqrt{\omega_n}\Phi_{n-1} + \alpha_{n+1}\Phi_n) \\ &= P_{n+1} + \alpha_{n+1} P_n + \omega_n P_{n-1}, \end{aligned}$$

the Jacobi coefficient of $\{P_n\}$ coincides with the given $(\{\omega_n\}, \{\alpha_n\})$. □

1.6 The Accardi–Bożejko Formula

Given $(\{\omega_n\}, \{\alpha_n\}) \in \mathfrak{J}$, let $\Gamma_{\{\omega_n\}} = (\Gamma, \{\Phi_n\}, B^+, B^-)$ be the interacting Fock space associated with $\{\omega_n\}$ and B° the diagonal operator defined by

$$B^\circ = \alpha_{N+1}.$$

We are interested in the moment sequence of the real random variable $B^+ + B^- + B^\circ$ in the algebraic probability space $(L(\Gamma), \Phi_0)$:

$$M_m = \langle \Phi_0, (B^+ + B^- + B^\circ)^m \Phi_0 \rangle, \qquad m = 1, 2, \dots. \tag{1.43}$$

Expanding the right-hand side, we obtain

$$M_m = \sum_{\epsilon} \langle \Phi_0, B^{\epsilon_m} \cdots B^{\epsilon_2} B^{\epsilon_1} \Phi_0 \rangle, \tag{1.44}$$

where $\epsilon = (\epsilon_1, \dots, \epsilon_m)$ runs over $\{+, -, \circ\}^m$.

In order to observe the action of $B^{\epsilon_m} \cdots B^{\epsilon_2} B^{\epsilon_1}$ to the vacuum vector Φ_0 it is convenient to associate a sequence of points (i.e., a path) in $\mathbb{Z}^2$ starting at $(0,0)$ as follows. Given $\epsilon = (\epsilon_1, \dots, \epsilon_m) \in \{+, -, \circ\}^m$ we associate a sequence of points in $\mathbb{Z}^2$ defined by

$$(0,0),\ (1, \epsilon_1),\ (2, \epsilon_1 + \epsilon_2), \dots, (m, \epsilon_1 + \epsilon_2 + \cdots + \epsilon_m),$$

where numbers $+1, -1, 0$ are assigned to ϵ_i according to $\epsilon_i = +, -, \circ$. It is more instructive to draw edges connecting these points in order (see Fig. 1.1). Let $\mathcal{E}_m^+$ denote the set of paths which end at $(m, 0)$ and pass only in the upper half plane, namely,

$$\mathcal{E}_m^+ = \left\{ (\epsilon_1, \dots, \epsilon_m) \in \{+, -, \circ\}^m ; \begin{array}{l} \epsilon_1 + \cdots + \epsilon_k \ge 0, \quad k = 1, 2, \dots, m-1 \\ \epsilon_1 + \cdots + \epsilon_{m-1} + \epsilon_m = 0 \end{array} \right\}. \tag{1.45}$$

In view of the action of B^ϵ we easily see that

$$\langle \Phi_0, B^{\epsilon_m} \cdots B^{\epsilon_2} B^{\epsilon_1} \Phi_0 \rangle = 0, \quad (\epsilon_1, \dots, \epsilon_m) \in \{+, -, \circ\}^m \setminus \mathcal{E}_m^+.$$

Hence (1.44) becomes

$$M_m = \sum_{\epsilon \in \mathcal{E}_m^+} \langle \Phi_0, B^{\epsilon_m} \cdots B^{\epsilon_2} B^{\epsilon_1} \Phi_0 \rangle. \tag{1.46}$$

Definition 1.57. Let S be a finite set. A collection ϑ of non-empty subsets of S is called a *partition* of S if

$$S = \bigcup_{v \in \vartheta} v, \qquad v \cap v' = \emptyset, \quad v \neq v'.$$

A partition ϑ is called (i) a *pair partition* if $|v| = 2$ for all $v \in \vartheta$; (ii) a *pair partition with singletons* if $|v| = 2$ or $|v| = 1$ for all $v \in \vartheta$. An element $v \in \vartheta$ is called a *singleton* if $|v| = 1$.

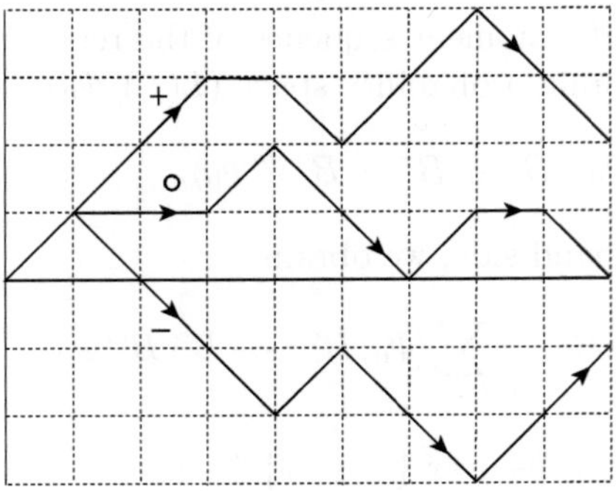

Fig. 1.1. Paths in $\{+,-,\circ\}^m$ and $\mathcal{E}_m^+$

We next associate with each $\epsilon \in \mathcal{E}_m^+$ a partition $\vartheta(\epsilon)$ of $\{1,2,\dots,m\}$. In general, $\epsilon \in \{+,-,\circ\}^m$ being regarded as a map $\epsilon : \{1,2,\dots,m\} \to \{+,-,\circ\}$, we obtain a partition:

$$\{1,2,\dots,m\} = \epsilon^{-1}(\circ) \cup \epsilon^{-1}(+) \cup \epsilon^{-1}(-).$$

Let $\epsilon \in \mathcal{E}_m^+$. Since $|\epsilon^{-1}(+)| = |\epsilon^{-1}(-)|$ we may set

$$\epsilon^{-1}(\circ) = \{s_1 < \dots < s_j\}, \quad \epsilon^{-1}(\{+,-\}) = \{t_1 < \dots < t_{2k}\},$$

where $j + 2k = m$. We shall divide $\{t_1 < \dots < t_{2k}\}$ into a union of pairs. First we take $1 \le \alpha \le 2k$ such that

$$\epsilon(t_1) = \dots = \epsilon(t_\alpha) = +, \quad \epsilon(t_{\alpha+1}) = -.$$

Note that such an α always exists whenever $\epsilon^{-1}(\{+,-\}) \neq \varnothing$. Then we make a pair $\{t_\alpha < t_{\alpha+1}\}$. Setting

$$\{t'_1 < \dots < t'_{2k-2}\} = \{t_1 < \dots < t_{2k}\} \setminus \{t_\alpha < t_{\alpha+1}\},$$

and applying a similar argument, we make the second pair. Repeating this procedure, we obtain a pair partition

$$\{t_1 < \dots < t_{2k}\} = \{l_1 < r_1\} \cup \dots \cup \{l_k < r_k\},$$

where $\epsilon(l_1) = \dots = \epsilon(l_k) = +$ and $\epsilon(r_1) = \dots = \epsilon(r_k) = -$. Finally we define a partition $\vartheta(\epsilon)$ by

$$\vartheta(\epsilon) = \{\{s_1\},\dots,\{s_j\},\{l_1 < r_1\},\dots,\{l_k < r_k\}\}, \tag{1.47}$$

which is a pair partition with singleton (see Fig. 1.2).

Definition 1.58. Let ϑ be a pair partition with singleton of $\{1,2,\dots,m\}$, say,

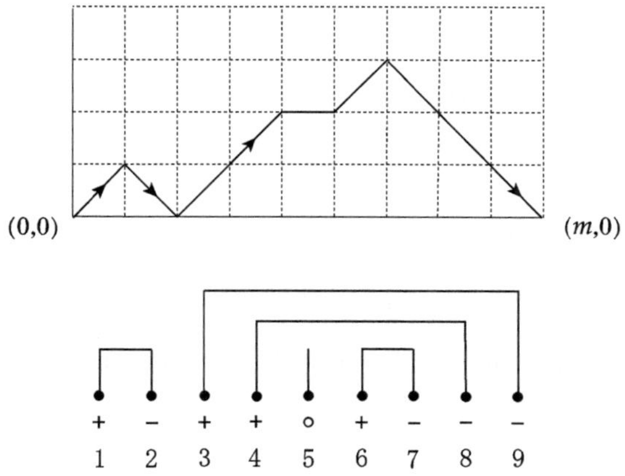

Fig. 1.2. Path in $\mathcal{E}_m^+$ and partition in $\mathcal{P}_{\rm NCPS}(m)$

$$\vartheta = \{\{s_1\}, \dots, \{s_j\}, \{l_1, r_1\}, \dots, \{l_k, r_k\}\},$$

where we may assume without loss of generality that

$$s_1 < \cdots < s_j, \quad l_1 < \cdots < l_k, \quad l_1 < r_1, \dots, \quad l_k < r_k.$$

We say that ϑ is *non-crossing* if for any $1 \le \alpha, \beta \le k$,

$$[l_\alpha, r_\alpha] \subset [l_\beta, r_\beta] \quad \text{or} \quad [l_\beta, r_\beta] \subset [l_\alpha, r_\alpha] \quad \text{or} \quad [l_\alpha, r_\alpha] \cap [l_\beta, r_\beta] = \varnothing$$

occurs, where $[l, r] = \{u \,;\, l \le u \le r\}$. Let $\mathcal{P}_{\rm NCP}(m)$ and $\mathcal{P}_{\rm NCPS}(m)$ denote the set of non-crossing pair partitions of $\{1, 2, \dots, m\}$ and that of non-crossing pair partitions with singletons, respectively.

Lemma 1.59. *Let $\epsilon \in \mathcal{E}_m^+$ and $\vartheta(\epsilon)$ the pair partition with singleton of $\{1, 2, \dots, m\}$ defined as in (1.47). Then $\vartheta(\epsilon)$ is non-crossing. Moreover, the map $\epsilon \mapsto \vartheta(\epsilon)$ is a bijection from $\mathcal{E}_m^+$ onto $\mathcal{P}_{\rm NCPS}(m)$.*

Proof. It is obvious from construction that $\vartheta(\epsilon)$ is non-crossing and that $\epsilon \mapsto \vartheta(\epsilon)$ is injective. Suppose we are given $\vartheta \in \mathcal{P}_{\rm NCPS}(m)$. Set

$$\vartheta = \{\{s_1\}, \dots, \{s_j\}, \{l_1, r_1\}, \dots, \{l_k, r_k\}\}$$

and assume that

$$s_1 < \cdots < s_j, \quad l_1 < \cdots < l_k, \quad l_1 < r_1, \quad \dots, \quad l_k < r_k. \tag{1.48}$$

Define $\epsilon \in \{+, -, \circ\}^m$ by

$$\epsilon(s_t) = \circ, \quad \epsilon(l_u) = +, \quad \epsilon(r_u) = -. \tag{1.49}$$

It is apparent that $\epsilon(1) + \cdots + \epsilon(m) = 0$. We shall prove that $\epsilon \in \mathcal{E}_m^+$, i.e.,

$$\epsilon(1) + \cdots + \epsilon(i) \geq 0, \qquad i = 1, 2, \ldots, m. \tag{1.50}$$

Given i, we choose u such that

$$l_1 < \cdots < l_u \leq i < l_{u+1} < \cdots < l_k.$$

Then, by (1.48) we have

$$\{r_1, \ldots, r_k\} \cap [1, i] \subset \{r_1, \ldots, r_u\}.$$

Hence on the left-hand side of (1.50), $(+1)$ appears u times and (-1) at most u times, which shows that (1.50) holds. Finally, we need to prove that for ϵ defined in (1.49), $\vartheta(\epsilon) = \vartheta$. Set

$$\{l_1, \ldots, l_k, r_1, \ldots, r_k\} = \{w_1 < \cdots < w_{2k}\}.$$

The first step of constructing the partition $\vartheta(\epsilon)$ is to find $1 \leq \alpha \leq 2k$ such that

$$\epsilon(w_1) = \cdots = \epsilon(w_\alpha) = +, \quad \epsilon(w_{\alpha+1}) = -.$$

Obviously,

$$w_1 = l_1, \ldots, \quad w_\alpha = l_\alpha,$$

and by the non-crossing condition we have

$$w_{\alpha+1} = r_\alpha.$$

Thus, $\vartheta(\epsilon)$ contains a pair $\{l_\alpha, r_\alpha\}$. Repeating this argument, we conclude that $\vartheta(\epsilon) = \vartheta$. □

Definition 1.60. Let $\vartheta \in \mathcal{P}_{\rm NCPS}(m)$. The *depth* of $v \in \vartheta$ is defined by

$$d_\vartheta(v) = \begin{cases} |\{\{a < b\} \in \vartheta \,;\, a < s < b\}| + 1, & \text{if } v = \{s\}, \\ |\{\{a < b\} \in \vartheta \,;\, a < l < r < b\}| + 1, & \text{if } v = \{l < r\}. \end{cases}$$

For example, for ϑ in Fig. 1.2 it holds that

$$d_\vartheta(\{1,2\}) = 1, \quad d_\vartheta(\{4,8\}) = 2, \quad d_\vartheta(\{5\}) = 3.$$

The next result is easy to see.

Lemma 1.61. *Let $\vartheta \in \mathcal{P}_{\rm NCPS}(m)$ be corresponding to $\epsilon = (\epsilon_1, \ldots, \epsilon_m) \in \mathcal{E}_m^+$. Then*

$$d_\vartheta(v) = \begin{cases} \displaystyle\sum_{i=1}^{s-1} \epsilon_i + 1, & \text{if } v = \{s\}, \\ \displaystyle\sum_{i=1}^{l-1} \epsilon_i + 1 = \sum_{i=1}^{r-1} \epsilon_i \,, & \text{if } v = \{l < r\}. \end{cases}$$

With these notations we continue the calculation of (1.46) and obtain a combinatorial expression of (1.43).

Proposition 1.62. *Let $(\Gamma_{\{\omega_n\}}, B^+, B^-)$ be an interacting Fock space and $B^\circ = \alpha_{N+1}$ the diagonal operator, where $\{\alpha_n\}$ is a real sequence. Then,*

$$\begin{aligned}&\langle \Phi_0, (B^+ + B^- + B^\circ)^m \Phi_0 \rangle \\ &\quad = \sum_{\vartheta \in \mathcal{P}_{\mathrm{NCPS}}(m)} \prod_{\substack{v \in \vartheta \\ |v|=1}} \alpha(d_\vartheta(v)) \prod_{\substack{v \in \vartheta \\ |v|=2}} \omega(d_\vartheta(v)), \qquad m = 1, 2, \dots . \end{aligned} \tag{1.51}$$

In particular,

$$\begin{cases} \langle \Phi_0, (B^+ + B^-)^{2m-1} \Phi_0 \rangle = 0, \\[2mm] \langle \Phi_0, (B^+ + B^-)^{2m} \Phi_0 \rangle = \displaystyle\sum_{\vartheta \in \mathcal{P}_{\mathrm{NCP}}(2m)} \prod_{v \in \vartheta} \omega(d_\vartheta(v)). \end{cases} \tag{1.52}$$

Proof. From (1.46) we already know that

$$\langle \Phi_0, (B^+ + B^- + B^\circ)^m \Phi_0 \rangle = \sum_{\epsilon \in \mathcal{E}_m^+} \langle \Phi_0, B^{\epsilon_m} \cdots B^{\epsilon_2} B^{\epsilon_1} \Phi_0 \rangle.$$

We shall calculate $B^{\epsilon_m} \cdots B^{\epsilon_2} B^{\epsilon_1} \Phi_0$ for $\epsilon = (\epsilon_1, \dots, \epsilon_m) \in \mathcal{E}_m^+$. Denote by $\vartheta = \vartheta(\epsilon) \in \mathcal{P}_{\mathrm{NCPS}}(m)$ the corresponding partition and set

$$\vartheta(\epsilon) = \{\{s_1\}, \dots, \{s_j\}, \{l_1, r_1\}, \dots, \{l_k, r_k\}\}.$$

First consider a singleton $s = s_i$. Since $B^{\epsilon_{s-1}} \cdots B^{\epsilon_1} \Phi_0 \in \mathbb{C}\Phi_{\epsilon_1 + \cdots + \epsilon_{s-1}}$ and $B^{\epsilon_s} = B^\circ$, we obtain by virtue of Lemma 1.61

$$\begin{aligned} B^{\epsilon_s} B^{\epsilon_{s-1}} \cdots B^{\epsilon_1} \Phi_0 &= \alpha(\epsilon_1 + \cdots + \epsilon_{s-1} + 1) B^{\epsilon_{s-1}} \cdots B^{\epsilon_1} \Phi_0 \\ &= \alpha(d_\vartheta(\{s\})) B^{\epsilon_{s-1}} \cdots B^{\epsilon_1} \Phi_0. \end{aligned}$$

Applying the above argument to all the singletons, we obtain

$$B^{\epsilon_m} \cdots B^{\epsilon_1} \Phi_0 = \left\{ \prod_{i=1}^{j} \alpha(d_\vartheta(\{s_i\})) \right\} [[B^{\epsilon_m} \cdots B^{\epsilon_1}]] \Phi_0, \tag{1.53}$$

where $[[B^{\epsilon_m} \cdots B^{\epsilon_1}]]$ stands for omission of B°. Then $[[B^{\epsilon_m} \cdots B^{\epsilon_1}]]$ is a product of k creation operators B^+ and k annihilation operators B^- which form a non-crossing pair partition. Hence there exists $\{l, r\} = \{l_i, r_i\}$ such that B^{ϵ_r} and B^{ϵ_l} are consecutive. In that case

$$[[B^{\epsilon_m} \cdots B^{\epsilon_r} B^{\epsilon_l} \cdots B^{\epsilon_1}]] \Phi_0 = [[B^{\epsilon_m} \cdots B^- B^+ \cdots B^{\epsilon_1}]] \Phi_0.$$

Since the action of B° does not change the level of the number vectors, in the above expression $[[\cdots B^{\epsilon_1}]] \Phi_0 \in \mathbb{C}\Phi_{\epsilon_1 + \cdots + \epsilon_{l-1}}$ so that the action of $B^- B^+$ on it becomes a scalar $\omega(\epsilon_1 + \cdots + \epsilon_{l-1} + 1) = \omega(d_\vartheta(\{l,r\}))$, where Lemma 1.61 is taken into account. Thus, we have

$$[[B^{\epsilon_m} \cdots B^{\epsilon_r} B^{\epsilon_l} \cdots B^{\epsilon_1}]]\Phi_0 = \omega(d_\vartheta(\{l,r\}))[[B^{\epsilon_m} \cdots \check{B}^{\epsilon_r} \check{B}^{\epsilon_l} \cdots B^{\epsilon_1}]]\Phi_0,$$

where $\check{B}^{\epsilon_r}\check{B}^{\epsilon_l}$ means that $B^{\epsilon_r}B^{\epsilon_l}$ is omitted. Repeating this argument, we obtain

$$[[B^{\epsilon_m} \cdots B^{\epsilon_1}]]\Phi_0 = \left\{ \prod_{i=1}^{k} \omega(d_\vartheta(\{l_i, r_i\})) \right\} \Phi_0. \tag{1.54}$$

Now formula (1.51) follows immediately from (1.53) and (1.54). Formula (1.52) follows from (1.51). □

Theorem 1.63 (Accardi–Bożejko formula). *For $\mu \in \mathfrak{P}_{\mathrm{fm}}(\mathbb{R})$ let $\{M_m\}$ be its moment sequence and $(\{\omega_n\}, \{\alpha_n\})$ its Jacobi coefficient. Then it holds that*

$$M_m = \sum_{\vartheta \in \mathcal{P}_{\mathrm{NCPS}}(m)} \prod_{\substack{v \in \vartheta \\ |v|=1}} \alpha(d_\vartheta(v)) \prod_{\substack{v \in \vartheta \\ |v|=2}} \omega(d_\vartheta(v)), \quad m = 1, 2, \dots. \tag{1.55}$$

Moreover, if μ is symmetric,

$$\begin{cases} M_{2m-1} = 0, \\ M_{2m} = \displaystyle\sum_{\vartheta \in \mathcal{P}_{\mathrm{NCP}}(2m)} \prod_{v \in \vartheta} \omega(d_\vartheta(v)), \quad m = 1, 2, \dots. \end{cases} \tag{1.56}$$

Proof. Consider an algebraic probability space $(L(\mathcal{P}(\mathbb{R}, \mu)), P_0)$ and a real random variable M_X (the multiplication operator by X). Let $\Gamma_{\{\omega_n\}}$ be the interacting Fock space associated with $\{\omega_n\}$ and consider the diagonal operator $B^\circ = \alpha_{N+1}$. It then follows from Corollary 1.53 that

$$M_X \stackrel{\mathrm{s}}{=} B^+ + B^- + B^\circ,$$

namely,

$$\langle M_X^m \rangle = \langle \Phi_0, (B^+ + B^- + B^\circ)^m \Phi_0 \rangle, \qquad m = 1, 2, \dots.$$

Note that the left-hand side coincides with the mth moment of μ, while the right-hand side was calculated in Proposition 1.62. Thus we obtain (1.55). If μ is symmetric, we have $\alpha_n = 0$ for all n, see Proposition 1.47. Hence in (1.55) the terms including singletons vanish and (1.56) follows. □

In (1.26) we defined a map $\mathfrak{M} \to \mathfrak{J}$ and in Proposition 1.56 we proved it to be surjective. As an application of the Accardi–Bożejko formula, we prove the following:

Theorem 1.64. *The map $\mathfrak{M} \to \mathfrak{J}$ defined in* (1.26) *is bijective. Moreover, the inverse is given by* (1.55).

Proof. We only need to prove the injectivity. Suppose that $\{M_m\}, \{M'_m\} \in \mathfrak{M}$ have the same image $(\{\omega_n\}, \{\alpha_n\}) \in \mathfrak{J}$. Take $\mu, \mu' \in \mathfrak{P}_{\mathrm{fm}}(\mathbb{R})$ whose moment sequences are $\{M_m\}, \{M'_m\}$, respectively. It then follows from Theorem 1.63 that $M_m = M'_m$ for all $m = 1, 2, \dots$. □

We are now in a good position to discuss uniqueness of the quantum decomposition stated in Corollary 1.55.

Corollary 1.65. *Let $(\mathcal{A}, \varphi)$ be an algebraic probability space. For a real random variable $a \in \mathcal{A}$ there exists an interacting Fock space $\Gamma_{\{\omega_n\}} = (\Gamma, \{\Phi_n\}, B^+, B^-)$ and a diagonal operator $B^\circ = \alpha_{N+1}$ such that*

$$a \stackrel{\mathrm{s}}{=} B^+ + B^- + B^\circ.$$

This quantum decomposition is unique in the sense that the Jacobi coefficient $(\{\omega_n\}, \{\alpha_n\})$ is uniquely specified by a.

Proof. The first half is just a repetition of Corollary 1.55. Let

$$a \stackrel{\mathrm{s}}{=} C^+ + C^- + C^\circ$$

be another quantum decomposition of a obtained from another Jacobi coefficient $(\{\omega'_n\}, \{\alpha'_n\})$, where $(\Gamma_{\{\omega'_n\}}, \{\Psi_n\}, C^+, C^-)$ and $C^\circ = \alpha'_{N+1}$. Since $B^+ + B^- + B^\circ \stackrel{\mathrm{s}}{=} C^+ + C^- + C^\circ$, we have

$$\sum_{\vartheta \in \mathcal{P}_{\mathrm{NCPS}}(m)} \prod_{\substack{v \in \vartheta \\ |v|=1}} \alpha(d_\vartheta(v)) \prod_{\substack{v \in \vartheta \\ |v|=2}} \omega(d_\vartheta(v))$$

$$= \sum_{\vartheta \in \mathcal{P}_{\mathrm{NCPS}}(m)} \prod_{\substack{v \in \vartheta \\ |v|=1}} \alpha'(d_\vartheta(v)) \prod_{\substack{v \in \vartheta \\ |v|=2}} \omega'(d_\vartheta(v)), \quad m = 1, 2, \dots.$$

But $\mathfrak{M} \to \mathfrak{J}$ being bijective by Theorem 1.64, $(\{\omega_n\}, \{\alpha_n\}) = (\{\omega'_n\}, \{\alpha'_n\})$ follows. □

Recall that the map $\mathfrak{P}_{\mathrm{fm}}(\mathbb{R}) \to \mathfrak{M}$ is not injective (determinate moment problem). Therefore, $\mathfrak{P}_{\mathrm{fm}}(\mathbb{R}) \to \mathfrak{J}$ is not either.

Theorem 1.66 (Carleman's condition). *Let $(\{\omega_n\}, \{\alpha_n\})$ be a Jacobi coefficient of $\mu \in \mathfrak{P}_{\mathrm{fm}}(\mathbb{R})$. Then, μ is the solution of a determinate moment problem if*

$$\sum_{n=1}^{\infty} \frac{1}{\sqrt{\omega_n}} = +\infty. \tag{1.57}$$

Remark 1.67. If $\omega_n = 0$ occurs, we understand that (1.57) is fulfilled automatically. In that case, Theorem 1.66 says that μ is the solution of a determinate moment problem. Indeed, the Jacobi coefficient is of finite type so that μ is a finite sum of δ-measures.

Remark 1.68. Carleman's moment test (Theorem 1.36) and Theorem 1.66 are not equivalent. By using the Accardi–Bożejko formula we may construct an example which satisfies the condition of Theorem 1.66 but does not satisfy that of Theorem 1.36.

Finally, it may be worthwhile to mention a few words about how to deal with a classical random variable. Let X be a classical $\mathbb{R}$-valued random variable defined on a probability space $(\Omega, \mathcal{F}, P)$. Let μ be the distribution of X and assume that $\mu \in \mathfrak{P}_{\mathrm{fm}}(\mathbb{R})$, that is, $\mathbb{E}(|X|^m) < \infty$ for all $m = 1, 2, \dots$. Then, taking the Jacobi coefficient $(\{\omega_n\}, \{\alpha_n\})$ of μ, we obtain

$$\mathbb{E}(X^m) = \int_{-\infty}^{+\infty} x^m \mu(dx) = \langle \Phi_0, (B^+ + B^- + B^\circ)^m \Phi_0 \rangle, \qquad m = 1, 2, \dots.$$

We thereby write

$$X \stackrel{\mathrm{s}}{=} B^+ + B^- + B^\circ$$

and call it the *quantum decomposition* of a classical random variable X. The quantum decomposition brings a classical variable X into a non-commutative paradigm where X is studied by means of its quantum components.

1.7 Fermion, Free and Boson Fock Spaces

Theorem 1.69. *For the Fermion Fock space* $(\Gamma_{\mathrm{Fermion}}, \{\Phi_n\}, B^+, B^-)$,

$$\langle \Phi_0, (B^+ + B^-)^m \Phi_0 \rangle = \frac{1}{2} \int_{-\infty}^{+\infty} x^m (\delta_{-1} + \delta_{+1})(dx), \qquad m = 1, 2, \dots. \tag{1.58}$$

Proof. Let $\{M_m\}$ be the right-hand side of (1.58). Obviously, we have

$$M_{2m-1} = 0, \qquad M_{2m} = 1, \qquad m = 1, 2, \dots.$$

By using Proposition 1.62 with $\omega_1 = 1$, $\omega_2 = \omega_3 = \cdots = 0$ and $\alpha_n = 0$ for all n, we obtain

$$\langle \Phi_0, (B^+ + B^-)^{2m-1} \Phi_0 \rangle = 0 = M_{2m-1},$$

which shows that (1.58) holds for odd m. Similarly, we start with the formula

$$\langle \Phi_0, (B^+ + B^-)^{2m} \Phi_0 \rangle = \sum_{\vartheta \in \mathcal{P}_{\mathrm{NCP}}(2m)} \prod_{v \in \vartheta} \omega(d_\vartheta(v)) \tag{1.59}$$

and observe that on the right-hand side a non-zero term appears only when $d_\vartheta(v) = 1$ for all $v \in \vartheta$. Since such a pair partition ϑ is unique, (1.59) is equal to 1 and hence to M_{2m}. Namely, (1.58) holds also for even m. □

The probability measure appearing on the right-hand side of (1.58) is called the *Bernoulli distribution.* This is the probability distribution of a Bernoulli random variable X such that $P(X = +1) = P(X = -1) = 1/2$, which is also called a coin-tossing. Hence it follows from Theorem 1.69 that $X \stackrel{s}{=} B^+ + B^-$, which is sometimes called a *quantum coin-tossing*. In an explicit form we have

$$\text{(coin-tossing)} \stackrel{s}{=} \begin{pmatrix} 0 & 1 \\ 0 & 0 \end{pmatrix} + \begin{pmatrix} 0 & 0 \\ 1 & 0 \end{pmatrix},$$

where the right-hand side is regarded as a sum of random variables in the vector state $\binom{0}{1}$.

We next discuss the free Fock space.

Definition 1.70. Set $\mathcal{C}_m = \mathcal{E}_{2m}^+ \cap \{+,-\}^{2m}$. An element in $\mathcal{C}_m$ is called a *Catalan path* of length $2m$. In other words, $(\epsilon_1, \epsilon_2, \dots, \epsilon_{2m}) \in \{+,-\}^{2m}$ is a Catalan path if and only if

$$\epsilon_1 + \cdots + \epsilon_k \geq 0, \quad k = 1, 2, \dots, 2m-1,$$
$$\epsilon_1 + \cdots + \epsilon_{2m} = 0.$$

We call $|\mathcal{C}_m|$ the mth *Catalan number.*

Lemma 1.71. *For $m = 1, 2, \dots$ the mth Catalan number is given by*

$$|\mathcal{C}_m| = |\mathcal{P}_{\mathrm{NCP}}(2m)| = \frac{(2m)!}{m!(m+1)!} = \frac{1}{m+1}\binom{2m}{m}.$$

Proof. The bijection between $\mathcal{E}_{2m}^+$ and $\mathcal{P}_{\mathrm{NCPS}}(2m)$ established in Lemma 1.59 gives rise to a bijective correspondence between $\mathcal{C}_m$ and $\mathcal{P}_{\mathrm{NCP}}(2m)$. We need to count the number of Catalan paths in $\mathcal{C}_m$. Set

$$\tilde{\mathcal{C}}_m = \left\{\epsilon = (\epsilon_1, \epsilon_2, \dots, \epsilon_{2m}) \in \{+,-\}^{2m}\,;\, \epsilon_1 + \cdots + \epsilon_{2m} = 0\right\}.$$

Obviously, $\mathcal{C}_m \subset \tilde{\mathcal{C}}_m$. Each $\epsilon \in \tilde{\mathcal{C}}_m$ corresponds to a path connecting the vertices

$$(0,0),\ (1,\epsilon_1),\ (2,\epsilon_1+\epsilon_2),\ \dots,\ (2m, \epsilon_1+\epsilon_2+\cdots+\epsilon_{2m}) = (2m, 0)$$

in order. Since we know

$$|\tilde{\mathcal{C}}_m| = \binom{2m}{m} = \frac{(2m)!}{m!m!},$$

for $|\mathcal{C}_m|$ it is sufficient to count the number of paths in $\tilde{\mathcal{C}}_m \setminus \mathcal{C}_m$. By definition a path $\epsilon = (\epsilon_1, \epsilon_2, \dots, \epsilon_{2m})$ in $\tilde{\mathcal{C}}_m \setminus \mathcal{C}_m$ has one or more vertices with negative ordinates. Let k be the abscissa of the first such vertex. Then $1 \leq k \leq 2m-1$. If $k = 1$ we have $\epsilon_1 = -1$. Otherwise,

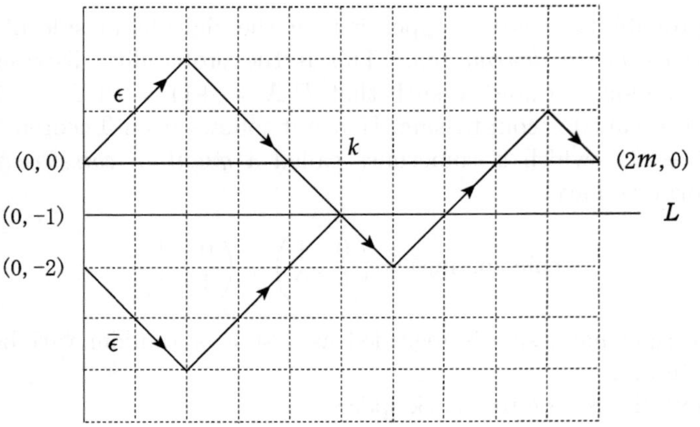

Fig. 1.3. Counting the Catalan number

$$\epsilon_1 \geq 0, \quad \epsilon_1 + \epsilon_2 \geq 0, \quad \dots, \quad \epsilon_1 + \dots + \epsilon_{k-1} = 0,$$
$$\epsilon_1 + \dots + \epsilon_{k-1} + \epsilon_k = -1.$$

Let L be the horizontal line passing through $(0,-1)$. Then ϵ has one or more vertices which lie on L and $(k,-1)$ is the first one. Define $\bar{\epsilon}$ to be the path obtained from ϵ by reflecting the first part of ϵ up to $(k,-1)$ with respect to L (see Fig. 1.3). Then $\bar{\epsilon}$ becomes a path from $(0,-2)$ to $(2m,0)$ passing through $(k,-1)$ as the first meeting point with L. It is easily verified that $\epsilon \leftrightarrow \bar{\epsilon}$ is a one-to-one correspondence between $\tilde{\mathcal{C}}_m \setminus \mathcal{C}_m$ and the set of paths connecting $(0,-2)$ and $(2m,0)$. Obviously, the number of such paths is

$$\binom{2m}{m+1} = \frac{(2m)!}{(m+1)!(m-1)!} = |\tilde{\mathcal{C}}_m \setminus \mathcal{C}_m|.$$

Hence

$$|\mathcal{C}_m| = \frac{(2m)!}{m!m!} - \frac{(2m)!}{(m+1)!(m-1)!} = \frac{(2m)!}{m!(m+1)!},$$

which completes the proof. □

The method used in the above proof is well known under the name of reflection principle.

Theorem 1.72. *For the free Fock space $(\Gamma_{\text{free}}, \{\Phi_n\}, B^+, B^-)$ it holds that*

$$\langle \Phi_0, (B^+ + B^-)^m \Phi_0 \rangle = \frac{1}{2\pi} \int_{-2}^{+2} x^m \sqrt{4 - x^2}\, dx, \quad m = 1, 2, \dots. \tag{1.60}$$

Definition 1.73. The probability measure whose density function is given by

$$\rho(x) = \begin{cases} \dfrac{1}{2\pi}\sqrt{4-x^2}\,, & |x| \le 2, \\ 0, & \text{otherwise}, \end{cases}$$

is called the (normalized) *Wigner semicircle law.*

Proof. Let $\{M_m\}$ denote the moment sequence of the Wigner semicircle law. By an elementary calculus one may easily check that the odd moments vanish and the even moments are given by

$$M_{2m} = \frac{1}{2\pi}\int_{-2}^{+2} x^{2m}\sqrt{4-x^2}\,dx = \frac{(2m)!}{m!(m+1)!}, \qquad m = 1, 2, \dots .$$

On the other hand, from Proposition 1.62 we see that

$$\langle \Phi_0, (B^+ + B^-)^{2m-1}\Phi_0\rangle = 0 = M_{2m-1},$$

which shows that (1.60) is valid for odd m. As for the even moments, since $\omega_n = 1$ for all n we obtain

$$\langle \Phi_0, (B^+ + B^-)^{2m}\Phi_0\rangle = \sum_{\vartheta\in\mathcal{P}_{\mathrm{NCP}}(2m)} \prod_{v\in\vartheta} \omega(d_\vartheta(v)) = |\mathcal{P}_{\mathrm{NCP}}(2m)|,$$

which is the Catalan number (Lemma 1.71) and coincides with M_{2m}. □

The orthogonal polynomials associated with the Wigner semicircle law are essentially the Chebyshev polynomials of the second kind. In accordance with the tradition we give the following:

Definition 1.74. The polynomials $T_n(x)$ defined by

$$T_n(\cos\theta) = \cos n\theta, \quad n = 0, 1, 2, \dots,$$

are called the *Chebyshev polynomials of the first kind.* Similarly, the polynomials $U_n(x)$ defined by

$$U_n(\cos\theta) = \frac{\sin(n+1)\theta}{\sin\theta}, \quad n = 0, 1, 2, \dots,$$

are called the *Chebyshev polynomials of the second kind.*

Theorem 1.75. *The orthogonal polynomials associated with the Wigner semicircle law are given by*

$$\tilde{U}_n(x) = U_n\left(\frac{x}{2}\right), \quad n = 0, 1, 2, \dots,$$

where $\{U_n(x)\}$ are the Chebyshev polynomials of the second kind.

Proof. The three-term recurrence relation for $\{\tilde{U}_n(x)\}$ is easily obtained by means of the elementary formulae for the trigonometric functions. We then see that the Jacobi coefficient is $(\{\omega_n \equiv 1\}, \{\alpha_n \equiv 0\})$. See also Exercise 1.14. □

For the Boson Fock space we need some combinatorial preparation. Let $\mathcal{P}_{\mathrm{P}}(2m)$ denote the set of pair partitions of $\{1, 2, \dots, 2m\}$.

Lemma 1.76. *For* $\vartheta = \{\{l_1 < r_1\}, \dots, \{l_m < r_m\}\} \in \mathcal{P}_{\mathrm{P}}(2m)$ *we define*

$$\epsilon_i = \begin{cases} +, & i \in \{l_1, \dots, l_m\}, \\ -, & i \in \{r_1, \dots, r_m\}. \end{cases} \tag{1.61}$$

Then $(\epsilon_1, \dots, \epsilon_{2m}) \in \mathcal{C}_m$.

The proof is obvious. Then, by using the bijection between $\mathcal{C}_m$ and $\mathcal{P}_{\mathrm{NCP}}(2m)$, with each $\vartheta \in \mathcal{P}_{\mathrm{P}}(2m)$ we may associate a non-crossing pair partition which is denoted by $\pi(\vartheta) \in \mathcal{P}_{\mathrm{NCP}}(2m)$.

Lemma 1.77. *The map* $\pi : \mathcal{P}_{\mathrm{P}}(2m) \to \mathcal{P}_{\mathrm{NCP}}(2m)$ *defined above is surjective and is the identity on* $\mathcal{P}_{\mathrm{NCP}}(2m)$. *Moreover,*

$$|\pi^{-1}(\vartheta)| = \prod_{v \in \vartheta} d_\vartheta(v), \qquad \vartheta \in \mathcal{P}_{\mathrm{NCP}}(2m). \tag{1.62}$$

Proof. We only prove (1.62). Let $\vartheta \in \mathcal{P}_{\mathrm{NCP}}(2m)$ such that

$$\vartheta = \{\{l_1 < r_1\}, \dots, \{l_m < r_m\}\}, \quad r_1 < \dots < r_m = 2m,$$

and define $(\epsilon_1, \dots, \epsilon_{2m}) \in \mathcal{C}_m$ as in (1.61). By definition any element in $\pi^{-1}(\vartheta)$ is of the form

$$\vartheta' = \{\{l'_1 < r_1\}, \dots, \{l'_m < r_m\}\}, \quad \{l'_1, \dots, l'_m\} = \{l_1, \dots, l_m\} \equiv L.$$

We study how to choose $l'_1, \dots, l'_m$ (see Fig. 1.4). First l'_1 should be chosen from

$$\{1, 2, \dots, r_1 - 1\} \cap L.$$

Since the above set does not contain r_i's, the number of choice of l'_1 is

$$r_1 - 1 = \sum_{i=1}^{r_1 - 1} \epsilon_i = d_\vartheta(\{l_1, r_1\}).$$

Suppose $l'_1, \dots, l'_{k-1}$ is chosen. Then l'_k is chosen from

$$\{1, 2, \dots, r_k - 1\} \setminus \{l'_1, r_1, \dots, l'_{k-1}, r_{k-1}\}.$$

The number of choice is

$$r_k - 1 - 2(k-1) = \sum_{i=1}^{r_k - 1} \epsilon_i = d_\vartheta(\{l_k, r_k\}).$$

Repeating this argument, we see that the number of choice of $l'_1, \dots, l'_m$ is $\prod_{k=1}^m d_\vartheta(\{l_k, r_k\})$, which proves (1.62). □

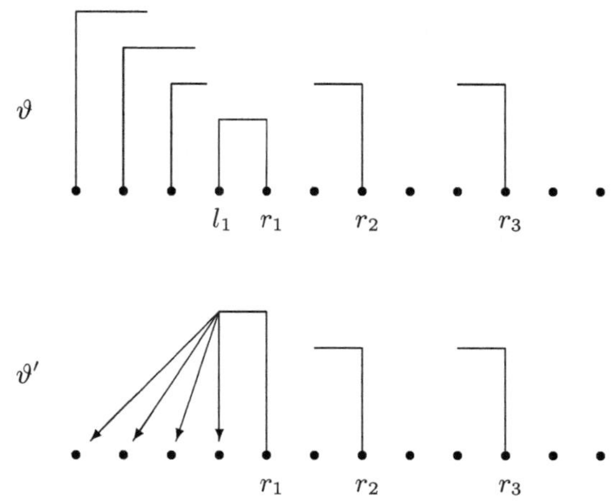

Fig. 1.4. Constructing $\vartheta' \in \pi^{-1}(\vartheta)$

Theorem 1.78. *For the Boson Fock space* $\Gamma_{\mathrm{Boson}} = (\Gamma, \{\Phi_n\}, B^+, B^-)$ *it holds that*

$$\langle \Phi_0, (B^+ + B^-)^m \Phi_0 \rangle = \frac{1}{\sqrt{2\pi}} \int_{-\infty}^{+\infty} x^m e^{-x^2/2} dx, \qquad m = 1, 2, \dots. \tag{1.63}$$

Definition 1.79. The probability measure with density function

$$\frac{1}{\sqrt{2\pi}} e^{-x^2/2}, \qquad x \in \mathbb{R},$$

is called the *standard Gaussian distribution.*

Proof. The moment sequence $\{M_m\}$ of the standard Gaussian distribution is easily computed:

$$M_{2m-1} = 0, \qquad M_{2m} = \frac{(2m)!}{2^m m!}, \qquad m = 1, 2, \dots.$$

On the other hand,

$$\langle \Phi_0, (B^+ + B^-)^{2m-1} \Phi_0 \rangle = 0, \qquad m = 1, 2, \dots,$$

is obvious so that (1.63) holds for an odd m. Recall that the Boson Fock space is by definition the interacting Fock space associated with the Jacobi sequence $\{\omega_n = n\}$. Hence we see from Proposition 1.62 that

$$\langle \Phi_0, (B^+ + B^-)^{2m} \Phi_0 \rangle = \sum_{\vartheta \in \mathcal{P}_{\mathrm{NCP}}(2m)} \prod_{v \in \vartheta} d_\vartheta(v). \tag{1.64}$$

Applying Lemma 1.77, we see that (1.64) becomes

$$\sum_{\vartheta \in \mathcal{P}_{\mathrm{NCP}}(2m)} |\pi^{-1}(\vartheta)| = |\mathcal{P}_{\mathrm{P}}(2m)| = \frac{(2m)!}{2^m m!}.$$

Thus,

$$\langle \Phi_0, (B^+ + B^-)^{2m} \Phi_0 \rangle = \frac{(2m)!}{2^m m!} = M_{2m},$$

which shows that (1.63) holds for an even m too. □

As is easily verified, for each $n = 0, 1, 2, \dots$, there exists a polynomial $H_n(x)$ of degree n such that

$$e^{-z^2+2xz} = \sum_{n=0}^{\infty} \frac{z^n}{n!} H_n(x). \tag{1.65}$$

Definition 1.80. The above $\{H_n(x)\}$ are called the *Hermite polynomials.*

Theorem 1.81. *The orthogonal polynomials associated with the standard Gaussian measure are given by*

$$\tilde{H}_n(x) = \frac{1}{\sqrt{2^n}} H_n\left(\frac{x}{\sqrt{2}}\right), \qquad n = 0, 1, 2, \dots,$$

where $\{H_n(x)\}$ are the Hermite polynomials.

Proof. By simple manipulation of (1.65) we obtain the recurrence relation

$$H_0(x) = 1, \quad H_1(x) = 2x, \quad H_n(x) = 2xH_{n-1}(x) - 2(n-1)H_{n-2}(x).$$

The assertion is then immediate. □

Remark 1.82. The Bernoulli distribution, the Wigner semicircle law and the Gaussian distribution are the solutions of determinate moment problems.

1.8 Theory of Finite Jacobi Matrices

First we recall the notion of a continued fraction. In general, expressions of the forms

$$\cfrac{a_1}{b_1 + \cfrac{a_2}{b_2 + \cfrac{a_3}{b_3 + \ddots + \cfrac{a_n}{b_n}}}} = \frac{a_1}{b_1} + \frac{a_2}{b_2} + \frac{a_3}{b_3} + \cdots + \frac{a_n}{b_n} \tag{1.66}$$

and

$$\cfrac{a_1}{b_1 + \cfrac{a_2}{b_2 + \cfrac{a_3}{b_3 + \ddots}}} = \frac{a_1}{b_1} + \frac{a_2}{b_2} + \frac{a_3}{b_3} + \cdots \tag{1.67}$$

are called *continued fractions.* Since the expressions on the left-hand sides are space-consuming, we hereafter adopt the ones on the right-hand sides. We only need to consider complex numbers $\{a_k\}$ and $\{b_k\}$. To be slightly more precise, we define a linear fractional transformation τ_k by

$$\tau_k(w) = \frac{a_k}{b_k + w}, \qquad w \in \mathbb{C} \cup \{\infty\}, \qquad k = 1, 2, \dots. \tag{1.68}$$

Then the continued fraction (1.66) means by definition

$$\frac{a_1}{b_1} + \frac{a_2}{b_2} + \frac{a_3}{b_3} + \cdots + \frac{a_n}{b_n} = \tau_1 \tau_2 \cdots \tau_n(0). \tag{1.69}$$

For the infinite continued fraction (1.67), if $\tau_1 \cdots \tau_n(0)$ takes a value in $\mathbb{C}$ except finitely many n and $\lim_{n\to\infty} \tau_1 \cdots \tau_n(0)$ exists, we say that the infinite fraction converges and define

$$\frac{a_1}{b_1} + \frac{a_2}{b_2} + \frac{a_3}{b_3} + \cdots = \lim_{n\to\infty} \tau_1 \cdots \tau_n(0).$$

In other words, the value of the infinite continued fraction (1.67) is defined as the limit of the *nth approximant*:

$$\frac{a_1}{b_1} + \frac{a_2}{b_2} + \frac{a_3}{b_3} + \cdots = \lim_{n\to\infty} \frac{a_1}{b_1} + \frac{a_2}{b_2} + \frac{a_3}{b_3} + \cdots + \frac{a_n}{b_n}.$$

We now prove the fundamental recurrence relations satisfied by the coefficients of the linear fractional transformation $\tau_1 \cdots \tau_n$.

Proposition 1.83. *Let $\{a_n\}$ and $\{b_n\}$ be two sequences of complex numbers. Define $\{A_n\}$ and $\{B_n\}$ respectively by the following recurrence relations:*

$$\begin{cases} A_{-1} = 1, \quad A_0 = 0, \\ A_n = b_n A_{n-1} + a_n A_{n-2}, & n = 1, 2, \dots, \end{cases}$$

$$\begin{cases} B_{-1} = 0, \quad B_0 = 1, \\ B_n = b_n B_{n-1} + a_n B_{n-2}, & n = 1, 2, \dots \end{cases}$$

Then, for the linear fractional transformations defined in (1.68) *we have*

$$\tau_1 \cdots \tau_n(w) = \frac{A_n + A_{n-1} w}{B_n + B_{n-1} w}, \qquad n = 1, 2, \dots. \tag{1.70}$$

In particular,

$$\frac{a_1}{b_1} + \frac{a_2}{b_2} + \frac{a_3}{b_3} + \cdots + \frac{a_n}{b_n} = \frac{A_n}{B_n}. \tag{1.71}$$

Proof. By induction. First note that

$$A_1 = b_1 A_0 + a_1 A_{-1} = a_1, \qquad B_1 = b_1 B_0 + a_1 B_{-1} = b_1.$$

Hence

$$\tau_1(w) = \frac{a_1}{b_1 + w} = \frac{A_1 + A_0 w}{B_1 + B_0 w},$$

which means that (1.70) is true for $n = 1$. Suppose the assertion holds up to $n \geq 1$. Then,

$$\begin{aligned}
\tau_1 \cdots \tau_n \tau_{n+1}(w) &= \tau_1 \cdots \tau_n \left(\frac{a_{n+1}}{b_{n+1} + w} \right) \\
&= \frac{A_{n-1} a_{n+1} + A_n (b_{n+1} + w)}{B_{n-1} a_{n+1} + B_n (b_{n+1} + w)} \\
&= \frac{(b_{n+1} A_n + a_{n+1} A_{n-1}) + A_n w}{(b_{n+1} B_n + a_{n+1} B_{n-1}) + B_n w} \\
&= \frac{A_{n+1} + A_n w}{B_{n+1} + B_n w},
\end{aligned}$$

so that (1.71) is true for $n + 1$. □

The nth approximant of the continued fraction being given in (1.71), we call A_n and B_n the *nth numerator* and the *nth denominator*, respectively.

A simple application of Proposition 1.83 yields the following:

Proposition 1.84. *Let $\alpha_1, \dots, \alpha_n \in \mathbb{R}$ and $\omega_1 > 0, \dots, \omega_{n-1} > 0$ be constant numbers. Let $\{P_k(z)\}$ and $\{Q_k(z)\}$ are monic polynomials defined respectively by the following recurrence relations:*

$$\begin{cases} P_0(z) = 1, \quad P_1(z) = z - \alpha_1, \\ P_k(z) = (z - \alpha_k) P_{k-1}(z) - \omega_{k-1} P_{k-2}(z), \quad k = 2, 3, \dots, n, \end{cases} \tag{1.72}$$

$$\begin{cases} Q_0(z) = 1, \quad Q_1(z) = z - \alpha_2, \\ Q_k(z) = (z - \alpha_{k+1}) Q_{k-1}(z) - \omega_k Q_{k-2}(z), \; k = 2, 3, \dots, n-1. \end{cases} \tag{1.73}$$

Then it holds that

$$\frac{1}{z - \alpha_1} \, \frac{}{-} \, \frac{\omega_1}{z - \alpha_2} \, \frac{}{-} \, \frac{\omega_2}{z - \alpha_3} \, \frac{}{-} \cdots \frac{}{-} \, \frac{\omega_{k-1}}{z - \alpha_k} = \frac{Q_{k-1}(z)}{P_k(z)}, \tag{1.74}$$

for $k = 1, 2, \dots, n$.

Throughout the rest of this section, letting $n \geq 1$ be a fixed integer, we assume that we are given two sequences

$$\alpha_1, \dots, \alpha_n \in \mathbb{R}, \quad \text{and} \quad \omega_1 > 0, \dots, \omega_{n-1} > 0.$$

In other words, we start with Jacobi coefficients of finite type. It is natural to associate a tridiagonal matrix defined by

$$T = \begin{pmatrix} \alpha_1 & \sqrt{\omega_1} & & & & & \\ \sqrt{\omega_1} & \alpha_2 & \sqrt{\omega_2} & & & & \\ & \sqrt{\omega_2} & \alpha_3 & \sqrt{\omega_3} & & & \\ & & \ddots & \ddots & \ddots & & \\ & & & \ddots & \ddots & \ddots & \\ & & & & \sqrt{\omega_{n-2}} & \alpha_{n-1} & \sqrt{\omega_{n-1}} \\ & & & & & \sqrt{\omega_{n-1}} & \alpha_n \end{pmatrix},$$

which is called a *Jacobi matrix (of finite type)*. The eigenvalues of T are of interest and we shall establish a relation with the continued fraction (1.74). Let $\{e_0, e_1, \dots, e_{n-1}\}$ be the canonical basis of $\mathbb{R}^n$, that is,

$$e_0 = \begin{pmatrix} 1 \\ 0 \\ 0 \\ \vdots \\ 0 \end{pmatrix}, \qquad e_1 = \begin{pmatrix} 0 \\ 1 \\ 0 \\ \vdots \\ 0 \end{pmatrix}, \qquad \dots, \qquad e_{n-1} = \begin{pmatrix} 0 \\ 0 \\ 0 \\ \vdots \\ 1 \end{pmatrix}.$$

Lemma 1.85. *Let $z \in \mathbb{R}$ and $f \in \mathbb{R}^n$. If $(z-T)f = 0$ and $f \neq 0$, then $\langle e_0, f\rangle \neq 0$ and $\langle e_{n-1}, f\rangle \neq 0$.*

Proof. For simplicity we set $f_i = \langle e_i, f\rangle$ and $f = f_0 e_0 + f_1 e_1 + \cdots + f_{n-1} e_{n-1}$. Then the equation $(z-T)f = 0$ is equivalent to

$$(z-\alpha_1) f_0 - \sqrt{\omega_1}\, f_1 = 0, \tag{1.75}$$

$$-\sqrt{\omega_i}\, f_{i-1} + (z-\alpha_{i+1}) f_i - \sqrt{\omega_{i+1}}\, f_{i+1} = 0, \qquad i = 1, 2, \dots, n-2, \tag{1.76}$$

$$-\sqrt{\omega_{n-1}}\, f_{n-2} + (z-\alpha_n) f_{n-1} = 0. \tag{1.77}$$

If $f_0 = \langle e_0, f\rangle = 0$, from (1.75) we obtain $f_1 = 0$. Then, using (1.76) repeatedly, we obtain $f_i = 0$ for all $i = 2, \dots, n-1$. This contradicts the assumption of $f \neq 0$. If $f_{n-1} = \langle e_{n-1}, f\rangle = 0$, a similar argument starting from (1.77) again yields contradiction. □

Proposition 1.86. *Every eigenvalue of T is simple.*

Proof. Suppose that $f, g \in \mathbb{R}^n$ are eigenvectors associated with a common eigenvalue λ. Choosing $a, b \in \mathbb{R}$ such that

$$a\langle e_0, f\rangle + b\langle e_0, g\rangle = 0, \qquad (a,b) \neq (0,0),$$

we set $h = af + bg$. Then

$$(\lambda - T)h = a(\lambda - T)f + b(\lambda - T)g = 0$$

and

$$\langle e_0, h\rangle = a\langle e_0, f\rangle + b\langle e_0, g\rangle = 0.$$

Applying Lemma 1.85 we see that $h = 0$, that is, f and g are linearly dependent. Consequently, each eigenspace is of one dimension. □

Let $\operatorname{Spec} T$ denote the set of eigenvalues of T. Proposition 1.86 says that $\operatorname{Spec} T$ consists of n different real numbers. Given $z \notin \operatorname{Spec} T$, we consider an equation

$$(z - T)f = e_0. \tag{1.78}$$

Obviously, equation (1.78) has a unique solution $f = (z - T)^{-1}e_0$, for which we shall study several expressions in order to derive important identities.

Lemma 1.87.

$$\langle e_0, (z - T)^{-1}e_0\rangle = \frac{1}{z - \alpha_1} \, \frac{}{-} \, \frac{\omega_1}{z - \alpha_2} \, \frac{}{-} \, \frac{\omega_2}{z - \alpha_3} \, \frac{}{-} \cdots \frac{}{-} \, \frac{\omega_{n-1}}{z - \alpha_n}. \tag{1.79}$$

Proof. We write the solution to (1.78) as $f = f_0 e_0 + f_1 e_1 + \cdots + f_{n-1} e_{n-1}$. Then the left-hand side of (1.79) is

$$\langle e_0, (z - T)^{-1}e_0\rangle = \langle e_0, f\rangle = f_0.$$

On the other hand, equation (1.78) is equivalent to

$$(z - \alpha_1)f_0 - \sqrt{\omega_1}\, f_1 = 1, \tag{1.80}$$

$$-\sqrt{\omega_i}\, f_{i-1} + (z - \alpha_{i+1})f_i - \sqrt{\omega_{i+1}}\, f_{i+1} = 0, \qquad i = 1, 2, \dots, n-2, \tag{1.81}$$

$$-\sqrt{\omega_{n-1}}\, f_{n-2} + (z - \alpha_n)f_{n-1} = 0. \tag{1.82}$$

From (1.80) we obtain

$$f_0 \left\{ (z - \alpha_1) - \sqrt{\omega_1}\, \frac{f_1}{f_0} \right\} = 1,$$

and hence

$$f_0 = \frac{1}{z - \alpha_1 - \sqrt{\omega_1}\, \dfrac{f_1}{f_0}}. \tag{1.83}$$

Similarly, from (1.81) we obtain

$$-\sqrt{\omega_i}\, f_{i-1} + f_i \left\{ (z - \alpha_{i+1}) - \sqrt{\omega_{i+1}}\, \frac{f_{i+1}}{f_i} \right\} = 0,$$

and therefore

$$\sqrt{\omega_i}\,\frac{f_i}{f_{i-1}} = \frac{\omega_i}{z - \alpha_{i+1} - \sqrt{\omega_{i+1}}\,\dfrac{f_{i+1}}{f_i}}. \tag{1.84}$$

Finally, from (1.82) we have

$$\sqrt{\omega_{n-1}}\,\frac{f_{n-1}}{f_{n-2}} = \frac{\omega_{n-1}}{z - \alpha_n}. \tag{1.85}$$

Combining (1.83)–(1.85), we obtain

$$f_0 = \frac{1}{z-\alpha_1} - \frac{\omega_1}{z-\alpha_2} - \frac{\omega_2}{z-\alpha_3} - \cdots - \frac{\omega_{n-1}}{z-\alpha_n},$$

from which (1.79) follows. □

Lemma 1.88. *Let $P_n(z)$ and $Q_{n-1}(z)$ be monic polynomials defined by the recurrence relations* (1.72) *and* (1.73), *respectively. Then*

$$\langle e_0, (z-T)^{-1} e_0 \rangle = \frac{Q_{n-1}(z)}{P_n(z)}. \tag{1.86}$$

Proof. Immediate from Lemma 1.87 and Proposition 1.84. □

Proposition 1.89 (Determinantal formula). *For $k = 1, 2, \ldots, n$ it holds that*

$$P_k(z) = \det \begin{pmatrix} z-\alpha_1 & -\sqrt{\omega_1} & & & & \\ -\sqrt{\omega_1} & z-\alpha_2 & -\sqrt{\omega_2} & & & \\ & -\sqrt{\omega_2} & z-\alpha_3 & -\sqrt{\omega_3} & & \\ & & \ddots & \ddots & \ddots & \\ & & & -\sqrt{\omega_{k-2}} & z-\alpha_{k-1} & -\sqrt{\omega_{k-1}} \\ & & & & -\sqrt{\omega_{k-1}} & z-\alpha_k \end{pmatrix}.$$

In particular,

$$P_n(z) = \det(z - T).$$

For $k = 2, 3, \ldots, n$ it holds that

$$Q_{k-1}(z) = \det \begin{pmatrix} z-\alpha_2 & -\sqrt{\omega_2} & & & \\ -\sqrt{\omega_2} & z-\alpha_3 & -\sqrt{\omega_3} & & \\ & \ddots & \ddots & \ddots & \\ & & -\sqrt{\omega_{k-2}} & z-\alpha_{k-1} & -\sqrt{\omega_{k-1}} \\ & & & -\sqrt{\omega_{k-1}} & z-\alpha_k \end{pmatrix}.$$

Proof. By expanding the determinants in the last column one can easily check that these determinants satisfy the recurrence relations (1.72) and (1.73). □

Proposition 1.90. *Every zero of $P_n(z)$ is real and simple. Moreover,*

$$\operatorname{Spec} T = \{\lambda \in \mathbb{C}\,;\, P_n(\lambda) = 0\}. \tag{1.87}$$

Proof. (1.87) follows from Proposition 1.89. Then, the first assertion is immediate from Proposition 1.86. □

Lemma 1.91. *Let $\lambda \in \operatorname{Spec} T$. Then,*

$$f(\lambda) = \begin{pmatrix} P_0(\lambda) \\ P_1(\lambda)/\sqrt{\omega_1} \\ \vdots \\ P_{n-1}(\lambda)/\sqrt{\omega_{n-1}\cdots\omega_1} \end{pmatrix} \tag{1.88}$$

is an eigenvector associated with it. Moreover,

$$\|f(\lambda)\|^2 = \sum_{j=0}^{n-1} \frac{P_j(\lambda)^2}{\omega_j \cdots \omega_1}. \tag{1.89}$$

Proof. In view of (1.72) we obtain

$$\begin{aligned}
&P_0(\lambda) = 1, \\
&P_1(\lambda) = \lambda - \alpha_1, \\
&P_k(\lambda) = (\lambda - \alpha_k)P_{k-1}(\lambda) - \omega_{k-1}P_{k-2}(\lambda), \quad k = 2, 3, \dots, n-1, \\
&0 = (\lambda - \alpha_n)P_{n-1}(\lambda) - \omega_{n-1}P_{n-2}(\lambda).
\end{aligned}$$

The last identity comes from $P_n(\lambda) = \det(\lambda - T) = 0$. Then a simple computation yields

$$\begin{aligned}
&\sqrt{\omega_1}\,\frac{P_1(\lambda)}{\sqrt{\omega_1}} = \lambda - \alpha_1 = (\lambda - \alpha_1)P_0(\lambda), \\
&\sqrt{\omega_k}\,\frac{P_k(\lambda)}{\sqrt{\omega_k \cdots \omega_1}} = (\lambda - \alpha_k)\frac{P_{k-1}(\lambda)}{\sqrt{\omega_{k-1}\cdots\omega_1}} - \sqrt{\omega_{k-1}}\,\frac{P_{k-2}(\lambda)}{\sqrt{\omega_{k-2}\cdots\omega_1}},
\end{aligned}$$

for $k = 2, 3, \dots, n-1$, and

$$0 = (\lambda - \alpha_n)\frac{P_{n-1}(\lambda)}{\sqrt{\omega_{n-1}\cdots\omega_1}} - \sqrt{\omega_{n-1}}\,\frac{P_{n-2}(\lambda)}{\sqrt{\omega_{n-2}\cdots\omega_1}}.$$

The above relations are combined into a single identity: $(\lambda - T)f(\lambda) = 0$. Since $f(\lambda) \neq 0$ for $\langle e_0, f(\lambda)\rangle = P_0(\lambda) = 1$, we see that $f(\lambda)$ is an eigenvector associated with λ. □

Lemma 1.92. *Define a measure μ on $\mathbb{R}$ by*

$$\mu = \sum_{\lambda \in \operatorname{Spec} T} \|f(\lambda)\|^{-2} \delta_\lambda, \tag{1.90}$$

where the weights are given in (1.89). *Then,* $\mu \in \mathfrak{P}_{\mathrm{fm}}(\mathbb{R})$ *and*

$$\langle e_0, (z-T)^{-1} e_0 \rangle = \int_{-\infty}^{+\infty} \frac{\mu(dx)}{z-x}. \tag{1.91}$$

Proof. Since every eigenvalue of T is simple (Proposition 1.86), we see from Lemma 1.91 that $\{f(\lambda)/\|f(\lambda)\|\,;\, \lambda \in \operatorname{Spec} T\}$ becomes a complete orthonormal basis of $\mathbb{C}^n$. The spectral decomposition of T is given by

$$Tv = \sum_{\lambda \in \operatorname{Spec} T} \lambda \left\langle \frac{f(\lambda)}{\|f(\lambda)\|}, v \right\rangle \frac{f(\lambda)}{\|f(\lambda)\|}, \qquad v \in \mathbb{C}^n. \tag{1.92}$$

It then follows that

$$(z-T)^{-1} v = \sum_{\lambda \in \operatorname{Spec} T} (z-\lambda)^{-1} \left\langle \frac{f(\lambda)}{\|f(\lambda)\|}, v \right\rangle \frac{f(\lambda)}{\|f(\lambda)\|}, \qquad v \in \mathbb{C}^n,$$

and hence

$$\langle e_0, (z-T)^{-1} e_0 \rangle = \sum_{\lambda \in \operatorname{Spec} T} (z-\lambda)^{-1} \frac{|\langle f(\lambda), e_0 \rangle|^2}{\|f(\lambda)\|^2}.$$

Using $\langle f(\lambda), e_0 \rangle = P_0(\lambda) = 1$, which is immediate from (1.88), and the definition of μ, we obtain

$$\langle e_0, (z-T)^{-1} e_0 \rangle = \sum_{\lambda \in \operatorname{Spec} T} \frac{\|f(\lambda)\|^{-2}}{z-\lambda} = \int_{-\infty}^{+\infty} \frac{\mu(dx)}{z-x},$$

which proves (1.91).

We need to show that $\mu(\mathbb{R}) = 1$. This may be proved by observing asymptotics of both sides of (1.91). In fact, with the help of Lemma 1.88 we see that

$$\lim_{\substack{z \to \infty \\ \operatorname{Re} z = 0}} z \langle e_0, (z-T)^{-1} e_0 \rangle = \lim_{\substack{z \to \infty \\ \operatorname{Re} z = 0}} \frac{z Q_{n-1}(z)}{P_n(z)} = 1, \tag{1.93}$$

where we applied the fact that both $zQ_{n-1}(z)$ and $P_n(z)$ are monic polynomials of degree n. On the other hand,

$$\lim_{\substack{z \to \infty \\ \operatorname{Re} z = 0}} z \int_{-\infty}^{+\infty} \frac{\mu(dx)}{z-x} = \int_{-\infty}^{+\infty} \mu(dx) = \mu(\mathbb{R}) \tag{1.94}$$

by the dominated convergence theorem. It then follows from (1.93) and (1.94) that $\mu(\mathbb{R}) = 1$. □

Lemma 1.93. *It holds that*

$$P_0(T)e_0 = e_0, \quad P_k(T)e_0 = \sqrt{\omega_k \cdots \omega_1}\, e_k, \qquad k = 1, 2, \dots, n-1.$$

Proof. Obviously, $P_0(T)e_0 = e_0$. For $k = 1$, using $P_1(z) = z - \alpha_1$ and the explicit form of T, we have

$$P_1(T)e_0 = (T - \alpha_1)e_0 = (\alpha_1 e_0 + \sqrt{\omega_1}\, e_1) - \alpha_1 e_0 = \sqrt{\omega_1}\, e_1.$$

Let $k \ge 2$ and assume that the assertion is true up to $k-1$. Then, by using the recurrence formula (1.72) and by the assumption of induction, we have

$$\begin{aligned} P_k(T)e_0 &= (T - \alpha_k)P_{k-1}(T)e_0 - \omega_{k-1}P_{k-2}(T)e_0 \\ &= (T - \alpha_k)\sqrt{\omega_{k-1}\cdots\omega_1}\, e_{k-1} - \omega_{k-1}\sqrt{\omega_{k-2}\cdots\omega_1}\, e_{k-2} \\ &= \sqrt{\omega_{k-1}\cdots\omega_1}\left\{(T - \alpha_k)e_{k-1} - \sqrt{\omega_{k-1}}\, e_{k-2}\right\}. \end{aligned}$$

Since

$$(T - \alpha_k)e_{k-1} - \sqrt{\omega_{k-1}}\, e_{k-2} = \sqrt{\omega_k}\, e_k,$$

which follows by the explicit action of T, the assertion is true for k. □

Proposition 1.94. *Let* $\alpha_1, \dots, \alpha_n \in \mathbb{R}$ *and* $\omega_1 > 0, \dots, \omega_{n-1} > 0$. *Then the polynomials* $P_0(z), P_1(z), \dots, P_{n-1}(z)$ *defined by the recurrence relation* (1.72) *are the orthogonal polynomials associated with* μ *defined in* (1.90).

Proof. For a polynomial $p \in \mathbb{C}[X]$ let us calculate the norm of $p(T)e_0$. Using the spectral decomposition (1.92), we have

$$p(T)e_0 = \sum_{\lambda \in \mathrm{Spec}\, T} p(\lambda) \left\langle \frac{f(\lambda)}{\|f(\lambda)\|}, e_0 \right\rangle \frac{f(\lambda)}{\|f(\lambda)\|}.$$

In view of $\langle e_0, f(\lambda)\rangle = 1$ we obtain

$$\begin{aligned} \|p(T)e_0\|^2 &= \sum_{\lambda \in \mathrm{Spec}\, T} |p(\lambda)|^2 \left|\left\langle \frac{f(\lambda)}{\|f(\lambda)\|}, e_0 \right\rangle\right|^2 = \sum_{\lambda \in \mathrm{Spec}\, T} \frac{|p(\lambda)|^2}{\|f(\lambda)\|^2} \\ &= \int_{-\infty}^{+\infty} |p(x)|^2 \mu(dx) = \langle p, p\rangle_\mu, \qquad p \in \mathbb{C}[X]. \end{aligned}$$

Hence the map $p \mapsto p(T)e_0$ gives rise to an isometry from $\mathcal{P}(\mathbb{R}, \mu)$ into $\mathbb{C}^n$. In fact, this is surjective due to Lemma 1.93 and, hence, a unitary isomorphism from $\mathcal{P}(\mathbb{R}, \mu)$ onto $\mathbb{C}^n$. In particular,

$$\langle P_j, P_k\rangle_\mu = \langle P_j(T)e_0, P_k(T)e_0\rangle = \omega_j \cdots \omega_1 \langle e_j, e_k\rangle$$

by Lemma 1.93. Together with the fact that $P_0(z), P_1(z), \dots, P_{n-1}(z)$ are monic polynomials, which follows by definition, we conclude that they are the orthogonal polynomials associated with μ. □

The above argument is summarized as follows.

Theorem 1.95. *Let* $(\{\omega_1, \dots, \omega_{n-1}\}, \{\alpha_1, \dots, \alpha_n\})$ *be a Jacobi coefficient of finite type. Define polynomials* $P_0(z), P_1(z), \dots, P_n(z)$ *by the recurrence relation* (1.72) *and a measure* μ *on* $\mathbb{R}$ *by*

$$\mu = \sum_{\lambda:\, P_n(\lambda)=0} \|f(\lambda)\|^{-2} \delta_\lambda,$$

where the weights are given in (1.89). *Then,* μ *becomes a probability measure on* $\mathbb{R}$ *with finite support* $|\operatorname{supp} \mu| = n$ *and* $\{P_0(z), P_1(z), \dots, P_{n-1}(z)\}$ *form the orthogonal polynomials associated with* μ. *Moreover,*

$$\frac{1}{z-\alpha_1} \,\overline{\,-\,}\, \frac{\omega_1}{z-\alpha_2} \,\overline{\,-\,}\, \frac{\omega_2}{z-\alpha_3} \,\overline{\,-\,} \cdots \overline{\,-\,}\, \frac{\omega_{n-1}}{z-\alpha_n} = \int_{-\infty}^{+\infty} \frac{\mu(dx)}{z-x}$$

holds for $z \in \mathbb{C} \setminus \{\lambda\,;\, P_n(\lambda) = 0\}$.

1.9 Stieltjes Transform and Continued Fractions

For a probability measure $\mu \in \mathfrak{P}(\mathbb{R})$ (not necessarily having finite moments) the *Stieltjes transform* or the *Cauchy transform* is defined by

$$G_\mu(z) = \int_{-\infty}^{+\infty} \frac{\mu(dx)}{z-x}. \tag{1.95}$$

The integral exists for all $z \in \mathbb{C} \setminus \operatorname{supp} \mu$ since the distance between such a z and $\operatorname{supp} \mu$ is positive. We list some fundamental properties, the proofs of which are straightforward.

Proposition 1.96. *Let* $G(z) = G_\mu(z)$ *be the Stieltjes transform of a probability measure* $\mu \in \mathfrak{P}(\mathbb{R})$.

(1) $G(z)$ *is analytic on* $\mathbb{C} \setminus \operatorname{supp} \mu$.
(2) $\operatorname{Im} G(z) < 0$ *for* $\operatorname{Im} z > 0$ *and* $\operatorname{Im} G(z) > 0$ *for* $\operatorname{Im} z < 0$.
(3) $|G(z)| \le |\operatorname{Im} z|^{-1}$ *for* $\operatorname{Im} z \neq 0$.
(4) $G(\bar{z}) = \overline{G(z)}$. *In particular,* $G(z)$ *is completely determined by its values on the upper half plane* $\{\operatorname{Im} z > 0\}$.

We are interested in a continued fraction expansion of $G_\mu(z)$. For μ having a finite support the result was already established in Theorem 1.95. For a general $\mu \in \mathfrak{P}_{\mathrm{fm}}(\mathbb{R})$ we shall prove the following:

Theorem 1.97. *Let* $\mu \in \mathfrak{P}_{\mathrm{fm}}(\mathbb{R})$ *and* $(\{\omega_n\}, \{\alpha_n\})$ *its Jacobi coefficient. If* μ *is the solution of a determinate moment problem, the Stieltjes transform of* μ *is expanded into a continued fraction*

$$G_\mu(z) = \frac{1}{z-\alpha_1} \,\overline{\,-\,}\, \frac{\omega_1}{z-\alpha_2} \,\overline{\,-\,}\, \frac{\omega_2}{z-\alpha_3} \,\overline{\,-\,}\, \frac{\omega_3}{z-\alpha_4} \,\overline{\,-\,} \cdots, \tag{1.96}$$

which converges in $\{\operatorname{Im} z \neq 0\}$.

Proof. The assertion for $\mu \in \mathfrak{P}_{\mathrm{fm}}(\mathbb{R})$ having a finite support, or equivalently for $(\{\omega_n\}, \{\alpha_n\})$ being a Jacobi coefficient of finite type, has already been proved in Theorem 1.95. So we assume that μ has an infinite support and its Jacobi coefficient $(\{\omega_n\}, \{\alpha_n\})$ of infinite type.

Define polynomials $P_0(x), P_1(x), \dots, P_n(x), \dots$ by the recurrence relation (1.72). For each $n = 1, 2, \dots$ let μ_n be the unique probability measure whose Jacobi coefficient is

$$(\{\omega_1, \omega_2, \dots, \omega_{n-1}\}, \{\alpha_1, \alpha_2, \dots, \alpha_n\}).$$

It then follows from Theorem 1.95 that $\{P_0(x), P_1(x), \dots, P_{n-1}(x)\}$ form the orthogonal polynomials associated with μ_n and

$$\frac{1}{z-\alpha_1} - \frac{\omega_1}{z-\alpha_2} - \frac{\omega_2}{z-\alpha_3} - \dots - \frac{\omega_{n-1}}{z-\alpha_n} = \int_{-\infty}^{+\infty} \frac{\mu_n(dx)}{z-x} \tag{1.97}$$

holds on $\{\mathrm{Im}\, z \neq 0\}$.

Let us consider the mth moment of μ_n. As is seen from the Accardi–Bożejko formula (Theorem 1.63), for $M_m(\mu_n)$ we need only the first $\lceil m/2 \rceil$ terms of the Jacobi coefficient of μ_n. Hence for a fixed m, the sequence $M_m(\mu_n)$ stays constant for all large n. Since the Jacobi coefficient of μ_n is obtained by cutting off the Jacobi coefficient of μ, the constant coincides with $M_m(\mu)$. Therefore we have

$$\lim_{n\to\infty} M_m(\mu_n) = M_m(\mu), \qquad m = 1, 2, \dots. \tag{1.98}$$

Since μ is the solution of a determinate moment problem, thanks to the standard result on weak convergence of probability measures (see Proposition 1.98 below), we see that

$$\lim_{n\to\infty} \int_{-\infty}^{+\infty} \frac{\mu_n(dx)}{z-x} = \int_{-\infty}^{+\infty} \frac{\mu(dx)}{z-x}, \qquad \mathrm{Im}\, z \neq 0.$$

Hence, using (1.97) we obtain

$$\int_{-\infty}^{+\infty} \frac{\mu(dx)}{z-x} = \lim_{n\to\infty} \frac{1}{z-\alpha_1} - \frac{\omega_1}{z-\alpha_2} - \frac{\omega_2}{z-\alpha_3} - \dots - \frac{\omega_{n-1}}{z-\alpha_n},$$

where the right-hand side converges on $\{\mathrm{Im}\, z \neq 0\}$. This completes the proof. □

In the above proof we used the following:

Proposition 1.98. *Let $\mu, \mu_1, \mu_2, \dots \in \mathfrak{P}_{\mathrm{fm}}(\mathbb{R})$ and assume that μ is the solution of a determinate moment problem. If*

$$\lim_{n\to\infty} M_m(\mu_n) = M_m(\mu), \qquad m = 1, 2, \dots,$$

then μ_n converges weakly to μ, i.e.,

$$\lim_{n\to\infty}\int_{-\infty}^{+\infty} f(x)\mu_n(dx) = \int_{-\infty}^{+\infty} f(x)\mu(dx)$$

holds for any bounded continuous function $f(x)$ on $\mathbb{R}$.

In the rest of this section we shall discuss how the probability measure μ is recovered from the Stieltjes transform $G_\mu(z)$. We start with the following:

Theorem 1.99 (Stieltjes inversion formula). *Let $G(z)$ be the Stieltjes transform of $\mu \in \mathfrak{P}(\mathbb{R})$. Then for any pair of real numbers $s < t$,*

$$-\frac{2}{\pi}\lim_{y\to+0}\int_s^t \operatorname{Im} G(x+iy)\,dx = \mu(\{s\}) + \mu(\{t\}) + 2\mu((s,t)).$$

Proof. Let $x \in \mathbb{R}$ and $y > 0$. By definition we have

$$G(x+iy) = \int_{-\infty}^{+\infty} \frac{\mu(d\xi)}{x+iy-\xi} = \int_{-\infty}^{+\infty} \frac{(x-\xi)-iy}{(x-\xi)^2+y^2}\,\mu(d\xi)$$

so that

$$\begin{aligned} -\int_s^t \operatorname{Im} G(x+iy)\,dx &= \int_s^t dx \int_{-\infty}^{+\infty} \frac{y}{(x-\xi)^2+y^2}\,\mu(d\xi) \\ &= \int_{-\infty}^{+\infty} \mu(d\xi)\int_s^t \frac{y}{(x-\xi)^2+y^2}\,dx, \end{aligned} \tag{1.99}$$

where we applied the Fubini theorem after observing integrability of the integrand with respect to the product measure $\mu(d\xi)dx$. The inner integral of (1.99) is calculated immediately as

$$\int_s^t \frac{y}{(x-\xi)^2+y^2}\,dx = \arctan\frac{t-\xi}{y} - \arctan\frac{s-\xi}{y}.$$

For simplicity we set

$$I(t,y) = \int_{-\infty}^{+\infty} \arctan\frac{t-\xi}{y}\,\mu(d\xi) \tag{1.100}$$

so that (1.99) becomes

$$-\int_s^t \operatorname{Im} G(x+iy)\,dx = I(t,y) - I(s,y). \tag{1.101}$$

We now consider the limit as $y \to +0$. Going back to (1.100), we first see that the integrand is bounded and integrable. Hence the integral $I(t,y)$ is divided into three parts according to $\mathbb{R} = (-\infty, t) \cup \{t\} \cup (t, +\infty)$. In fact, since the integrand vanishes at $\xi = t$, we obtain

$$I(t, y) = \int_{(-\infty,t)} \arctan \frac{t-\xi}{y}\, \mu(d\xi) + \int_{(t,+\infty)} \arctan \frac{t-\xi}{y}\, \mu(d\xi).$$

Then, applying the bounded convergence theorem, we have

$$\begin{aligned}\lim_{y\to+0} I(t, y) &= \int_{(-\infty,t)} \frac{\pi}{2}\, \mu(d\xi) + \int_{(t,+\infty)} \left(-\frac{\pi}{2}\right) \mu(d\xi) \\ &= \frac{\pi}{2} \{\mu((-\infty, t)) - \mu((t, +\infty))\}. \end{aligned} \tag{1.102}$$

Combining (1.101) and (1.102), we obtain

$$\begin{aligned}&-\frac{2}{\pi} \lim_{y\to+0} \int_s^t \operatorname{Im} G(x + iy)\, dx \\ &= \mu((-\infty, t)) - \mu((t, +\infty)) - \mu((-\infty, s)) + \mu((s, +\infty)). \end{aligned} \tag{1.103}$$

The assertion then follows by immediate modification. □

The Stieltjes inversion formula is presented conventionally in terms of the distribution function. Let F be the distribution function of μ, i.e.,

$$F(x) = \mu((-\infty, x]), \qquad x \in \mathbb{R}. \tag{1.104}$$

Obviously, $F(x)$ possesses the following properties:

(i) $F(x)$ is non-decreasing, i.e., $F(x_1) \le F(x_2)$ for $x_1 \le x_2$;
(ii) $\lim_{x\to-\infty} F(x) = 0$ and $\lim_{x\to+\infty} F(x) = 1$;
(iii) $F(x)$ is right continuous, i.e., $\lim_{\epsilon\to+0} F(x + \epsilon) = F(x)$ for all $x \in \mathbb{R}$.

Conversely, for a function $F(x)$ satisfying the above conditions (i)–(iii) there exists a unique probability measure $\mu \in \mathfrak{P}(\mathbb{R})$ whose distribution function is $F(x)$. Moreover, for a bounded continuous function $f(x)$ we have

$$\int_{-\infty}^{+\infty} f(x)\, \mu(dx) = \int_{-\infty}^{+\infty} f(x)\, dF(x),$$

where the right-hand side is the so-called Stieltjes integral.

Now Theorem 1.99 is rephrased as follows.

Theorem 1.100. *Let $G(z)$ be the Stieltjes transform of $\mu \in \mathfrak{P}(\mathbb{R})$. Then for any pair of real numbers $s < t$, we have*

$$-\frac{2}{\pi} \lim_{y\to+0} \int_s^t \operatorname{Im} G(x + iy)\, dx = F(t) + F(t - 0) - F(s) - F(s - 0),$$

where F is the distribution function defined in (1.104).

Proof. The assertion easily follows from (1.103) together with

$$\mu((-\infty,t)) = \lim_{\epsilon\to+0} \mu((-\infty,t-\epsilon]) = \lim_{\epsilon\to+0} F(t-\epsilon) = F(t-0),$$
$$\mu((t,+\infty)) = 1 - F(t),$$

which are immediate from (1.104) and the succeedingly listed properties. □

Corollary 1.101. *The Stieltjes transform is injective, i.e., for* $\mu,\nu \in \mathfrak{P}(\mathbb{R})$, $G_\mu(z) = G_\nu(z)$ *on* $\{\operatorname{Im} z > 0\}$ *implies* $\mu = \nu$.

Proof. Let F and E denote the distribution functions of μ and ν, respectively. It follows from Theorem 1.100 that

$$F(t) + F(t-0) - F(s) - F(s-0) = E(t) + E(t-0) - E(s) - E(s-0),$$

for $s < t$. Then letting $s \to -\infty$, we obtain

$$F(t) + F(t-0) = E(t) + E(t-0), \quad t \in \mathbb{R}, \tag{1.105}$$

where we took into account that $F(s-0) \le F(s) \to 0$ as $s \to -\infty$ and a similar property of E. In particular, (1.105) becomes $F(t) = E(t)$ if both F and E are continuous at $t \in \mathbb{R}$. For an arbitrary $t \in \mathbb{R}$ we may choose a sequence $t_1 > t_2 > \cdots \to t$ such that both F and E are continuous at $t_n \in \mathbb{R}$. Such a sequence exists because the number of points at which F or E has a jump is at most countable. Then we have

$$F(t) = \lim_{n\to\infty} F(t_n) = \lim_{n\to\infty} E(t_n) = E(t),$$

from which we infer that $\mu = \nu$ since a distribution function determines a probability measure uniquely. □

Theorem 1.102. *Let* $G(z)$ *be the Stieltjes transform of* $\mu \in \mathfrak{P}(\mathbb{R})$. *If the distribution function* $F(x)$ *of* μ *is differentiable at* x, *we have*

$$F'(x) = -\frac{1}{\pi} \lim_{y\to+0} \operatorname{Im} G(x+iy). \tag{1.106}$$

Proof. Throughout the proof we fix a point $x \in \mathbb{R}$ at which $F(x)$ is differentiable. Let us start with

$$\operatorname{Im} G(x+iy) = -y \int_{-\infty}^{+\infty} \frac{\mu(d\xi)}{(x-\xi)^2 + y^2} = -y \int_{-\infty}^{+\infty} \frac{dF(\xi)}{(x-\xi)^2 + y^2}.$$

Since the integrand is differentiable and vanishes at $\xi \to \pm\infty$, we apply integration by parts to get

$$\operatorname{Im} G(x+iy) = y \int_{-\infty}^{+\infty} F(\xi) \frac{2(x-\xi)}{((x-\xi)^2 + y^2)^2}\, d\xi,$$

which becomes

$$2y \int_{-\infty}^{+\infty} F(x-\xi) \frac{\xi}{(\xi^2+y^2)^2} \, d\xi \tag{1.107}$$

by changing variables. Moreover, (1.107) is also equal to

$$2y \int_{-\infty}^{+\infty} F(x+\xi) \frac{-\xi}{(\xi^2+y^2)^2} \, d\xi. \tag{1.108}$$

Taking the mean value of (1.107) and (1.108), we obtain

$$\operatorname{Im} G(x+iy) = -y \int_{-\infty}^{+\infty} (F(x+\xi) - F(x-\xi)) \frac{\xi}{(\xi^2+y^2)^2} \, d\xi. \tag{1.109}$$

Here we use the assumption that $F(x)$ is differentiable at x. Given $\epsilon > 0$, we choose $\delta > 0$ such that

$$\left| \frac{F(x+\xi) - F(x-\xi)}{2\xi} - F'(x) \right| < \epsilon, \qquad |\xi| < \delta.$$

We divide the integral (1.109) into two parts. Set

$$I_1(y) = -y \int_{|\xi| \ge \delta} (F(x+\xi) - F(x-\xi)) \frac{\xi}{(\xi^2+y^2)^2} \, d\xi,$$

$$I_2(y) = -y \int_{|\xi| < \delta} (F(x+\xi) - F(x-\xi)) \frac{\xi}{(\xi^2+y^2)^2} \, d\xi.$$

We start with $I_1(y)$. Since $y > 0$ and $|F(x+\xi) - F(x-\xi)| \le 1$ we have

$$|I_1(y)| \le 2y \int_{\delta}^{+\infty} \frac{\xi}{(\xi^2+y^2)^2} \, d\xi \le 2y \int_{\delta}^{+\infty} \frac{\xi}{\xi^4} \, d\xi = \frac{y}{\delta^2},$$

and therefore,

$$\lim_{y \to +0} I_1(y) = 0. \tag{1.110}$$

As for $I_2(y)$ we rewrite

$$\begin{aligned} I_2(y) &= -y \int_{-\delta}^{+\delta} \frac{F(x+\xi) - F(x-\xi)}{2\xi} \frac{2\xi^2}{(\xi^2+y^2)^2} \, d\xi \\ &= -y \int_{-\delta}^{+\delta} \left\{ \frac{F(x+\xi) - F(x-\xi)}{2\xi} - F'(x) \right\} \frac{2\xi^2}{(\xi^2+y^2)^2} \, d\xi \\ &\quad - y \int_{-\delta}^{+\delta} F'(x) \frac{2\xi^2}{(\xi^2+y^2)^2} \, d\xi \\ &\equiv I_{21}(y) + I_{22}(y). \end{aligned}$$

By virtue of

$$y\int_{-\delta}^{+\delta} \frac{2\xi^2}{(\xi^2+y^2)^2}\,d\xi = 4\int_0^{\delta/y} \frac{s^2}{(s^2+1)^2}\,ds \uparrow \pi \quad \text{as } y \to +0,$$

which follows by elementary computation, we see that

$$|I_{21}(y)| \le y\int_{-\delta}^{+\delta} \epsilon\,\frac{2\xi^2}{(\xi^2+y^2)^2}\,d\xi \le \epsilon\pi, \quad y>0, \tag{1.111}$$

and

$$\lim_{y\to+0} I_{22}(y) = -F'(x)\lim_{y\to+0} y\int_{-\delta}^{+\delta} \frac{2\xi^2}{(\xi^2+y^2)^2}\,d\xi = -\pi F'(x). \tag{1.112}$$

Consequently, combining (1.110), (1.111) and (1.112), we obtain

$$\begin{aligned}&\lim_{y\to+0} |\operatorname{Im} G(x+iy) + \pi F'(x)| \\ &\quad = \lim_{y\to+0} |I_1(y) + I_{21}(y) + I_{22}(y) + \pi F'(x)| \le \epsilon\pi,\end{aligned}$$

from which (1.106) follows. □

Corollary 1.103. *Let $G(z)$ be the Stieltjes transform of $\mu \in \mathfrak{P}(\mathbb{R})$. Then the limit*

$$-\frac{1}{\pi}\lim_{y\to+0} \operatorname{Im} G(x+iy) \tag{1.113}$$

exists for a.e. $x \in \mathbb{R}$. Define $\rho(x)$ to be the above limit (1.113) if the limit exists and 0 otherwise. Then $\rho(x)dx$ is the absolutely continuous part of μ.

Proof. Let $F(x)$ be the distribution function of $\mu \in \mathfrak{P}(\mathbb{R})$. Because non-decreasing, $F(x)$ is differentiable at a.e. $x \in \mathbb{R}$ with respect to the Lebesgue measure dx. It then follows from Theorem 1.102 that the limit (1.113) exists for a.e. $x \in \mathbb{R}$. The rest is clear since $F'(x)dx$ is the absolutely continuous part of μ. □

The discrete or singular continuous part of μ is more complicated to obtain from its Stieltjes transform. For our later application we only need the following:

Proposition 1.104. *Let $\mu \in \mathfrak{P}(\mathbb{R})$. Then its Stieltjes transform $G(z)$ has a simple pole at $z = a \in \mathbb{R}$ if and only if a is an isolated point of* $\operatorname{supp}\mu$, *i.e., μ is a convex combination of δ_a and a probability measure $\nu \in \mathfrak{P}(\mathbb{R})$ such that* $\operatorname{supp}\nu \cap \{a\} = \varnothing$ *in such a way that*

$$\mu = c\delta_a + (1-c)\nu, \qquad 0 < c \le 1.$$

In that case, $c = \operatorname{Res}_{z=a} G(z)$.

Proof. The 'if part' follows easily from Proposition 1.96 (1). We shall prove 'only if part.' By assumption we choose $R > 0$ such that

$$G(z) = \frac{c}{z-a} + H(z), \tag{1.114}$$

where $H(z)$ is analytic in the disc $\{|z-a| < R\}$. Since $G(\bar{z}) = \overline{G(z)}$, we get

$$\frac{c}{z-a} + H(z) = \frac{\bar{c}}{z-a} + \overline{H(\bar{z})}, \qquad |z-a| < R;$$

hence $c \in \mathbb{R}$ and $H(z) = \overline{H(\bar{z})}$. Thus from (1.114) we see that

$$\operatorname{Im} G(x+iy) = -\frac{cy}{(x-a)^2 + y^2} + \operatorname{Im} H(x+iy), \quad y > 0.$$

Integrating by x over $[a-r, a+r]$, where $0 < r < R$, we obtain

$$\int_{a-r}^{a+r} \operatorname{Im} G(x+iy)\, dx = -c \int_{-r/y}^{r/y} \frac{dx}{x^2+1} + \int_{a-r}^{a+r} \operatorname{Im} H(x+iy)\, dx.$$

It follows from $H(z) = \overline{H(\bar{z})}$ that $H(x)$ is real for $a - r \le x \le a + r$. Then, by the dominated convergence theorem we see that

$$\lim_{y \to +0} \int_{a-r}^{a+r} \operatorname{Im} H(x+iy)\, dx = 0,$$

and therefore

$$\begin{aligned} -\frac{2}{\pi} &\lim_{y \to +0} \int_{a-r}^{a+r} \operatorname{Im} G(x+iy)\, dx \\ &= -\frac{2}{\pi} \lim_{y \to +0} (-c) \int_{-r/y}^{r/y} \frac{dx}{x^2+1} = 2c. \end{aligned} \tag{1.115}$$

On the other hand, the left-hand side of (1.115) is equal to

$$\mu(\{a-r\}) + \mu(\{a+r\}) + 2\mu((a-r, a+r))$$

by Theorem 1.99. Thus (1.115) becomes

$$\mu(\{a-r\}) + \mu(\{a+r\}) + 2\mu((a-r, a+r)) = 2c, \qquad 0 < r < R.$$

Since the right-hand side is independent of r, we see that

$$\mu((a-r, a)) = \mu((a, a+r)) = 0, \quad \mu(\{a\}) = c.$$

Consequently, $\mu - c\delta_a$ is a finite measure (possibly zero) and the assertion is now proved. □

Remark 1.105. An analytic function $H(z)$ defined on the upper half plane $\{\operatorname{Im} z > 0\}$ with $\operatorname{Im} H(z) \geq 0$ is called a *Herglotz function* or *Pick function*. For such a function there exist $a \geq 0$, $b \in \mathbb{R}$ and a finite Borel measure ν on $\mathbb{R}$ (possibly $\nu(\mathbb{R}) = 0$) such that

$$H(z) = az + b - \int_{-\infty}^{+\infty} \frac{1 + zx}{z - x}\, \nu(dx), \qquad \operatorname{Im} z > 0.$$

Moreover, the triple (a, b, ν) is uniquely determined by $H(z)$. Note that $\operatorname{Im} H(z_0) = 0$ occurs at some $z_0 \in \{\operatorname{Im} z > 0\}$ if and only if $H(z)$ is a real constant. The above representation is useful in the study of the Stieltjes transform $G_\mu(z)$.

Exercises

1.1. Prove that given a state φ on $M(n, \mathbb{C})$ there exists a unique density matrix ρ such that $\varphi(a) = \operatorname{tr}(a\rho)$, $a \in M(n, \mathbb{C})$. [Example 1.6]

1.2. Prove that there exists a one-to-one correspondence between the set of density matrices of rank 1 and the set of vector states modulo constant factors. [Example 1.7]

1.3. Let G be a discrete group. Show that $\mathbb{C}[G]$ is commutative if and only if G is commutative.

1.4. Let G be a discrete group. Show that the vacuum state φ_e corresponding to the unit $e \in G$ is tracial, i.e.,

$$\varphi_e(ab) = \varphi_e(ba), \qquad a, b \in \mathbb{C}[G].$$

1.5. Let $(\Gamma, \{\Phi_n\}, B^+, B^-)$ be the interacting Fock space associated with a Jacobi sequence $\{\omega_n\}$. Show that $B^\pm$ is bounded (in the sense of operators on Hilbert space obtained by completing Γ) if and only if $\{\omega_n\}$ is a bounded sequence.

1.6. Prove that if a probability measure $\mu \in \mathfrak{P}(\mathbb{R})$ has a finite moment of order m, $m \geq 2$, then it has a finite moment of order k with $1 \leq k \leq m-1$. Show an example of a probability measure which has a finite moment of order m but does not have a finite moment of order $m+1$.

1.7 (Stieltjes' example). For $-1 \leq c \leq 1$ define

$$\rho_c(x) = \begin{cases} x^{-\log x}(1 + c\sin(2\pi \log x)), & x > 0, \\ 0, & x \leq 0. \end{cases}$$

(1) Prove that $\rho_c(x)dx$ is a finite measure on $\mathbb{R}$ having finite moments of all orders.

(2) Prove that the moment sequence

$$M_m = \int_{-\infty}^{+\infty} x^m \rho_c(x)\, dx, \qquad m = 1, 2, \dots,$$

is independent of c.

(3) Show that

$$\sum_{m=1}^{\infty} M_{2m}^{-\frac{1}{2m}} < \infty.$$

1.8. Let $\rho_c(x)$ be as in Exercise 1.7. Choose $-1 \le c_1, c_2 \le 1$ arbitrarily and we set

$$\rho(x) = K(\rho_{c_1}(x) + \rho_{c_2}(-x)), \qquad x \in \mathbb{R},$$

where $K > 0$ is chosen in such a way that $\mu(dx) = \rho(x)dx$ is a probability measure. Prove that the moments of odd orders all vanish though μ is not necessarily symmetric.

1.9. Prove that the polynomial functions span a dense subspace of $L^2(\mathbb{R}, \mu)$ if $\mu \in \mathfrak{P}_{\mathrm{fm}}(\mathbb{R})$ has a compact support. [Remark 1.42]

1.10. Let $C_m = |\mathcal{C}_m|$ be the Catalan number and set

$$F(z) = 1 + \sum_{m=1}^{\infty} C_m z^m.$$

Show by graphical consideration that

$$C_m = \sum_{k=1}^{m} C_{k-1} C_{m-k}, \qquad m = 1, 2, \dots,$$

and derive the identity

$$F(z) - 1 = zF(z)^2.$$

Solving the above functional equation, find the explicit form of the Catalan numbers.

1.11. Let $\{M_m\}$ and $(\{\omega_n\}, \{\alpha_n\})$ be in correspondence through the canonical map between $\mathfrak{M}$ and $\mathfrak{J}$. Then $M_{2m-1} = 0$ for all $m = 1, 2, \dots$ if and only if $\alpha_n = 0$ for all $n = 1, 2, \dots$.

1.12. Let X be a Bernoulli random variable such that

$$P(X = +1) = p, \quad P(X = -1) = q, \quad p, q > 0, \quad p + q = 1.$$

Show that a quantum decomposition of X is given by

$$X \stackrel{\mathrm{s}}{=} \begin{pmatrix} 0 & 2\sqrt{pq} \\ 0 & 0 \end{pmatrix} + \begin{pmatrix} 0 & 0 \\ 2\sqrt{pq} & 0 \end{pmatrix} + \begin{pmatrix} q-p & 0 \\ 0 & p-q \end{pmatrix}.$$

1.13. Examine the following integral formulae:

$$\frac{1}{2\pi}\int_{-2}^{+2} x^{2m}\sqrt{4-x^2}\,dx = \frac{(2m)!}{m!(m+1)!},$$

$$\frac{1}{\sqrt{2\pi}}\int_{-\infty}^{+\infty} x^{2m}e^{-x^2/2}dx = \frac{(2m)!}{2^m m!},$$

where $m = 1, 2, \dots$. [Theorems 1.72 and 1.78]

1.14. Let $\{T_n(x)\}$ and $\{U_n(x)\}$ be the Chebyshev polynomials of the first and second kind, respectively. Prove the following recurrence relations:

$$T_0(x) = 1, \quad T_1(x) = x, \quad T_{n+1}(x) - 2xT_n(x) + T_{n-1}(x) = 0,$$
$$U_0(x) = 1, \quad U_1(x) = 2x, \quad U_{n+1}(x) - 2xU_n(x) + U_{n-1}(x) = 0.$$

[Theorem 1.75]

1.15. Derive directly from definition (1.65) the orthogonality relation for the Hermite polynomials

$$\int_{-\infty}^{+\infty} H_m(x)H_n(x)e^{-x^2}dx = \sqrt{\pi}\,2^n n!\delta_{mn}.$$

[Theorem 1.81]

1.16. Prove in detail the determinantal formulae for $\{P_k\}$ and $\{Q_k\}$ in Proposition 1.89.

1.17. Let $(\Gamma_{\{\omega_n\}}, \{\Phi_n\}, B^+, B^-)$ be an interacting Fock space and B° the diagonal operator associated with a real sequence $\{\alpha_n\}$. Let $\{P_n\}$ be the orthogonal polynomials associated with the Jacobi coefficient $(\{\omega_n\}, \{\alpha_n\})$. Show the following formula:

$$P_n(B^+ + B^- + B^\circ)\Phi_0 = \sqrt{\omega_n \cdots \omega_1}\,\Phi_n, \qquad n = 1, 2, \dots.$$

1.18. Let $G(z)$ be the Stieltjes transform of $\mu \in \mathfrak{P}(\mathbb{R})$. Then $G(z)$ is analytic at $z = \infty$ if and only if μ has a compact support. In that case, $G(z)$ admits an expansion

$$G(z) = \frac{1}{z}\sum_{m=0}^{\infty}\frac{M_m}{z^m}, \qquad |z| > L, \qquad L = \sup\{|x|\,;\, x \in \operatorname{supp}\mu\},$$

where $\{M_m\}$ is the moment sequence of μ.

Notes

The most basic ideas of an algebraic probability space and an algebraic random variable trace back to von Neumann [219]. The rapid and vast development of quantum probability during the last quarter century cannot be covered by a single monograph. Some fundamentals are found in the books Accardi–Lu–Volovich [9], Hiai–Petz [99], Meyer [158], Parthasarathy [176], Voiculescu–Dykema–Nica [216], and references cited therein.

The GNS-representation (Gelfand–Naimark–Segal representation) is standard, see, e.g., Takesaki [204, Chap. I], Pedersen [177, Chap. 3]. As we do not go into topological argument (C^*-algebras, von Neumann algebras, etc.) in this book, our GNS-representation remains at an algebraic level.

The notion of an interacting Fock space has a physical origin, where an interacting Fock space over an infinite-dimensional Hilbert space $\mathcal{H}$ is more essential, see, Accardi–Lu–Volovich [9] and references cited therein. An interacting Fock space studied in this chapter is the simplest one in the sense of $\mathcal{H} = \mathbb{C}$ and is often called *one-mode*.

The fundamental link between interacting Fock spaces and orthogonal polynomials (Theorems 1.51 and 1.52) was discovered by Accardi–Bożejko [3] and was used to prove the Accardi–Bożejko formula (Theorem 1.63). Cabanal-Duvillard–Ionescu [54] obtained the same formula in the particular case of $\alpha_n \equiv 0$. The reflection principle used in the proof of the Accardi–Bożejko formula is well known in the theory of random walks, see, e.g., Feller [77, Section 3.2] where the Catalan number appears (without this name).

The classical moment problem has a long history tracing back to the investigation of Chebyshev in the early 1870s. Stieltjes (1894) solved the question in the case of μ being supported by the half line $[0, +\infty)$, and Hamburger (1920–21) in the case of the real line $\mathbb{R} = (-\infty, +\infty)$. There are quite a few standard references, among which we refer to Chihara [56], Deift [69], Shohat–Tamarkin [190], Szegö [202]. The recent book by Simon [193] contains enormous materials. Concerning Remark 1.42, a necessary and sufficient condition for $\mathcal{P}(\mathbb{R}, \mu)$ being dense in $L^2(\mathbb{R}, \mu)$ is found in Akhiezer [11, Theorem 2.3.3]. For Carleman's condition (Theorem 1.66) we refer to Shohat–Tamarkin [190, Sect. 2.17].

The bulk of Sect. 1.8 depends on Deift [69]. The proof of Proposition 1.98 is found in Chung [58, Sect. 4.5], Durrett [76, Sect. 2.3].

More on Remark 1.105 is found in Donoghue [73, Chap. II] and Wall [222, Chap. XV].

The Riesz–Markov representation theorem (Theorem 1.3) is well known, see, e.g., Dunford–Schwartz [75, Theorem IV.6.3], Reed–Simon [181, Theorem IV.14]. A *Baire measure* on a compact Hausdorff space X is by definition a finite measure defined on the σ-field generated by the compact G_δ-sets in X. A Baire measure admits many Borel extensions in general, but unique regular Borel extension. Thus, the Riesz–Markov representation theorem claims equivalently that there is a one-to-one correspondence between the states on

$C(X)$ and the Baire probability measures on X. If X is metrizable, every compact set is a G_δ-set so that every Borel probability measure on X is regular.

Let $(\{\omega_n\}, \{\alpha_n\})$ be a pair of arbitrary infinite sequences of complex numbers and define a sequence of polynomials $\{P_n(x)\}$ recurrently by

$$\begin{aligned} P_0(x) &= 1, \\ P_1(x) &= x - \alpha_1, \\ xP_n(x) &= P_{n+1}(x) + \alpha_{n+1}P_n(x) + \omega_n P_{n-1}(x), \qquad n = 1, 2, \dots. \end{aligned}$$

Then, there exists a probability measure $\mu \in \mathfrak{P}_{\mathrm{fm}}(\mathbb{R})$ with which $\{P_n\}$ is the orthogonal polynomials if and only if $(\{\omega_n\}, \{\alpha_n\}) \in \mathfrak{J}$. There is now a natural question whether or not $\{P_n\}$ are still considered as 'orthogonal polynomials' in an extended sense. An answer is given by Favard's theorem (though the result was implicitly known before).

Theorem 1.106 (Favard). *Notations and assumptions being as above, there exists a unique linear functional* $\mathcal{F} : \mathbb{C}[X] \to \mathbb{C}$ *such that*

$$\mathcal{F}(P_0) = \omega_1, \qquad \mathcal{F}(P_m P_n) = 0 \quad \text{for } m \neq n, \quad m, n = 0, 1, 2, \dots.$$

Moreover, $\mathcal{F}$ *is quasi-definite (i.e., the Hanckel determinant* $\Delta_n \neq 0$ *for all* $n = 0, 1, 2, \dots$*) if and only if* $\omega_n \neq 0$ *for all* $n = 1, 2, \dots$*, while* $\mathcal{F}$ *is positive-definite (i.e.,* $\mathcal{F}(P) > 0$ *for all polynomials* P *such that* $P \not\equiv 0$ *and* $P(x) \geq 0$ *for all* $x \in \mathbb{R}$*) if and only if* $\omega_n > 0$ *and* $\alpha_n \in \mathbb{R}$ *for all* $n = 1, 2, \dots$.

2

Adjacency Matrices

In this chapter we study the adjacency matrix of a graph and an appropriate algebraic probability space in which it lives. Especially the vacuum state and their deformation on the adjacency algebra are treated. In Sect. 2.4 we introduce the *quantum decomposition* of the adjacency matrix, which will play a key role throughout this book.

2.1 Notions in Graph Theory

A *graph* is a pair $\mathcal{G} = (V, E)$, where V is a non-empty set and E is a subset of the set of unordered pairs of V, i.e., $E \subset \{\{x, y\}\,;\, x, y \in V,\, x \neq y\}$. An element of V is called a *vertex* and an element of E an *edge*. We say that two vertices x and y are *adjacent* and write $x \sim y$ if $\{x, y\} \in E$. A graph $\mathcal{G} = (V, E)$ is called *finite* if V is a finite set.

It is natural to think a picture of a graph $\mathcal{G} = (V, E)$. First, we assign a planar point to each vertex in V. Then, if two planar points correspond to an edge in E, we draw an arc connecting those points. We need to keep in mind that appearance of a picture of a graph varies largely.

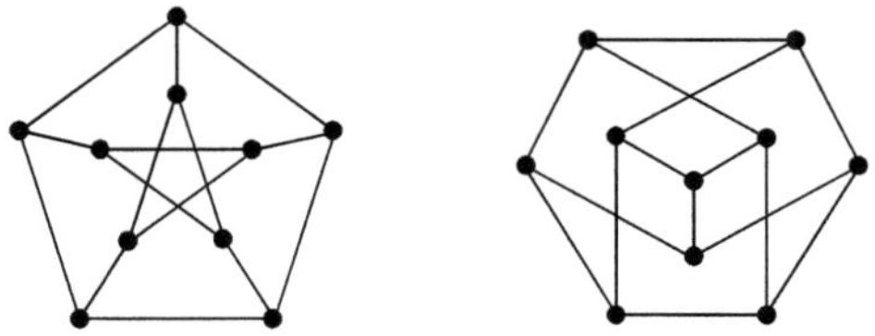

Fig. 2.1. Two pictures of the Petersen graph

A sequence $x_0, x_1, \dots, x_n \in V$, $n \geq 1$, is called a *walk of length* n or an *n-step walk* if $x_i \sim x_{i+1}$, for all $i = 0, 1, \dots, n-1$. A graph is called *connected* if any pair of distinct vertices $x, y \in V$ are connected by a walk.

A. Hora and N. Obata: Adjacency Matrices. In: A. Hora and N. Obata, Quantum Probability and Spectral Analysis of Graphs, Theoretical and Mathematical Physics, 65–83 (2007)
DOI 10.1007/3-540-48863-4_2

For a connected graph we define $\partial(x, y)$ to be the length of the shortest walk connecting x and y. By definition $\partial(x, x) = 0$ for $x \in V$. It is easy to show that $\partial(x, y)$ satisfies the axioms of a distance function. We call $\partial(x, y)$ the *graph distance*. Obviously, $x \sim y$ if and only if $\partial(x, y) = 1$. The *diameter* of a connected graph is defined by

$$\operatorname{diam}(\mathcal{G}) = \sup\{\partial(x, y)\,;\, x, y \in V\}. \tag{2.1}$$

To avoid notational confusion some more definitions are given. Note first that a walk is allowed to visit a same vertex repeatedly. A walk $x_0 \sim x_1 \sim \cdots \sim x_n$ is called a *path* if the vertices x_i, $0 \le i \le n$, are distinct from each other. A walk $x_0 \sim x_1 \sim \cdots \sim x_n$ is called a *cycle* if $n \ge 3$, $x_0 = x_n$, and the vertices x_i, $0 \le i < n$, are distinct from each other. A connected graph is called a *tree* if it possesses no cycle.

Consider two graphs $\mathcal{G} = (V, E)$ and $\mathcal{G}' = (V', E')$. A map $\alpha : V \to V'$ is called an *isomorphism* if (i) α is bijective and (ii) $x \sim y$ if and only if $\alpha(x) \sim \alpha(y)$. Obviously, α^{-1} is also an isomorphism. If there exists an isomorphism between two graphs, they are called *isomorphic* and denoted by $\mathcal{G} \cong \mathcal{G}'$. An isomorphism between $\mathcal{G}$ and itself is called an *automorphism*. The set of all automorphisms of $\mathcal{G}$ is denoted by $\operatorname{Aut}(\mathcal{G})$. As is easily verified, $\operatorname{Aut}(\mathcal{G})$ forms a group and is called the *automorphism group* of $\mathcal{G}$.

The *degree (or valency)* of $x \in V$ is defined by

$$\kappa(x) = |\{y \in V\,;\, y \sim x\}|.$$

A graph $\mathcal{G} = (V, E)$ is called *locally finite* if $\kappa(x) < \infty$ for all $x \in V$. Graphs considered in this book are not necessarily finite (i.e., $|V| < \infty$) but locally finite. In fact, many interesting examples satisfy the stronger condition:

$$\sup\{\kappa(x)\,;\, x \in V\} < \infty. \tag{2.2}$$

Such a graph is called *uniformly locally finite* and is characterized by boundedness of the adjacency matrix (Exercise 2.4). If the degree of each vertex is constant, i.e., $\kappa(x) = \kappa < \infty$ for all $x \in V$, the graph is called *regular* with degree κ.

Convention. Hereafter, unless otherwise specified, a *graph* is always assumed to be locally finite and connected.

We shall deal mostly with two classes of regular graphs: distance-regular graphs (the definition will be given in Sect. 3.1) and Cayley graphs of discrete groups.

Definition 2.1. Let G be a discrete group and Σ a finite subset of G. The pair (G, Σ) is called a *Cayley graph* if

(i) $g \in \Sigma \Leftrightarrow g^{-1} \in \Sigma$;
(ii) the unit $e \in G$ does not belong to Σ;

(iii) Σ generates G, i.e., every element of G is a product of ones in Σ.

A Cayley graph gives rise to a (locally finite and connected) graph (V, E), where

$$V = G, \qquad E = \{\{x, y\}\,;\, x, y \in G,\ xy^{-1} \in \Sigma\}.$$

By definition $x \sim y$ if and only if $x = gy$ for some $g \in \Sigma$. The next result is obvious.

Proposition 2.2. *A Cayley graph (G, Σ) is regular with degree $|\Sigma|$.*

We give two examples. More examples will be discussed later.

Example 2.3 (Integer lattice). The additive group $\mathbb{Z}^N$ furnished with the standard generators

$$e_{\pm 1} = (\pm 1, 0, \dots, 0), \quad \dots, \quad e_{\pm N} = (0, \dots, 0, \pm 1),$$

is the *N-dimensional integer lattice.* Obviously, the degree is $2N$.

Example 2.4 (Homogeneous tree with even degree). A tree is called *homogeneous* if it is regular. A homogeneous tree is uniquely determined by its degree. Let F_N be the free group on N free generators $g_1, \dots, g_N$. For simplicity, we write $g_{-i} = g_i^{-1}$. The Cayley graph $(F_N, \{g_{\pm 1}, \dots, g_{\pm N}\})$ is a homogeneous tree with degree $2N$. Note that a homogeneous tree with degree two is one-dimensional integer lattice.

2.2 Adjacency Matrices and Adjacency Algebras

For a graph $\mathcal{G} = (V, E)$, its *adjacency matrix* $A = (A_{xy})$ is defined by

$$A_{xy} = \begin{cases} 1, & \text{if } x \sim y, \\ 0, & \text{otherwise.} \end{cases} \tag{2.3}$$

Thus A is a (possibly infinite) matrix with rows and columns indexed by V. The adjacency matrix contains full information of a given graph.

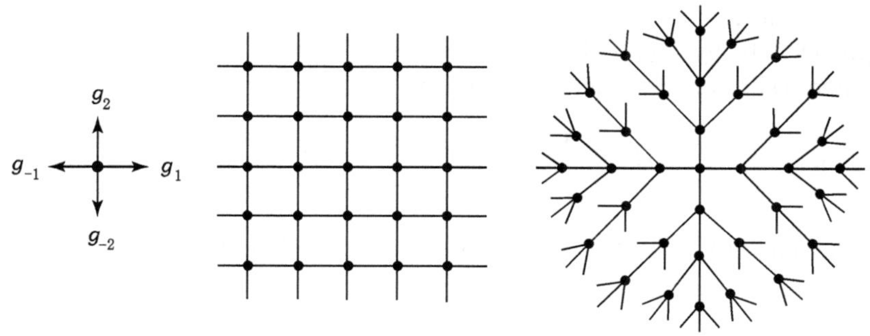

Fig. 2.2. Cayley graphs: $\mathbb{Z}^2$ and F_2

Lemma 2.5. *(In this statement a graph is not assumed to be locally finite nor to be connected.) The adjacency matrix $A = (A_{xy})$ of a graph $\mathcal{G} = (V, E)$ has the following properties:*

(i) *A_{xy} takes values in $\{0, 1\}$;*
(ii) *A is symmetric, i.e., $A_{xy} = A_{yx}$ for all $x, y \in V$;*
(iii) *The diagonal of A vanishes, i.e., $A_{xx} = 0$ for all $x \in V$.*

Conversely, if a matrix $A = (A_{xy})$, where x, y run over a non-empty set V, satisfies the above conditions (i)–(iii), A is the adjacency matrix of a graph.

Proof. It is straightforward to check that the adjacency matrix of a graph satisfies conditions (i)–(iii). Conversely, given a matrix $A = (A_{xy})$, where x, y run over a non-empty set V, satisfying conditions (i)–(iii), we define E to be the set of unordered pairs $\{x, y\}$ such that $A_{xy} = 1$. Then, E is well-defined and (V, E) is a graph whose adjacency matrix is A. □

Spectral properties of the adjacency matrix A are our subjects to study. If $\mathcal{G} = (V, E)$ is a finite graph, A becomes a $|V| \times |V|$ real symmetric matrix so that the spectrum of A is nothing else the array

$$\begin{matrix} \lambda_1 & \lambda_2 & \dots & \lambda_s \\ w_1 & w_2 & \dots & w_s, \end{matrix} \tag{2.4}$$

where $\lambda_1 < \lambda_2 < \dots < \lambda_s$ are the set of eigenvalues of A and w_i is the multiplicity of λ_i. The array (2.4) is called the *spectrum* of a (finite) graph $\mathcal{G} = (V, E)$. Instead of (2.4) we may consider a probability measure defined by

$$\mu = \frac{1}{|V|} \sum_{i=1}^{s} w_i \delta_{\lambda_i}. \tag{2.5}$$

This is called the *eigenvalue distribution* or *spectral distribution* of A (or of a graph $\mathcal{G}$). In this book we also treat infinite graphs and, more seriously, growing graphs, for which spectral properties cannot be expressed as in the array (2.4). More analytic and probabilistic ideas should be employed, where the expression (2.5) is more convenient.

Let us formulate our problems. Let $\mathcal{G} = (V, E)$ be a (finite or infinite) graph. Denote by $\ell^2(V)$ the Hilbert space of square-summable functions on V, where the inner product is defined by

$$\langle f, g \rangle = \sum_{x \in V} \overline{f(x)}\, g(x), \qquad f, g \in \ell^2(V),$$

and the norm by $\|f\| = \langle f, f \rangle^{1/2}$. For $x \in V$ we define $\delta_x \in \ell^2(V)$ by

$$\delta_x(y) = \begin{cases} 1, & \text{if } y = x, \\ 0, & \text{otherwise.} \end{cases}$$

The collection $\{\delta_x \,;\, x \in V\}$ forms a complete orthonormal basis of $\ell^2(V)$. Let $C_0(V)$ be a dense subspace of $\ell^2(V)$ spanned by $\{\delta_x \,;\, x \in V\}$. In other words, $C_0(V)$ consists of $\mathbb{C}$-valued functions on V with finite supports.

Note that $C_0(V)$ is a pre-Hilbert space. According to our notation introduced in Sect. 1.1, let $L(C_0(V))$ be the $*$-algebra of linear operators on $C_0(V)$. By definition, $T \in L(C_0(V))$ admits the adjoint operator from $C_0(V)$ into itself. For $T \in L(C_0(V))$ we define its matrix elements by

$$T_{xy} = \langle \delta_x, T\delta_y \rangle, \qquad x, y \in V. \tag{2.6}$$

Obviously, $T \in L(C_0(V))$ is uniquely determined by its matrix elements. Since $T^* \in L(C_0(V))$ by definition, we see that

$$|\{y \in V \,;\, T_{xy} \neq 0\}| < \infty \quad \text{for any } x \in V,$$
$$|\{x \in V \,;\, T_{xy} \neq 0\}| < \infty \quad \text{for any } y \in V.$$

A complex matrix (T_{xy}) indexed by V is called *locally finite* if the above condition is satisfied. The next assertion is clear.

Proposition 2.6. *The set of locally finite matrices becomes a $*$-algebra with the usual matrix operations. Moreover, the map $T \mapsto (T_{xy})$ is a $*$-isomorphism from $L(C_0(V))$ onto the $*$-algebra of locally finite matrices indexed by V.*

The adjacency matrix A is a locally finite matrix. In fact, we have

$$\kappa(x) = \sum_{y\in V} A_{xy} = \sum_{y\in V} A_{yx}, \qquad x \in V.$$

(Recall that the graph is always assumed to be locally finite.) Thus, the adjacency matrix A is identified with an operator in $L(C_0(V))$. The *adjacency algebra* of $\mathcal{G}$ is defined to be the $*$-subalgebra of $L(C_0(V))$ generated by A and is denoted by $\mathcal{A} = \mathcal{A}(\mathcal{G})$. Since $A = A^*$, every element $a \in \mathcal{A}$ is a polynomial in A, namely, is expressible in the form:

$$a = \lambda_0 1 + \lambda_1 A + \lambda_2 A^2 + \cdots + \lambda_n A^n, \qquad \lambda_i \in \mathbb{C}.$$

In particular, $\mathcal{A}$ is a commutative $*$-algebra.

Proposition 2.7. *If $\mathcal{G}$ is a finite graph, then* $\operatorname{diam}(\mathcal{G}) + 1 \le \dim \mathcal{A}(\mathcal{G}) < \infty$. *If $\mathcal{G}$ is an infinite graph, then* $\dim \mathcal{A}(\mathcal{G}) = \infty$.

Proof. Choose distinct vertices $x, y \in V$ and set $\partial(x, y) = n$. Note that

$$1_{xy} = A_{xy} = (A^2)_{xy} = \cdots = (A^{n-1})_{xy} = 0, \qquad (A^n)_{xy} \neq 0.$$

Hence by induction $\{1, A, A^2, \dots, A^n\}$ is linearly independent. For a finite graph, choosing vertices $x, y \in V$ such that $\partial(x, y) = \operatorname{diam}(\mathcal{G})$, we see that $\mathcal{A}(\mathcal{G})$ contains a linearly independent subset $\{1, A, A^2, \dots, A^n\}$ with $n = \operatorname{diam}(\mathcal{G})$, which means that $\dim \mathcal{A}(\mathcal{G}) \ge \operatorname{diam}(\mathcal{G}) + 1$. The rest of the assertion is clear. □

As soon as a state $\langle\cdot\rangle$ is given to the adjacency algebra $\mathcal{A}(\mathcal{G})$, we consider A as an algebraic random variable in an algebraic probability space $(\mathcal{A}(\mathcal{G}), \langle\cdot\rangle)$. In particular, as a spectral property of A, we are interested in a probability measure $\mu \in \mathfrak{P}_{\mathrm{fm}}(\mathbb{R})$ satisfying

$$\langle A^m \rangle = \int_{-\infty}^{+\infty} x^m \mu(dx), \quad m = 1, 2, \dots . \tag{2.7}$$

We call μ the *spectral distribution* of A in the state $\langle\cdot\rangle$. Remind that μ is not uniquely determined in general (known as the determinate moment problem).

In fact, we are more interested in asymptotic spectral properties of a growing graph (or a very large graph if a static situation is preferable). To be more precise, let $\mathcal{G}^{(\nu)} = (V^{(\nu)}, E^{(\nu)})$ be a growing graph, where a growing parameter ν runs over a directed set. Let A_ν denote the adjacency matrix of $\mathcal{G}^{(\nu)}$ and suppose that each adjacency algebra $\mathcal{A}(\mathcal{G}^{(\nu)})$ is given a state $\langle\cdot\rangle_\nu$. Then there exists a probability measure $\mu_\nu \in \mathfrak{P}_{\mathrm{fm}}(\mathbb{R})$ such that

$$\langle A_\nu^m \rangle_\nu = \int_{-\infty}^{+\infty} x^m \mu_\nu(dx), \qquad m = 1, 2, \dots .$$

Our interest lies in the limit of μ_ν as the graph grows. However, as is suggested by limit theorems in probability theory, such a limit does not exist in general without suitable scaling. A natural normalization is given by

$$\frac{A_\nu - \langle A_\nu \rangle}{\Sigma_\nu}, \qquad \Sigma_\nu^2 = \langle (A_\nu - \langle A_\nu \rangle)^2 \rangle. \tag{2.8}$$

(The suffix ν is often dropped when there is no danger of confusion.) Our aim is to find a probability measure $\mu \in \mathfrak{P}_{\mathrm{fm}}(\mathbb{R})$ satisfying

$$\lim_{\nu} \left\langle \left(\frac{A_\nu - \langle A_\nu \rangle}{\Sigma_\nu} \right)^m \right\rangle = \int_{-\infty}^{+\infty} x^m \mu(dx), \qquad m = 1, 2, \dots .$$

The above μ is called the *asymptotic spectral distribution* of A_ν in the state $\langle\cdot\rangle_\nu$. The uniqueness does not hold in general.

2.3 Vacuum and Deformed Vacuum States

By analogy of an interacting Fock space we give the following:

Definition 2.8. Let $\mathcal{G} = (V, E)$ be a graph and $\mathcal{A}(\mathcal{G})$ its adjacency algebra. The *vacuum state* at a fixed origin $o \in V$ is defined by

$$\langle a \rangle_o = \langle \delta_o, a\delta_o \rangle, \qquad a \in \mathcal{A}(\mathcal{G}). \tag{2.9}$$

It is noted that $\langle A^m \rangle_o$ is the number of m-step walks from $o \in V$ to itself. More generally, we have the following:

Proposition 2.9. *Let $x, y \in V$ and $m = 1, 2, \dots$. Then*

$$(A^m)_{xy} = \langle \delta_x, A^m \delta_y \rangle$$

coincides with the number of m-step walks connecting y and x.

Proof. We first note that

$$(A^m)_{xy} = \sum_{z_1,\dots,z_{m-1} \in V} A_{xz_1} A_{z_1 z_2} \cdots A_{z_{m-1} y}$$

by definition. Since $A_{xz_1} A_{z_1 z_2} \cdots A_{z_{m-1} y} = 1$ or $= 0$ according as $x \sim z_1 \sim z_2 \sim \cdots \sim z_{m-1} \sim y$ or not, the assertion follows immediately. □

We are also interested in deformation of the vacuum state. Given a function $t : V \to \mathbb{C}$ with $t(o) = 1$, we consider

$$\varphi(a) = \sum_{x \in V} \overline{t(x)} \langle \delta_x, a\delta_o \rangle, \qquad a \in \mathcal{A}(\mathcal{G}). \tag{2.10}$$

The right-hand side is in fact a finite sum because $a \in \mathcal{A}(\mathcal{G})$ is locally finite. Then, $\varphi : \mathcal{A}(\mathcal{G}) \to \mathbb{C}$ is linear and normalized ($\varphi(1) = 1$); however, positivity ($\varphi(a^* a) \geq 0$ for all $a \in \mathcal{A}(\mathcal{G})$) does not hold in general. If $t \in \ell^2(V)$, then

$$t = \sum_{x \in V} t(x) \delta_x \tag{2.11}$$

converges in the sense of norm of $\ell^2(V)$ and (2.10) becomes

$$\varphi(a) = \langle t, a\delta_o \rangle, \qquad a \in \mathcal{A}(\mathcal{G}), \tag{2.12}$$

where the right-hand side is the inner product of $\ell^2(V)$. For simplicity, the notation (2.12) will be often adopted even when t does not belong to $\ell^2(V)$. More precisely, the inner product $\langle \cdot, \cdot \rangle$ on $\ell^2(V)$ is extended to the canonical sesquilinear form on $C_0(V)^* \times C_0(V)$, where $C_0(V)^*$ is the space of formal series as in (2.11).

We are interested in a particular one-parameter deformation of the vacuum state. For $q \in \mathbb{R}$ (one may consider $q \in \mathbb{C}$ though our interesting case happens only when $-1 \leq q \leq 1$, see Proposition 2.13), we define a matrix $Q = Q_q$, called the *Q-matrix* of a graph $\mathcal{G} = (V, E)$, by

$$Q = Q_q = (q^{\partial(x,y)})_{x,y \in V}. \tag{2.13}$$

For $q = 0$ we understand that $0^0 = 1$ and $Q_0 = 1$ (the identity matrix). Viewing Q as an element of $L(C_0(V))$, we have

$$Q\delta_o = \sum_{x \in V} q^{\partial(x,o)} \delta_x.$$

Note that the right-hand side is in general an infinite sum and does not necessarily belong to $\ell^2(V)$. Nevertheless, as a special case of (2.10), we may define

$$\langle a\rangle_q = \sum_{x\in V} q^{\partial(x,o)}\langle \delta_x, a\delta_o\rangle, \qquad a\in\mathcal{A}(\mathcal{G}). \tag{2.14}$$

Let us adopt a shorthand notation as explained in the last paragraph:

$$\langle a\rangle_q = \langle Q\delta_o, a\delta_o\rangle, \qquad a\in\mathcal{A}(\mathcal{G}). \tag{2.15}$$

Obviously, $\mathcal{A}(\mathcal{G}) \ni a \mapsto \langle a\rangle_q$ is a normalized linear function on $\mathcal{A}(\mathcal{G})$. Being slightly free from the strict wording of 'state', we give the following:

Definition 2.10. A normalized linear function defined in (2.15) is called a *deformed vacuum state* on $\mathcal{A}(\mathcal{G})$.

Thus, a deformed vacuum state is not necessarily a state. We shall give a simple sufficient condition for the positivity. The following general terminology is adequate.

Definition 2.11. Let X be a set and $C_0(X)$ the space of $\mathbb{C}$-valued function on X with finite supports. A complex matrix $K = (K_{xy})$ indexed by X is called a *positive definite kernel* on X if

$$\sum_{x,y\in X} \overline{f(x)}\, K_{xy} f(y) \ge 0, \qquad f\in C_0(X). \tag{2.16}$$

Lemma 2.12. *The normalized linear function $\langle\cdot\rangle_q$ defined by (2.15) is positive, hence a state on $\mathcal{A}(\mathcal{G})$ if the following two conditions are fulfilled:*

(i) *Q is a positive definite kernel on V;*
(ii) *$QA = AQ$. (Note that Q is not necessarily locally finite but A is. Therefore the matrix elements of both sides are well-defined.)*

Proof. Let $a\in\mathcal{A}(\mathcal{G})$. Since a is a polynomial in A, we have $Qa = aQ$. Then, by the definition (2.15) we have

$$\langle a^*a\rangle_q = \langle Q\delta_o, a^*a\delta_o\rangle = \langle aQ\delta_o, a\delta_o\rangle = \langle Qa\delta_o, a\delta_o\rangle \ge 0,$$

which proves the assertion. □

As for the positivity of $Q = (q^{\partial(x,y)})$, we first note the following:

Proposition 2.13. *Let $\mathcal{G} = (V,E)$ be a graph with $|V|\ge 2$. If $Q = (q^{\partial(x,y)})$ is a positive definite kernel on V, then $-1\le q\le 1$.*

Proof. By assumption there is a pair of $a,b\in V$ such that $\partial(a,b)=1$. Since $Q = (q^{\partial(x,y)})$ is a positive definite kernel on V, taking $f = \alpha\delta_a + \beta\delta_b$ in $C_0(V)$, we obtain

$$\left\langle \begin{pmatrix}\alpha\\ \beta\end{pmatrix}, \begin{pmatrix}1 & q\\ q & 1\end{pmatrix}\begin{pmatrix}\alpha\\ \beta\end{pmatrix}\right\rangle \ge 0, \qquad \alpha,\beta\in\mathbb{C}, \tag{2.17}$$

where $\langle\cdot,\cdot\rangle$ is the usual Hermitian inner product of $\mathbb{C}^2$. Therefore, the 2×2 matrix appearing in (2.17) is positive definite. Hence $q\in\mathbb{R}$ and $1-q^2\ge 0$. □

Proposition 2.14 (Bożejko's quadratic embedding test). *If a graph $\mathcal{G} = (V, E)$ admits a quadratic embedding, i.e., if there is a map F from V into a real Hilbert space $\mathcal{H}$ such that*

$$\|F(x) - F(y)\|^2 = \partial(x, y), \qquad x, y \in V,$$

then $Q = (q^{\partial(x,y)})$ is positive definite for all $0 \le q \le 1$.

Proof. For $q = 0$ by definition we have $Q = 1$ (the identity matrix), which is positive definite. For $q = 1$ we have $q^{\partial(x,y)} = 1$ for all $x, y \in V$ so that

$$\sum_{x,y \in V} \overline{f(x)} f(y)\, q^{\partial(x,y)} = \sum_{x,y \in V} \overline{f(x)} f(y) = \left| \sum_{x \in V} f(x) \right|^2 \ge 0. \tag{2.18}$$

Assume that $0 < q < 1$ and we set $q = \mathrm{e}^{-t}$ with $0 < t < \infty$. Then

$$q^{\partial(x,y)} = \mathrm{e}^{-t\partial(x,y)} = \mathrm{e}^{-t\|F(x)-F(y)\|^2}, \qquad x, y \in V.$$

We need to prove that

$$\sum_{x,y \in V} \overline{f(x)}\, f(y) \mathrm{e}^{-t\|F(x)-F(y)\|^2} \ge 0, \qquad f \in C_0(V). \tag{2.19}$$

Let $\mathcal{H}_0 \subset \mathcal{H}$ be the subspace spanned by $\{F(x)\,;\, x \in \operatorname{supp} f\} \subset \mathcal{H}$. Since f has a finite support, $\mathcal{H}_0$ is finite dimensional and is identified with a Euclidean space, say $\mathbb{R}^n$. Now recall the n-dimensional Gaussian distribution

$$\gamma(u)du = (4\pi t)^{-n/2} \mathrm{e}^{-\|u\|^2/(4t)} du, \qquad u \in \mathbb{R}^n$$

and its Fourier transform

$$\mathrm{e}^{-t\|v\|^2} = \int_{\mathbb{R}^n} \mathrm{e}^{i\langle v,u\rangle} \gamma(u)\, du.$$

Then the left-hand side of (2.19) becomes

$$\begin{aligned}
&\sum_{x,y \in V} \overline{f(x)}\, f(y)\, \mathrm{e}^{-t\|F(x)-F(y)\|^2} \\
&\quad = \sum_{x,y \in V} \overline{f(x)}\, f(y) \int_{\mathbb{R}^n} \mathrm{e}^{i\langle F(x)-F(y),u\rangle} \gamma(u)\, du \\
&\quad = \int_{\mathbb{R}^n} \sum_{x,y \in V} \overline{f(x) \mathrm{e}^{-i\langle F(x),u\rangle}}\, f(y)\, \mathrm{e}^{-i\langle F(y),u\rangle} \gamma(du) \\
&\quad = \int_{\mathbb{R}^n} \left| \sum_{x \in V} f(x)\, \mathrm{e}^{-i\langle F(x),u\rangle} \right|^2 \gamma(du) \ge 0.
\end{aligned}$$

This completes the proof. □

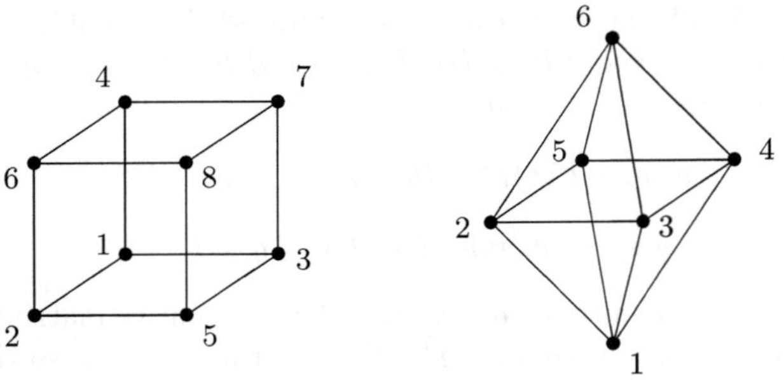

Fig. 2.3. Cube and octahedron

Remark 2.15. Replacing the Gaussian distribution by a stable one, we can extend the quadratic exponent to p such that $0 < p \le 2$.

Example 2.16 (Cube). For a cube the matrix Q is positive definite if and only if $-1 \le q \le 1$. This can be checked as follows: Allocate a number to each vertex of a cube as in Fig. 2.3 and consider $Q = (q^{\partial(i,j)})$, where i, j run over $\{1, 2, \dots, 8\}$. For $k = 1, 2, \dots, 8$ consider the $k \times k$ matrix defined by $(q^{\partial(i,j)})$ with i, j running over $\{1, 2, \dots, k\}$ and its determinant D_k. Then we have

$$D_1 = 1, \quad D_2 = 1 - q^2, \quad D_3 = (1-q^2)^2, \quad D_4 = (1-q^2)^3,$$
$$D_5 = (1-q^2)^5, \quad D_6 = (1-q^2)^7, \quad D_7 = (1-q^2)^9, \quad D_8 = (1-q^2)^{12}.$$

We see that $D_k > 0$ for all $k = 1, 2, \dots, 8$ if and only if $-1 < q < 1$. This condition is equivalent to that Q is strictly positive definite. In that case every principal minor of Q is positive (note that $\{D_k\}$ covers a part of principal minors). By continuity, for all $-1 \le q \le 1$ every principal minor of Q is non-negative so that Q is positive definite. An alternative proof is by computing the eigenvalues of Q directly. With no difficulty we obtain

$$\begin{aligned}\det(Q - \lambda) &= ((1-q^2)(1+q) - \lambda)^3 \\ &\quad \times ((1-q^2)(1-q) - \lambda)^3((1-q)^3 - \lambda)((1+q)^3 - \lambda),\end{aligned}$$

from which the desired result follows.

Example 2.17 (Octahedron). As in Example 2.16, by computing all the principal minor matrices of Q or by determining all the eigenvalues. In fact, we obtain easily that

$$\det(Q - \lambda) = (1 + 4q + q^2 - \lambda)(1 - q^2 - \lambda)^3((1-q)^2 - \lambda)^2.$$

Hence the matrix Q is positive definite if and only if $-2 + \sqrt{3} \le q \le 1$.

In order to derive a sufficient condition for the equality $QA = AQ$ we consider a geometric property of a graph. A graph $\mathcal{G} = (V, E)$ is called *quasi-distance-regular* if

$$\left|\left\{z \in V\,;\, \begin{array}{c}\partial(z,x) = n \\ \partial(z,y) = 1\end{array}\right\}\right| = \left|\left\{z \in V\,;\, \begin{array}{c}\partial(z,x) = 1 \\ \partial(z,y) = n\end{array}\right\}\right| \tag{2.20}$$

holds for any choice of $x, y \in V$ and $n = 0, 1, 2, \dots$. Here the number defined by (2.20) may depend on the choice of $x, y \in V$.

Remark 2.18. The equality (2.20) holds always for $n = 0, 1$ and $x, y \in V$. It is seen immediately by definition that a distance-regular graph (Definition 3.1) is quasi-distance-regular. On the other hand, if (2.20) depends only on $\partial(x, y)$, the graph $\mathcal{G}$ becomes distance-regular (Exercise 3.6).

Lemma 2.19. *If a graph is quasi-distance-regular, then $QA = AQ$ for all $q \in \mathbb{R}$. Conversely, if $QA = AQ$ holds for q running over a non-empty open interval, then the graph is quasi-distance-regular.*

Proof. Let $x, y \in V$. Then

$$\begin{aligned}(QA)_{xy} &= \sum_{z \in V} q^{\partial(x,z)} A_{zy} = \sum_{z \sim y} q^{\partial(x,z)} \\ &= \sum_{n=0}^{\infty} q^n |\{z \in V\,;\, \partial(z,x) = n,\ \partial(z,y) = 1\}|,\end{aligned} \tag{2.21}$$

which is in fact a finite sum. Similarly, we have

$$(AQ)_{xy} = \sum_{n=0}^{\infty} q^n |\{z \in V\,;\, \partial(z,x) = 1,\ \partial(z,y) = n\}|. \tag{2.22}$$

Hence, if the graph is quasi-distance-regular, the coefficients of q^n in (2.21) and (2.22) coincide and we obtain $(QA)_{xy} = (AQ)_{xy}$ for all $x, y \in V$. The converse assertion is readily clear. □

Proposition 2.20. *Let $\mathcal{G} = (V, E)$ be a graph. If for any choice of $x, y \in V$ there exists an automorphism $\alpha \in \mathrm{Aut}\,(\mathcal{G})$ such that $\alpha(x) = y$ and $\alpha(y) = x$, then we have $QA = AQ$ for all $q \in \mathbb{R}$.*

Proof. A graph having the property mentioned in the statement is quasi-distance-regular. In fact, for fixed $x, y \in V$, the automorphism α induces a one-to-one correspondence between $\{z \in V\,;\partial(z,x) = n,\ \partial(z,y) = 1\}$ and $\{z \in V\,;\, \partial(z,x) = 1,\ \partial(z,y) = n\}$. The assertion is then an immediate consequence from Lemma 2.19. □

2.4 Quantum Decomposition of an Adjacency Matrix

The problems mentioned in Sect. 2.2 belong to the classical region for $\mathcal{A}(\mathcal{G})$ being commutative. Our strategy is to introduce a non-commutative extension of the adjacency algebra by means of the quantum decomposition of the

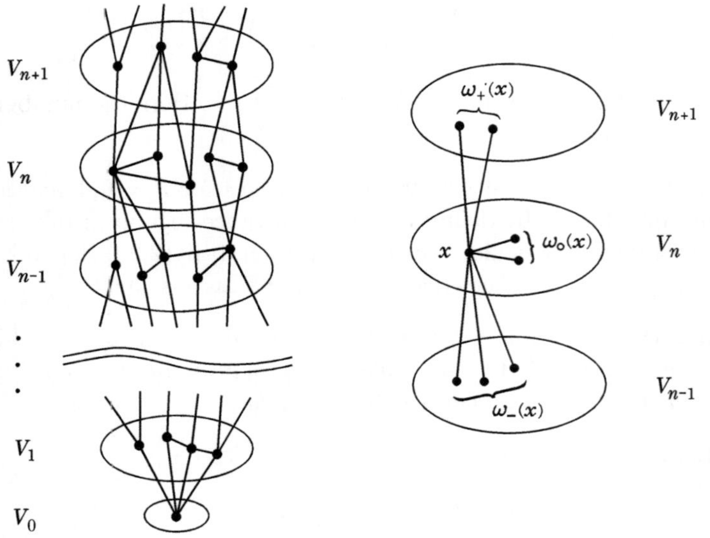

Fig. 2.4. Stratification $V = \bigcup_{n=0}^{\infty} V_n$ and $\kappa(x) = \omega_+(x) + \omega_-(x) + \omega_\circ(x)$

adjacency matrix and to employ techniques that are unique and typical in quantum probability theory.

Let $\mathcal{G} = (V, E)$ be a graph with a fixed origin $o \in V$. The graph is stratified into a disjoint union of strata:

$$V = \bigcup_{n=0}^{\infty} V_n, \qquad V_n = \{x \in V\,;\, \partial(o,x) = n\}. \tag{2.23}$$

This is called the *stratification* or the *distance partition* associated with $o \in V$, see Fig. 2.4. Apparently, $V_n = \varnothing$ happens for some $n = 1, 2, \dots$ if and only if the graph is finite. In that case $V_n = \varnothing$ for all n after a certain number.

Lemma 2.21. *Let $x, y \in V$. If $x \in V_n$ and $x \sim y$, then $y \in V_{n-1} \cup V_n \cup V_{n+1}$.*

Proof. By the triangle inequality we have

$$|\partial(o,x) - \partial(x,y)| \le \partial(o,y) \le \partial(o,x) + \partial(x,y),$$

from which we see that $n - 1 \le \partial(o,y) \le n + 1$. □

For $x \in V$ and $\epsilon \in \{+, -, \circ\}$ we define

$$\omega_\epsilon(x) = |\{y \in V\,;\, y \sim x,\ \partial(o,y) = \partial(o,x) + \epsilon\}|, \tag{2.24}$$

see Fig. 2.4. Obviously,

$$\kappa(x) = \omega_+(x) + \omega_-(x) + \omega_\circ(x), \qquad x \in V. \tag{2.25}$$

Also note the following:

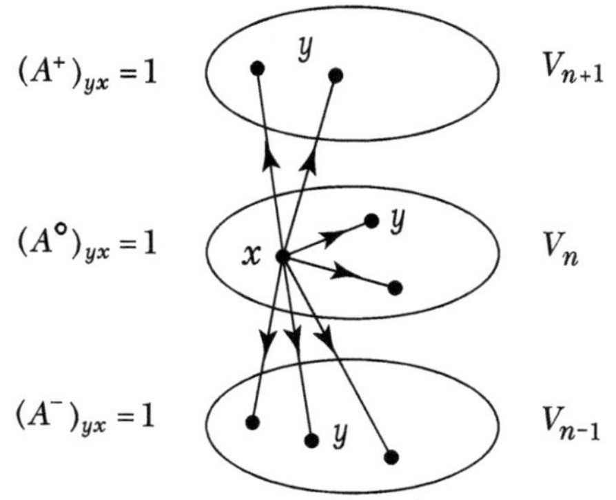

Fig. 2.5. Quantum decomposition $A = A^+ + A^- + A^\circ$

Lemma 2.22 (Matching identity). *It holds that*

$$\sum_{x\in V_n} \omega_+(x) = \sum_{y\in V_{n+1}} \omega_-(y), \qquad n = 0, 1, 2, \dots. \tag{2.26}$$

Proof. In fact, (2.26) is the number of edges between V_n and V_{n+1}. □

Given a stratification (2.23), we define three matrices A^ϵ, $\epsilon \in \{+, -, \circ\}$, indexed by V as follows: For $x \in V_n$, $n = 0, 1, 2, \dots$,

$$(A^\epsilon)_{yx} = \begin{cases} A_{yx}, & \text{if } y \in V_{n+\epsilon}, \\ 0, & \text{otherwise}, \end{cases} \qquad \epsilon \in \{+, -, \circ\},$$

where $n + \epsilon = n + 1, n - 1, n$ according as $\epsilon = +, -, \circ$. Obviously, A^ϵ is again locally finite. The following result is essential but the proof is obvious.

Lemma 2.23. *The adjacency matrix A is decomposed into a sum of three matrices:*

$$A = A^+ + A^- + A^\circ. \tag{2.27}$$

Moreover,

$$(A^+)^* = A^-, \quad (A^-)^* = A^+, \quad (A^\circ)^* = A^\circ. \tag{2.28}$$

Taking into account the $*$-isomorphism between $L(C_0(V))$ and the locally finite matrices indexed by V, we see that (2.28) is equivalent to

$$\langle A^+ f, g\rangle = \langle f, A^- g\rangle, \quad \langle A^\circ f, g\rangle = \langle f, A^\circ g\rangle, \qquad f, g \in C_0(V).$$

Definition 2.24. The expression (2.27) is called the *quantum decomposition* of A associated with the stratification (2.23) and A^+, A^-, A° the *quantum components*.

Recall that the adjacency matrix A acts on $C_0(V)$ in the canonical manner:

$$Af(x) = \sum_{y \in V} A_{xy} f(y) = \sum_{y \sim x} f(y), \qquad f \in C_0(V),$$

or equivalently,

$$A\delta_x = \sum_{y \sim x} \delta_y, \qquad x \in V.$$

Similarly, the action of A^ϵ is given by

$$A^\epsilon \delta_x = \sum_{\substack{y \sim x \\ y \in V_{n+\epsilon}}} \delta_y, \qquad x \in V_n. \tag{2.29}$$

Let $\tilde{\mathcal{A}}(\mathcal{G})$ be the $*$-algebra generated by the quantum components A^+, A^-, A° of the adjacency matrix A. Note that $\tilde{\mathcal{A}}(\mathcal{G})$ is non-commutative unless the graph $\mathcal{G}$ consists of a single vertex. Except such a trivial case, the quantum decomposition yields a non-commutative extension of $\mathcal{A}(\mathcal{G})$:

$$\mathcal{A}(\mathcal{G}) \subset \tilde{\mathcal{A}}(\mathcal{G}),$$

which plays an essential role in our study.

According to the stratification we shall introduce a particular subspace of $C_0(V)$. First note that $|V_n| < \infty$ for all n, which is easily verified by induction. Then for each $n = 0, 1, 2, \dots$, we define

$$\Phi_n = |V_n|^{-1/2} \sum_{x \in V_n} \delta_x, \tag{2.30}$$

whenever $V_n \neq \varnothing$. If the graph is finite, taking the smallest number $m_0 \geq 1$ with $V_{m_0} = \varnothing$, we obtain a finite sequence $\{\Phi_0, \Phi_1, \dots, \Phi_{m_0-1}\}$. Otherwise, we obtain an infinite sequence $\{\Phi_0, \Phi_1, \dots\}$. At any rate let $\Gamma(\mathcal{G}) \subset C_0(V)$ denote the subspace spanned by the sequence $\{\Phi_n\}$. Note that $\Gamma(\mathcal{G})$ is determined by the stratification of the graph $\mathcal{G}$ hence by the choice of an origin $o \in V$. A function in $\Gamma(\mathcal{G})$ is *radial* in the sense that its values depend only on the distance from the origin.

Remark 2.25. For some purposes it would be more natural to consider the completion of $\Gamma(\mathcal{G})$ in $\ell^2(V)$; however, in accordance with the definition of an interacting Fock space (Definition 1.27), we avoid the completion.

Theorem 2.26. *Let $\mathcal{G} = (V, E)$ be a graph with a fixed origin $o \in V$. Let $A = A^+ + A^- + A^\circ$ be the quantum decomposition of the adjacency matrix and $\Gamma(\mathcal{G})$ the space spanned by $\{\Phi_n\}$ defined in* (2.30). *Then we have*

$$A^+\Phi_n = |V_n|^{-1/2} \sum_{y\in V_{n+1}} \omega_-(y)\,\delta_y, \tag{2.31}$$

$$A^-\Phi_n = |V_n|^{-1/2} \sum_{y\in V_{n-1}} \omega_+(y)\,\delta_y\,, \tag{2.32}$$

$$A^\circ\Phi_n = |V_n|^{-1/2} \sum_{y\in V_n} \omega_\circ(y)\,\delta_y\,. \tag{2.33}$$

Proof. It follows from definition that

$$|V_n|^{1/2}A^+\Phi_n = \sum_{x\in V_n} A^+\delta_x = \sum_{y\in V_{n+1}} \omega_-(y)\,\delta_y,$$

from which (2.31) follows. The rest is proved similarly. □

It is noted from (2.31)–(2.33) that $\Gamma(\mathcal{G})$ is not necessarily invariant under the actions of the quantum components of A. The method of quantum decomposition will be best effective when

(i) $\Gamma(\mathcal{G})$ is invariant under the quantum components or
(ii) $\Gamma(\mathcal{G})$ is 'asymptotically' invariant under the quantum components.

There will also be some discussion when $\Gamma(\mathcal{G})$ is not invariant under the quantum components.

We end this section with the following observation.

Proposition 2.27. *Notations being as above, $\Gamma(\mathcal{G})$ is invariant under the quantum components A^+, A^-, A° if and only if $\omega_+(y), \omega_-(y), \omega_\circ(y)$ are constant on V_n for all $n = 0, 1, 2, \dots$. In that case, $(\Gamma(\mathcal{G}), \{\Phi_n\}, A^+, A^-)$ becomes an interacting Fock space and A° a diagonal operator. The associated Jacobi coefficient is given by*

$$\omega_n = \frac{|V_n|}{|V_{n-1}|}\,\omega_-(y)^2, \quad y \in V_n\,,$$

$$\alpha_n = \omega_\circ(y), \qquad y \in V_{n-1}, \qquad n = 1, 2, \dots.$$

In particular,

$$\omega_1 = \kappa(o), \qquad \alpha_1 = 0. \tag{2.34}$$

Proof. The first assertion is obvious from Eqs. (2.31)–(2.33). Assume that $\Gamma(\mathcal{G})$ is invariant under the actions of A^+, A^-, A°. Then $A^+\Phi_n$ is a constant multiple of Φ_{n+1} for $n = 0, 1, \dots$. By (2.31) the constant coincides with

$$|V_{n+1}|^{1/2}|V_n|^{-1/2}\omega_-(y), \qquad y \in V_{n+1},$$

which is independent of $y \in V_{n+1}$. Setting

$$\omega_{n+1} = \frac{|V_{n+1}|}{|V_n|}\,\omega_-(y)^2, \qquad y \in V_{n+1}, \qquad n = 0, 1, 2, \dots,$$

we have

$$A^+\Phi_n = \sqrt{\omega_{n+1}}\Phi_{n+1}, \qquad n = 0, 1, 2, \dots. \tag{2.35}$$

Since A^+ and A^- are mutually adjoint, we have

$$\sqrt{\omega_{n+1}} = \langle A^+\Phi_n, \Phi_{n+1}\rangle = \langle \Phi_n, A^-\Phi_{n+1}\rangle.$$

Since $A^-\Phi_{n+1}$ is a constant multiple of Φ_n by assumption, we obtain

$$A^-\Phi_{n+1} = \sqrt{\omega_{n+1}}\,\Phi_n. \tag{2.36}$$

It follows from (2.35) and (2.36) that $(\Gamma(\mathcal{G}), \{\Phi_n\}, A^+, A^-)$ is an interacting Fock space associated with a Jacobi sequence $\{\omega_n\}$ defined above. One may check easily that $\{\omega_n\}$ is of infinite type if the graph is infinite and of finite type otherwise. A similar argument is applied for A°. Finally, for (2.34) we note that

$$\langle \Phi_0, A\Phi_0\rangle = |\{\text{1-step walks connecting } o \text{ and } o\}| = 0,$$
$$\langle \Phi_0, A^2\Phi_0\rangle = |\{\text{2-step walks connecting } o \text{ and } o\}| = \kappa(o).$$

Then $A^\circ\Phi_0 = 0$ and $A^+\Phi_0 = \sqrt{\kappa(o)}\,\Phi_1$ follows easily. □

We note that the vacuum state corresponding to the fixed origin $o \in V$ becomes

$$\langle a\rangle_o = \langle \delta_o, a\delta_o\rangle = \langle \Phi_0, a\Phi_0\rangle, \qquad a \in \mathcal{A}(\mathcal{G}).$$

Hence, Proposition 2.27 says that the theory of an interacting Fock space established in Chap. 1 is directly applicable to the spectral analysis of $A = A^+ + A^- + A^\circ$ in the vacuum state. Concrete examples and calculations will be shown later. As for the deformed vacuum state (Definition 2.10), we have an alternative expression:

$$\langle a\rangle_q = \sum_{n=0}^{\infty} q^n |V_n|^{1/2} \langle \Phi_n, a\Phi_0\rangle, \qquad a \in \mathcal{A}(\mathcal{G}), \tag{2.37}$$

which is also useful.

Remark 2.28. It may happen that $\Gamma(\mathcal{G})$ is invariant under A^+ but not under A^-, or conversely that $\Gamma(\mathcal{G})$ is invariant under A^- but not under A^+ (see Exercise 2.15).

Exercises

2.1. Let K_N be a complete graph with N vertices. Show in a concrete form the adjacency matrix and determine the spectrum. [A finite graph is called *complete* if every pair of vertices is connected by an edge.]

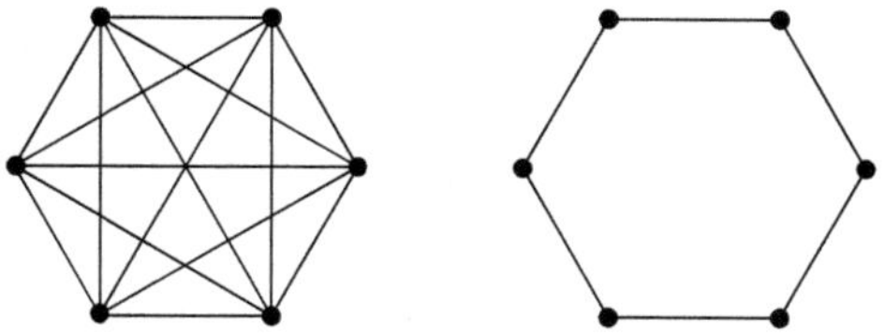

Fig. 2.6. Complete graph K_6 and cyclic graph C_6

2.2. Let C_N be a cyclic graph with N vertices. Show in a concrete form the adjacency matrix and determine the spectrum. [A finite graph is called *cyclic* or a *polygon* if there is a unique cycle on which every vertex lies.]

2.3. Prove that the following conditions for a graph (always assumed to be locally finite and connected) are equivalent:

(i) $\mathcal{G}$ is a finite graph;
(ii) $\operatorname{diam}(\mathcal{G}) < \infty$;
(iii) in the stratification $V_n = \varnothing$ happens for some $n \geq 1$;
(iv) $\dim \Gamma(\mathcal{G}) < \infty$;
(v) $\dim \mathcal{A}(\mathcal{G}) < \infty$.

2.4. Let $\mathcal{G} = (V, E)$ be a graph and A its adjacency matrix. Then A is a bounded operator on $\ell^2(V)$ if and only if the graph is uniformly locally finite. Prove also that

$$\sqrt{\kappa} \leq \|A\| \leq \kappa, \qquad \kappa = \sup\{\kappa(x)\,;\, x \in V\}.$$

2.5. Let $\mathcal{G}$ be a finite graph and define $\kappa = \max\{\kappa(x)\,;\, x \in V\}$. Show that every eigenvalue of the adjacency matrix A lies in the interval $[-\kappa, \kappa]$. Moreover, κ is an eigenvalue of A if and only if $\mathcal{G}$ is regular.

2.6. Let A be the adjacency matrix of a finite graph having at least one edge. Then A has at least one positive eigenvalue and at least one negative eigenvalue.

2.7. Let X be a set and k a $\mathbb{C}$-valued function on X. Prove that

$$K_{xy} = \overline{k(x)}\, k(y), \qquad x, y \in X$$

is a positive definite kernel on X. [Such a kernel is called *Gram kernel.*]

2.8. Let $\mathcal{G} = (V, E)$ be a regular graph of degree κ. Show that

$$Q\delta_o = \sum_{x \in V} q^{\partial(x,o)} \delta_x$$

belongs to $\ell^2(V)$ for $|q| < \kappa^{-1/2}$.

2.9. Let K_N be a complete graph with $N \geq 2$ vertices. Show that $Q = (q^{\partial(x,y)})$ is a positive definite kernel on K_N if and only if $-1/(N-1) \leq q \leq 1$. [Hint: Show $\det Q = (1-q)^{N-1}(1+(N-1)q)$ and mimic the argument in Example 2.16.]

2.10. Let $\mathcal{G} = (V, E)$ be a graph containing a complete graph K_N, $N \geq 2$. Show that $Q = (q^{\partial(x,y)})$ can be a positive definite kernel on V only when $-1/(N-1) \leq q \leq 1$. In particular, if $\mathcal{G}$ contains a triangle, $Q = (q^{\partial(x,y)})$ can be a positive definite kernel on V only when $-1/2 \leq q \leq 1$.

2.11. Let C_N be a cyclic graph with $N \geq 3$ vertices and consider the matrix $Q = (q^{\partial(x,y)})$. Prove the following statements. [Hint: Find all the eigenvalues of Q explicitly.]

(1) If N is even, Q is a positive definite kernel on C_N for all $-1 \leq q \leq 1$.
(2) If N is odd, there exists $-1 < r_N < 0$ such that Q is a positive definite kernel on C_N if and only if $r_N \leq q \leq 1$.
(3) $-1/2 = r_3 > r_5 > r_7 > \cdots \to -1$.

2.12. Show that every graph $\mathcal{G} = (V, E)$ with $|V| \leq 4$ admits a quadratic embedding. [Proposition 2.14]

2.13. Show that the graph illustrated in Fig. 2.7 does not admit a quadratic embedding. [Bożejko's obstruction.]

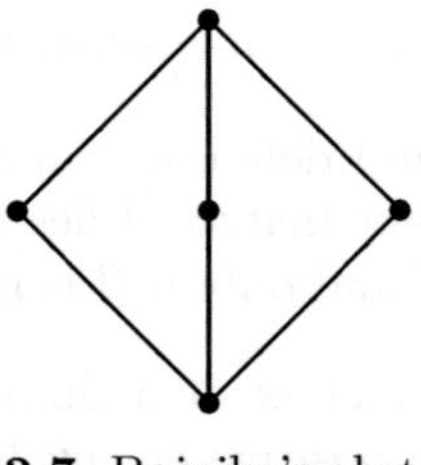

Fig. 2.7. Bożejko's obstruction

2.14. Let $N \geq 2$.

(1) Show that we may choose N points $v_1, \dots, v_N$ in $\mathbb{R}^{N-1}$ satisfying

$$\|v_1\| = \cdots = \|v_N\| < 1, \qquad \|v_i - v_j\| = 1, \quad i \neq j.$$

(2) Show that a complete graph K_N admits a quadratic embedding.

2.15. It may occur that $\Gamma(\mathcal{G})$ is invariant under A^+ but not under A^-, or conversely that $\Gamma(\mathcal{G})$ is invariant under A^- but not under A^+. Examine this by using the graphs illustrated in Fig. 2.8.

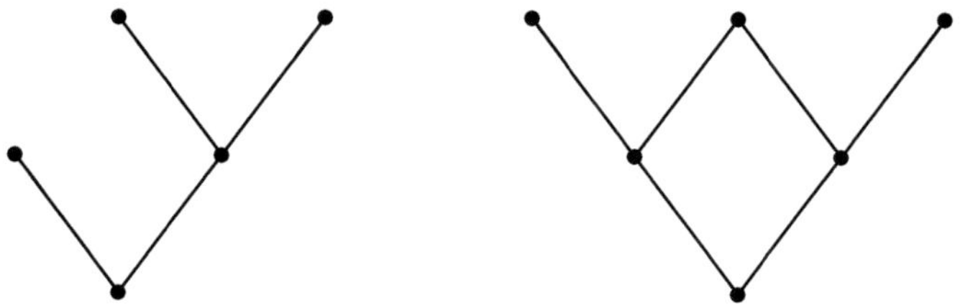

Fig. 2.8. Exercise 2.15

Notes

There are a huge number of literatures on graph theory. For very basic notions we referred to Balakrishnan–Ranganathan [18], Bollobás [32], Diestel [72]. An algebraic method for graph theory, i.e., an approach based on the adjacency matrix, is concisely overviewed in Biggs [30]. The study of spectrum of a (finite) graph has a long history, see Cvetković–Doob–Sachs [62], Cvetković–Rowlinson–Simić [63], Godsil–Royle [87] and references cited therein. While, spectral analysis of infinite graphs has been a subject in functional analysis, harmonic analysis and probability theory, see e.g., Woess [226] and references cited therein.

The matrix $Q = (q^{\partial(x,y)})$ plays a key role in harmonic analysis on free groups. An attention has been called to the positivity problem for Q by Bożejko in his Heidelberg lectures [38] (more notes can be found in Chap. 4). The graph geometric property concerning $AQ = QA$ is hardly found in literatures. The term 'quasi-distance-regular' is our provisional usage.

The idea of quantum decomposition for the study of asymptotic spectral distribution of a growing graph appeared first in Hashimoto–Obata–Tabei [97], where a growing Hamming graph was studied. The explicit use of the wording 'quantum decomposition' in this context is due to Hashimoto [94]. However, in quantum probability theory the notion of quantum decomposition traces back to Hudson–Parthasarathy [113], where the Brownian motion is decomposed into a sum of the annihilation and creation processes: $B_t = A_t^+ + A_t^-$. This aspect is the essence of quantum stochastic calculus developed considerably during the last quarter century, see Meyer [158], Parthasarathy [176]. Also see Ji–Obata [119] for quantum white noise theory.

In some literatures a graph $\mathcal{G} = (V, E)$ may possess a *loop*, i.e., an edge connecting a vertex with itself. In this case the adjacency matrix may have non-zero (in fact, the value is one) diagonal elements. We may consider a *multiedge*, namely, two vertices are connected by two or more edges. More generally, each edge may be given a certain value. Such an object is called a *network* and the corresponding 'adjacency matrix' is no longer $\{0, 1\}$-valued. In another context, we consider a *directed graph*, where each edge is given an orientation. Our approach developed in this book is applicable to some extent to these generalizations of graphs.

3

Distance-Regular Graphs

We focus on a class of highly symmetric graphs enjoying what we call distance-regularity, to which the method of quantum decomposition is applied quite efficiently. This chapter is devoted to developing the general theory while concrete examples will be discussed in the following chapters.

3.1 Definition and Some Properties

Definition 3.1. Let $\mathcal{G} = (V, E)$ be a graph. Let i, j, k be non-negative integers. A graph $\mathcal{G} = (V, E)$ is called *distance-regular* if for any choice of $x, y \in V$ with $\partial(x, y) = k$ the number of vertices $z \in V$ such that $\partial(x, z) = i$ and $\partial(y, z) = j$ is independent of the choice of x, y. Thus, taking $x, y \in V$ with $\partial(x, y) = k$ the number

$$p_{ij}^k = |\{z \in V\,;\ \partial(x, z) = i,\ \partial(y, z) = j\}|$$

is defined only depending on i, j, k. These constants are called the *intersection numbers* of $\mathcal{G} = (V, E)$.

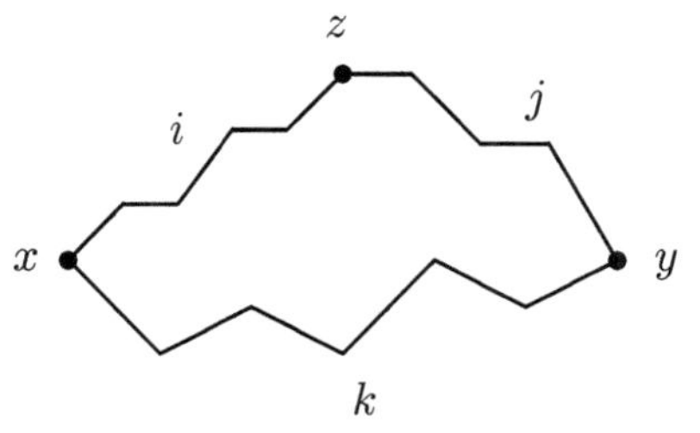

Fig. 3.1. Distance-regularity

The main examples are homogeneous trees (Chap. 4), Hamming graphs (Chap. 5), Johnson graphs, odd graphs (Chap. 6), for which the asymptotic spectral distributions will be studied in detail.

A. Hora and N. Obata: Distance-Regular Graphs. In: A. Hora and N. Obata, Quantum Probability and Spectral Analysis of Graphs, Theoretical and Mathematical Physics, 85–103 (2007)
DOI 10.1007/3-540-48863-4_3

Lemma 3.2. *A distance-regular graph is regular with degree* p^0_{11}.

Lemma 3.3. *The intersection numbers* $\{p^k_{ij}\}$ *of a distance-regular graph possess the following properties:*

(1) $p^k_{ij} = p^k_{ji}$.

(2) $p^k_{ij} = 0$ *if* $i+j<k$ *or* $j+k<i$ *or* $k+i<j$.

(3) $p^0_{ij} = 0$ *for* $i \neq j$.

(4) *If the graph is finite,* $p^k_{ij} = 0$ *unless* $i,j,k \le \mathrm{diam}(\mathcal{G})$, *where* $\mathrm{diam}(\mathcal{G})$ *is the diameter of a graph, see* (2.1).

The proofs of the above assertions are straightforward and omitted. In general, for a graph $\mathcal{G} = (V,E)$ we define the kth *distance matrix* (or kth *adjacency matrix*) A_k by

$$(A_k)_{xy} = \begin{cases} 1, & \partial(x,y) = k, \\ 0, & \text{otherwise}. \end{cases} \tag{3.1}$$

Obviously, the 0th distance matrix is the identity matrix $A_0 = 1$ and the 1st is what we call the *adjacency matrix* so that $A_1 = A$. It is noted that A_k is locally finite for all $k = 0,1,2,\dots$. Denoting by J the matrix of which entries are all one, we have

$$\sum_k A_k = J.$$

Moreover, if the graph $\mathcal{G}$ is finite, $A_k = 0$ for all $k > \mathrm{diam}\,(\mathcal{G})$. For a distance-regular graph these distance matrices are useful.

Lemma 3.4 (Linearization formula). *Let* $\mathcal{G} = (V,E)$ *be a distance-regular graph with intersection numbers* $\{p^k_{ij}\}$. *Then the distance matrices* $\{A_k\}$ *satisfy*

$$A_i A_j = \sum_{k=|i-j|}^{i+j} p^k_{ij} A_k. \tag{3.2}$$

Proof. We first show that

$$A_i A_j = \sum_k p^k_{ij} A_k. \tag{3.3}$$

Take $x, y \in V$ and set $\partial(x,y) = d$. Then, by definition we have

$$\begin{aligned} (A_i A_j)_{xy} &= \sum_{z \in V} (A_i)_{xz} (A_j)_{zy} \\ &= |\{z \in V\,;\, \partial(x,z) = i,\ \partial(y,z) = j\}| = p^d_{ij}. \end{aligned}$$

On the other hand, by (3.1) the xy-entry of the right-hand side of (3.3) is p^d_{ij} so that (3.3) holds. It follows from Lemma 3.3 (2) that $p^k_{ij} = 0$ when $k < |i-j|$ or $k > i+j$. Therefore the sum in (3.3) is taken over $|i-j| \le k \le i+j$. □

Recall that the adjacency algebra $\mathcal{A}(\mathcal{G})$ is by definition the $*$-algebra generated by a single element $A = A_1$. By repeated application of (3.2) we obtain the following:

Proposition 3.5. *The adjacency algebra $\mathcal{A}(\mathcal{G})$ of a distance-regular graph $\mathcal{G}$ is a linear space with a linear basis $A_0 = 1$ (the identity matrix), $A_1 = A$ (adjacency matrix), $A_2, \dots$. In particular, if $\mathcal{G}$ is finite, we have* $\dim \mathcal{A}(\mathcal{G}) = \mathrm{diam}(\mathcal{G}) + 1$.

In view of Proposition 2.7 we know that for a finite distance-regular graph the adjacency algebra has the minimum possible dimension.

It follows from Proposition 3.5 that, for a distance-regular graph, $\mathcal{A}(\mathcal{G})$ coincides with the $*$-algebra generated by $\{A_0, A_1, \dots, A_k, \dots\}$. In fact, this is a characteristic property of a distance-regular graph.

Proposition 3.6. *A graph $\mathcal{G} = (V, E)$ is distance-regular if and only if for any $k = 0, 1, 2, \dots$, the kth distance matrix A_k is expressible in a polynomial of A of degree k whenever $A_k \neq 0$.*

Proof. It suffices to express A^k as a linear combination of $A_0, A_1, \dots, A_k$ with a positive coefficient of A_k. We use induction on k. Let $A^k = \sum_{i=0}^{k} c_i A_i$ with $c_k > 0$. Multiplying A on the both sides and taking into account (3.2), we see that A^{k+1} has a desired expression with $c_k p_{1k}^{k+1} > 0$ as the coefficient of A_{k+1}. □

We give a simple criterion for a graph to be distance-regular. In general, a graph is called *distance-transitive* if for any $x, x', y, y' \in V$ such that $\partial(x, y) = \partial(x', y')$ there exists $\alpha \in \mathrm{Aut}(\mathcal{G})$ such that $\alpha(x) = x'$, $\alpha(y) = y'$. The next result is easy to see (Exercise 3.3).

Proposition 3.7. *A distance-transitive graph is distance-regular.*

Remark 3.8. A distance-regular graph possesses high symmetry; however, the Cayley graph of a group is not necessarily distance-regular. The two-dimensional integer lattice $\mathbb{Z}^2$ is such an example.

Although we do not go into detailed discussion, our consideration for distance-regular graphs can cover without essential change a more general algebraic system. Let X be a finite set and $\{R_0, R_1, \dots\}$ a partition of $X \times X$, i.e.,

$$X \times X = \bigcup R_i, \quad R_i \neq \varnothing, \quad R_i \cap R_j = \varnothing \quad \text{for } i \neq j.$$

Then $\{R_i\}$ is called an *association scheme of Bose–Mesner type* if

(i) $R_0 = \{(x, x)\,;\, x \in X\}$;
(ii) $(x, y) \in R_i \iff (y, x) \in R_i$;
(iii) Let $i, j, k \geq 0$. For $(x, y) \in R_k$ the number

$$|\{z \in X\,;\, (x, z) \in R_i,\ (z, y) \in R_j\}| \equiv p_{ij}^k$$

is independent of the choice of $(x, y) \in R_k$.

For an association scheme of Bose–Mesner type we define the ith adjacency matrix by

$$(A_i)_{xy} = \begin{cases} 1, & \text{if } (x,y) \in R_i\,, \\ 0, & \text{otherwise.} \end{cases}$$

Then the linearization formula holds:

$$A_i A_j = \sum_k p_{ij}^k A_k. \tag{3.4}$$

Transposing (3.4) and using condition (ii), we have

$$p_{ij}^k = p_{ji}^k.$$

The $*$-algebra generated by $A_0 = 1, A_1, A_2, \dots$ is called a *Bose–Mesner algebra.* For combinatorial interests X is usually assumed to be a finite set; however, the definition works unless X is a finite set. A distance-regular graph $\mathcal{G} = (V, E)$ becomes an association scheme of Bose–Mesner type by setting

$$X = V, \qquad R_i = \{(x, y) \in V \times V\,;\, \partial(x, y) = i\}.$$

3.2 Spectral Distributions in the Vacuum States

Let $\mathcal{G} = (V, E)$ be a distance-regular graph with intersection numbers $\{p_{ij}^k\}$ and A its adjacency matrix. Let us consider the spectral distribution of A in the vacuum state, i.e., a probability measure $\mu \in \mathfrak{P}_{\mathrm{fm}}(\mathbb{R})$ satisfying

$$\langle \delta_o, A^m \delta_o \rangle = \int_{-\infty}^{+\infty} x^m \mu(dx), \qquad m = 1, 2, \dots, \tag{3.5}$$

where $o \in V$ is a fixed origin of the graph. We apply the general method established in Sect. 2.4. As before, we set

$$V = \bigcup_n V_n, \quad V_n = \{x \in V\,;\, \partial(o, x) = n\}, \tag{3.6}$$

$$\omega_\epsilon(x) = |\{y \in V_{n+\epsilon}\,;\, y \sim x\}|, \quad x \in V_n\,, \quad \epsilon \in \{+, -, \circ\}. \tag{3.7}$$

Lemma 3.9. *For any $n = 0, 1, 2, \dots$ and $\epsilon \in \{+, -, \circ\}$ we have*

$$\omega_\epsilon(x) = p_{1,n+\epsilon}^n, \qquad x \in V_n.$$

Proof. Let $x \in V_n$ and $\epsilon = +$. Then we have

$$\omega_+(x) = |\{y \in V\,;\, \partial(o, y) = n + 1,\ \partial(x, y) = 1\}|,$$

which is constant independent of $x \in V_n$ and is the definition of $p_{1,n+1}^n$. The rest is similar. □

Lemma 3.10. (1) $|V_n| = p^0_{nn}$.
(2) $p^n_{1,n+1}|V_n| = p^{n+1}_{1,n}|V_{n+1}|$.
(3) $p^n_{1,n+1}p^0_{nn} = p^{n+1}_{1,n}p^0_{n+1,n+1}$.

Proof. (1) is obvious by definition. For (2) we only need the matching identity:

$$\sum_{x\in V_n} \omega_+(x) = \sum_{y\in V_{n+1}} \omega_-(y), \qquad n = 0, 1, 2, \dots,$$

which is, in fact, the number of edges between V_n and V_{n+1}, see Lemma 2.22. The assertion is then immediate from Lemma 3.9. Lastly, (3) follows immediately from (1) and (2). □

According to the stratification (3.6) we introduce the quantum decomposition of the adjacency matrix:

$$A = A^+ + A^- + A^\circ.$$

Define a unit vector $\Phi_n \in \ell^2(V)$ by

$$\Phi_n = |V_n|^{-1/2} \sum_{x\in V_n} \delta_x$$

and denote by $\Gamma(\mathcal{G})$ the subspace spanned by $\{\Phi_n\}$. Note that $\delta_o = \Phi_0$.

Theorem 3.11. *Let $\mathcal{G}$ be a distance-regular graph with intersection numbers $\{p^k_{ij}\}$ and A the adjacency matrix. Then $\Gamma(\mathcal{G})$ is invariant under the action of the quantum components A^ϵ, $\epsilon \in \{+, -, \circ\}$. Moreover,*

$$A^+\Phi_n = \sqrt{p^{n+1}_{1,n}p^n_{1,n+1}}\,\Phi_{n+1}\,, \quad n = 0, 1, 2, \dots, \tag{3.8}$$

$$A^-\Phi_0 = 0, \quad A^-\Phi_n = \sqrt{p^n_{1,n-1}p^{n-1}_{1,n}}\,\Phi_{n-1}\,, \quad n = 1, 2, \dots, \tag{3.9}$$

$$A^\circ\Phi_n = p^n_{1,n}\Phi_n, \quad n = 0, 1, 2, \dots. \tag{3.10}$$

Proof. We start with a general formula:

$$A^+\Phi_n = |V_n|^{-1/2} \sum_{y\in V_{n+1}} \omega_-(y)\,\delta_y,$$

see Theorem 2.26. Applying Lemma 3.9, we see that

$$A^+\Phi_n = |V_n|^{-1/2} \sum_{y\in V_{n+1}} p^{n+1}_{1,n}\,\delta_y = p^{n+1}_{1,n}|V_n|^{-1/2}|V_{n+1}|^{1/2}\Phi_{n+1}.$$

Then, in view of Lemma 3.10 (2) we obtain (3.8). The rest is proved in a similar manner. □

According to Theorem 3.11 $(\Gamma(\mathcal{G}), \{\Phi_n\}, A^+, A^-)$ is an interacting Fock space associated with a Jacobi sequence

$$\omega_n = p^n_{1,n-1} p^{n-1}_{1,n}, \qquad n = 1, 2, \dots, \tag{3.11}$$

and the quantum component A° is the diagonal operator defined by the sequence

$$\alpha_n = p^{n-1}_{1,n-1}, \qquad n = 1, 2, \dots. \tag{3.12}$$

Then by an immediate application of the theory of an interacting Fock space we may state the following:

Theorem 3.12. *Let $\mathcal{G} = (V, E)$ be a distance-regular graph and A its adjacency matrix. Let μ be a spectral distribution of A in the vacuum state at an origin $o \in V$ fixed arbitrarily. Then the pair of sequence $(\{\omega_n\}, \{\alpha_n\})$ given in* (3.11) *and* (3.12) *is the Jacobi coefficient of μ.*

As we mentioned in Sect. 1.9, if μ is the solution of a determinate moment problem, e.g., under Theorem 1.66, then the Stieltjes transform of μ admits a continued fraction expansion:

$$\int_{-\infty}^{+\infty} \frac{\mu(dx)}{z - x} = \frac{1}{z - \alpha_1} - \frac{\omega_1}{z - \alpha_2} - \frac{\omega_2}{z - \alpha_3} - \frac{\omega_3}{z - \alpha_4} - \cdots, \tag{3.13}$$

where the right-hand side converges in $\{\mathrm{Im}\, z \neq 0\}$. We see from (3.11) and (3.12) that

$$\omega_1 = p^0_{11} = \kappa, \qquad \alpha_1 = 0.$$

Thus the spectral distribution of A in the vacuum state has mean zero and variance p^0_{11}.

Corollary 3.13. *Let $\mathcal{G} = (V, E)$ be a distance-regular graph. For any $m = 1, 2, \dots,$ the number of m-step walks from $x \in V$ to itself is independent of $x \in V$.*

Proof. Let $o \in V$ be an arbitrarily chosen origin and μ a spectral distribution of A in the vacuum state at the origin $o \in V$. Then we have

$$\langle \delta_o, A^m \delta_o \rangle = \int_{-\infty}^{+\infty} x^m \mu(dx), \qquad m = 1, 2, \dots.$$

The integral in the right-hand side is expressible in terms of the Jacobi coefficient $(\{\omega_n\}, \{\alpha_n\})$, which is independent of the choice of $o \in V$ by Theorem 3.12. And so is $\langle \delta_o, A^m \delta_o \rangle$. □

Remark 3.14. The polynomials in Proposition 3.6 are orthogonal with respect to the spectral distribution μ of A in the vacuum state. In fact, let $A_k = P_k(A)$ in Proposition 3.6. Equation (3.5) yields

$$\langle \delta_o, A_i A_j \delta_o \rangle = \langle \delta_o, P_i(A) P_j(A) \delta_o \rangle = \int_{-\infty}^{+\infty} P_i(x) P_j(x) \mu(dx).$$

On the other hand, the left-hand side is expressed by (3.2) as

$$\sum_k p_{ij}^k \langle \delta_o, A_k \delta_o \rangle = p_{ij}^0 = \delta_{ij} |V_i|.$$

3.3 Finite Distance-Regular Graphs

Let us consider a finite distance-regular graph $\mathcal{G} = (V, E)$. Let A be its adjacency matrix and μ the spectral distribution in the vacuum state at $o \in V$. By Corollary 3.13, for $m = 1, 2, \dots,$

$$\sum_{x \in V} \langle \delta_x, A^m \delta_x \rangle = |V| \langle \delta_o, A^m \delta_o \rangle = |V| \int_{-\infty}^{+\infty} x^m \mu(dx),$$

that is,

$$\frac{1}{|V|} \mathrm{tr}\,(A^m) = \int_{-\infty}^{+\infty} x^m \mu(dx), \tag{3.14}$$

where $\mathrm{tr}\,(\cdot)$ is the usual trace of a finite matrix. On the other hand, the *eigenvalue distribution* of A is the probability measure σ defined by

$$\sigma = \frac{1}{|V|} \sum_{j=1}^{s} w_j \delta_{\lambda_j},$$

where $\lambda_1 < \lambda_2 < \cdots < \lambda_s$ are the eigenvalues of A and w_j the multiplicity of λ_j (see also Sect. 2.2). Then,

$$\frac{1}{|V|} \mathrm{tr}\,(A^m) = \frac{1}{|V|} \sum_{j=1}^{s} w_j \lambda_j^m = \int_{-\infty}^{+\infty} x^m \sigma(dx). \tag{3.15}$$

It follows from (3.14) and (3.15) that $\mu = \sigma$. (Apparently, σ is the solution of a determinate moment problem.) Summing up, we claim the following:

Theorem 3.15. *Let A be the adjacency matrix of a finite distance-regular graph $\mathcal{G} = (V, E)$. Then the spectral distribution of A in the vacuum state at an arbitrary origin coincides with the eigenvalue distribution of A.*

Remark 3.16. As is seen from the above argument, Theorem 3.15 remains valid for a finite graph for which the number of m-step walks from $x \in V$ to itself is independent of the choice of $x \in V$.

Corollary 3.17. *Let $\mathcal{G} = (V, E)$ be a finite distance-regular graph and A its adjacency matrix. The number of distinct eigenvalues of A coincides with* $\dim \mathcal{A}(\mathcal{G}) = \mathrm{diam}\,(\mathcal{G}) + 1$.

Proof. Let μ be a spectral distribution of A in the vacuum state at an arbitrary chosen origin $o \in V$. The Jacobi coefficient of μ is given by $(\{\omega_n\}, \{\alpha_n\})$ defined in (3.11) and (3.12). Since the graph is finite, denoting $d = \mathrm{diam}\,(\mathcal{G})$ we have

$$\omega_d = p^d_{1,d-1} p^{d-1}_{1,d} > 0, \qquad \omega_{d+1} = p^{d+1}_{1,d} p^d_{1,d+1} = 0.$$

Thus the Jacobi matrix corresponding to μ is a $(d+1) \times (d+1)$ matrix so that $|\mathrm{supp}\,\mu| = d+1$ by Proposition 1.90. It then follows from Theorem 3.15 that the support of the eigenvalue distribution of A consists of exactly $d+1$ points. □

We shall give an alternative proof of Theorem 3.15 without using Corollary 3.13. Let $\mathcal{G} = (V, E)$ be a finite distance-regular graph with $\mathrm{diam}\,(\mathcal{G}) = d$. In general, for a matrix $a = (a_{xy}) \in \mathcal{A}(\mathcal{G})$ the *normalized trace* is defined by

$$\varphi_{\mathrm{tr}}(a) = \frac{1}{|V|} \mathrm{tr}\,(a) = \frac{1}{|V|} \sum_{x \in V} a_{xx}.$$

The normalized trace is a state on $\mathcal{A}(\mathcal{G})$. We consider the GNS-representation of the algebraic probability space $(\mathcal{A}(\mathcal{G}), \varphi_{\mathrm{tr}})$. It was proved in Proposition 3.5 that $\{A_0, A_1, \dots, A_d\}$ forms a linear basis of $\mathcal{A}(\mathcal{G})$.

Lemma 3.18. $\varphi_{\mathrm{tr}}(A_i A_j) = \delta_{ij} p^0_{ii}$.

Proof. Recall the linearization formula (Lemma 3.4) and consider the trace of both sides of (3.2). Since the diagonal entries are zero for adjacency matrices A_k except $k = 0$ and $A_0 = 1$ is the identity matrix,

$$\mathrm{tr}\,(A_i A_j) = \mathrm{tr}\,(p^0_{ij} A_0) = p^0_{ij} |V| = \delta_{ij} p^0_{ii} |V|.$$

This is what we wanted to see. □

Define an inner product on $\mathcal{A}(\mathcal{G})$ by

$$\langle a, b \rangle_{\mathcal{A}} = \varphi_{\mathrm{tr}}(a^* b), \qquad a, b \in \mathcal{A}(\mathcal{G}).$$

In fact, $\mathrm{tr}(a^* a) = 0$ implies $a = 0$. It follows from Lemma 3.18 that

$$\Psi_n = \frac{A_n}{\sqrt{p^0_{nn}}} = \frac{A_n}{\sqrt{|V_n|}}, \qquad n = 0, 1, 2, \dots, d,$$

becomes a complete orthonormal basis of $\mathcal{A}(\mathcal{G})$. Let π be the regular representation of $\mathcal{A}(\mathcal{G})$, that is, $\pi : \mathcal{A}(\mathcal{G}) \to L(\mathcal{A}(\mathcal{G}))$ is the $*$-homomorphism defined by

$$\pi(a) b = ab, \qquad a, b \in \mathcal{A}(\mathcal{G}).$$

By definition $(\pi, \mathcal{A}(\mathcal{G}), \Psi_0)$ is a GNS-representation of $(\mathcal{A}(\mathcal{G}), \varphi_{\mathrm{tr}})$.

Lemma 3.19. *Define a linear map* $U : \Gamma(\mathcal{G}) \to \mathcal{A}(\mathcal{G})$ *by*

$$U\Phi_n = \Psi_n, \qquad n = 0, 1, 2, \ldots, d.$$

Then, U *is a bijection which preserves the inner product. Moreover,* $Ua = \pi(a)U$ *for all* $a \in \mathcal{A}(\mathcal{G})$.

Proof. Since $\{\Phi_n\}$ and $\{\Psi_n\}$ are orthonormal basis of $\Gamma(\mathcal{G})$ and $\mathcal{A}(\mathcal{G})$, respectively, U is a bijection which preserves the inner product. We prove that $UA = \pi(A)U$. By the linearization formula (Lemma 3.4) we have

$$\begin{aligned}\pi(A)\Psi_n &= \frac{1}{\sqrt{|V_n|}} AA_n = \frac{1}{\sqrt{|V_n|}} (p_{1n}^{n-1} A_{n-1} + p_{1n}^n A_n + p_{1n}^{n+1} A_{n+1}) \\ &= \sqrt{p_{1,n-1}^n p_{1n}^{n-1}}\, \Psi_{n-1} + p_{1n}^n \Psi_n + \sqrt{p_{1,n}^{n+1} p_{1,n+1}^n}\, \Psi_{n+1}, \qquad (3.16)\end{aligned}$$

where Lemma 3.10 was taken into account. On the other hand, it follows from Theorem 3.11 that

$$A\Phi_n = \sqrt{p_{1,n-1}^n p_{1n}^{n-1}}\, \Phi_{n-1} + p_{1n}^n \Phi_n + \sqrt{p_{1,n}^{n+1} p_{1,n+1}^n}\, \Phi_{n+1}. \qquad (3.17)$$

We see from (3.16) and (3.17) that $UA = \pi(A)U$. Since $\mathcal{A}(\mathcal{G})$ is the $*$-algebra generated by A, we also have $Ua = \pi(a)U$ for all $a \in \mathcal{A}(\mathcal{G})$. □

Lemma 3.20. $\langle \delta_o, a\delta_o \rangle = \varphi_{\mathrm{tr}}(a)$ *for* $a \in \mathcal{A}(\mathcal{G})$. *In other words, the vacuum state at* $o \in V$ *and the trace* φ_{tr} *coincide.*

Proof. By Lemma 3.19 we have

$$\langle \delta_o, a\delta_o \rangle = \langle \Phi_0, a\Phi_0 \rangle = \langle \Psi_0, \pi(a)\Psi_0 \rangle_{\mathcal{A}}.$$

Since Ψ_0 is the identity matrix, the last expression becomes

$$= \varphi_{\mathrm{tr}}(\Psi_0^*(a\Psi_0)) = \varphi_{\mathrm{tr}}(a),$$

which completes the proof. □

It follows from Lemma 3.20 that

$$\int_{-\infty}^{+\infty} x^m \mu(dx) = \langle \delta_o, A^m \delta_o \rangle = \varphi_{\mathrm{tr}}(A^m), \qquad m = 1, 2, \ldots.$$

Thus (3.14) is obtained and hence Theorem 3.15 follows.

3.4 Asymptotic Spectral Distributions

Let $\mathcal{G}^{(\nu)} = (V^{(\nu)}, E^{(\nu)})$ be a growing distance-regular graph. Let $\{p_{ij}^k(\nu)\}$ be the intersection numbers and $\kappa(\nu) = p_{11}^0(\nu)$ the degree. Each graph $\mathcal{G}^{(\nu)}$ has a fixed origin $o_\nu \in V^{(\nu)}$, associated with which we have the stratification:

$$V^{(\nu)} = \bigcup_{n=0}^{\infty} V_n^{(\nu)}, \qquad V_n^{(\nu)} = \{x \in V^{(\nu)}\,;\, \partial(x, o_\nu) = n\},$$

the subspace $\Gamma(\mathcal{G}^{(\nu)}) \subset \ell^2(V^{(\nu)})$ spanned by the unit vectors:

$$\Phi_n^{(\nu)} = |V_n^{(\nu)}|^{-1/2} \sum_{x \in V_n^{(\nu)}} \delta_x,$$

and the quantum decomposition of the adjacency matrix:

$$A_\nu = A_\nu^+ + A_\nu^- + A_\nu^\circ.$$

We first study the asymptotic spectral distribution of A_ν in the vacuum states (the vector state corresponding to the origin o_ν or equivalently to $\Phi_0^{(\nu)}$). Hereafter the suffix ν is often dropped for simple notations. The mean and the variance of A_ν are given, respectively, by

$$\langle \Phi_0, A_\nu \Phi_0 \rangle = 0, \qquad \langle \Phi_0, A_\nu^2 \Phi_0 \rangle = \kappa(\nu) = p_{11}^0(\nu).$$

Hence the normalized adjacency matrix becomes

$$\frac{A_\nu}{\sqrt{\kappa(\nu)}} = \frac{A_\nu^+}{\sqrt{\kappa(\nu)}} + \frac{A_\nu^-}{\sqrt{\kappa(\nu)}} + \frac{A_\nu^\circ}{\sqrt{\kappa(\nu)}}.$$

The actions of the normalized quantum components are readily known. For $n = 1, 2, \dots$ we set

$$\bar{\omega}_n(\nu) = \frac{p_{1,n-1}^n(\nu) p_{1,n}^{n-1}(\nu)}{\kappa(\nu)}, \tag{3.18}$$

$$\bar{\alpha}_n(\nu) = \frac{p_{1,n-1}^{n-1}(\nu)}{\sqrt{\kappa(\nu)}}. \tag{3.19}$$

It then follows immediately from Theorem 3.11 that

$$\frac{A_\nu^+}{\sqrt{\kappa(\nu)}} \Phi_n = \sqrt{\bar{\omega}_{n+1}(\nu)}\, \Phi_{n+1}\,, \quad n = 0, 1, 2, \dots, \tag{3.20}$$

$$\frac{A_\nu^-}{\sqrt{\kappa(\nu)}} \Phi_0 = 0, \quad \frac{A_\nu^-}{\sqrt{\kappa(\nu)}} \Phi_n = \sqrt{\bar{\omega}_n(\nu)}\, \Phi_{n-1}\,, \quad n = 1, 2, \dots, \tag{3.21}$$

$$\frac{A_\nu^\circ}{\sqrt{\kappa(\nu)}} \Phi_n = \bar{\alpha}_{n+1}(\nu) \Phi_n, \quad n = 0, 1, 2, \dots. \tag{3.22}$$

Thus we need to consider the limits of (3.18) and (3.19). For $n = 1, 2, \dots$ we set

$$\omega_n = \lim_{\nu\to\infty} \bar{\omega}_n(\nu) = \lim_{\nu\to\infty} \frac{p^n_{1,n-1}(\nu) p^{n-1}_{1,n}(\nu)}{\kappa(\nu)}, \tag{3.23}$$

$$\alpha_n = \lim_{\nu\to\infty} \bar{\alpha}_n(\nu) = \lim_{\nu\to\infty} \frac{p^{n-1}_{1,n-1}(\nu)}{\sqrt{\kappa(\nu)}}. \tag{3.24}$$

Since $\bar{\omega}_1(\nu) = 1$, for all ν we have $\omega_1 = 1$. Note also that $\alpha_1 = 0$. However, for the rest there is no guarantee that the limits exist. We consider the condition:

(DR) (i) for all $n = 1, 2, \dots$ the limits ω_n and α_n exist with $\omega_n > 0$ or (ii) there exists $n = 1, 2, \dots$ such that the limits $\omega_1, \dots, \omega_n$ and $\alpha_1, \dots, \alpha_n$ exist with

$$\omega_1 = 1,\ \omega_2 > 0,\ \dots,\ \omega_{n-1} > 0,\ \omega_n = 0.$$

If condition (DR) is fulfilled, $(\{\omega_n\}, \{\alpha_n\})$ becomes a Jacobi coefficient (condition (i) for an infinite type and (ii) for a finite type). We are now in a position to state the *quantum central limit theorem* for a growing distance-regular graph.

Theorem 3.21 (QCLT for a growing DRG). *Let $\mathcal{G}^{(\nu)} = (V^{(\nu)}, E^{(\nu)})$ be a growing distance-regular graph with the adjacency matrix A_ν. The degree is denoted by $\kappa(\nu)$. Assume that the condition (DR) is fulfilled. Let $\Gamma_{\{\omega_n\}} = (\Gamma, \{\Psi_n\}, B^+, B^-)$ be an interacting Fock space associated with $\{\omega_n\}$ and $B^\circ = \alpha_{N+1}$ the diagonal operator defined by $\{\alpha_n\}$, N being the number operator. Then we have*

$$\lim_{\nu\to\infty} \frac{A^\epsilon_\nu}{\sqrt{\kappa(\nu)}} = B^\epsilon, \qquad \epsilon \in \{+, -, \circ\}, \tag{3.25}$$

in the sense of stochastic convergence with respect to the vacuum states, i.e.,

$$\lim_{\nu\to\infty} \left\langle \Phi^{(\nu)}_0, \frac{A^{\epsilon_m}_\nu}{\sqrt{\kappa(\nu)}} \cdots \frac{A^{\epsilon_1}_\nu}{\sqrt{\kappa(\nu)}} \Phi^{(\nu)}_0 \right\rangle = \langle \Psi_0, B^{\epsilon_m} \cdots B^{\epsilon_1} \Psi_0 \rangle \tag{3.26}$$

for any $\epsilon_1, \dots, \epsilon_m \in \{+, -, \circ\}$ and $m = 1, 2, \dots$.

Proof. We see from (3.20)–(3.22) that

$$\frac{A^{\epsilon_m}_\nu}{\sqrt{\kappa(\nu)}} \cdots \frac{A^{\epsilon_1}_\nu}{\sqrt{\kappa(\nu)}} \Phi^{(\nu)}_0$$

is a constant multiple of $\Phi^{(\nu)}_{\epsilon_1+\dots+\epsilon_m}$ and the constant is given by a finite product of $\bar{\omega}_n(\nu)$ and $\bar{\alpha}_n(\nu)$, $n = 1, 2, \dots$. Hence the limit in the left-hand side of (3.26) exists by assumption. Moreover, since the actions of A^ϵ_ν and B^ϵ on the number vectors are given by the Jacobi coefficients $(\{\bar{\omega}_n\}, \{\bar{\alpha}_n\})$ and $(\{\omega_n\}, \{\alpha_n\})$, respectively, one may easily verify that the limit coincides with $\langle \Psi_0, B^{\epsilon_m} \cdots B^{\epsilon_1} \Psi_0 \rangle$. □

Corollary 3.22 (CLT for a growing DRG). *Notations and assumptions being as in Theorem 3.21, let μ be a probability measure of which the Jacobi coefficient is $(\{\omega_n\}, \{\alpha_n\})$. Then we have*

$$\lim_{\nu\to\infty} \left\langle \left(\frac{A_\nu}{\sqrt{\kappa(\nu)}} \right)^m \right\rangle_o = \int_{-\infty}^{+\infty} x^m \, \mu(dx), \qquad m = 1, 2, \dots .$$

The following result is slightly more general than Theorem 3.21 though the proof is similar.

Theorem 3.23. *Notations and assumptions being as in Theorem 3.21, it holds that*

$$\lim_{\nu} \left\langle \Phi_j^{(\nu)}, \frac{A_\nu^{\epsilon_m}}{\sqrt{\kappa(\nu)}} \cdots \frac{A_\nu^{\epsilon_1}}{\sqrt{\kappa(\nu)}} \Phi_k^{(\nu)} \right\rangle = \langle \Psi_j, B^{\epsilon_m} \cdots B^{\epsilon_1} \Psi_k \rangle,$$

for any $\epsilon_1, \dots, \epsilon_m \in \{+, -, \circ\}$, $m = 1, 2, \dots$ and $j, k = 0, 1, 2, \dots$.

Below we mention a simple example for illustrating the above consideration. More examples will be discussed in Chaps. 4–6.

Example 3.24 (Cyclic graph). Consider a cyclic graph C_{2N+1} with $2N+1$ vertices, where N is taken to be the growing parameter. The intersection numbers necessary to the limits (3.23) and (3.24) are obtained by simple observation as follows:

$$p^n_{1,n-1}(N) = \begin{cases} 1, & n = 1, 2, \dots, N, \\ 0, & \text{otherwise}, \end{cases} \qquad p^{n-1}_{1,n}(N) = \begin{cases} 2, & n = 1, \\ 1, & n = 2, 3, \dots, N, \\ 0, & \text{otherwise}, \end{cases}$$

$$p^{n-1}_{1,n-1}(N) = \begin{cases} 1, & n = N+1, \\ 0, & \text{otherwise}, \end{cases}$$

In particular, $\kappa(N) = p^0_{11}(N) = 2$. Hence

$$\omega_n(N) = \begin{cases} 1, & n = 1, \\ 1/2, & n = 2, 3, \dots, N, \\ 0, & n \ge N+1, \end{cases} \qquad \alpha_n(N) = \begin{cases} 1/\sqrt{2}, & n = N+1, \\ 0, & \text{otherwise}. \end{cases}$$

Obviously, (DR) is fulfilled with $\{\omega_n\} = \{1, 1/2, 1/2, \dots\}$ and $\{\alpha_n \equiv 0\}$, which is the Jacobi coefficient of the asymptotic spectral distribution μ of the normalized adjacency matrix in the vacuum state. The explicit form of μ is obtained from the continued fraction expansion corresponding to the Jacobi coefficient. Here the detailed computation is omitted for a more general result will be obtained in Sect. 4.1. As a result, we come to

$$\lim_{N\to\infty} \left\langle \left(\frac{A_N}{\sqrt{2}} \right)^m \right\rangle_o = \frac{1}{\pi} \int_{-\sqrt{2}}^{+\sqrt{2}} \frac{x^m}{\sqrt{2 - x^2}} \, dx, \qquad m = 1, 2, \dots .$$

The probability measure in the right-hand side is called the (normalized) arcsine law (see Definition 4.6).

We next consider the deformed vacuum state (Definition 2.10) defined by

$$\langle a\rangle_q = \langle Q\delta_o, a\delta_o\rangle = \sum_{n=0}^{\infty} q^n |V_n|^{1/2} \langle \Phi_n, a\Phi_0\rangle, \quad a \in \mathcal{A}(\mathcal{G}), \tag{3.27}$$

where $Q = (q^{\partial(x,y)})$ with $q \in \mathbb{R}$. Keeping the same notations, let $\mathcal{G} = (V, E)$ be a distance-regular graph with intersection numbers $\{p_{ij}^k\}$. The degree is given by $\kappa = p_{11}^0$. For normalization of the adjacency matrix we prepare the following:

Lemma 3.25. *The mean and the variance of the adjacency matrix A in the deformed vacuum state are respectively given as follows:*

$$\langle A\rangle_q = q\kappa, \tag{3.28}$$

$$\Sigma_q^2(A) = \langle (A - \langle A\rangle_q)^2\rangle_q = \kappa(1-q)(1+q+qp_{11}^1). \tag{3.29}$$

Proof. Since $A = A^+ + A^- + A^\circ$, we see from Lemma 3.10 and Theorem 3.11 that for $n = 0, 1, 2, \dots,$

$$\begin{aligned} A\Phi_n &= \sqrt{p_{1n}^{n+1} p_{1,n+1}^n}\, \Phi_{n+1} + p_{1n}^n \Phi_n + \sqrt{p_{1,n-1}^n p_{1n}^{n-1}}\, \Phi_{n-1} \\ &= p_{1n}^{n+1} \left(\frac{|V_{n+1}|}{|V_n|}\right)^{1/2} \Phi_{n+1} + p_{1n}^n \Phi_n + p_{1,n-1}^n \left(\frac{|V_n|}{|V_{n-1}|}\right)^{1/2} \Phi_{n-1}. \end{aligned}$$

Using the above formula, we easily obtain

$$\begin{aligned} A\Phi_0 &= |V_1|^{1/2}\Phi_1, \\ A^2\Phi_0 &= p_{11}^0 |V_0|^{1/2}\Phi_0 + p_{11}^1 |V_1|^{1/2}\Phi_1 + p_{11}^2 |V_2|^{1/2}\Phi_2. \end{aligned}$$

Taking (3.27) into account, we obtain

$$\langle A\rangle_q = \langle Q\delta_o, A\delta_o\rangle = \langle q|V_1|^{1/2}\Phi_1, |V_1|^{1/2}\Phi_1\rangle = q|V_1| = q\kappa,$$

which proves (3.28). Similarly,

$$\begin{aligned} \langle A^2\rangle_q &= \langle Q\delta_o, A^2\delta_o\rangle \\ &= \langle \Phi_0, p_{11}^0 |V_0|^{1/2}\Phi_0\rangle \\ &\quad + \langle q|V_1|^{1/2}\Phi_1, p_{11}^1|V_1|^{1/2}\Phi_1\rangle + \langle q^2|V_2|^{1/2}\Phi_2, p_{11}^2|V_2|^{1/2}\Phi_2\rangle \\ &= p_{11}^0 + qp_{11}^1|V_1| + q^2 p_{11}^2 |V_2| \\ &= p_{11}^0 + qp_{11}^1|V_1| + q^2 p_{12}^1 |V_1| \\ &= \kappa(1 + qp_{11}^1 + q^2 p_{12}^1), \end{aligned}$$

where $p_{11}^2|V_2| = p_{12}^1|V_1|$ was used (see Lemma 3.10). Thus we obtain

$$\langle (A - \langle A\rangle_q)^2\rangle_q = \langle A^2\rangle_q - \langle A\rangle_q^2 = \kappa(1 + qp_{11}^1 + q^2p_{12}^1) - (q\kappa)^2.$$

Finally, with the help of the identity $p_{11}^1 + p_{12}^1 + 1 = \kappa$, we obtain (3.29). □

Lemma 3.26. *For a distance-regular graph it holds that $QA = AQ$.*

Proof. By definition, a distance-regular graph is quasi-distance-regular. Hence the assertion is apparent by Lemma 2.19. It is also immediate to check that $(QA)_{xy} = (AQ)_{xy}$ for all $x, y \in V$. □

Theorem 3.27. *Let $\mathcal{G} = (V, E)$ be a distance-regular graph. If $Q = (q^{\partial(x,y)})$ is a positive definite kernel on V, the deformed vacuum state $\langle\cdot\rangle_q$ is positive (i.e., a state in a strict sense) on the adjacency algebra $\mathcal{A}(\mathcal{G})$.*

Proof. The assertion follows by combining Lemmas 2.12 and 3.26. □

Here we do not go into the question whether or not Q is a positive definite kernel. In order to outline our idea, it is sufficient to assume that $\Sigma_q^2(A) > 0$.

Now let us consider a growing distance-regular graph $\mathcal{G}^{(\nu)} = (V^{(\nu)}, E^{(\nu)})$. Suppose that each $\mathcal{G}^{(\nu)}$ is given a deformed vacuum state $\langle\cdot\rangle_q$, where q may depend on ν. By virtue of Lemma 3.25 the normalized adjacency matrix is given by

$$\frac{A_\nu - \langle A_\nu\rangle_q}{\Sigma_q(A_\nu)}, \tag{3.30}$$

where $\langle A_\nu\rangle_q$ and $\Sigma_q^2(A)$ are given by (3.28) and (3.29), respectively. Taking the quantum decomposition $A_\nu = A_\nu^+ + A_\nu^- + A_\nu^\circ$ into account, we obtain

$$\frac{A_\nu - \langle A_\nu\rangle_q}{\Sigma_q(A_\nu)} = \frac{A_\nu^+}{\Sigma_q(A_\nu)} + \frac{A_\nu^-}{\Sigma_q(A_\nu)} + \frac{A_\nu^\circ - q\kappa(\nu)}{\Sigma_q(A_\nu)}. \tag{3.31}$$

We shall observe the explicit actions of the above quantum components. For $n = 1, 2, \dots$ we set

$$\bar{\omega}_n(\nu, q) = \frac{p_{1,n-1}^n(\nu) p_{1,n}^{n-1}(\nu)}{\Sigma_q^2(A_\nu)}, \tag{3.32}$$

$$\bar{\alpha}_n(\nu, q) = \frac{p_{1,n-1}^{n-1}(\nu) - q\kappa(\nu)}{\Sigma_q(A_\nu)}. \tag{3.33}$$

Using Theorem 3.11 we obtain

$$\frac{A_\nu^+}{\Sigma_q(A_\nu)} \Phi_n = \sqrt{\bar{\omega}_{n+1}(\nu, q)}\, \Phi_{n+1}, \quad n = 0, 1, 2, \dots, \tag{3.34}$$

$$\frac{A_\nu^-}{\Sigma_q(A_\nu)} \Phi_0 = 0, \qquad \frac{A_\nu^-}{\Sigma_q(A_\nu)} \Phi_n = \sqrt{\bar{\omega}_n(\nu, q)}\, \Phi_{n-1}, \quad n = 1, 2, \dots, \tag{3.35}$$

$$\frac{A_\nu^\circ - q\kappa(\nu)}{\Sigma_q(A_\nu)} \Phi_n = \bar{\alpha}_{n+1}(\nu, q)\, \Phi_n, \qquad n = 0, 1, 2, \dots. \tag{3.36}$$

Then, the situation is quite similar to the case of the vacuum state. We consider the following limits:

$$\omega_n = \lim_{\nu,q} \bar{\omega}_n(\nu, q) = \lim_{\nu,q} \frac{p^n_{1,n-1}(\nu) p^{n-1}_{1,n}(\nu)}{\Sigma^2_q(A_\nu)}, \tag{3.37}$$

$$\alpha_n = \lim_{\nu,q} \bar{\alpha}_n(\nu, q), = \lim_{\nu,q} \frac{p^{n-1}_{1,n-1}(\nu) - q\kappa(\nu)}{\Sigma_q(A_\nu)} \tag{3.38}$$

when they exist under a good scaling balance of ν and q. If (DR) is fulfilled for (3.37) and (3.38), we obtain a Jacobi coefficient $(\{\omega_n\}, \{\alpha_n\})$. Let $\Gamma_{\{\omega_n\}} = (\Gamma, \{\Psi_n\}, B^+, B^-)$ be an interacting Fock space associated with $\{\omega_n\}$ and B° the diagonal operator defined by $\{\alpha_n\}$.

Lemma 3.28. *Notations and assumptions being as above, we set*

$$\tilde{A}^\pm_\nu = A^\pm_\nu, \qquad \tilde{A}^\circ_\nu = A^\circ_\nu - q\kappa(\nu).$$

It then holds that

$$\lim_{\nu,q} \left\langle \Phi^{(\nu)}_j, \frac{\tilde{A}^{\epsilon_m}_\nu}{\Sigma_q(A_\nu)} \cdots \frac{\tilde{A}^{\epsilon_1}_\nu}{\Sigma_q(A_\nu)} \Phi^{(\nu)}_k \right\rangle = \langle \Psi_j, B^{\epsilon_m} \cdots B^{\epsilon_1} \Psi_k \rangle,$$

for any $\epsilon_1, \dots, \epsilon_m \in \{+, -, \circ\}$, $m = 1, 2, \dots$ *and* $j, k = 0, 1, 2, \dots$.

Proof. Apparently, the same argument as in the proof of Theorems 3.21 is applicable. □

Theorem 3.29 (QCLT for a growing DRG in the deformed vacuum state). *Let* $\mathcal{G}^{(\nu)} = (V^{(\nu)}, E^{(\nu)})$ *be a growing distance-regular graph with* A_ν *being the adjacency matrix, and each* $\mathcal{A}(\mathcal{G}^{(\nu)})$ *be given a deformed vacuum state* $\langle\cdot\rangle_q$. *Assume that condition (DR) is fulfilled for* (3.37) *and* (3.38), *and that the limit*

$$c_n = \lim_{\nu,q} q^n |V^{(\nu)}_n|^{1/2} = \lim_{\nu,q} q^n \sqrt{p^0_{nn}(\nu)} \tag{3.39}$$

exists for all n for which $\{\alpha_n\}$ *is defined. Let* $\Gamma_{\{\omega_n\}} = (\Gamma, \{\Psi_n\}, B^+, B^-)$ *be an interacting Fock space associated with* $\{\omega_n\}$, B° *the diagonal operator defined by* $\{\alpha_n\}$, *and* Υ *the formal sum of vectors defined by*

$$\Upsilon = \sum_{n=0}^{\infty} c_n \Psi_n.$$

Then, we have

$$\lim_{\nu,q} \left\langle \frac{\tilde{A}^{\epsilon_m}_\nu}{\Sigma_q(A_\nu)} \cdots \frac{\tilde{A}^{\epsilon_1}_\nu}{\Sigma_q(A_\nu)} \right\rangle_q = \langle \Upsilon, B^{\epsilon_m} \cdots B^{\epsilon_1} \Psi_0 \rangle, \tag{3.40}$$

for any $\epsilon_1, \dots, \epsilon_m \in \{+, -, \circ\}$ *and* $m = 1, 2, \dots$.

Proof. By definition of the deformed vacuum state we have

$$\left\langle \frac{\tilde{A}_\nu^{\epsilon_m}}{\Sigma_q(A_\nu)} \cdots \frac{\tilde{A}_\nu^{\epsilon_1}}{\Sigma_q(A_\nu)} \right\rangle_q = \sum_{n=0}^{\infty} q^n |V_n^{(\nu)}|^{1/2} \left\langle \Phi_n, \frac{\tilde{A}_\nu^{\epsilon_m}}{\Sigma_q(A_\nu)} \cdots \frac{\tilde{A}_\nu^{\epsilon_1}}{\Sigma_q(A_\nu)} \Phi_0 \right\rangle.$$

In view of the up–down actions of A^ϵ the right-hand side is in fact a finite sum up to $n = m$ at most. Then, applying Lemma 3.28 and (3.39), we obtain

$$\lim_{\nu,q} \left\langle \frac{\tilde{A}_\nu^{\epsilon_m}}{\Sigma_q(A_\nu)} \cdots \frac{\tilde{A}_\nu^{\epsilon_1}}{\Sigma_q(A_\nu)} \right\rangle_q = \sum_{n=0}^{\infty} c_n \langle \Psi_n, B^{\epsilon_m} \cdots B^{\epsilon_1} \Psi_0 \rangle,$$

from which (3.40) follows. □

Remark 3.30. In Theorem 3.29 we do not assume that the deformed vacuum state $\langle \cdot \rangle_q$ is positive, but assume that $\Sigma_q^2(A_\nu) > 0$ for normalization. If each deformed vacuum state $\langle \cdot \rangle_q$ on $\mathcal{A}(\mathcal{G}^{(\nu)})$ is positive, there exists a probability measure $\mu \in \mathfrak{P}_{\mathrm{fm}}(\mathbb{R})$ such that

$$\langle \Upsilon, (B^+ + B^- + B^\circ)^m \Psi_0 \rangle = \int_{-\infty}^{+\infty} x^m \mu(dx), \qquad m = 1, 2, \dots. \tag{3.41}$$

This μ is the asymptotic spectral distribution of A_ν in the deformed vacuum state that we are interested in. However, derivation of an explicit form of μ from (3.41) seems to be a difficult problem in general. We shall discuss only some special cases later.

Remark 3.31. In Theorem 3.29 there are two basic assumptions: (i) condition (DR) is fulfilled for (3.37) and (3.38) and (ii) the limit (3.39) exists for all n for which $\{\alpha_n\}$ is defined. These conditions are naturally satisfied by various examples; it is a problem, on the contrary, to find a proper scaling balance of ν and q in such a way that (i) and (ii) hold. In Chap. 7 we shall introduce a widely applicable sufficient condition for (i) and (ii) when $\kappa(\nu) \to \infty$. We note, however, that Theorem 3.29 does not require the condition $\kappa(\nu) \to \infty$, see also Exercise 3.10.

3.5 Coherent States in General

At the end of the previous section we come to the expression

$$\langle \Upsilon, (B^+ + B^- + B^\circ)^m \Psi_0 \rangle, \tag{3.42}$$

where $\Gamma_{\{\omega_n\}} = (\Gamma, \{\Phi_n\}, B^+, B^-)$ is an interacting Fock space, B° a diagonal operator, and Υ a formal sum of vectors defined by

$$\Upsilon = \sum_{n=0}^{\infty} c_n \Psi_n.$$

We note that (3.42) defines a normalized linear function on the $*$-algebra generated by $\{B^+, B^-, B^\circ\}$. In this section we consider a special case.

Definition 3.32. Let $\Gamma_{\{\omega_n\}} = (\Gamma, \{\Psi_n\}, B^+, B^-)$ be an interacting Fock space associated with a Jacobi sequence $\{\omega_n\}$. The *coherent vector* with parameter $z \in \mathbb{C}$ is a formal sum of vectors defined by

$$\Omega_z = \Psi_0 + \sum_{n=1}^{\infty} \frac{z^n}{\sqrt{\omega_n \cdots \omega_2 \omega_1}} \Psi_n \,, \tag{3.43}$$

where the right-hand side becomes a finite sum when the Jacobi sequence is of finite type.

Definition 3.33. Let $\Gamma_{\{\omega_n\}} = (\Gamma, \{\Psi_n\}, B^+, B^-)$ be an interacting Fock space associated with a Jacobi sequence $\{\omega_n\}$ and B° a diagonal operator. The normalized linear function

$$\langle b \rangle_z = \langle \Omega_z, b\Psi_0 \rangle, \tag{3.44}$$

where b runs over the $*$-algebra generated by $\{B^+, B^-, B^\circ\}$, is called the *coherent state* with parameter $z \in \mathbb{C}$.

Note that a coherent state is not necessarily positive. This lax wording is similar to a deformed vacuum state on an adjacency algebra. In the following chapters, a coherent state will emerge in various contexts of limit theorems and play an interesting role in computing asymptotic spectral distributions.

Here we only note the following result, the proof of which is direct.

Proposition 3.34. *Let $\Gamma_{\{\omega_n\}} = (\Gamma, \{\Phi_n\}, B^+, B^-)$ be an interacting Fock space of infinite type. Then a coherent vector Ω_z is a generalized eigenvector of B^- with an eigenvalue z, i.e., $B^-\Omega_z = z\Omega_z$. More precisely,*

$$\langle \Omega_z, B^+\Phi \rangle = z\langle \Omega_z, \Phi \rangle, \qquad \Phi \in \Gamma.$$

Exercises

3.1. Show that the cyclic graph C_N with $N(\geq 3)$ vertices is distance-regular.

3.2. Show that the two-dimensional integer lattice $\mathbb{Z}^2$ is not distance-regular.

3.3. Prove that a distance-transitive graph is distance-regular. [Proposition 3.7]

3.4. A graph is called *vertex-transitive* if for any pair of vertices $x, y \in V$ there exists $\alpha \in \mathrm{Aut}\,(\mathcal{G})$ such that $\alpha(x) = y$. Show that for a vertex-transitive graph the number of m-step walks from $x \in V$ to itself is independent of the choice of $x \in V$. [Remark 3.16]

3.5. The direct product of two cyclic graphs $C_M \times C_N$ $(M, N \geq 3)$ is vertex-transitive but not distance-regular.

3.6. Let $\mathcal{G} = (V, E)$ be a quasi-distance-regular graph, i.e.,

$$\left|\left\{z \in V\,;\; \begin{matrix} \partial(z,x) = n \\ \partial(z,y) = 1 \end{matrix}\right\}\right| = \left|\left\{z \in V\,;\; \begin{matrix} \partial(z,x) = 1 \\ \partial(z,y) = n \end{matrix}\right\}\right|$$

is satisfied for any choice of $x, y \in V$ and $n = 0, 1, 2, \dots$. Prove that if the above number depends only on $\partial(x, y)$, then $\mathcal{G}$ becomes distance-regular.

3.7. Prove that the complete graph K_N with $N \geq 2$ vertices is distance-regular with intersection numbers

$$p^0_{00} = 1, \quad p^0_{11} = N - 1, \quad p^1_{01} = p^1_{10} = 1, \quad p^1_{11} = N - 2,$$

and all other $p^k_{ij} = 0$. In particular, the degree $\kappa = p^0_{11} = N - 1$.

3.8. Let A_N be the adjacency matrix of the complete graph K_N. Show that condition (DR) is not fulfilled for the limits defined by (3.23) and (3.24) as $N \to \infty$. Hence, the asymptotic spectral distribution of A_N in the vacuum state does not exist. Moreover, show that there is no scaling balance of $N \to \infty$ and $q \to 0$ such that the asymptotic spectral distribution of A_N in the deformed vacuum state exists.

3.9. For the cyclic graph C_{2N} with $2N(\geq 4)$ vertices derive the asymptotic spectral distribution of the adjacency matrix in the vacuum state. [See Example 3.24]

3.10. For $N \geq 3$ let A_N be the adjacency matrix of the cyclic graph C_N. Compute the limits (3.37), (3.38) and (3.39) as $N \to \infty$ with $-1 < q < 1$ being fixed. Then derive an explicit form of the asymptotic spectral distribution of A_N in the deformed vacuum state.

3.11. Let $(\Gamma, \{\Phi_n\}, B^+, B^-)$ be an interacting Fock space associated with a Jacobi sequence $\{\omega_n\}$ of finite type. Show that 0 is the unique eigenvalue of B^- and the eigenspace is spanned by the vacuum vector Φ_0.

3.12. Let $\Gamma_{\{\omega_n\}} = (\Gamma, \{\Phi_n\}, B^+, B^-)$ be an interacting Fock space of infinite type and define $0 \leq R \leq \infty$ by

$$R = \liminf_{n\to\infty} (\omega_n \cdots \omega_1)^{\frac{1}{2n}}.$$

Let $\bar{\Gamma}$ be the Hilbert space spanned by $\{\Phi_n\}$, that is, the completion of Γ. Show that the coherent vector Ω_z belongs to $\bar{\Gamma}$ for all $z \in \mathbb{C}$ with $|z| < R$.

3.13. Show that R defined in Exercise 3.12 coincides with

$$R = \lim_{n\to\infty} \sqrt{\omega_n},$$

whenever the limit exists.

Notes

(Finite) distance-regular graphs have been considerably discussed in graph theory and algebraic combinatorics. A standard textbook is Bannai–Ito [17] for basic knowledge. For more information we refer to Brouwer–Cohen–Neumaier [50], which includes 800 references.

The quantum probabilistic framework was first introduced by Hora [102] for asymptotic spectral analysis of growing distance-regular graphs, where concrete examples included Hamming graphs, Johnson graphs and bipartite halves, q-analogue of Johnson graphs. However, the method worked only when the spectrum is known explicitly for each graph in question. Nevertheless, the discussion was extended to the case of deformed vacuum states by Hora [104], where the Johnson graphs were studied in detail.

Our method requires neither combinatorial struggle nor explicit knowledge of spectrum before taking the limit. We only need to extract the 'leading terms' surviving in the limit, which is reasonably reachable by means of quantum decomposition. The method presented in this chapter originated in the concrete study of Hamming graphs by Hashimoto–Obata–Tabei [97] and Johnson graphs by Hashimoto–Hora–Obata [96]. The concrete results, discussed in Chaps. 5 and 6, are now fully reproduced by our framework with high transparency.

The coherent state is a widely known concept in quantum physics, e.g., Klauder [133], Klauder–Skagerstam [134]. See also Ali–Antoine–Gazeau [13]. In the context of an interacting Fock space the coherent state was introduced by Das [66], see also Das [64,65] for a restricted case. Our definition coincides with his and seems to be reasonable because the coherent vector is characterized as an eigenvector of the annihilation operator. This is a well-known characteristic property of the coherent vectors in Boson Fock space.

4

Homogeneous Trees

We shall go further through the general results developed in the preceding chapter to investigate spectral properties of a homogeneous tree and its variant in a quite explicit manner. The central limit distributions obtained from such growing graphs include the Wigner semicircle law, free Poisson distributions and free Meixner laws.

4.1 Kesten Distribution

Definition 4.1. A graph (always assumed to be connected) is called a *tree* if it has no cycle. A tree is called *homogeneous* if it is regular.

A homogeneous tree is always an infinite graph. A homogeneous tree of even degree, say $2N$, appears as the Cayley graph $(F_N, \{g_{\pm 1}, \dots, g_{\pm N}\})$, where F_N is a free group on N generators $g_1, \dots, g_N$. Note that a homogeneous tree is determined uniquely by its degree. We denote by T_κ a homogeneous tree with degree $\kappa \geq 2$ and by $A = A_\kappa$ its adjacency matrix.

As is easily verified, a homogeneous tree T_κ is distance-transitive and hence distance-regular. By straightforward observation we easily obtain the intersection numbers:

$$\begin{aligned} &p_{11}^{0} = \kappa, \\ &p_{1,n}^{n-1} = \kappa - 1, \quad n = 2, 3, \dots, \\ &p_{1,n-1}^{n} = 1, \quad n = 1, 2, 3, \dots, \\ &p_{1,n}^{n} = 0, \quad n = 0, 1, 2, \dots. \end{aligned} \tag{4.1}$$

Fix an origin o of T_κ arbitrarily and consider the quantum decomposition:

$$A = A^+ + A^-.$$

In fact, $A^\circ = 0$ by geometric observation. We are interested in a probability measure $\mu \in \mathfrak{P}_{\mathrm{fm}}(\mathbb{R})$ such that

A. Hora and N. Obata: Homogeneous Trees. In: A. Hora and N. Obata, Quantum Probability and Spectral Analysis of Graphs, Theoretical and Mathematical Physics, 105–130 (2007)
DOI 10.1007/3-540-48863-4_4

$$\langle A^m \rangle_o = \langle \delta_o, A^m \delta_o \rangle = \int_{-\infty}^{+\infty} x^m \mu(dx), \qquad m = 1, 2, \ldots.$$

By Theorem 3.12 the Jacobi coefficient of the above μ is given by

$$\omega_n = p^n_{1,n-1} p^{n-1}_{1,n} = \begin{cases} \kappa, & \text{if } n = 1, \\ \kappa - 1, & \text{if } n = 2, 3, \ldots, \end{cases} \tag{4.2}$$

$$\alpha_n = p^{n-1}_{1,n-1} = 0, \qquad n = 1, 2, \ldots.$$

In other words, letting $\Gamma_{\{\omega_n\}} = (\Gamma, \{\Phi_n\}, B^+, B^-)$ be the interacting Fock space associated with $\{\omega_n\}$ given in (4.2), we have

$$\langle \Phi_0, (A^+ + A^-)^m \Phi_0 \rangle = \int_{-\infty}^{+\infty} x^m \mu(dx), \qquad m = 1, 2, \ldots.$$

Then μ is uniquely determined (Theorem 1.66) and is obtained from the Stieltjes transform (Theorem 1.97):

$$\int_{-\infty}^{+\infty} \frac{\mu(dx)}{z - x} = \frac{1}{z} \frac{}{-} \frac{\kappa}{z} \frac{}{-} \frac{\kappa - 1}{z} \frac{}{-} \frac{\kappa - 1}{z} \frac{}{-} \cdots. \tag{4.3}$$

More generally, we give the following:

Definition 4.2. For $p > 0$ and $q \geq 0$, the probability measure $\mu = \mu_{p,q}$ specified by

$$\int_{-\infty}^{+\infty} \frac{\mu(dx)}{z - x} = \frac{1}{z} \frac{}{-} \frac{p}{z} \frac{}{-} \frac{q}{z} \frac{}{-} \frac{q}{z} \frac{}{-} \cdots \tag{4.4}$$

is called the *Kesten distribution* with parameters p, q. The Kesten distribution $\mu_{p,q}$ has mean 0 and variance p.

Let us compute the density function of the Kesten distribution.

Proposition 4.3. *For $p > 0$ and $q \geq 0$ we define $\rho_{p,q}(x)$ by*

$$\rho_{p,q}(x) = \begin{cases} \dfrac{p}{2\pi} \dfrac{\sqrt{4q - x^2}}{p^2 - (p - q)x^2}, & \textit{if } |x| \leq 2\sqrt{q}, \\ 0, & \textit{otherwise.} \end{cases} \tag{4.5}$$

Then the Kesten distribution $\mu_{p,q}$ is given by

$$\mu_{p,q}(dx) = \begin{cases} \rho_{p,q}(x)dx, & \textit{if } 0 < p \leq 2q, \\ \rho_{p,q}(x)dx + \dfrac{p - 2q}{2(p - q)} (\delta_{-p/\sqrt{p-q}} + \delta_{p/\sqrt{p-q}}), & \textit{if } 0 \leq 2q < p. \end{cases}$$

Proof. The case of $q = 0$ being trivial, we only discuss the case of $q > 0$. We need to compute the continued fraction:

$$G(z) = \frac{1}{z} \frac{}{-} \frac{p}{z} \frac{}{-} \frac{q}{z} \frac{}{-} \frac{q}{z} \frac{}{-} \cdots . \tag{4.6}$$

We first consider

$$H = \frac{q}{z} \frac{}{-} \frac{q}{z} \frac{}{-} \frac{q}{z} \frac{}{-} \frac{q}{z} \frac{}{-} \cdots = \frac{q}{z - H},$$

from which we obtain

$$H = \frac{q}{z} \frac{}{-} \frac{q}{z} \frac{}{-} \frac{q}{z} \frac{}{-} \frac{q}{z} \frac{}{-} \cdots = \frac{z - \sqrt{z^2 - 4q}}{2}, \tag{4.7}$$

where the minus sign in front of the square root is for convenience and the branch will be chosen later. Inserting (4.7) into (4.6), we obtain

$$G(z) = \cfrac{1}{z - \cfrac{p}{q} H} = -\frac{1}{2} \frac{(p - 2q)z + p\sqrt{z^2 - 4q}}{p^2 - (p-q)z^2}. \tag{4.8}$$

In order to determine the branch of the square root we recall that $\operatorname{Im} G(z) < 0$ for $\operatorname{Im} z > 0$ and $\operatorname{Im} G(z) > 0$ for $\operatorname{Im} z < 0$, see Proposition 1.96. Examining this property, for example, by taking $z = iy$, we conclude that $\sqrt{z^2 - 4q}$ is a holomorphic function defined on $\mathbb{C}\backslash[-2\sqrt{q}, +2\sqrt{q}]$ and takes positive values on $(+2\sqrt{q}, +\infty)$ and negative values on $(-\infty, -2\sqrt{q})$. The absolutely continuous part of $\mu_{p,q}$ is obtained by calculating the limit

$$-\frac{1}{\pi} \lim_{y \downarrow 0} \operatorname{Im} G(x + iy),$$

see also Corollary 1.103. In fact, the limit is easily shown to be equal to $\rho_{p,q}(x)dx$ defined in (4.5). For the singular part of $\mu_{p,q}$ we first note the integral formula:

$$\frac{p}{2\pi} \int_{-2\sqrt{q}}^{+2\sqrt{q}} \frac{\sqrt{4q - x^2}}{p^2 - (p-q)x^2} \, dx = \begin{cases} 1, & \text{if } 0 < p \le 2q, \\ \dfrac{q}{p-q}, & \text{if } 0 < 2q \le p, \end{cases} \tag{4.9}$$

the verification of which is elementary. Hence, for $0 < p \le 2q$ there is no singular part and $\mu_{p,q}(dx) = \rho_{p,q}(x)dx$. Suppose that $0 < 2q < p$. Then by (4.9) we see that

$$\int_{-\infty}^{+\infty} \rho_{p,q}(x)\, dx = \frac{q}{p-q} < 1, \tag{4.10}$$

which means that $\mu_{p,q}$ has a singular part. Note that $G(z)$ has simple poles at $\pm p/\sqrt{p-q}$ and

$$\operatorname*{Res}_{z=\pm p/\sqrt{p-q}} G(z) = \frac{p-2q}{2(p-q)}$$

because of the above branch for $\sqrt{z^2-4q}$. It then follows from Proposition 1.104 that $\mu_{p,q}$ has two atoms at $\pm p/\sqrt{p-q}$ with weight $(p-2q)/2(p-q)$. Moreover, we see from (4.10) that

$$\rho_{p,q}(x)dx + \frac{p-2q}{2(p-q)}(\delta_{-p/\sqrt{p-q}} + \delta_{p/\sqrt{p-q}}) \tag{4.11}$$

is a probability measure. Consequently, $\mu_{p,q}(dx)$ is given by (4.11). □

Going back to the homogeneous tree, we claim the following:

Theorem 4.4 (Kesten). *Let A be the adjacency matrix of a homogeneous tree of degree $\kappa \geq 2$. Then,*

$$\langle \delta_o, A^m \delta_o \rangle = \frac{\kappa}{2\pi} \int_{-2\sqrt{\kappa-1}}^{+2\sqrt{\kappa-1}} x^m \frac{\sqrt{4(\kappa-1)-x^2}}{\kappa^2-x^2}\, dx, \qquad m = 1, 2, \dots,$$

where the probability measure on the right-hand side is the Kesten distribution $\mu_{\kappa,\kappa-1}(dx) = \rho_{\kappa,\kappa-1}(x)dx$ (see Fig. 4.1).

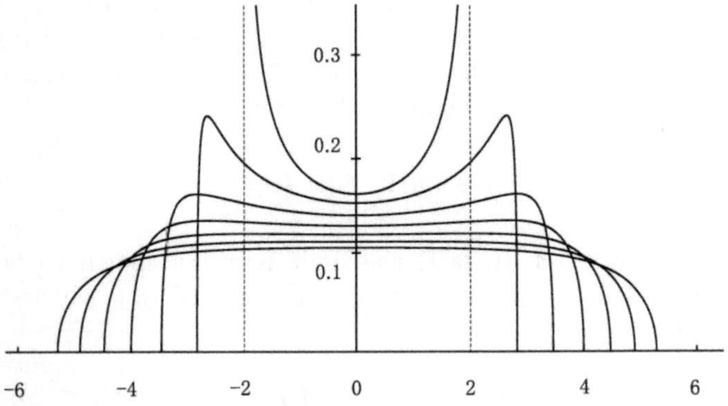

Fig. 4.1. Kesten distributions $\mu_{\kappa,\kappa-1}$ with $\kappa = 2, 3, 4, 5, 6, 7, 8$

Proposition 4.5. *Let $p > 0$.*

(1) *The Kesten distribution $\mu_{p,p}$ is the semicircle law which has mean 0 and variance p.*
(2) *The Kesten distribution $\mu_{p,p/2}$ is the arcsine law which has mean 0 and variance p.*

The proof is obvious by definition (see below).

Definition 4.6. Let $a > 0$. The probability measure with the density function

$$\rho(x) = \begin{cases} \dfrac{1}{\pi\sqrt{2a^2 - x^2}}, & |x| < \sqrt{2}\,a, \\ 0, & \text{otherwise}, \end{cases}$$

is called the *arcsine law* with mean 0 and variance a^2.

A homogeneous tree of degree $\kappa = 2$ is the one-dimensional integer lattice. The spectral distribution of the adjacency matrix in the vacuum state is the Kesten distribution $\mu_{2,1}$, which coincides with the arcsine law with mean 0 and variance 2. Thus,

Corollary 4.7. *The spectral distribution of the adjacency matrix of the one-dimensional integer lattice* $\mathbb{Z}$ *in the vacuum state at an origin is the arcsine law with mean* 0 *and variance* 2.

Remark 4.8. We showed in Example 3.24 that the asymptotic spectral distribution of the cyclic graphs C_{2N+1} as $N \to \infty$ is the (normalized) arcsine law, which coincides (up to normalization) with the spectral distribution of the adjacency matrix of $\mathbb{Z}$. This reflects our intuition that the geometric sight of C_{2N+1} from the origin becomes similar to that of $\mathbb{Z}$ as $N \to \infty$. In Example 3.24 we consider cyclic graphs of odd degrees. This restriction is not essential and the same result follows for C_N as $N \to \infty$.

4.2 Asymptotic Spectral Distributions in the Vacuum State (Free CLT)

Having studied in the previous section the spectral distribution in the vacuum state of the adjacency matrix A_κ of a homogeneous tree T_κ, we are now interested in its asymptotic behaviour as $\kappa \to \infty$. Since T_κ is distance-regular, we can simply apply the general result in Theorem 3.21. We need a Jacobi coefficient $(\{\omega_n\}, \{\alpha_n\})$ which describes the limit of the normalized adjacency matrix:

$$\frac{A_\kappa}{\sqrt{\kappa}} = \frac{A_\kappa^+ + A_\kappa^-}{\sqrt{\kappa}}.$$

In fact, we obtain $\{\omega_n = 1\}$ from the following simple calculations. For $n = 1$ we have

$$\omega_1 = \lim_{\kappa\to\infty} \frac{p^1_{1,0}(\kappa)p^0_{1,1}(\kappa)}{\kappa} = \lim_{\kappa\to\infty} \frac{1\cdot\kappa}{\kappa} = 1.$$

For $n \geq 2$,

$$\omega_n = \lim_{\kappa\to\infty} \frac{p^n_{1,n-1}(\kappa)p^{n-1}_{1,n}(\kappa)}{\kappa} = \lim_{\kappa\to\infty} \frac{1\cdot(\kappa-1)}{\kappa} = 1.$$

Moreover, since

$$\alpha_n = \lim_{\kappa \to \infty} \frac{p_{1,n-1}^{n-1}(\kappa)}{\sqrt{\kappa}} = 0, \quad n = 1, 2, \dots,$$

no diagonal operator appears in the limit. Recall that an interacting Fock space associated with $\{\omega_n \equiv 1\}$ is the free Fock space.

Theorem 4.9 (QCLT for homogeneous trees). *Let $\Gamma_{\mathrm{free}} = (\Gamma, \{\Psi_n\}, B^+, B^-)$ be a free Fock space. Then, for the quantum components $A_\kappa^\pm$ of the adjacency matrix of a homogeneous tree T_κ we have*

$$\lim_{\kappa \to \infty} \frac{A_\kappa^\pm}{\sqrt{\kappa}} = B^\pm \tag{4.12}$$

in the sense of stochastic convergence with respect to the vacuum states.

We then come to the following:

Theorem 4.10 (CLT for homogeneous trees). *Let A_κ be the adjacency matrix of a homogeneous tree of degree $\kappa \geq 2$. Then,*

$$\lim_{\kappa \to \infty} \left\langle \left(\frac{A_\kappa}{\sqrt{\kappa}} \right)^m \right\rangle_o = \frac{1}{2\pi} \int_{-2}^{+2} x^m \sqrt{4 - x^2}\, dx, \qquad m = 1, 2, \dots,$$

where the probability measure on the right-hand side is the Wigner semicircle law.

Proof. By Theorem 4.9 we have

$$\begin{aligned}\lim_{\kappa \to \infty} \left\langle \delta_o, \left(\frac{A_\kappa}{\sqrt{\kappa}} \right)^m \delta_o \right\rangle &= \lim_{\kappa \to \infty} \left\langle \Phi_0, \left(\frac{A_\kappa^+ + A_\kappa^-}{\sqrt{\kappa}} \right)^m \Phi_0 \right\rangle \\ &= \langle \Psi_0, (B^+ + B^-)^m \Psi_0 \rangle,\end{aligned}$$

which is the mth moment of the Wigner semicircle law, see Theorem 1.72. □

Theorem 4.10 is a prototype of the celebrated *free central limit theorem* (free CLT). In fact, this wording is adequate because A_κ is decomposed into a sum of 'free independent' random variables, see Theorem 8.21.

One may prove Theorem 4.10 directly by means of the concrete expression of Kesten distributions obtained in Theorem 4.4 (Exercise 4.2).

4.3 The Haagerup State

We start with the following famous result. The proof will be deferred to Sect. 4.6, where we shall prove a more general result (Theorem 4.35). Note that the 'only if' part follows from Proposition 2.13.

Theorem 4.11 (Haagerup). *Let $T_\kappa = (V, E)$ be a homogeneous tree of degree $\kappa \ge 2$. Then $Q = (q^{\partial(x,y)})$ is a positive definite kernel on V, i.e.,*

$$\langle f, Qf\rangle = \sum_{x,y\in V} \overline{f(x)}\, q^{\partial(x,y)} f(y) \ge 0, \qquad f \in C_0(V),$$

if and only if $-1 \le q \le 1$.

Together with $AQ = QA$, which follows from T_κ being distance-regular, we see that the deformed vacuum state $\langle\cdot\rangle_q$ defined by

$$\langle a\rangle_q = \langle Q\delta_o, a\delta_o\rangle = \langle Q\Phi_0, a\Phi_0\rangle, \qquad a \in \mathcal{A}(T_\kappa),$$

is positive (i.e., a state in the strict sense) whenever $-1 \le q \le 1$. This is called the *Haagerup state*. The mean and the variance of the adjacency matrix A_κ are easily obtained from Lemma 3.25:

$$\langle A_\kappa\rangle_q = \kappa q, \qquad \Sigma_q^2(A_\kappa) = \langle (A_\kappa - \langle A_\kappa\rangle_q)^2\rangle_q = \kappa(1-q^2).$$

The normalized adjacency matrix becomes

$$\frac{A_\kappa - \langle A_\kappa\rangle_q}{\Sigma_q(A_\kappa)} = \frac{A_\kappa - \kappa q}{\sqrt{\kappa(1-q^2)}} \tag{4.13}$$

and we are interested in the limit in the deformed vacuum states $\langle\cdot\rangle_q$. As was discussed in Sect. 3.4, to describe the limit we need three sequences $\{\omega_n\}, \{\alpha_n\}, \{c_n\}$ defined in (3.37), (3.38) and (3.39), respectively.

Recall first that

$$\omega_n = \lim_{\kappa\to\infty} \frac{p^n_{1,n-1}(\kappa)p^{n-1}_{1,n}(\kappa)}{\Sigma_q^2(A_\kappa)},$$

which describes the interacting Fock space in the limit. In view of (4.1), for $n = 1$ we have

$$\omega_1 = \lim_{\kappa\to\infty} \frac{1\cdot\kappa}{\kappa(1-q^2)} = \lim_{\kappa\to\infty} \frac{1}{1-q^2}.$$

For $n \ge 2$ we come to the same expression:

$$\omega_n = \lim_{\kappa\to\infty} \frac{1\cdot(\kappa-1)}{\kappa(1-q^2)} = \lim_{\kappa\to\infty} \frac{1}{1-q^2}. \tag{4.14}$$

(Remind that q may depend on κ.) For the diagonal operator we need

$$\alpha_n = \lim_{\kappa\to\infty} \frac{p^{n-1}_{1,n-1}(\kappa) - q\kappa}{\Sigma_q(A_\nu)} = \lim_{\kappa\to\infty} \frac{-q\kappa}{\sqrt{\kappa(1-q^2)}} = \lim_{\kappa\to\infty} \frac{-q\sqrt{\kappa}}{\sqrt{1-q^2}}. \tag{4.15}$$

In order that both limits (4.14) and (4.15) exist we need to find a suitable balance between q and κ. In fact, we obtain with no difficulty the following:

$$\lim_{\kappa\to\infty} q = 0, \qquad \lim_{\kappa\to\infty} q\sqrt{\kappa} = \gamma. \tag{4.16}$$

Here we also note that $\gamma \in \mathbb{R}$ can be arbitrarily chosen. Under these circumstances we obtain

$$\omega_n = 1, \quad \alpha_n = -\gamma, \quad n = 1, 2, \dots,$$

so that the limit is described by the free Fock space $\Gamma_{\text{free}} = (\Gamma, \{\Psi_n\}, B^+, B^-)$. We also need to compute

$$c_n = \lim_{\substack{q\sqrt{\kappa}\to\gamma \\ \kappa\to\infty, q\to 0}} q^n \sqrt{p^0_{nn}(\kappa)}.$$

Since $p^0_{nn} = \kappa(\kappa-1)^{n-1}$ for $n = 1, 2, \dots$, we have

$$c_n = \gamma^n, \qquad n = 0, 1, 2, \dots.$$

The corresponding (formal) sum of vectors is given by

$$\sum_{n=0}^{\infty} c_n \Psi_n = \sum_{n=0}^{\infty} \gamma^n \Psi_n = \Omega_\gamma,$$

which is the coherent vector of the free Fock space. The coherent state is denoted by $\langle\cdot\rangle_\gamma$ for simplicity. We thus come to the following result, which is just a specialization of Theorem 3.29.

Theorem 4.12 (QCLT for homogeneous trees in Haagerup states). *Let A_κ be the adjacency matrix of a homogeneous tree T_κ of degree $\kappa \geq 2$ and the adjacency algebra $\mathcal{A}(T_\kappa)$ be equipped with a Haagerup state $\langle\cdot\rangle_q$. Let $(\Gamma_{\text{free}}, \{\Psi_n\}, B^+, B^-)$ be the free Fock space. Then for any $\gamma \in \mathbb{R}$ we have*

$$\lim_{\substack{q\sqrt{\kappa}\to\gamma \\ \kappa\to\infty, q\to 0}} \frac{A_\kappa^\pm}{\Sigma_q(A_\kappa)} = B^\pm, \qquad \lim_{\substack{q\sqrt{\kappa}\to\gamma \\ \kappa\to\infty, q\to 0}} \frac{-\langle A_\kappa\rangle_q}{\Sigma_q(A_\kappa)} = -\gamma,$$

in the sense of stochastic convergence with respect to the Haagerup state $\langle\cdot\rangle_q$ on the left-hand sides and the coherent state $\langle\cdot\rangle_\gamma$ on the right-hand sides.

As a classical reduction we obtain

Theorem 4.13. *Notations and assumptions being as in Theorem 4.12, we have*

$$\lim_{\substack{q\sqrt{\kappa}\to\gamma \\ \kappa\to\infty, q\to 0}} \left\langle \left(\frac{A_\kappa - \langle A_\kappa\rangle_q}{\Sigma_q(A_\kappa)} \right)^m \right\rangle_q = \langle (B^+ + B^- - \gamma)^m \rangle_\gamma, \tag{4.17}$$

for $m = 1, 2, \dots$, where $\langle\cdot\rangle_\gamma$ is the coherent state of the free Fock space.

It is not a priori evident that the coherent state $\langle\cdot\rangle_\gamma$ is positive on the $*$-algebra generated by $B^+ + B^-$. The positivity follows from the fact that the Haagerup state is positive. Therefore there exists a probability measure $\mu \in \mathfrak{P}_{\mathrm{fm}}(\mathbb{R})$ satisfying

$$\lim_{\substack{q\sqrt{\kappa}\to\gamma \\ \kappa\to\infty, q\to 0}} \left\langle \left(\frac{A_\kappa - \langle A_\kappa\rangle_q}{\Sigma_q(A_\kappa)} \right)^m \right\rangle_q = \int_{-\infty}^{+\infty} x^m \mu(dx), \quad m = 1, 2, \dots, \tag{4.18}$$

which is the asymptotic spectral distribution of A_κ in the Haagerup state. By virtue of Theorem 4.13 the problem is reduced to finding a probability measure $\mu \in \mathfrak{P}_{\mathrm{fm}}(\mathbb{R})$ satisfying

$$\langle (B^+ + B^- - \gamma)^m \rangle_\gamma = \int_{-\infty}^{+\infty} x^m \mu(dx), \qquad m = 1, 2, \dots. \tag{4.19}$$

The answer to the question (4.19) for $\gamma = 0$ is readily known. The left-hand side becomes the moment of $B^+ + B^-$ in the vacuum state of the free Fock space so that we obtain

$$\begin{aligned} \lim_{\substack{q\sqrt{\kappa}\to 0 \\ \kappa\to\infty, q\to 0}} \left\langle \left(\frac{A_\kappa - \langle A_\kappa\rangle_q}{\Sigma_q(A_\kappa)} \right)^m \right\rangle_q &= \langle \Psi_0, (B^+ + B^-)^m \Psi_0 \rangle \\ &= \frac{1}{2\pi} \int_{-2}^{+2} x^m \sqrt{4 - x^2}\, dx, \end{aligned}$$

for $m = 1, 2, \dots$. To find a probability measure μ in (4.19) for a general $\gamma \neq 0$ we need a few steps.

The Catalan path and Catalan number introduced in Definition 1.70 are generalized as follows.

Definition 4.14. A finite sequence $\epsilon = (\epsilon_1, \epsilon_2, \dots, \epsilon_m) \in \{+, -\}^m$, $m \geq 1$, is called a *Catalan path* if

$$\epsilon_1 + \cdots + \epsilon_k \geq 0, \qquad k = 1, 2, \dots, m.$$

Let $0 \leq n \leq m$ be integers. A Catalan path $\epsilon = (\epsilon_1, \epsilon_2, \dots, \epsilon_{m+n})$ is said to be of *type* (m, n) if

$$\epsilon_1 + \cdots + \epsilon_{m+n} = m - n,$$

or equivalently if

$$m = |\{1 \leq i \leq m+n\,;\, \epsilon_i = +\}|, \quad n = |\{1 \leq i \leq m+n\,;\, \epsilon_i = -\}|.$$

Let $\mathcal{C}_{m,n}$ denote the set of Catalan paths of type (m, n). The number of Catalan paths of type (m, n) is called the (m, n)-*Catalan number* and is denoted by $C_{m,n}$.

A Catalan path of length $2m$ in Definition 1.70 is a Catalan path of type (m, m). Note also that $\mathcal{C}_{m,m} = \mathcal{C}_m$ and $C_{m,m} = C_m$. It is known that

$$C_{m,n} = |\mathcal{C}_{m,n}| = \binom{m+n}{m} - \binom{m+n}{m+1}. \tag{4.20}$$

The proof is similar to that of Lemma 1.71 (Exercise 4.3).

Throughout the rest of this section, let $\Gamma_{\text{free}} = (\Gamma, \{\Psi_n\}, B^+, B^-)$ be the free Fock space and P the vacuum projection, i.e., the projection onto the one-dimensional space spanned by the vacuum vector Ψ_0.

Proposition 4.15. *For any $z \in \mathbb{C}$ and $m = 1, 2, \ldots,$*

$$\langle \Omega_{\bar{z}}, (B^+ + B^-)^m \Psi_0 \rangle = \langle \Psi_0, (B^+ + B^- + zP)^m \Psi_0 \rangle, \tag{4.21}$$

where $\Omega_{\bar{z}}$ is the coherent vector with parameter $\bar{z}$.

Proof. We start with the obvious identity:

$$\langle \Omega_{\bar{z}}, (B^+ + B^-)^m \Psi_0 \rangle = \sum_{\epsilon_1, \ldots, \epsilon_m \in \{+,-\}} \langle \Omega_{\bar{z}}, B^{\epsilon_m} \cdots B^{\epsilon_1} \Psi_0 \rangle. \tag{4.22}$$

Clearly, for the sum we only need to take Catalan paths. Let $\mathcal{D}_{m,n}$ denote the set of Catalan paths $\epsilon = (\epsilon_1, \epsilon_2, \ldots, \epsilon_m)$ satisfying $\epsilon_1 + \cdots + \epsilon_m = n$. Then, (4.22) becomes

$$\langle \Omega_{\bar{z}}, (B^+ + B^-)^m \Psi_0 \rangle = \sum_{n=0}^{\infty} \sum_{\epsilon \in \mathcal{D}_{m,n}} \langle \Omega_{\bar{z}}, B^{\epsilon_m} \cdots B^{\epsilon_1} \Psi_0 \rangle. \tag{4.23}$$

For $\epsilon = (\epsilon_1, \epsilon_2, \ldots, \epsilon_m) \in \mathcal{D}_{m,n}$ we have

$$\langle \Omega_{\bar{z}}, B^{\epsilon_m} \cdots B^{\epsilon_1} \Psi_0 \rangle = \langle \Omega_{\bar{z}}, \Psi_n \rangle = z^n,$$

so that (4.23) becomes

$$\langle \Omega_{\bar{z}}, (B^+ + B^-)^m \Psi_0 \rangle = \sum_{n=0}^{\infty} z^n |\mathcal{D}_{m,n}|. \tag{4.24}$$

We shall obtain an alternative expression of $|\mathcal{D}_{m,n}|$. For a Catalan path $\epsilon = (\epsilon_1, \epsilon_2, \ldots, \epsilon_m) \in \mathcal{D}_{m,n}$ we have

$$|\{1 \le i \le m\,;\, \epsilon_i = +\}| = \frac{m+n}{2}, \quad |\{1 \le i \le m\,;\, \epsilon_i = -\}| = \frac{m-n}{2}.$$

Now, with each $\epsilon \in \mathcal{D}_{m,n}$ we associate a 'remainder sequence'

$$1 \le i_1 < i_2 < \cdots < i_n \le m \tag{4.25}$$

as follows: if ϵ consists of only $+$, the remainder sequence is $1 < 2 < \cdots < m$. Suppose that ϵ contains at least one $-$. Since $m \geq 2$ and $\epsilon_1 = +$, we may choose the smallest $i \geq 2$ such that $\epsilon_i = -$. Then we extract $\epsilon_{i-1}, \epsilon_i$ from $\epsilon = (\epsilon_1, \epsilon_2, \dots, \epsilon_m)$ to obtain a shorter Catalan path. After repeating this procedure until all $-$ are extracted from ϵ, we obtain a subsequence ϵ' of ϵ consisting of only $+$. The remainder sequence (4.25) is defined to be the sequence of suffixes of ϵ'. It may happen that the remainder sequence is empty.

Now, let $C(i_1, \dots, i_n)$ be the set of Catalan paths $\epsilon \in \mathcal{D}_{m,n}$ of which the remainder sequence is $i_1 < \cdots < i_n$. Then

$$|\mathcal{D}_{m,n}| = \sum_{1 \leq i_1 < i_2 < \cdots < i_n \leq m} |C(i_1, \dots, i_n)|.$$

By construction, $\{1, 2, i_1 - 1\}, \{i_1 + 1, \dots, i_2 - 1\}, \dots, \{i_n + 1, \dots, m\}$ consist of even elements unless empty sets. We denote the numbers by

$$2j_1 = i_1 - 1, \quad 2j_2 = i_2 - i_1 - 1, \dots, \quad 2j_{n+1} = m - i_n.$$

Moreover, each of these sets is extracted from ϵ by making a non-crossing pair partition. Therefore,

$$|C(i_1, \dots, i_n)| = C_{j_1} \cdots C_{j_{n+1}},$$

where C_j is the jth Catalan number. On the other hand, since

$$C_j = \langle \Psi_0, (B^+ + B^-)^j \Psi_0 \rangle,$$

we obtain

$$\begin{aligned}
&|C(i_1, \dots, i_n)| \\
&\quad = \langle \Psi_0, (B^+ + B^-)^{j_{n+1}} \Psi_0 \rangle \cdots \\
&\qquad \times \langle \Psi_0, (B^+ + B^-)^{j_2} \Psi_0 \rangle \langle \Psi_0, (B^+ + B^-)^{j_1} \Psi_0 \rangle \\
&\quad = \langle \Psi_0, (B^+ + B^-)^{j_{n+1}} P \cdots P (B^+ + B^-)^{j_2} P (B^+ + B^-)^{j_1} \Psi_0 \rangle,
\end{aligned}$$

where the vacuum projection P appears n times. Then

$$\begin{aligned}
&\sum_{n=0}^{\infty} z^n |\mathcal{D}_{m,n}| \\
&\quad = \sum_{n=0}^{\infty} z^n \sum_{1 \leq i_1 < i_2 < \cdots < i_n \leq m} |C(i_1, \dots, i_n)| \\
&\quad = \sum_{n=0}^{\infty} \sum_{1 \leq i_1 < i_2 < \cdots < i_n \leq m} \langle \Psi_0, (B^+ + B^-)^{j_{n+1}} (zP) \cdots \\
&\qquad \times (zP)(B^+ + B^-)^{j_2} (zP)(B^+ + B^-)^{j_1} \Psi_0 \rangle.
\end{aligned}$$

It turns out that the last expression coincides with the expansion of

$$\langle \Psi_0, (B^+ + B^- + zP)^m \Psi_0 \rangle.$$

Consequently, combining (4.24) we obtain (4.21) as desired. □

Lemma 4.16. $B^+B^- = 1 - P$.

The proof is obvious. We are now in a position to obtain an alternative expression of the left-hand side of (4.19).

Proposition 4.17. *For any $z \in \mathbb{C}$ and $m = 1, 2, \dots$ we have*

$$\langle \Omega_{\bar{z}}, (B^+ + B^- - z)^m \Psi_0 \rangle = \langle \Psi_0, (B^+ + B^- - zB^+B^-)^m \Psi_0 \rangle. \tag{4.26}$$

Proof. With the help of Proposition 4.15, we obtain

$$\begin{aligned}
&\langle \Omega_{\bar{z}}, (B^+ + B^- - z)^m \Psi_0 \rangle \\
&= \sum_{n=0}^{m} \binom{m}{n} (-z)^{m-n} \langle \Omega_{\bar{z}}, (B^+ + B^-)^n \Psi_0 \rangle \\
&= \sum_{n=0}^{m} \binom{m}{n} (-z)^{m-n} \langle \Psi_0, (B^+ + B^- + zP)^n \Psi_0 \rangle \\
&= \langle \Psi_0, (B^+ + B^- + zP - z)^m \Psi_0 \rangle.
\end{aligned}$$

By Lemma 4.16 the last expression becomes

$$= \langle \Psi_0, (B^+ + B^- - zB^+B^-)^m \Psi_0 \rangle,$$

which proves (4.26). □

Theorem 4.18 (CLT for homogeneous trees in Haagerup states). *Let A_κ be the adjacency matrix of a homogeneous tree T_κ of degree $\kappa \geq 2$ and the adjacency algebra $\mathcal{A}(T_\kappa)$ be equipped with a Haagerup state $\langle \cdot \rangle_q$. Then for $\gamma \in \mathbb{R}$ there exists a unique probability measure μ such that*

$$\lim_{\substack{q\sqrt{\kappa} \to \gamma \\ \kappa \to \infty, q \to 0}} \left\langle \left(\frac{A_\kappa - \langle A_\kappa \rangle_q}{\Sigma_q(A_\kappa)} \right)^m \right\rangle_q = \int_{-\infty}^{+\infty} x^m \mu(dx), \quad m = 1, 2, \dots. \tag{4.27}$$

Moreover, μ is given by

$$\mu(dx) = \begin{cases} \rho_\gamma(x)dx, & 0 \leq |\gamma| \leq 1, \\ \rho_\gamma(x)dx + \left(1 - \dfrac{1}{\gamma^2}\right) \delta_{1/\gamma}, & |\gamma| \geq 1, \end{cases} \tag{4.28}$$

where

$$\rho_\gamma(x) = \begin{cases} \dfrac{\sqrt{4 - (x+\gamma)^2}}{2\pi(1 - \gamma x)}, & |x + \gamma| \leq 2, \\ 0, & \textit{otherwise}. \end{cases} \tag{4.29}$$

(See Fig. 4.2)

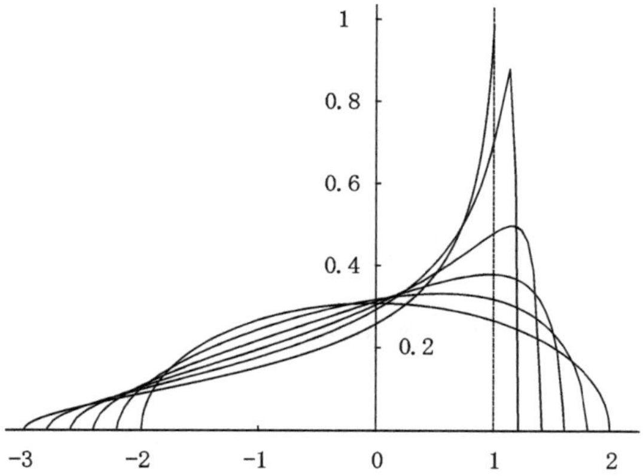

Fig. 4.2. $\rho_\gamma(x)$ with $\gamma = 0, 0.2, 0.4, 0.6, 0.8, 1$ (Theorem 4.18)

Proof. Combining Theorem 4.13 and Proposition 4.17, we obtain

$$\lim_{\substack{q\sqrt{\kappa}\to\gamma \\ \kappa\to\infty, q\to 0}} \left\langle \left(\frac{A_\kappa - \langle A_\kappa \rangle_q}{\Sigma_q(A_\kappa)} \right)^m \right\rangle_q = \langle \Omega_\gamma, (B^+ + B^- - \gamma)^m \Psi_0 \rangle$$
$$= \langle \Psi_0, (B^+ + B^- - \gamma B^+ B^-)^m \Psi_0 \rangle.$$

Then (4.27) becomes

$$\langle \Psi_0, (B^+ + B^- - \gamma B^+ B^-)^m \Psi_0 \rangle = \int_{-\infty}^{+\infty} x^m \mu(dx), \quad m = 1, 2, \dots . \tag{4.30}$$

Note that $-\gamma B^+ B^-$ is a diagonal operator defined by the sequence:

$$\alpha_1 = 0, \quad \alpha_2 = \alpha_3 = \dots = -\gamma.$$

Hence the Jacobi coefficient of μ in (4.30) is $(\{\omega_n \equiv 1\}, \{\alpha_n\})$. From the growth rate μ is uniquely determined and its Stieltjes transform is given by

$$G(z) = \int_{-\infty}^{+\infty} \frac{\mu(dx)}{z - x} = \frac{1}{z -} \, \frac{1}{z + \gamma -} \, \frac{1}{z + \gamma -} \, \frac{1}{z + \gamma - \cdots}$$
$$= \frac{z - \gamma - \sqrt{(z+\gamma)^2 - 4}}{2(1 - \gamma z)},$$

where $\sqrt{(z+\gamma)^2 - 4}$ is holomorphic on $\mathbb{C} \setminus [-2-\gamma, 2-\gamma]$ and takes positive values on $(2-\gamma, +\infty)$. The density function (4.29) is easily obtained by applying Stieltjes' inversion formula. For $0 \le |\gamma| \le 1$, elementary integral calculus shows that $\rho_\gamma(x)dx$ is a probability measure so that $\mu(dx) = \rho_\gamma(x)dx$. For $|\gamma| \ge 1$, observing that

$$\int_{-\infty}^{+\infty} \rho_\gamma(x)\, dx = \frac{1}{\gamma^2}, \qquad \operatorname*{Res}_{z=1/\gamma} G(z) = 1 - \frac{1}{\gamma^2},$$

we obtain the expression for μ as in (4.28). □

4.4 Free Poisson Distribution

In the last part of the previous section an essential role was played by a real random variable

$$B^+ + B^- - \gamma B^+ B^- = -\gamma \left(B^+ - \frac{1}{\gamma}\right)\left(B^- - \frac{1}{\gamma}\right) + \frac{1}{\gamma}$$

with respect to the vacuum state Ψ_0. We are in a good position to give the following:

Definition 4.19. Let $\Gamma_{\mathrm{free}} = (\Gamma, \{\Psi_n\}, B^+, B^-)$ be the free Fock space and $\lambda > 0$ a constant. The vacuum spectral distribution of $(B^+ + \sqrt{\lambda})(B^- + \sqrt{\lambda})$ is called the *free Poisson distribution* or *Marchenko–Pastur distribution* with parameter λ. In other words, the free Poisson distribution with parameter λ is a probability measure μ uniquely specified by

$$\langle \Psi_0, ((B^+ + \sqrt{\lambda})(B^- + \sqrt{\lambda}))^m \Psi_0 \rangle = \int_{-\infty}^{+\infty} x^m \mu(dx), \tag{4.31}$$

for $m = 1, 2, \dots$.

The density function of the free Poisson distribution is given explicitly.

Proposition 4.20. *For $\lambda > 0$ we set*

$$\rho_\lambda(x) = \begin{cases} \dfrac{\sqrt{4\lambda - (x-1-\lambda)^2}}{2\pi x}, & (1-\sqrt{\lambda})^2 \le x \le (1+\sqrt{\lambda})^2, \\ 0, & \textit{otherwise}. \end{cases}$$

The free Poisson distribution with parameter λ is given by

$$\begin{cases} (1-\lambda)\delta_0 + \rho_\lambda(x)dx, & 0 < \lambda < 1, \\ \rho_\lambda(x)dx, & \lambda \ge 1. \end{cases}$$

The mean and the variance are both λ.

Proof. From (4.31) we obtain

$$\left\langle \Psi_0, \left(B^+ + B^- + \frac{B^+B^-}{\sqrt{\lambda}} + \sqrt{\lambda}\right)^m \Psi_0 \right\rangle = \int_{-\infty}^{+\infty} \left(\frac{x}{\sqrt{\lambda}}\right)^m \mu(dx). \tag{4.32}$$

We note that

$$B^\circ \equiv \frac{B^+B^-}{\sqrt{\lambda}} + \sqrt{\lambda}$$

is the diagonal operator associated with a sequence $\{\alpha_n\}$ defined by

$$\alpha_1 = \sqrt{\lambda}, \quad \alpha_2 = \alpha_3 = \cdots = \sqrt{\lambda} + \frac{1}{\sqrt{\lambda}}.$$

Let ν be a probability measure such that

$$\langle \Psi_0, (B^+ + B^- + B^\circ)^m \Psi_0 \rangle = \int_{-\infty}^{+\infty} x^m \nu(dx), \quad m = 1, 2, \dots. \tag{4.33}$$

We then see that ν is uniquely determined by the Jacobi coefficient $(\{\omega_n \equiv 1\}, \{\alpha_n\})$. The Stieltjes transform $G(z)$ of ν is computed easily by continued fraction and then, applying Stieltjes' inversion formula, we obtain the absolutely continuous part:

$$\tilde{\rho}(x) = \begin{cases} \dfrac{\sqrt{4\lambda - (\sqrt{\lambda}x - \lambda - 1)^2}}{2\pi x}, & (\sqrt{\lambda} - 1)^2 \le \sqrt{\lambda}x \le (\sqrt{\lambda} + 1)^2, \\ 0, & \text{otherwise.} \end{cases}$$

Checking the residues of $G(z)$, we obtain

$$\nu(dx) = \begin{cases} \tilde{\rho}(x)\, dx + (1 - \lambda)\delta_0(dx), & 0 < \lambda < 1, \\ \tilde{\rho}(x)\, dx, & \lambda \ge 1. \end{cases}$$

From (4.32) and (4.33) we see that μ is obtained from ν by simple scaling and we come to the desired assertion. □

The probability measure obtained from the central limit theorem for homogeneous trees in Haagerup states (Theorem 4.18) is related to a free Poisson distribution.

Proposition 4.21. *The probability measure in (4.28) for $\gamma \neq 0$ is obtained from the free Poisson distribution with parameter γ^{-2} by reflection with respect to $x = 0$ and normalization.*

The proof is easy and omitted. Finally we record the following:

Proposition 4.22. *The Jacobi coefficient of the free Poisson distribution with parameter λ is given by*

$$\omega_1 = \omega_2 = \cdots = \lambda, \quad \alpha_1 = \lambda,\ \alpha_2 = \alpha_3 = \cdots = \lambda + 1.$$

In particular, the mean and variance are both λ.

Remark 4.23. Given an interacting Fock space $\Gamma_{\{\omega_n\}} = (\Gamma, \{\Psi_n\}, B^+, B^-)$, an algebraic random variable $(B^+ + \sqrt{\lambda})(B^- + \sqrt{\lambda})$ is interesting to study. In the case of the free Fock space it gives rise to a free Poisson distribution. As we shall see later (Theorem 5.12), in the case of the Boson Fock space a (classical) Poisson distribution will appear.

Remark 4.24. As another justification of using the term 'Poisson distribution', we remark that the free cumulants of the free Poisson distribution with parameter λ are all λ. See Sect. 11.5 for free cumulants of a probability measure.

4.5 Spidernets and Free Meixner Law

A tree has no edge lying in the same stratum in its stratification. As generalization we discuss a particular graph obtained by adding some edges in each stratum. As usual, for $x \in V$ and $\epsilon \in \{+, -, \circ\}$ we set

$$\omega_\epsilon(x) = |\{y \in V\,;\, y \sim x,\ \partial(o, y) = \partial(o, x) + \epsilon\}|.$$

Definition 4.25. Let a, b, c be integers such that $a \ge 1$, $b \ge 2$ and $1 \le c \le b - 1$. A *spidernet* is a graph which satisfies the following conditions:

$$\begin{cases} \omega_+(o) = a, \\ \omega_-(o) = 0, \\ \omega_\circ(o) = 0, \end{cases} \qquad \begin{cases} \omega_+(x) = c, \\ \omega_-(x) = 1, \\ \omega_\circ(x) = b - 1 - c, \end{cases} \qquad \text{for } x \neq o. \tag{4.34}$$

Such a spidernet is denoted by $S(a, b, c)$.

Examples are shown in Fig. 4.3. It is noted that a spidernet $S(a, b, c)$ is not uniquely determined by (a, b, c) in general. Note also that a spidernet $S(a, b, c)$ with $c = b-1$ is a tree. A spidernet $S(a, b, c)$ is not necessarily a regular graph, much less a distance-regular graph. But it is still highly symmetric and our method is effective. Note that

$$\kappa(x) = \begin{cases} a, & \text{if } x = o, \\ b, & \text{otherwise,} \end{cases}$$

which follows by $\kappa(x) = \omega_+(x) + \omega_-(x) + \omega_\circ(x)$ with (4.34).

Example 4.26. An infinite distance-transitive graph serves as an example of a spidernet. For a pair of integers $a, b \ge 2$, let $\Gamma(a, b)$ denote the graph constructed by gluing a copies of the complete graph K_b at each vertex in a tree-like way. For instance, $\Gamma(a, 2)$ is the homogeneous tree with degree a. It is easily verified that $\Gamma(a, b) = S(a(b-1), a(b-1), (a-1)(b-1))$ holds with any fixed origin. Due to MacPherson's classification $\{\Gamma(a, b)\,;\, a, b \ge 2\}$ exhausts infinite, locally finite and distance-transitive graphs.

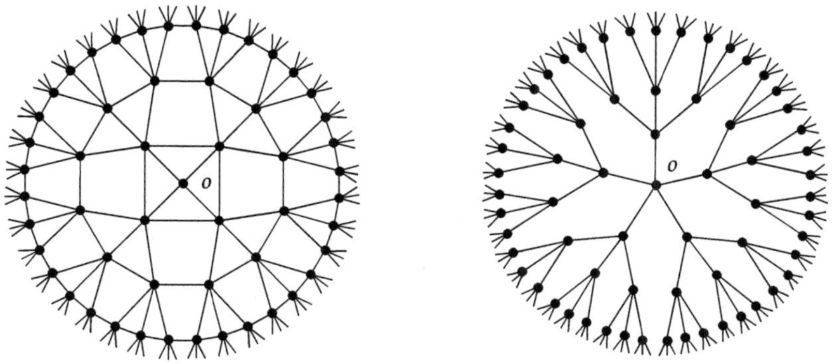

Fig. 4.3. Spidernets $S(4,6,3)$ and $S(5,4,3)$

Proposition 4.27. *Let $S = S(a,b,c)$ be a spidernet and A its adjacency matrix. Then $\Gamma(S)$ is invariant under the action of the quantum components A^ϵ, $\epsilon \in \{+,-,\circ\}$. Moreover, we have*

$$A^+\Phi_0 = \sqrt{a}\,\Phi_1, \quad A^+\Phi_n = \sqrt{c}\,\Phi_{n+1}, \quad n = 1,2,\dots, \tag{4.35}$$

$$A^-\Phi_0 = 0, \quad A^-\Phi_1 = \sqrt{a}\,\Phi_0, \quad A^-\Phi_n = \sqrt{c}\,\Phi_{n-1}, \quad n = 2,3,\dots, \tag{4.36}$$

$$A^\circ\Phi_0 = 0, \quad A^\circ\Phi_n = (b-1-c)\Phi_n, \quad n = 1,2,\dots. \tag{4.37}$$

Proof. It is easily derived from (4.34) that

$$|V_0| = 1, \qquad |V_n| = ac^{n-1}, \qquad n = 1,2,\dots.$$

For $n = 0,1,2,\dots$ we have

$$\sqrt{|V_n|}\,A^+\Phi_n = \sum_{x\in V_n} A^+\delta_x = \sum_{y\in V_{n+1}} \delta_y = \sqrt{|V_{n+1}|}\,\Phi_{n+1}.$$

This proves (4.35). The rest is similarly proved. □

Proposition 4.27 says that $(\Gamma(S), \{\Phi_n\}, A^+, A^-)$ is an interacting Fock space with Jacobi sequence

$$\omega_1 = a, \quad \omega_2 = \omega_3 = \cdots = c \tag{4.38}$$

and A° is a diagonal operator defined by $A^\circ\Phi_n = \alpha_{n+1}\Phi_n$ with

$$\alpha_1 = 0, \quad \alpha_2 = \alpha_3 = \cdots = b-1-c. \tag{4.39}$$

Then a probability measure μ such that

$$\langle \delta_o, A^m\delta_o\rangle = \int_{-\infty}^{+\infty} x^m \mu(dx), \qquad m = 1,2,\dots, \tag{4.40}$$

is uniquely determined by the Jacobi coefficient $(\{\omega_n\},\{\alpha_n\})$ given by (4.38) and (4.39). To be slightly more general, we give the following:

Definition 4.28. For three real numbers $p > 0$, $q \geq 0$, $a \in \mathbb{R}$, a probability measure $\mu_{p,q,a}$ uniquely specified by

$$\int_{-\infty}^{+\infty} \frac{\mu_{p,q,a}(dx)}{z-x} = \frac{1}{z} - \frac{p}{z-a} - \frac{q}{z-a} - \frac{q}{z-a} - \frac{q}{z-a-\cdots}, \tag{4.41}$$

is called the *free Meixner law* with parameter (p, q, a). In other words, $\mu_{p,q,a}$ is characterized by a Jacobi coefficient $(\{p, q, q, \dots\}, \{0, a, a, \dots\})$.

Clearly, $\mu_{p,q,a}$ has mean zero and variance p. The density function will be given explicitly at the end of this section. Now the following claim is immediate from the above argument.

Theorem 4.29. *Let $S(a,b,c)$ be a spidernet and $A = A_{a,b,c}$ its adjacency matrix. It then holds that*

$$\langle A^m \rangle_o = \int_{-\infty}^{+\infty} x^m \mu_{a,c,b-1-c}(dx), \qquad m = 1, 2, \dots,$$

where $\mu_{a,c,b-1-c}$ is the free Meixner law with parameter $(a, c, b-1-c)$.

We next consider growing spidernets $S(a,b,c)$ as $a \to \infty$. Since $\langle \delta_o, A\delta_o \rangle = 0$ and $\langle \delta_o, A^2 \delta_o \rangle = a$, the proper normalization of the adjacency matrix is $A/\sqrt{a}$. In view of Proposition 4.27 we obtain

$$\frac{A^+}{\sqrt{a}} \Phi_0 = \Phi_1, \quad \frac{A^+}{\sqrt{a}} \Phi_n = \sqrt{\frac{c}{a}}\, \Phi_{n+1}, \quad n = 1, 2, \dots,$$

$$\frac{A^-}{\sqrt{a}} \Phi_0 = 0, \quad \frac{A^-}{\sqrt{a}} \Phi_1 = \Phi_0, \quad \frac{A^-}{\sqrt{a}} \Phi_n = \sqrt{\frac{c}{a}}\, \Phi_{n-1}, \quad n = 2, 3, \dots,$$

$$\frac{A^\circ}{\sqrt{a}} \Phi_0 = 0, \quad \frac{A^\circ}{\sqrt{a}} \Phi_n = \frac{b-1-c}{\sqrt{a}} \Phi_n, \quad n = 1, 2, \dots.$$

Then, with no difficulty we come to the following:

Theorem 4.30 (QCLT for spidernets). *Let $q \geq 0$ and $r \geq 0$ be real numbers, and define a Jacobi coefficient $(\{\omega_n\}, \{\alpha_n\})$ by*

$$\omega_1 = 1, \quad \omega_2 = \omega_3 = \cdots = q,$$
$$\alpha_1 = 0, \quad \alpha_2 = \alpha_3 = \cdots = r.$$

Let $(\Gamma, \{\Psi_n\}, B^+, B^-)$ be the interacting Fock space associated with $\{\omega_n\}$ and B° the diagonal operator defined by $\{\alpha_n\}$. Let $A = A_{a,b,c}$ be the adjacency matrix of a spidernet $S(a,b,c)$. Then we have

$$\lim \frac{A^\epsilon_{a,b,c}}{\sqrt{a}} = B^\epsilon, \qquad \epsilon \in \{+, -, \circ\},$$

in the sense of stochastic convergence with respect to the vacuum states, where the limit is taken as

$$a \to \infty, \quad \frac{c}{a} \to q, \quad \frac{b-c}{\sqrt{a}} \to r. \tag{4.42}$$

Theorem 4.31 (CLT for spidernets). *Notations and assumptions being as in Theorem 4.30, we have*

$$\lim_{(4.42)} \left\langle \left(\frac{A_{a,b,c}}{\sqrt{a}} \right)^m \right\rangle_o = \int_{-\infty}^{+\infty} x^m \mu_{1,q,r}(dx), \quad m = 1, 2, \dots,$$

where $\mu_{1,q,r}$ is the free Meixner law with parameter $(1, q, r)$.

The rest of this section will be devoted to computation of the density function of the free Meixner law $\mu_{p,q,a}$ for $p > 0$, $q \geq 0$, $a \in \mathbb{R}$. According to Definition 4.28, we start with

$$G(z) = \frac{1}{z} - \frac{p}{z - a} - \frac{q}{z - a} - \frac{q}{z - a} - \frac{q}{z - a} - \cdots$$

$$= \frac{(2q - p)z + pa - p\sqrt{(z - a)^2 - 4q}}{2(q - p)z^2 + 2paz + 2p^2}, \tag{4.43}$$

where the square root is a holomorphic function on $\mathbb{C} \setminus [a - 2\sqrt{q}, a + 2\sqrt{q}\,]$ and the branch is chosen in such a way that $\sqrt{(z - a)^2 - 4q} > 0$ for $z \in \mathbb{R}$ with $z > a + 2\sqrt{q}$. The absolutely continuous part of $\mu_{p,q,a}(dx)$, denoted by $\rho_{p,q,a}(x)$, is obtained by Stieltjes' inversion formula:

$$\rho_{p,q,a}(x) = \frac{p}{2\pi} \frac{\sqrt{4q - (x - a)^2}}{(q - p)x^2 + pax + p^2}, \qquad |x - a| \leq 2\sqrt{q}. \tag{4.44}$$

We note that $\rho_{p,q,-a}(x) = \rho_{p,q,a}(-x)$. For simplicity, we set

$$g(z) = (q - p)z^2 + paz + p^2,$$

which appears in the denominator of (4.44). Note that the possible real zeros of $g(z)$ lie outside the open interval $(a - 2\sqrt{q}, a + 2\sqrt{q}\,)$. We consider the following five cases.

Case 1. $q = p > 0$ and $a = 0$. Then the density function becomes

$$\rho_{p,p,0}(x) = \frac{1}{2\pi p} \sqrt{4p - x^2}, \qquad |x| \leq 2\sqrt{p},$$

which is the Wigner semicircle law with variance p. There is no atom and

$$\mu_{p,p,0}(dx) = \rho_{p,p,0}(x)dx.$$

Case 2. $q = p > 0$ and $a \neq 0$. The density function becomes

$$\rho_{p,p,a}(x) = \frac{1}{2\pi} \frac{\sqrt{4p - (x - a)^2}}{ax + p}, \qquad |x - a| \leq 2\sqrt{p}.$$

Moreover, $\mu_{p,p,a}$ may possess one atom at the zero of $g(z)$, i.e., at $x = \lambda_0 = -p/a$. The weight is obtained from the residue of $G(z)$. For example, if $a > 0$, noting the signature of $\sqrt{(z-a)^2 - 4q}$ for $z = \lambda_0 \le a - 2\sqrt{p}$, we have

$$\lim_{z \to \lambda_0} (z - \lambda_0) G(z) = \frac{1}{2a} \left(a - \frac{p}{a} + \left| a - \frac{p}{a} \right| \right),$$

which is $1 - p/a^2$ for $a^2 > p$, and 0 for $a^2 \le p$. After similar computation for $a < 0$ we obtain

$$\mu_{p,p,a}(dx) = \begin{cases} \rho_{p,p,a}(x)dx, & \text{for } a^2 \le p, \\ \rho_{p,p,a}(x)dx + \left(1 - \dfrac{p}{a^2}\right) \delta_{-p/a}, & \text{for } a^2 > p. \end{cases}$$

In fact, the above result covers Case 1 too. It is easily verified that $\mu_{p,p,a}$ is an affine transformation of the free Poisson law with parameter p/a^2 (see Sect. 4.4).

We next consider the case where $q \ne p$. Then $g(z)$ is a quadratic function and we need its discriminant:

$$D = a^2 - 4(q - p).$$

Case 3. $D < 0$, that is, $0 \le a^2 < 4(q-p)$. Then $g(z)$ has no real zeros so that

$$\mu_{p,q,a}(dx) = \rho_{p,q,a}(x)dx. \tag{4.45}$$

Case 4. $D = 0$, that is, $0 < a^2 = 4(q-p)$. Then $g(z)$ has a real multiple zero outside $[a - 2\sqrt{q}, a + 2\sqrt{q}]$, nevertheless $\mu_{p,q,a}$ has no atom and (4.45) remains valid.

Case 5. $D > 0$, that is, $4(q-p) < a^2$. Then $g(z)$ has two real zeros:

$$\lambda_\pm = \frac{p}{2(q-p)}(-a \pm \sqrt{D}),$$

and $\mu_{p,q,a}$ is of the form

$$\mu_{p,q,a}(dx) = \rho_{p,q,a}(x)dx + w_+ \delta_{\lambda_+} + w_- \delta_{\lambda_-}. \tag{4.46}$$

To describe $w_\pm$ we define

$$\nu_+ = \frac{1}{\sqrt{D}} \left(\frac{q\lambda_+}{p} - \frac{p}{\lambda_+} \right), \quad \nu_- = \frac{1}{\sqrt{D}} \left(\frac{q\lambda_-}{p} - \frac{p}{\lambda_-} \right). \tag{4.47}$$

Case 5-1. $0 < 4(q-p) < a^2$. Then $\lambda_- < \lambda_+$ and both lie in the same half line $(-\infty, a - 2\sqrt{q}]$ or $[a + 2\sqrt{q}, +\infty)$. The weights are given as follows:

$$w_+ = \begin{cases} 0, & a \le -2\sqrt{q-p}, \quad 2\sqrt{q-p} < a \le (2q-p)/\sqrt{q}, \\ \nu_+, & a \ge (2q-p)/\sqrt{q}, \end{cases}$$

$$w_- = \begin{cases} -\nu_-, & a \le -(2q-p)/\sqrt{q}, \\ 0, & -(2q-p)/\sqrt{q} \le a < -2\sqrt{q-p}, \quad 2\sqrt{q-p} < a. \end{cases}$$

Case 5-2. $0 \le q < 2q < p$. Note that $\lambda_+ < a - 2\sqrt{p}$ and $\lambda_- > a + 2\sqrt{p}$. The weights $w_\pm$ are given as follows:

$$w_+ = \begin{cases} 0, & a \le -(p-2q)/\sqrt{q}, \\ \nu_+, & a \ge -(p-2q)/\sqrt{q}, \end{cases} \qquad w_- = \begin{cases} -\nu_-, & a \le (p-2q)/\sqrt{q}, \\ 0, & a \ge (p-2q)/\sqrt{q}. \end{cases}$$

Case 5-3. $0 \le q < p < 2q$. The situation is similar to Case 5-2 and the weights are given as follows:

$$w_+ = \begin{cases} 0, & a \le (2q-p)/\sqrt{q}, \\ \nu_+, & a \ge (2q-p)/\sqrt{q}, \end{cases} \qquad w_- = \begin{cases} -\nu_-, & a \le -(2q-p)/\sqrt{q}, \\ 0, & a \ge -(2q-p)/\sqrt{q}. \end{cases}$$

In fact, Case 5-2 and Case 5-3 can be unified:

$$w_+ = \frac{1}{2}(|\nu_+| + \nu_+), \qquad w_- = \frac{1}{2}(|\nu_-| - \nu_-).$$

4.6 Markov Product of Positive Definite Kernels

We shall prove Theorem 4.11 in a more general form by means of an interesting construction of positive definite kernels due to Bożejko.

We first note a simple fact. Let $K = (K_{xy}) = (K(x,y))$ be a positive definite kernel on a set X. Then,

$$K_{xx} \ge 0, \quad x \in X; \qquad K_{yx} = \overline{K_{xy}}, \quad x, y \in X.$$

Lemma 4.32. *Let $K = (K(x,y))$ be a positive definite kernel on a set X. Then for any $o \in X$ and $f \in C_0(X)$ we have*

$$\left| \sum_{x \in X} \overline{f(x)}\, K(x,o) \right|^2 \le K(o,o) \sum_{x,y \in X} \overline{f(x)}\, K(x,y) f(y). \tag{4.48}$$

Proof. Define a sesquilinear form on $C_0(X)$ by

$$\langle f, g \rangle_K = \sum_{x,y \in X} \overline{f(x)}\, K(x,y) g(y), \qquad f, g \in C_0(X).$$

By the assumption $\langle f, f \rangle_K \ge 0$ for all $f \in C_0(X)$. Hence the Schwarz inequality holds. In particular,

$$|\langle f, \delta_o \rangle_K|^2 \le \langle f, f \rangle_K \langle \delta_o, \delta_o \rangle_K,$$

which is equivalent to (4.48). □

Theorem 4.33 (Bożejko). *Let X be a set which is a union of two subsets X_1, X_2 whose intersection consists of a single point, say $o \in X$. Namely,*

$$X = X_1 \cup X_2, \qquad X_1 \cap X_2 = \{o\}.$$

For $i = 1, 2$ let K_i be a positive definite kernel on X_i and assume that $K_1(o,o) = K_2(o,o) = 1$. Define a $\mathbb{C}$-valued function K on $X \times X$ by

$$K(x,y) = \begin{cases} K_1(x,y), & \text{if } x, y \in X_1, \\ K_2(x,y), & \text{if } x, y \in X_2, \\ K_1(x,o)K_2(o,y), & \text{if } x \in X_1,\ y \in X_2, \\ K_2(x,o)K_1(o,y), & \text{if } x \in X_2,\ y \in X_1 . \end{cases}$$

Then K is a positive definite kernel on X.

Proof. Let $f \in C_0(X)$. We may write $f = f_1 + f_2$ with $f_i \in C_0(X_i)$ though uniqueness does not hold for $X_1 \cap X_2 = \{o\}$. Then,

$$\begin{aligned} \sum_{x,y \in X} \overline{f(x)}\, K(x,y) f(y) &= \sum_{x,y \in X_1} \overline{f_1(x)}\, K_1(x,y) f_1(y) \\ &\quad + \sum_{x,y \in X_2} \overline{f_2(x)}\, K_2(x,y) f_2(y) \\ &\quad + \sum_{x \in X_1, y \in X_2} \overline{f_1(x)}\, K(x,y) f_2(y) \\ &\quad + \sum_{x \in X_2, y \in X_1} \overline{f_2(x)}\, K(x,y) f_1(y). \end{aligned} \tag{4.49}$$

By Lemma 4.32 we obtain

$$\begin{aligned} \left| \sum_{x \in X_1} \overline{f_1(x)}\, K_1(x,o) \right|^2 &\le K_1(o,o) \sum_{x,y \in X_1} \overline{f_1(x)}\, K_1(x,y) f_1(y) \\ &= \sum_{x,y \in X_1} \overline{f_1(x)}\, K_1(x,y) f_1(y). \end{aligned} \tag{4.50}$$

Similarly,

$$\left| \sum_{y \in X_2} \overline{f_2(y)}\, K_2(y,o) \right|^2 \le \sum_{x,y \in X_2} \overline{f_2(x)}\, K_2(x,y) f_2(y). \tag{4.51}$$

On the other hand, by definition we have

$$\begin{aligned}
&\sum_{x\in X_1, y\in X_2} \overline{f_1(x)}\, K(x,y) f_2(y) \\
&\quad= \sum_{x\in X_1, y\in X_2} \overline{f_1(x)}\, K_1(x,o) K_2(o,y) f_2(y) \\
&\quad= \sum_{x\in X_1} \overline{f_1(x)}\, K_1(x,o) \sum_{y\in X_2} K_2(o,y) f_2(y) \\
&\quad= \sum_{x\in X_1} \overline{f_1(x)}\, K_1(x,o) \overline{\sum_{y\in X_2} \overline{f_2(y)}\, K_2(y,o)}.
\end{aligned} \tag{4.52}$$

Similarly,

$$\begin{aligned}
&\sum_{x\in X_2, y\in X_1} \overline{f_2(x)}\, K(x,y) f_1(y) \\
&\quad= \overline{\sum_{x\in X_1} \overline{f_1(x)}\, K_1(x,o)} \sum_{y\in X_2} \overline{f_2(y)}\, K_2(y,o).
\end{aligned} \tag{4.53}$$

Combining (4.50)–(4.53), we see that (4.49) becomes

$$\begin{aligned}
&\sum_{x,y\in X} \overline{f(x)}\, K(x,y) f(y) \\
&\quad\geq \left| \sum_{x\in X_1} \overline{f_1(x)}\, K_1(x,o) \right|^2 + \left| \sum_{y\in X_2} \overline{f_2(y)}\, K_2(y,o) \right|^2 \\
&\qquad + \sum_{x\in X_1} \overline{f_1(x)}\, K_1(x,o) \overline{\sum_{y\in X_2} \overline{f_2(y)}\, K_2(y,o)} \\
&\qquad + \overline{\sum_{x\in X_1} \overline{f_1(x)}\, K_1(x,o)} \sum_{y\in X_2} \overline{f_2(y)}\, K_2(y,o) \\
&\quad= \left| \sum_{x\in X_1} \overline{f_1(x)}\, K_1(x,o) + \sum_{y\in X_2} \overline{f_2(y)}\, K_2(y,o) \right|^2 \geq 0.
\end{aligned}$$

This completes the proof. □

Definition 4.34. The positive definite kernel K on X constructed as in Theorem 4.33 is called the *Markov product* of K_1 and K_2.

Theorem 4.11 is included in the following:

Theorem 4.35. *Let $\mathcal{G} = (V, E)$ be a tree, i.e., a (connected, locally finite) graph without cycles and $\partial(x,y)$ the graph distance. Then $Q = (q^{\partial(x,y)})$ is positive definite for $-1 \leq q \leq 1$.*

Proof. We need to prove that

$$\sum_{x,y\in V} \overline{f(x)}\, q^{\partial(x,y)} f(y) \ge 0, \qquad f \in C_0(V).$$

Choose finite subsets $V_0 \subset V$ and $E_0 \subset E$ such that $\operatorname{supp} f \subset V_0$ and $\mathcal{G}_0 = (V_0, E_0)$ becomes a connected graph, i.e., a tree. It is noted that the graph distance on $\mathcal{G}_0$ coincides with the restriction of ∂ to $\mathcal{G}_0$. Hence it is sufficient to prove the assertion for a finite tree and we prove it by induction on the number of vertices. In the case of $|V| = 2$, in a matrix expression we have

$$Q = \begin{pmatrix} 1 & q \\ q & 1 \end{pmatrix},$$

which is apparently positive definite for $-1 \le q \le 1$. Suppose that the claim is true for a finite tree, the number of vertices of which is $\le n$ for $n \ge 2$. Let $\mathcal{G} = (V, E)$ be a finite tree with $|V| = n + 1$. There exists $x \in V$ whose degree is 1 and let $o \in V$ be the vertex adjacent to x. Set

$$V_1 = V \setminus \{x\}, \qquad E_1 = E \setminus \{\{x, o\}\}.$$

Then (V_1, E_1) is a tree with $|V_1| \le n$. Similarly, the pair

$$V_2 = \{o, x\}, \qquad E_2 = \{\{x, o\}\}$$

is a tree with $|V_2| = 2 \le n$. By the assumption of induction we see that

$$Q_1(x, y) = q^{\partial(x,y)}, \qquad x, y \in V_1,$$
$$Q_2(x, y) = q^{\partial(x,y)}, \qquad x, y \in V_2,$$

are positive definite kernels on V_1 and V_2, respectively. Moreover, for $x \in V_1$ and $y \in V_2$ we have

$$Q_1(x, o)Q_2(o, y) = q^{\partial(x,o)} q^{\partial(o,y)} = q^{\partial(x,o)+\partial(o,y)} = q^{\partial(x,y)}.$$

Hence $(q^{\partial(x,y)})$ is a product of Q_1 and Q_2 in the sense of Theorem 4.33 and hence is positive definite. □

Exercises

4.1. Let ν_κ be the normalized probability measure obtained from the Kesten distribution $\mu_{\kappa,\kappa-1}$ and ν the Wigner semicircle law (see Theorem 4.4). Prove that the density function $\rho_\kappa(x)$ of ν_κ converges to that of ν uniformly.

4.2. Using the above result, prove Theorem 4.10.

4.3. Prove formula (4.20) for the Catalan number. [Hint: reflection principle, see also Lemma 1.71]

4.4. Let $\Gamma_{\text{free}} = (\Gamma, \{\Phi_n\}, B^+, B^-)$ be the free Fock space. Prove the following recurrence relation: for $m, n = 1, 2, \dots$,

$$\langle \Phi_n, (B^+ + B^-)^m \Phi_0 \rangle = \sum_{k=0}^{\left[\frac{m-1}{2}\right]} C_k \langle \Phi_{n-1}, (B^+ + B^-)^{m-1-2k} \Phi_0 \rangle,$$

where C_k is the Catalan number.

4.5. Let $\Gamma_{\text{free}} = (\Gamma, \{\Phi_n\}, B^+, B^-)$ be the free Fock space and $\gamma \in \mathbb{C}$. Define

$$M_m = \langle \Omega_\gamma, (B^+ + B^-)^m \Phi_0 \rangle, \qquad \Omega_\gamma = \sum_{n=0}^{\infty} \gamma^n \Phi_n.$$

Show that $\{M_m\}$ satisfies the following recurrence relation:

$$M_m = C_{m/2} + \gamma \sum_{k=0}^{\left[\frac{m-1}{2}\right]} C_k M_{m-1-2k},$$

where $C_{m/2} = 0$ for an odd m. [Hint: Exercise 4.4.]

4.6. Show that a finite tree possesses a vertex whose degree is 1.

4.7. Let $\mathcal{G} = (V, E)$ be a graph with a fixed origin $o \in V$. Then $\mathcal{G}$ is a tree if and only if $\omega_-(x) = 1$ for all $x \in V - \{o\}$ and $\omega_\circ(x) = 0$ for all $x \in V$.

Notes

The distribution formulated in Theorem 4.4 is a simple scaling transformation of the one obtained by Kesten [131] for the transition matrix of an isotropic random walk on a free group (a homogeneous tree of even degree).

Theorem 4.11 is due to Haagerup [89]. The original statement for a free group remains valid for any tree by embedding. The positivity of the Q-matrix on a tree has been discussed by many authors. The proof presented here is due to Bożejko [39], where the idea of the Markov product of a positive definite kernel (Sect. 4.6) was introduced. This idea not only makes the proof of Theorem 4.11 extremely simple but also covers many different types of graphs, especially star graphs in Chap. 8. A further attempt at the positivity problem of the Q-matrix is found in Obata [171].

The Catalan paths and Catalan numbers appear in various combinatorial problems. Here we only mention Hilton–Pedersen [100].

Theorem 4.18 describes the asymptotic spectral distribution of the adjacency matrix of a free group in the Haagerup states. This question was first studied by Hashimoto [92] and the limit distribution was obtained under some restriction of the range of γ. This restriction, caused by his analytic method based on the Fourier transform of the spectral distribution and the power series expansion of Bessel functions, is now removed thanks to the method of quantum decomposition. The limit measure obtained in Theorem 4.18 is in fact an affine transformation of a free Poisson distribution and is a special case of a free Poisson distribution with two parameters introduced by Hiai–Petz [99].

Speicher [195] formulated in the free probability theory an analogue of Poisson's law of small numbers and derived a combinatorial formula for the moments of the limit distribution (reasonably called the free Poisson distribution). Moreover, he obtained a realization using the free Fock space. Afterwards, the density function was computed explicitly by Bożejko–Leinert–Speicher [43].

The spidernets have been studied for their interesting spectral geometric properties, see e.g., Urakawa [208], where a spidernet is called a semi-regular graph. The classification of infinite, locally finite distance-transitive graphs in terms of $\Gamma(a, b)$, mentioned in Example 4.26, is due to MacPherson [154]. See Bloom–Heyer [31] and Voit [218] for related harmonic analysis.

The free Meixner laws were first introduced by Bożejko–Bryc [40]. Their free Meixner laws are normalized to have variance 1 and are parametrized by $a \in \mathbb{R}$ and $b \geq -1$. In fact, the free Meixner law with parameter (a, b) of Bożejko–Bryc coincides with $\mu_{1,b+1,a}$ in our definition. A particular subclass of free Meixner laws was also discussed by Bożejko–Wysoczański [49]. The density function of the free Meixner law $\mu_{p,q,a}$ ($p > 0$, $q \geq 0$, $a \in \mathbb{R}$) was computed by Cohen–Trenholme [60] and Saitoh–Yoshida [185] (with different parametrization).

Znoĭko [229] introduced the notion of free products of graphs. Gutkin [85] extended some results in harmonic analysis on free product groups to the case of the free product of graphs. His idea seems to be relevant to further development, especially to the free independence, see Chap. 8.

5

Hamming Graphs

The Hamming graph is one of the most important and familiar distance-regular graphs and has been studied in a wide range of pure and applied mathematics. As the central limit distributions for these growing graphs, both Gaussian and Poisson distributions emerge.

5.1 Definition and Some Properties

Let F be a non-empty set and consider the cartesian product of $d \geq 1$ copies of F:

$$F^d = \{x = (\xi_1, \ldots, \xi_d)\,;\, \xi_i \in F,\ 1 \leq i \leq d\}.$$

For $x = (\xi_1, \ldots, \xi_d), y = (\eta_1, \ldots, \eta_d) \in F^d$ define

$$\partial(x, y) = |\{1 \leq i \leq d\,;\, \xi_i \neq \eta_i\}|.$$

Then ∂ becomes a distance function on F^d, which is called the *Hamming distance.*

Definition 5.1. Let F be a finite set with $|F| = N \geq 2$ and $d \geq 1$ an integer. The pair

$$V = F^d, \qquad E = \{\{x, y\}\,;\, x, y \in F^d,\ \partial(x, y) = 1\}$$

is called a *Hamming graph* and is denoted by $H(d, N)$.

We avoid the trivial case of $N = 1$ since $H(d, 1)$ is a trivial graph, i.e., consists of a single vertex. $H(d, 2)$ is the *d-cube* and $H(2, N)$ is the $N \times N$-*grid.* Furthermore, $H(1, N) \cong K_N$ (the complete graph with N vertices). Some Hamming graphs $H(d, N)$ with small d, N are illustrated below.

Proposition 5.2. *The graph distance of a Hamming graph $H(d, N) = (F^d, E)$ coincides with the Hamming distance on F^d.*

A. Hora and N. Obata: Hamming Graphs. In: A. Hora and N. Obata, Quantum Probability and Spectral Analysis of Graphs, Theoretical and Mathematical Physics, 131–146 (2007)
DOI 10.1007/3-540-48863-4_5

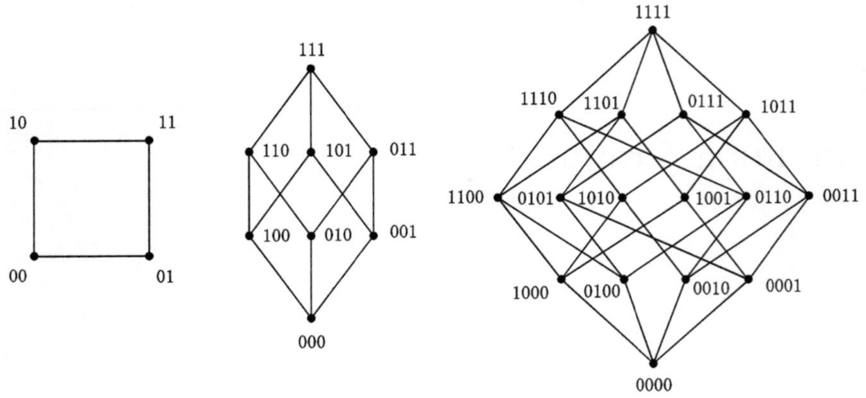

Fig. 5.1. Hamming graphs $H(2,2)$, $H(3,2)$ and $H(4,2)$

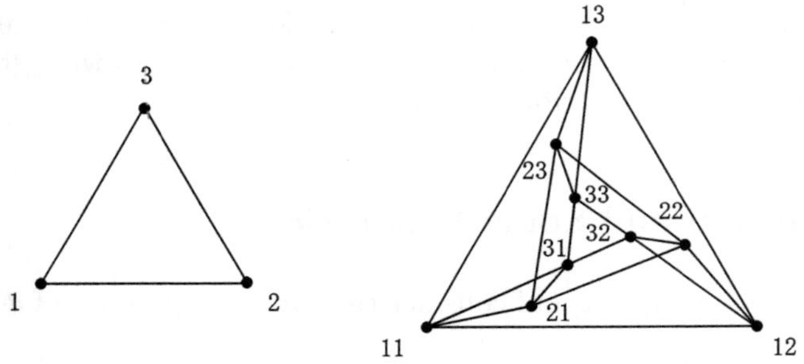

Fig. 5.2. Hamming graphs $H(1,3)$ and $H(2,3)$

Proposition 5.3. *A Hamming graph $H(d,N)$ is distance-transitive, hence distance-regular.*

The proofs of the above assertions are straightforward and omitted. The intersection number of a Hamming graph $H(d,N)$ is denoted by p_{ij}^k without indicating the parameters d,N to avoid cumbersome notation.

Lemma 5.4. *For the Hamming graph $H(d,N)$ we have*

$$p_{nn}^0 = \binom{d}{n}(N-1)^n, \quad n = 0,1,\dots,d. \tag{5.1}$$

In particular, the degree and the diameter are given as follows:

$$\kappa_{d,N} = p_{11}^0 = d(N-1), \qquad \operatorname{diam} H(d,N) = d.$$

Proof. p_{nn}^0 is the number of vertices $y \in F^d$, which has a distance n from a fixed vertex $x = (\xi_1,\dots,\xi_d)$. Such a y is obtained by selecting n components from $(\xi_1, \dots, \xi_d)$ and replacing each selected letter with one of the $N-1$ different letters. Hence (5.1) follows. By definition $\partial(x,y) \le d$ for any $x, y \in F^d$ and, in view of $N = |F| \ge 2$, the equality is attained for some x, y. Therefore $\operatorname{diam} H(d,N) = d$. □

Lemma 5.5. *For the Hamming graph $H(d,N)$ we have*

$$p_{1n}^n = n(N-2), \quad p_{1n}^{n-1} = (d-n+1)(N-1), \quad p_{1,n-1}^n = n, \tag{5.2}$$

for $n = 1, 2, \dots, d$.

Proof. We prove the first identity. For simplicity we set $F = \{1, 2, \dots, N\}$. Take two vertices $x, y \in V$ with $\partial(x,y) = n$. Without loss of generality we may set

$$x = (\underbrace{1, \dots, 1}_{n}, 1, \dots, 1), \qquad y = (\underbrace{2, \dots, 2}_{n}, 1, \dots, 1).$$

A vertex $z \in V$ with $\partial(x,z) = 1$ is of the form

$$z = (1, \dots, i, \dots, 1),$$

where $i \neq 1$ occurs at an arbitrary position. Among such z we need to determine ones with $\partial(y,z) = n$. Consider first the case where i occurs in the first n positions. Then, comparing

$$\begin{aligned} z &= (1, \dots, i, \dots, 1, 1, \dots, 1), \\ y &= (\underbrace{2, \dots, 2, \dots, 2}_{n}, 1, \dots, 1), \end{aligned}$$

we see that $\partial(y,z) = n$ if $i \neq 2$. Since such an i is chosen from $\{3, 4, \dots, N\}$ and there are n positions, the number of y satisfying the above conditions is $n(N-2)$. Consider next the case where i occurs in the last $d-n$ positions. Then we have

$$\begin{aligned} z &= (1, \dots, 1, 1, \dots, i, \dots, 1), \\ y &= (\underbrace{2, \dots, 2}_{n}, 1, \dots, 1, \dots, 1), \end{aligned}$$

so that $\partial(y,z) = n+1$, which does not satisfy the requirement $\partial(y,z) = n$. Consequently, $p_{1n}^n = n(N-2)$. The rest of (5.2) is shown easily in a similar fashion. □

A Hamming graph being highly symmetric, we here mention some group theoretical treatment. Without loss of generality we take $F = \{1, 2, \dots, N\}$. Let $S(d)$ denote the symmetric group of degree d, i.e. the group of permutations of $\{1, 2, \dots, d\}$, which acts on $H(d,N) = (F^d, E)$ by

$$\sigma x = (\xi_{\sigma^{-1}(1)}, \dots, \xi_{\sigma^{-1}(d)}), \quad \sigma \in S(d).$$

On the other hand, the direct product group $S(N)^d$ acts on $H(d,N)$ by

$$\tau x = (\tau_1(\xi_1), \dots, \tau_d(\xi_d)), \quad \tau = (\tau_1, \dots, \tau_d) \in S(N)^d.$$

Both actions give rise to automorphisms of $H(d,N)$. By a simple observation we have

$$\sigma^{-1}\tau\sigma x = (\tau_{\sigma(1)}(\xi_1), \dots, \tau_{\sigma(d)}(\xi_d)) = \tau^\sigma x,$$

where

$$\tau^\sigma = (\tau_1, \dots, \tau_d)^\sigma = (\tau_{\sigma(1)}, \dots, \tau_{\sigma(d)}).$$

Thus, the actions of $S(d)$ and $S(N)^d$ generate a semidirect product group $S(d) \ltimes S(N)^d$. Its action on $H(d,N)$ is transitive and the isotropy group of a vertex $o = (N, N, \dots, N)$ is $S(d) \ltimes S(N-1)^d$. Hence

$$H(d,N) \cong S(d) \ltimes S(N)^d / S(d) \ltimes S(N-1)^d.$$

In fact, it is shown that $\mathrm{Aut}\,(H(d,N)) \cong S(d) \ltimes S(N)^d$.

5.2 Asymptotic Spectral Distributions in the Vacuum State

Let $H(d,N) = (V,E)$, $V = F^d$, be a Hamming graph with an arbitrarily fixed origin $o \in V$. Since $\mathrm{diam}\, H(d,N) = d$, the corresponding stratification has the form:

$$V = \bigcup_{n=0}^{d} V_n.$$

By Lemma 5.4 we see that

$$|V_n| = p^0_{nn} = \binom{d}{n}(N-1)^n, \quad n = 0, 1, \dots, d.$$

Let $A = A_{d,N}$ be the adjacency matrix of $H(d,N)$. According to the above stratification we have the quantum decomposition:

$$A = A^+ + A^- + A^\circ.$$

Having obtained in Lemma 5.5 the intersection numbers necessary to our purpose, we shall apply the general theory established in Sect. 3.4. In view of (3.18) and (3.19), for $n = 1, 2, \dots$ we set

$$\bar{\omega}_n(d,N) = \frac{p^n_{1,n-1} p^{n-1}_{1,n}}{\kappa} = \frac{n(d-n+1)(N-1)}{d(N-1)} = n\left(1 - \frac{n}{d} + \frac{1}{d}\right),$$

$$\bar{\alpha}_n(d,N) = \frac{p^{n-1}_{1,n-1}}{\sqrt{\kappa}} = \frac{(n-1)(N-2)}{\sqrt{d(N-1)}} = (n-1)\sqrt{\left(\frac{N}{d} - \frac{2}{d}\right)\frac{N-2}{N-1}}.$$

On taking the limit we need to find a good balance between d and N. Apparently, the condition is given by

$$d \to \infty, \qquad \frac{N}{d} \to \tau,$$

where $\tau \geq 0$ can be arbitrary. Then we come to

$$\omega_n = \lim_{\substack{N/d \to \tau \\ d \to \infty}} \bar{\omega}_n(d, N) = n,$$

$$\alpha_n = \lim_{\substack{N/d \to \tau \\ d \to \infty}} \bar{\alpha}_n(d, N) = \sqrt{\tau}\,(n-1),$$

which shows that the limit is described by the Boson Fock space, i.e., an interacting Fock space with $\{\omega_n = n\}$. Moreover, the diagonal operator defined by $\{\alpha_n\}$ is $\sqrt{\tau}\,N$, where $N = B^+B^-$ is the number operator. Then, by an immediate application of Theorem 3.21 we come to the following:

Theorem 5.6 (QCLT for Hamming graphs). *Let $A_{d,N}$ be the adjacency matrix of the Hamming graph $H(d,N)$ and let $\tau \geq 0$. Denote by $\Gamma_{\text{Boson}} = (\Gamma, \{\Psi_n\}, B^+, B^-)$ the Boson Fock space and define a diagonal operator B° by $B^\circ = \sqrt{\tau}\,B^+B^-$. Then, we have*

$$\lim_{\substack{N/d \to \tau \\ d \to \infty}} \frac{A^\epsilon_{d,N}}{\sqrt{\kappa_{d,N}}} = B^\epsilon, \qquad \epsilon \in \{+, -, \circ\}, \tag{5.3}$$

in the sense of stochastic convergence with respect to the vacuum states.

Corollary 5.7. *For each $\tau \geq 0$ there exists a unique probability measure μ_τ such that*

$$\lim_{\substack{N/d \to \tau \\ d \to \infty}} \left\langle \left(\frac{A_{d,N}}{\sqrt{\kappa_{d,N}}} \right)^m \right\rangle_o = \int_{-\infty}^{+\infty} x^m \mu_\tau(dx), \quad m = 1, 2, \dots. \tag{5.4}$$

Moreover, the Jacobi coefficient of μ_τ is given by

$$\omega_n = n, \qquad \alpha_n = \sqrt{\tau}\,(n-1), \qquad n = 1, 2, \dots.$$

In fact, by Theorem 5.6 we have

$$\lim_{\substack{N/d \to \tau \\ d \to \infty}} \left\langle \delta_o, \left(\frac{A_{d,N}}{\sqrt{\kappa_{d,N}}} \right)^m \delta_o \right\rangle = \langle \Psi_0, (B^+ + B^- + \sqrt{\tau}\,B^+B^-)^m \Psi_0 \rangle,$$

for $m = 1, 2, \dots$. For $\tau = 0$ the probability measure μ_τ in (5.4) is readily known, that is, the standard Gaussian distribution (Theorem 1.78). For $\tau > 0$ the probability measure μ_τ is an affine transformation of the classical Poisson distribution. We shall discuss this topic in the next section.

5.3 Poisson Distribution

Definition 5.8. The *(classical) Poisson distribution* with parameter $\lambda > 0$ is a discrete measure defined by

$$p_\lambda = e^{-\lambda} \sum_{n=0}^{\infty} \frac{\lambda^n}{n!} \delta_n. \tag{5.5}$$

The parameter λ coincides with the mean and variance.

By definition the moment sequence of p_λ is given by

$$M_m = e^{-\lambda} \sum_{n=1}^{\infty} \frac{\lambda^n}{n!} n^m, \quad m = 0, 1, 2, \dots. \tag{5.6}$$

In particular, $M_1 = \lambda$ (mean) and $M_2 = \lambda^2 + \lambda$. The characteristic function $\psi_\lambda(z)$ is given by

$$\psi_\lambda(z) = e^{-\lambda} \sum_{n=0}^{\infty} \frac{\lambda^n}{n!} e^{inz} = \exp\{\lambda(e^{iz} - 1)\}. \tag{5.7}$$

Lemma 5.9. *The moment sequence $\{M_m\}$ of the Poisson distribution with parameter $\lambda > 0$ obeys the recurrence relation:*

$$M_0 = 1, \qquad M_m = \lambda \sum_{k=0}^{m-1} \binom{m-1}{k} M_k, \quad m = 1, 2, \dots. \tag{5.8}$$

Proof. We see from (5.6) that

$$M_m = e^{-\lambda} \sum_{n=0}^{\infty} \frac{\lambda^{n+1}}{(n+1)!} (n+1)^m = e^{-\lambda} \lambda \sum_{n=0}^{\infty} \frac{\lambda^n}{n!} (n+1)^{m-1}.$$

Then the binomial expansion leads us to

$$M_m = e^{-\lambda} \lambda \sum_{n=0}^{\infty} \frac{\lambda^n}{n!} \sum_{k=0}^{m-1} \binom{m-1}{k} n^k = \lambda \sum_{k=0}^{m-1} \binom{m-1}{k} M_k,$$

which proves the assertion. □

Lemma 5.10. *The moment sequence $\{M_m\}$ of the Poisson distribution with parameter $\lambda > 0$ satisfies*

$$M_m \le (\lambda + m)^m, \quad m = 1, 2, \dots. \tag{5.9}$$

Proof. Apparently (5.9) holds for $m = 1$ since $M_1 = \lambda$. Suppose that (5.9) is true up to $m \geq 1$. In view of Lemma 5.9, we obtain

$$M_{m+1} = \lambda \sum_{k=0}^{m} \binom{m}{k} M_k \leq \lambda \sum_{k=0}^{m} \binom{m}{k} (\lambda + k)^k.$$

Taking into account the obvious inequality $\lambda + k \leq \lambda + m$, we obtain

$$M_{m+1} \leq \lambda \sum_{k=0}^{m} \binom{m}{k} (\lambda + m)^k = \lambda(\lambda + m + 1)^m \leq (\lambda + m + 1)^{m+1}.$$

Thus (5.9) is true for $m + 1$, and the proof is complete. □

Proposition 5.11. *The Poisson distribution with parameter $\lambda > 0$ is the solution of a determinate moment problem. In other words, it is a unique probability measure in $\mathfrak{P}_{\mathrm{fm}}(\mathbb{R})$ whose moment sequence coincides with the sequence $\{M_m\}$ defined by* (5.8).

Proof. It follows from Lemma 5.10 that

$$M_{2m}^{-\frac{1}{2m}} \geq \{(\lambda + 2m)^{2m}\}^{-\frac{1}{2m}} = (\lambda + 2m)^{-1}$$

so that

$$\sum_{m=1}^{\infty} M_{2m}^{-\frac{1}{2m}} \geq \sum_{m=1}^{\infty} \frac{1}{\lambda + 2m} = \infty.$$

Then the assertion follows immediately from Carleman's moment test (Theorem 1.36). □

We now give an algebraic realization of a classical Poisson random variable.

Theorem 5.12. *Let $\Gamma_{\mathrm{Boson}} = (\Gamma, \{\Psi_n\}, B^+, B^-)$ be the Boson Fock space. For $\lambda > 0$, the vacuum spectral distribution of $(B^+ + \sqrt{\lambda})(B^- + \sqrt{\lambda})$ is the Poisson distribution with parameter λ. In other words,*

$$\langle \Psi_0, ((B^+ + \sqrt{\lambda})(B^- + \sqrt{\lambda}))^m \Psi_0 \rangle = \int_{-\infty}^{+\infty} x^m p_\lambda(dx), \tag{5.10}$$

for $m = 1, 2, \ldots$. Moreover, the Poisson distribution is uniquely determined by (5.10).

Proof. For simplicity we set

$$C^+ = B^+ + \sqrt{\lambda}, \quad C^- = B^- + \sqrt{\lambda}.$$

Since $B^- B^+ = B^+ B^- + 1$, we have

$$C^- C^+ = B^- B^+ + \sqrt{\lambda}\,(B^+ + B^-) + \lambda = C^+ C^- + 1.$$

Hence for $m = 1, 2, \dots$ we have

$$(C^+C^-)^m = C^+(C^-C^+)^{m-1}C^- = C^+(C^+C^- + 1)^{m-1}C^-. \tag{5.11}$$

Let M_m denote the left-hand side of (5.10), namely, the moment sequence of the algebraic random variable C^+C^- in the vacuum state:

$$M_m = \langle \Psi_0, (C^+C^-)^m \Psi_0 \rangle, \qquad m = 1, 2, \dots.$$

Then, by (5.11) and the obvious identity $C^-\Psi_0 = \sqrt{\lambda}\,\Psi_0$, we come to

$$M_m = \langle C^-\Psi_0, (C^+C^- + 1)^{m-1}C^-\Psi_0 \rangle = \lambda \langle \Psi_0, (C^+C^- + 1)^{m-1}\Psi_0 \rangle.$$

By the binomial expansion the last expression becomes

$$M_m = \lambda \sum_{k=0}^{m-1} \binom{m-1}{k} M_k, \quad m = 1, 2, \dots.$$

It is obvious that $M_0 = 1$. Thus we see from Lemma 5.9 that the recurrence relation satisfied by $\{M_m\}$ coincides with that by the moment sequence of the Poisson distribution with parameter λ. The uniqueness follows from Proposition 5.11. □

Corollary 5.13. *The Jacobi coefficient* $(\{\omega_n\}, \{\alpha_n\})$ *of the Poisson distribution with parameter* $\lambda > 0$ *is given by*

$$\omega_n = \lambda n, \quad \alpha_n = n - 1 + \lambda, \qquad n = 1, 2, \dots. \tag{5.12}$$

Proof. Let $\Gamma_{\mathrm{Boson}} = (\Gamma, \{\Psi_n\}, B^+, B^-)$ be the Boson Fock space. Using the number operator $N = B^+B^-$, we write

$$\begin{aligned}(B^+ + \sqrt{\lambda})(B^- + \sqrt{\lambda}) &= \sqrt{\lambda}\,(B^+ + B^-) + N + \lambda \\ &= \sqrt{\lambda}\,(B^+ + B^- + B^\circ),\end{aligned}$$

where

$$B^\circ = \frac{N}{\sqrt{\lambda}} + \sqrt{\lambda}$$

is the diagonal operator associated with a sequence defined by

$$\alpha_n = \frac{n-1}{\sqrt{\lambda}} + \sqrt{\lambda}, \qquad n = 1, 2, \dots.$$

Then (5.10) becomes

$$\langle \Psi_0, (B^+ + B^- + B^\circ)^m \Psi_0 \rangle = \int_{-\infty}^{+\infty} \left(\frac{x}{\sqrt{\lambda}} \right)^m p_\lambda(dx), \quad m = 1, 2, \dots. \tag{5.13}$$

Let S^*_λ be the dilation introduced in Sect. 1.4. We then have

$$\langle \Psi_0, (B^+ + B^- + B^\circ)^m \Psi_0 \rangle = \int_{-\infty}^{+\infty} x^m (S^*_{1/\sqrt{\lambda}} p_\lambda)(dx), \qquad m = 1, 2, \ldots.$$

Thus, the Jacobi parameter of $S^*_{1/\sqrt{\lambda}} p_\lambda$ is given by

$$\omega_n = n, \qquad \alpha_n = \frac{n-1}{\sqrt{\lambda}} + \sqrt{\lambda}, \qquad n = 1, 2, \ldots,$$

hence, that of $p_\lambda = S^*_{\sqrt{\lambda}} S^*_{1/\sqrt{\lambda}} p_\lambda$ as in (5.12), see Proposition 1.49. □

By virtue of Theorem 1.66, (5.12) shows also that the corresponding moment problem is determinate.

The orthogonal polynomials associated with the Poisson distribution with parameter λ is called the *Charlier polynomials* with parameter λ. It follows from Corollary 5.13 that the Charlier polynomials $C_n^\lambda(x)$ obeys

$$\begin{aligned} C_0^\lambda(x) &= 1, \\ C_1^\lambda(x) &= x - \lambda, \\ xC_n^\lambda(x) &= C_{n+1}^\lambda(x) + (n+\lambda)C_n^\lambda(x) + \lambda n C_{n-1}^\lambda(x), \qquad n = 1, 2, \ldots. \end{aligned}$$

We are in a position to claim the central limit theorem for the adjacency matrices of the Hamming graphs in the vacuum state.

Theorem 5.14 (CLT for Hamming graphs). *Let $H(d,N)$ be a Hamming graph, $A_{d,N}$ its adjacency matrix and $\kappa_{d,N} = d(N-1)$ its degree. Then for any $\tau \geq 0$ there exists a unique probability measure μ_τ such that*

$$\lim_{\substack{N/d \to \tau \\ d \to \infty}} \left\langle \left(\frac{A_{d,N}}{\sqrt{\kappa_{d,N}}} \right)^m \right\rangle_o = \int_{-\infty}^{+\infty} x^m \mu_\tau(dx), \qquad m = 1, 2, \ldots.$$

Moreover, μ_τ is given as follows: for $\tau = 0$, $\mu_\tau = \mu_0$ is the standard Gaussian distribution

$$\mu_0(dx) = \frac{1}{\sqrt{2\pi}} e^{-x^2/2} dx,$$

and for $\tau > 0$, μ_τ is an affine transformation of the Poisson distribution and is specified by

$$\mu_\tau\left(k\sqrt{\tau} - \frac{1}{\sqrt{\tau}} \right) = e^{-1/\tau} \frac{\tau^{-k}}{k!}, \qquad k = 0, 1, 2, \ldots. \tag{5.14}$$

Proof. For $\tau = 0$ the assertion is already shown in Corollary 5.7 and the succeeding comment. Let $\tau > 0$. It follows from Theorem 5.6 that

$$\lim_{\substack{N/d \to \tau \\ d \to \infty}} \frac{A_{d,N}}{\sqrt{\kappa_{d,N}}} = B^+ + B^- + \sqrt{\tau}\, N.$$

Since $N = B^+B^-$ holds in the Boson Fock space,

$$B^+ + B^- + \sqrt{\tau}\, N = B^+ + B^- + \sqrt{\tau}\, B^+B^-$$
$$= \sqrt{\tau}\left(B^+ + \frac{1}{\sqrt{\tau}}\right)\left(B^- + \frac{1}{\sqrt{\tau}}\right) - \frac{1}{\sqrt{\tau}}.$$

Viewing that $(B^+ + 1/\sqrt{\tau})(B^- + 1/\sqrt{\tau})$ is an algebraic realization of the classical Poisson random variable with parameter $1/\tau$ (see Theorem 5.12), we obtain (5.14) by easy calculation. □

5.4 Asymptotic Spectral Distributions in the Deformed Vacuum States

We start with the following:

Lemma 5.15. *A Hamming graph $H(d, N)$ admits a quadratic embedding into $\mathbb{R}^{(N-1)d}$.*

Proof. Recall that a complete graph K^N admits a quadratic embedding into a sphere in $\mathbb{R}^{N-1}$ (Exercise 2.14). In fact, we may choose $v_1, \dots, v_N \in \mathbb{R}^{N-1}$ in such a way that

$$\|v_1\| = \cdots = \|v_N\| < 1, \qquad \|v_i - v_j\| = 1, \quad i \neq j.$$

We define a map $v : \{1, 2, \dots, N\}^d \to \mathbb{R}^{(N-1)d}$ by

$$v : x = (\xi_1, \dots, \xi_d) \mapsto v(x) = (v_{\xi_1}, \dots, v_{\xi_n}).$$

Then for $x = (\xi_1, \dots, \xi_d)$ and $y = (\eta_1, \dots, \eta_d)$,

$$\|v(x) - v(y)\|^2 = \sum_{k=1}^{d} \|v_{\xi_k} - v_{\eta_k}\|^2 = |\{1 \le k \le d\,;\, \xi_k \neq \eta_k\}| = \partial(x, y),$$

which means that v is a quadratic embedding of $H(d, N)$ into $\mathbb{R}^{(N-1)d}$. □

Proposition 5.16. *For $0 \le q \le 1$, the deformed vacuum state $\langle\cdot\rangle_q$ on the adjacency algebra of a Hamming graph $H(d, N)$ is positive.*

Proof. Since a Hamming graph admits a quadratic embedding (Lemma 5.15), it follows by Bożejko's quadratic embedding test (Proposition 2.14) that $Q = (q^{\partial(x,y)})$ is a positive definite kernel for $0 \le q \le 1$. Then the assertion follows from Theorem 3.27. □

Remark 5.17. Since $H(1, 3)$ is a triangle, the corresponding matrix Q is positive definite if and only if $-1/2 \le q \le 1$. For a cube $H(3, 2)$, as is verified in Example 2.16, the corresponding matrix Q is positive definite if and only if $-1 \le q \le 1$. Here we do not discuss the positivity of Q for negative q.

The mean and the variance of the adjacency matrix $A_{d,N}$ in $\langle\cdot\rangle_q$ are easily obtained:

$$\langle A_{d,N}\rangle_q = qd(N-1),$$
$$\Sigma_q^2(A_{d,N}) = d(N-1)\{1+q(N-2)-q^2(N-1)\},$$

see the general formulae (3.28) and (3.29). We are interested in the asymptotic behaviour of

$$\frac{A_{d,N} - \langle A_{d,N}\rangle_q}{\Sigma_q(A_{d,N})}$$

in the limit with suitable balance of d, N, q. Following the general argument in Sect. 3.4, we set

$$\bar{\omega}_n(d,N,q) = \frac{p^n_{1,n-1}p^{n-1}_{1,n}}{\Sigma_q^2(A_{d,N})} = \frac{n(d-n+1)(N-1)}{d(N-1)\{1+q(N-2)-q^2(N-1)\}},$$

$$\bar{\alpha}_n(d,N,q) = \frac{p^{n-1}_{1,n-1} - q\kappa}{\Sigma_q(A_{d,N})} = \frac{(n-1)(N-2)-qd(N-1)}{\sqrt{d(N-1)\{1+q(N-2)-q^2(N-1)\}}},$$

$$c_n(d,N,q) = q^n\sqrt{p^0_{nn}} = q^n\sqrt{\binom{d}{n}(N-1)^n}.$$

For the limits we need to find a suitable balance among d, N, q. From

$$\bar{\omega}_n(d,N,q) = \frac{n\left(1-\dfrac{n}{d}+\dfrac{1}{d}\right)}{1+q(N-2)-q^2(N-1)},$$

$$\bar{\alpha}_n(d,N,q) = \sqrt{\left\{(n-1)\left(\frac{N}{d}-\frac{2}{d}\right)-q(N-1)\right\}}$$
$$\times\sqrt{\frac{(n-1)(N-2)-qd(N-1)}{(N-1)\{1+q(N-2)-q^2(N-1)\}}},$$

$$c_n(d,N,q) = \sqrt{1\cdot\left(1-\frac{1}{d}\right)\left(1-\frac{2}{d}\right)\cdots\left(1-\frac{n-1}{d}\right)\frac{\{d(N-1)q^2\}^n}{n!}},$$

it turns out that the limit is to be taken as

$$d\to\infty, \quad q\to 0, \quad \frac{N}{d}\to\tau\ge 0, \quad qd\to\sigma\ge 0, \tag{5.15}$$

where σ and τ are constant numbers. The limits are given by

$$\omega_n = \lim_{(5.15)} \bar{\omega}_n(d, N, q) = \frac{n}{1+\sigma\tau}, \tag{5.16}$$

$$\alpha_n = \lim_{(5.15)} \bar{\alpha}_n(d, N, q) = \frac{\sqrt{\tau}\,(n-1-\sigma)}{\sqrt{1+\sigma\tau}}, \tag{5.17}$$

$$c_n = \lim_{(5.15)} c_n(d, N, q) = \frac{(\sigma\sqrt{\tau})^n}{\sqrt{n!}}. \tag{5.18}$$

The above $\{\omega_n\}$ in (5.16) being a constant multiple of the Jacobi sequence of the Boson Fock space, we shall formulate the quantum central limit theorem in terms of the Boson Fock space $\Gamma_{\mathrm{Boson}} = (\Gamma, \{\Psi_n\}, B^+, B^-)$. The diagonal operator defined by $\{\alpha_n\}$ is expressible in the number operator $N = B^+B^-$. Note also that the vector corresponding to (5.18) is the coherent vector of the Boson Fock space (Definition 3.32), i.e.,

$$\sum_{n=0}^{\infty} c_n \Psi_n = \sum_{n=0}^{\infty} \frac{(\sigma\sqrt{\tau})^n}{\sqrt{n!}} \Psi_n = \Omega_{\sigma\sqrt{\tau}}.$$

We are now in a position to apply Theorem 3.29 to the Hamming graphs.

Theorem 5.18 (QCLT for Hamming graphs in the deformed vacuum state). *Let $A_{d,N}$ be the adjacency matrix of a Hamming graph $H(d,N)$ and $\Gamma_{\mathrm{Boson}} = (\Gamma, \{\Psi_n\}, B^{\pm})$ the Boson Fock space. Let $\tau \geq 0$ and $\sigma \geq 0$. Then, taking the limits as in* (5.15)*, we have*

$$\lim_{(5.15)} \frac{A^{\pm}_{d,N}}{\Sigma_q(A_{d,N})} = \frac{B^{\pm}}{\sqrt{1+\sigma\tau}}, \tag{5.19}$$

$$\lim_{(5.15)} \frac{A^{\circ}_{d,N} - \langle A_{d,N}\rangle_q}{\Sigma_q(A_\nu)} = \frac{\sqrt{\tau}\,(B^+B^- - \sigma)}{\sqrt{1+\sigma\tau}}, \tag{5.20}$$

in the sense of stochastic convergence, where the left-hand sides are in the deformed vacuum state $\langle\cdot\rangle_q$ and the right-hand sides in the coherent state $\langle\cdot\rangle_{\sigma\sqrt{\tau}}$.

Then the following result is immediate.

Corollary 5.19. *Notations and assumptions being as in Theorem 5.18,*

$$\lim_{(5.15)} \left\langle \left(\frac{A_{d,N} - \langle A_{d,N}\rangle_q}{\Sigma(A_{d,N})} \right)^m \right\rangle_q$$

$$= \left\langle \left(\frac{B^+ + B^- + \sqrt{\tau}\,(B^+B^- - \sigma)}{\sqrt{1+\sigma\tau}} \right)^m \right\rangle_{\sigma\sqrt{\tau}}, \qquad m = 1, 2, \dots. \tag{5.21}$$

The rest of this section will be devoted to finding a probability measure $\mu \in \mathfrak{P}_{\rm fm}(\mathbb{R})$, of which the mth moment coincides with (5.21), namely, such that

$$\left\langle \Omega_{\sigma\sqrt{\tau}}, \left(\frac{B^+ + B^- + \sqrt{\tau}\,(B^+B^- - \sigma)}{\sqrt{1+\sigma\tau}}\right)^m \Psi_0 \right\rangle = \int_{-\infty}^{+\infty} x^m \mu(dx), \qquad (5.22)$$

for $m = 1, 2, \dots$. If $\sigma = 0$ or $\tau = 0$, the coherent vector $\Omega_{\sigma\sqrt{\tau}}$ becomes the vacuum vector so that the probability measure μ is readily known from Theorem 5.12. For a general case we need the following:

Lemma 5.20. *Let $\Gamma_{\rm Boson} = (\Gamma, \{\Psi_n\}, B^+, B^-)$ be the Boson Fock space and Ω_α the coherent vector. Then for any $\lambda > 0$ and $\alpha > -\lambda$ we have*

$$\begin{aligned}&\langle \Omega_\alpha, \{(B^+ + \lambda)(B^- + \lambda)\}^m \Psi_0\rangle \\ &\qquad = \left\langle \Psi_0, \left\{\left(B^+ + \sqrt{(\alpha+\lambda)\lambda}\right)\left(B^- + \sqrt{(\alpha+\lambda)\lambda}\right)\right\}^m \Psi_0\right\rangle, \qquad (5.23)\end{aligned}$$

for $m = 0, 1, 2, \dots$. Moreover, (5.23) coincides with the mth moment of the Poisson distribution with parameter $(\alpha + \lambda)\lambda$.

Proof. For simplicity we set

$$C^+ = B^+ + \lambda, \qquad C^- = B^- + \lambda$$

and

$$M_m = \langle \Omega_\alpha, \{(B^+ + \lambda)(B^- + \lambda)\}^m \Psi_0\rangle = \langle \Omega_\alpha, (C^+C^-)^m \Psi_0\rangle.$$

We shall obtain a recurrence relation satisfied by $\{M_m\}$ above. Note first that

$$C^-\Omega_\alpha = (B^- + \lambda)\Omega_\alpha = (\alpha + \lambda)\Omega_\alpha\,, \qquad (5.24)$$

$$C^-\Psi_0 = (B^- + \lambda)\Psi_0 = \lambda\Psi_0. \qquad (5.25)$$

Then for $m = 1, 2, \dots$ we have

$$\begin{aligned}M_m &= \langle \Omega_\alpha, (C^+C^-)^m \Psi_0\rangle \\ &= \langle \Omega_\alpha, C^+(C^-C^+)^{m-1} C^-\Psi_0\rangle \\ &= \langle C^-\Omega_\alpha, (C^-C^+)^{m-1} C^-\Psi_0\rangle \\ &= (\alpha+\lambda)\lambda \langle \Omega_\alpha, (C^-C^+)^{m-1}\Psi_0\rangle. \qquad (5.26)\end{aligned}$$

Since

$$(C^-C^+)^{m-1} = (C^+C^- + 1)^{m-1} = \sum_{k=0}^{m-1} \binom{m-1}{k} (C^+C^-)^k,$$

which follows from the canonical commutation relation $B^-B^+ - B^+B^- = 1$, we see that (5.26) becomes

$$M_m = (\alpha+\lambda)\lambda \sum_{k=0}^{m-1} \binom{m-1}{k} \langle \Omega_\alpha, (C^+C^-)^k \Psi_0 \rangle$$
$$= (\alpha+\lambda)\lambda \sum_{k=0}^{m-1} \binom{m-1}{k} M_k. \tag{5.27}$$

Taking into account the obvious identity $M_0 = \langle \Omega_\alpha, \Psi_0 \rangle = 1$ and applying Lemma 5.9, we see that (5.27) determines the moment sequence of the Poisson distribution with parameter $(\alpha+\lambda)\lambda$. Then (5.23) follows from Theorem 5.12. □

We now go back to (5.22). For simplicity we set

$$\xi = \sigma + \frac{1}{\tau}.$$

We first note that

$$\frac{B^+ + B^- + \sqrt{\tau}\,(N-\sigma)}{\sqrt{1+\sigma\tau}} = \frac{1}{\sqrt{\xi}} \left\{ \left(B^+ + \frac{1}{\sqrt{\tau}}\right)\left(B^- + \frac{1}{\sqrt{\tau}}\right) - \xi \right\}. \tag{5.28}$$

On the other hand, it follows from Lemma 5.20 that

$$\left\langle \left\{ \left(B^+ + \frac{1}{\sqrt{\tau}}\right)\left(B^- + \frac{1}{\sqrt{\tau}}\right) \right\}^m \right\rangle_{\sigma\sqrt{\tau}} = \int_{-\infty}^{+\infty} x^m p_\xi(dx),$$

where p_ξ is the Poisson distribution with parameter ξ. Hence in view of (5.28),

$$\left\langle \left(\frac{B^+ + B^- + \sqrt{\tau}\,(N-\sigma)}{\sqrt{1+\sigma\tau}} \right)^m \right\rangle_{\sigma\sqrt{\tau}}$$
$$= \left(\frac{1}{\sqrt{\xi}}\right)^m \left\langle \left\{ \left(B^+ + \frac{1}{\sqrt{\tau}}\right)\left(B^- + \frac{1}{\sqrt{\tau}}\right) - \xi \right\}^m \right\rangle_{\sigma\sqrt{\tau}}$$
$$= \left(\frac{1}{\sqrt{\xi}}\right)^m \int_{-\infty}^{+\infty} (x-\xi)^m p_\xi(dx),$$

Thus, the probability distribution μ in (5.22) is an affine transformation of the Poisson distribution with parameter ξ. In fact,

$$\mu\left(\frac{k}{\sqrt{\xi}} - \sqrt{\xi}\right) = \mathrm{e}^{-\xi}\frac{\xi^k}{k!}, \qquad k = 0, 1, 2, \dots. \tag{5.29}$$

Summing up,

Theorem 5.21 (CLT for Hamming graphs in the deformed vacuum state). *Let $A_{d,N}$ be the adjacency matrix of a Hamming graph $H(d,N)$. Let*

$\tau \geq 0$ and $\sigma \geq 0$. Let μ be an asymptotic spectral distribution of $A_{d,N}$ in the deformed vacuum state under the limit (5.15), *i.e.,*

$$\lim_{(5.15)} \left\langle \left(\frac{A_{d,N} - \langle A_{d,N} \rangle_q}{\Sigma(A_{d,N})} \right)^m \right\rangle_q = \int_{-\infty}^{+\infty} x^m \mu(dx), \quad m = 0, 1, 2, \dots. \tag{5.30}$$

If $\tau > 0$, then μ is an affine transformation of the Poisson distribution with parameter $\xi = \sigma + 1/\tau$, explicitly given by (5.29). *If $\tau = 0$, then μ is the standard Gaussian distribution.*

Exercises

5.1. Show that the graph distance on a Hamming graph coincides with the Hamming distance. [Proposition 5.2]

5.2. For a Hamming graph $H(d, 2) = (V, E)$ define $E' = \{(x, y)\,;\, \partial(x, y) = 2\}$, where $\partial(x, y)$ is the Hamming distance. Show that (V, E') is distance-regular. [(V, E') is called the *bipartite half* of $H(d, 2)$.]

5.3. Let p_{ij}^k be the intersection numbers of the Hamming graph $H(d, N)$. Show that

$$p_{1n}^{n-1} = (d - n + 1)(N - 1), \qquad p_{1,n-1}^{n} = n,$$

for $n = 1, 2, \dots, d$. [Lemma 5.5]

5.4. Prove that the generating function of the Charlier polynomials $C_n^\lambda(x)$ is given by

$$\mathrm{e}^{-\lambda z}(1 + z)^x = \sum_{n=0}^{\infty} C_n^\lambda(x) \frac{z^n}{n!}.$$

5.5. Show that the Charlier polynomial $C_n^\lambda(x)$ is given by

$$C_n^\lambda(x) = \sum_{k=0}^{n} \binom{n}{k} \binom{x}{k} k!(-\lambda)^{n-k}, \quad n = 0, 1, 2, \dots.$$

5.6. Let $\Gamma_{\mathrm{Boson}} = (\Gamma, \{\Phi_n\}, B^+, B^-)$ be the Boson Fock space. Show the following identities: for $m = 1, 2, \dots$ and $z \in \mathbb{C}$,

(1) $B^-(B^+ + B^-)^m = (B^+ + B^-)^m B^- + m(B^+ + B^-)^{m-1}$.

(2) $\langle \Omega_z, (B^+ + B^-)^m \Phi_0 \rangle = \langle \Phi_0, (B^+ + B^- + \bar{z})^m \Phi_0 \rangle$.

[Hint: For (2) check that both sides obey the same recurrence relation.]

Notes

More algebraic and combinatorial properties of Hamming graphs are collected in Bannai–Ito [17] and Brouwer–Cohen–Neumaier [50]. The spectrum of the Hamming graph $H(d, N)$ is known:

$$\lambda_j = N(d-j) - d, \quad w_j = \binom{d}{j}(N-1)^j, \quad j = 0, 1, \dots, d, \tag{5.31}$$

see, for example, Bannai–Ito [17, Sect. 3.2], Biggs [30, Chap. 21] and Brouwer–Cohen–Neumaier [50, Sect. 9.2]. The Hamming graph serves as a useful model in classical probability theory, that is, Ehrenfests' urn model is the simple random walk on a Hamming graph.

The asymptotic spectral distribution of Hamming graphs in the vacuum state was first obtained by Hora [102] by using the spectrum (5.31) directly. With this method Hora [104] obtained the asymptotic spectral distribution in the deformed vacuum state too. The application of quantum decomposition was due to Hashimoto–Obata–Tabei [97] though the method was not yet well-developed. In their paper the quantum component A° was further decomposed into a sum of two parts and $A = (A^+ + A^\circ_+) + (A^- + A^\circ_-)$ was taken as the quantum decomposition. The decomposition $A^\circ = A^\circ_+ + A^\circ_-$ is reached by means of Euler's unicursal theorem (see e.g., Bollobás [32], Diestel [72]) and seems interesting in itself.

6

Johnson Graphs

As a further example of growing distance-regular graphs, in this chapter we shall consider the Johnson graphs and the odd graphs. Both are graphs whose vertices are certain subsets of a given finite set. As the central limit distributions, we shall obtain the exponential distribution and the geometric distribution from the Johnson graphs, and the two-sided Rayleigh distribution from the odd graphs.

6.1 Definition and Some Properties

Definition 6.1. Let $v \geq 1$ and $S = \{1, 2, \dots, v\}$. For $0 \leq d \leq v$ define

$$V = \{x \subset S\,;\, |x| = d\}, \quad E = \{\{x, y\}\,;\, x, y \in V,\ |x \cap y| = d - 1\}.$$

The pair (V, E) is called a *Johnson graph* and is denoted by $J(v, d)$.

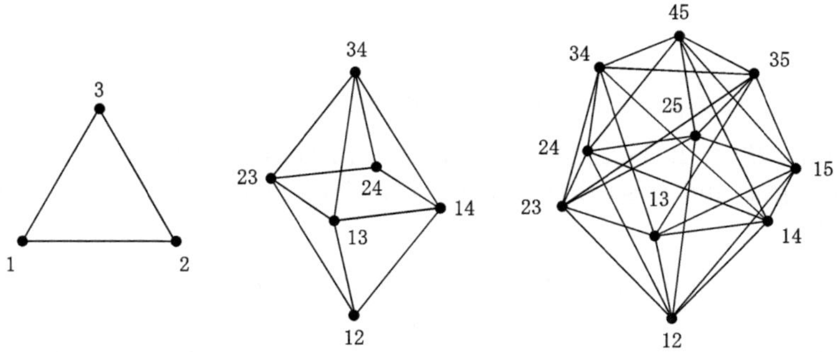

Fig. 6.1. Johnson graphs $J(3,2) \cong J(3,1)$, $J(4,2)$ and $J(5,2)$

A. Hora and N. Obata: Johnson Graphs. In: A. Hora and N. Obata, Quantum Probability and Spectral Analysis of Graphs, Theoretical and Mathematical Physics, 147–173 (2007)
DOI 10.1007/3-540-48863-4_6

Lemma 6.2. $J(v,d) \cong J(v,v-d)$.

Proof. The isomorphism is given by the map $x \mapsto S \setminus x$. □

Hence it is sufficient to consider v, d running over $1 \le d \le v/2$. By definition $J(v,0) \cong J(v,v)$ is a trivial graph (consisting of a single vertex) and $J(v,1) \cong J(v,v-1) \cong K_v$ is the complete graph with v vertices. $J(v,2)$ is sometimes called a *triangular graph.*

Lemma 6.3. *Let $\partial(x,y)$ denote the distance of a Johnson graph $J(v,d)$. Then $\partial(x,y) = k$ if and only if $|x \cap y| = d-k$.*

Proof. Assume that $|x \cap y| = d-k$, $0 \le k \le d$. Then we may write

$$x = \{\xi_1, \dots, \xi_k, \zeta_1, \dots, \zeta_{d-k}\}, \quad y = \{\eta_1, \dots, \eta_k, \zeta_1, \dots, \zeta_{d-k}\},$$

with

$$\{\xi_1, \dots, \xi_k\} \cap \{\eta_1, \dots, \eta_k\} = \varnothing.$$

For $i = 0, 1, \dots, k$ define

$$x_i = \{\underbrace{\eta_1, \dots, \eta_i}_{i}, \underbrace{\xi_{i+1}, \dots, \xi_k}_{k-i}, \zeta_1, \dots, \zeta_{d-k}\}.$$

Obviously,

$$x = x_0 \sim x_1 \sim x_2 \sim \cdots \sim x_k = y,$$

from which $\partial(x,y) \le k$ follows. We now take an arbitrary walk connecting x and y, say,

$$x = z_0 \sim z_1 \sim z_2 \sim \cdots \sim z_l = y. \tag{6.1}$$

In general, for three sets A, B, C it holds that $A \ominus B \subset (A \ominus C) \cup (C \ominus B)$, and hence $|A \ominus B| \le |A \ominus C| + |C \ominus B|$, where $A \ominus B = (A \cup B) - (A \cap B)$ is the symmetric difference. Using this formula repeatedly, we obtain

$$|x \ominus y| = |z_0 \ominus z_l| \le |z_0 \ominus z_1| + |z_1 \ominus z_2| + \cdots + |z_{l-1} \ominus z_l|.$$

Since

$$|x \ominus y| = |(x \cup y) - (x \cap y)| = 2k, \quad |z_i \ominus z_{i+1}| = 2,$$

we obtain $2k \le 2l$, i.e., $k \le l$. Therefore, (6.1) is one of the shortest walks connecting x and y so that $\partial(x,y) = k$. The converse assertion is shown similarly. □

A Johnson graph is highly symmetric. The permutation group $S(v)$ of $\{1, 2, \dots, v\}$ acts on $J(v,d)$ in a natural manner. The action is transitive. Since the isotropy group of $o = \{1, 2, \dots, d\}$ is $S(d) \times S(v-d)$, we have

$$J(v,d) \cong S(v)/(S(d) \times S(v-d)).$$

Thus, $J(v,d)$ is considered as a discrete analogue of a Grassmanian manifold.

Proposition 6.4. *A Johnson graph is distance-transitive and therefore distance-regular.*

Proof. Every $\sigma \in S(v)$ gives rise to a bijection on V defined by

$$z \mapsto \sigma(z) = \{\sigma(i)\,;\, i \in z\}, \qquad z \in V.$$

It is obvious that $\sigma \in \mathrm{Aut}(J(v,d))$. Suppose that $\partial(x,y) = \partial(x',y') = k$. We can set

$$x = \{\xi_1, \dots, \xi_k, \zeta_1, \dots, \zeta_{d-k}\}, \quad y = \{\eta_1, \dots, \eta_k, \zeta_1, \dots, \zeta_{d-k}\},$$
$$x' = \{\xi_1', \dots, \xi_k', \zeta_1', \dots, \zeta_{d-k}'\}, \quad y' = \{\eta_1', \dots, \eta_k', \zeta_1', \dots, \zeta_{d-k}'\},$$

where

$$\{\xi_1, \dots, \xi_k\} \cap \{\eta_1, \dots, \eta_k\} = \emptyset, \quad \{\xi_1', \dots, \xi_k'\} \cap \{\eta_1', \dots, \eta_k'\} = \emptyset.$$

Here it holds that $k \le d$ and $k \le v - d$ since $2k + d - k = d + k \le v$. We take $\alpha \in S(v)$ satisfying

$$\alpha(\xi_i) = \xi_i', \quad \alpha(\eta_i) = \eta_i', \quad \alpha(\zeta_i) = \zeta_i',$$

so as to have $\alpha(x) = x'$ and $\alpha(y) = y'$. This means that $J(v,d)$ is distance-transitive. It is a general fact that a distance-transitive graph is distance-regular (Proposition 3.7). □

For a Johnson graph $J(v,d)$ we denote by p_{ij}^k its intersection numbers, where we do not indicate the parameters v, d to avoid annoying notation. Let us compute those intersection numbers needed to discuss quantum central limit theorem. We see that all of them are obtained directly from the definition without difficulties.

Lemma 6.5. *The degree and the diameter of the Johnson graph $J(v,d)$ are*

$$\kappa_{v,d} = p_{11}^0 = d(v-d), \qquad \mathrm{diam}\, J(v,d) = \min\{d, v-d\},$$

respectively.

Proof. Take a vertex of $J(v,d)$, say $o = \{1, 2, \dots, d\}$. By definition a vertex x is adjacent to o if and only if x differs from o by a single element. Hence the number of such x's is $d(v-d)$. That $\mathrm{diam}\, J(v,d) = \min\{d, v-d\}$ follows immediately from Lemma 6.3. □

Lemma 6.6.

$$p_{n,n}^0 = \binom{d}{n}\binom{v-d}{n}, \qquad n = 0, 1, 2, \dots, \min\{d, v-d\}.$$

Proof. Set $o = \{1, 2, \dots, d\}$. It follows from Lemma 6.3 that

$$p^0_{nn} = |\{x\,;\, \partial(o, x) = n\}| = |\{x\,;\, |o \cap x| = d - n\}|.$$

Thus, p^0_{nn} is the number of vertices x that differ from $o = \{1, 2, \dots, d\}$ by n elements. Such a vertex x is obtained from o by replacing arbitrarily chosen n elements with the same number of elements arbitrarily chosen from $\{d + 1, d + 2, \dots, v\}$, and the number of such combination is given as in the statement. □

Lemma 6.7.

$$p^n_{1,n-1} = n^2, \qquad n = 1, \dots, \min\{d, v - d\}.$$

Proof. Without loss of generality we may take two vertices with distance n as follows:

$$x = \{1, 2, \dots, d - n, \xi_1, \dots, \xi_n\}, \quad y = \{1, 2, \dots, d - n, \eta_1, \dots, \eta_n\},$$
$$\{\xi_1, \dots, \xi_n\} \cap \{\eta_1, \dots, \eta_n\} = \varnothing.$$

We shall count the number of vertices z satisfying $\partial(x, z) = 1$ and $\partial(y, z) = n - 1$. Note that any z with $\partial(x, z) = 1$ is obtained from x replacing one element. We consider two cases according as the replacement happens in $\{1, 2, \dots, d-n\}$ or $\{\xi_1, \dots, \xi_n\}$, and study a form of z satisfying $\partial(y, z) = n-1$ or equivalently $|y \cap z| = d - n + 1$. The assertion follows by combining Case 1 and Case 2 below.

Case 1. z is obtained from x by replacing one element in $\{1, 2, \dots, d - n\}$, i.e.,

$$z = \{1, 2, \dots, i - 1, *, i + 1, \dots, d - n, \xi_1, \dots, \xi_n\}.$$

Then $|z \cap y| \le d - n$ and we see that

$$\partial(y, z) = d - |z \cap y| \ge d - (d - n) = n.$$

Namely, in this case $\partial(y, z) = n - 1$ never happens.

Case 2. z is obtained from x by replacing one element in $\{\xi_1, \dots, \xi_n\}$, i.e.,

$$z = \{1, 2, \dots, d - n, \xi_1, \dots, \xi_{i-1}, *, \xi_{i+1}, \dots, \xi_n\}.$$

Then $|y \cap z| = d - n + 1$ occurs if and only if $*$ belongs to y, that is, $*$ is one of $\{\eta_1, \dots, \eta_n\}$. Since there are n choices of i, the number of z satisfying our condition is n^2. □

Lemma 6.8.

$$p^{n-1}_{1,n} = (d - n + 1)(v - d - n + 1), \qquad n = 1, \dots, \min\{d, v - d\}.$$

Proof. We consider two points with distance $n-1$, say,

$$x = \{1, 2, \dots, d-n+1, \xi_1, \dots, \xi_{n-1}\},$$
$$y = \{1, 2, \dots, d-n+1, \eta_1, \dots, \eta_{n-1}\},$$
$$\{\xi_1, \dots, \xi_{n-1}\} \cap \{\eta_1, \dots, \eta_{n-1}\} = \varnothing.$$

In a similar manner as in the proof of Lemma 6.7, we consider two cases according to the form of z satisfying $\partial(x, z) = 1$. The assertion follows by combining Case 1 and Case 2.

Case 1. z is obtained from x by replacing one element in $\{1, 2, \dots, d-n+1\}$, i.e.,

$$z = \{1, 2, \dots, i-1, *, i+1, \dots, d-n+1, \xi_1, \dots, \xi_{n-1}\}.$$

Among such z's we count the number of z such that $\partial(y, z) = n$, or equivalently $|y \cap z| = d - n$. Since

$$y \cap z = \{1, 2, \dots, d-n+1, \eta_1, \dots, \eta_{n-1}\} \\ \cap \{1, 2, \dots, i-1, *, i+1, \dots, d-n+1\},$$

$|y \cap z| = d - n$ happens if and only if $*$ does not belong to y. The number of such a choice of $*$ is $v - d - (n-1)$. There are $(d-n+1)$ different ways of choosing i, the total number of the choice of z is $(d-n+1)(v-d-n+1)$.

Case 2. z is obtained from x by replacing one element in $\{\xi_1, \dots, \xi_{n-1}\}$, i.e.,

$$z = \{1, 2, \dots, d-n+1, \xi_1, \dots, \xi_{i-1}, *, \xi_{i+1}, \dots, \xi_{n-1}\}.$$

Then we have $|y \cap z| \ge d - n + 1$ and

$$\partial(y, z) = d - |y \cap z| \le d - (d-n+1) = n - 1.$$

Thus $\partial(y, z) = n$ never happens. □

Lemma 6.9.

$$p^n_{1,n} = n(v - 2n), \qquad n = 0, 1, \dots, \min\{d, v-d\}.$$

Proof. Take two points with distance n as follows:

$$x = \{1, 2, \dots, d-n, \xi_1, \dots, \xi_n\}, \quad y = \{1, 2, \dots, d-n, \eta_1, \dots, \eta_n\},$$
$$\{\xi_1, \dots, \xi_n\} \cap \{\eta_1, \dots, \eta_n\} = \varnothing.$$

Note that z with $\partial(x, z) = 1$ is obtained from x by replacing one element. We consider two cases according as the replacement happens in the first group of symbols or in the second group of symbols.

Case 1. z is obtained from x by replacing one element in $\{1, 2, \dots, d-n\}$, i.e.,

$$z = \{1, 2, \dots, i-1, *, i+1, \dots, d-n, \xi_1, \dots, \xi_n\}.$$

Note that $\partial(y, z) = n$ if and only if $|y \cap z| = d - n$. Since $\{\xi_1, \dots, \xi_n\}$ has no intersection with y, to have $|y \cap z| = d - n$, $*$ belongs to y. Such a $*$ is one of $\{\eta_1, \dots, \eta_n\}$. Therefore there are $(d-n)n$ possible choice of z.

Case 2. z is obtained from x by replacing one element in $\{\xi_1, \dots, \xi_n\}$, i.e.,

$$z = \{1, 2, \dots, d-n, \xi_1, \dots, \xi_{i-1}, *, \xi_{i+1}, \dots, \xi_n\}.$$

We shall determine z such that $|y \cap z| = d - n$. Since the first $d-n$ symbols of y and z are common, $*$ is chosen from the complement of $x \cup y$. The number of such choices is $(v - d - n)$. There are n different choices of i, and therefore the number of possible z is $n(v-d-n)$.

Combining Case 1 and Case 2, we come to

$$p^n_{1,n} = (d-n)n + n(v-d-n) = n(v-2n),$$

which completes the proof. □

6.2 Asymptotic Spectral Distributions in the Vacuum State

We now apply the general theory established in Sect. 3.4. According to (3.18) and (3.19) we set

$$\bar{\omega}_n(v,d) = \frac{p^n_{1,n-1} p^{n-1}_{1,n}}{\kappa} = \frac{n^2(d-n+1)(v-d-n+1)}{d(v-d)}$$
$$= n^2 \frac{\left(\dfrac{d}{v} - \dfrac{n-1}{v}\right)\left(1 - \dfrac{d}{v} - \dfrac{n-1}{v}\right)}{\dfrac{d}{v}\left(1 - \dfrac{d}{v}\right)},$$

$$\bar{\alpha}_n(v,d) = \frac{p^{n-1}_{1,n-1}}{\sqrt{\kappa}} = \frac{(n-1)(v-2(n-1))}{\sqrt{d(v-d)}}$$
$$= (n-1) \frac{1 - \dfrac{2(n-1)}{v}}{\sqrt{\dfrac{d}{v}\left(1 - \dfrac{d}{v}\right)}}.$$

Hereafter, we assume $2d \le v$ without loss of generality by virtue of Lemma 6.2. Then the proper scaling for the limit is given as

$$v \to \infty, \quad d \to \infty, \quad \frac{2d}{v} \to p \in (0, 1],$$

so that

$$\omega_n = \lim_{\substack{v,d\to\infty \\ 2d/v\to p}} \bar{\omega}_n(v, d) = n^2, \tag{6.2}$$

$$\alpha_n = \lim_{\substack{v,d\to\infty \\ 2d/v\to p}} \bar{\alpha}_n(v, d) = \frac{2(n-1)}{\sqrt{p(2-p)}}. \tag{6.3}$$

Thus we come to the following:

Theorem 6.10 (QCLT for Johnson graphs). *Let $J(v, d)$ be a Johnson graph and $A_{v,d}$ its adjacency matrix regarded as an algebraic random variable with respect to the vacuum state corresponding to a fixed origin in $J(v, d)$. Let $(\Gamma, \{\Psi_n\}, B^+, B^-)$ be the interacting Fock space associated with the Jacobi sequence $\{\omega_n = n^2\}$. For $0 < p \le 1$ define*

$$\alpha_n = \frac{2(n-1)}{\sqrt{p(2-p)}}, \qquad n = 1, 2, \dots,$$

and set $B^\circ = \alpha_{N+1}$, where N is the number operator defined by $N\Psi_n = n\Psi_n$. Then, for the quantum components $A^\epsilon_{v,d}$ we have

$$\lim_{\substack{d,v\to\infty \\ 2d/v\to p}} \frac{A^\epsilon_{v,d}}{\sqrt{d(v-d)}} = B^\epsilon, \qquad \epsilon \in \{+, -, \circ\},$$

in the sense of stochastic convergence. In particular,

$$\lim_{\substack{d,v\to\infty \\ 2d/v\to p}} \left\langle \left(\frac{A_{d,v}}{\sqrt{d(v-d)}} \right)^m \right\rangle_o$$
$$= \left\langle \Psi_0, \left(B^+ + B^- + \frac{2N}{\sqrt{p(2-p)}} \right)^m \Psi_0 \right\rangle, \quad m = 1, 2, \dots, \tag{6.4}$$

where o is an arbitrarily chosen origin of $J(v, d)$.

We are now interested in a probability distribution $\mu \in \mathfrak{P}_{\mathrm{fm}}(\mathbb{R})$, the mth moment of which coincides with (6.4). The Jacobi coefficient of μ is readily known:

$$\omega_n = n^2, \qquad \alpha_n = \frac{2(n-1)}{\sqrt{p(2-p)}}, \qquad n = 1, 2, \dots.$$

Therefore, the associated orthogonal polynomials $\{P_n(x)\}$ obey the following recurrence relation:

$$\begin{aligned} &P_0(x) = 1, \\ &P_1(x) = x, \\ &xP_n(x) = P_{n+1}(x) + \frac{2n}{\sqrt{p(2-p)}} P_n(x) + n^2 P_{n-1}(x), \quad n = 1, 2, \dots. \end{aligned} \tag{6.5}$$

We shall study these polynomials in the following sections.

Remark 6.11. The operators $\{B^+, B^-, 2N+1\}$ satisfy the commutation relations:

$$[B^-, B^+] = 2N + 1, \quad [2N+1, B^+] = 2B^+, \quad [2N+1, B^-] = -2B^-.$$

Therefore, the correspondences

$$\begin{pmatrix} 0 & 1 \\ 0 & 0 \end{pmatrix} \mapsto B^+, \quad \begin{pmatrix} 0 & 0 \\ -1 & 0 \end{pmatrix} \mapsto B^-, \quad \begin{pmatrix} 1 & 0 \\ 0 & -1 \end{pmatrix} \mapsto 2N+1,$$

give rise to an action of $\mathfrak{sl}(2, \mathbb{C})$ on Γ.

6.3 Exponential Distribution and Laguerre Polynomials

Definition 6.12. The *exponential distribution* with parameter $\lambda > 0$ is a probability distribution on $[0, +\infty)$ whose density function is given by $\lambda e^{-\lambda x}$ for $x \in [0, +\infty)$.

The mean of the exponential distribution with parameter $\lambda > 0$ is λ^{-1}. Unless otherwise stated, in the following the exponential distribution will mean the one with parameter $\lambda = 1$.

Connection of exponential distributions with the Laguerre polynomials is well known. We define the *Laguerre polynomial* $L_n(x)$ by

$$L_n(x) = \frac{1}{n!} e^x \frac{d^n}{dx^n} (x^n e^{-x}), \qquad n = 0, 1, 2, \dots. \tag{6.6}$$

The Leibniz formula yields

$$L_n(x) = \sum_{k=0}^{n} \binom{n}{k} \frac{(-x)^k}{k!}. \tag{6.7}$$

Using (6.6) and (6.7), we get through elementary computation the following characteristic properties of the Laguerre polynomials.

Lemma 6.13. *$L_n(x)$ is a polynomial of degree n and satisfies the orthogonality relation:*

$$\int_0^{+\infty} L_m(x) L_n(x) e^{-x} dx = \delta_{mn}, \qquad m, n = 0, 1, 2, \dots.$$

Lemma 6.14. *The generating function of the Laguerre polynomials* $\{L_n(x)\}$ *is given by*

$$\sum_{n=0}^{\infty} L_n(x) z^n = \frac{1}{1-z} \exp\left(-\frac{xz}{1-z}\right). \tag{6.8}$$

Lemma 6.15. *The Laguerre polynomials* $\{L_n(x)\}$ *obey the following recurrence relation:*

$$\begin{aligned} L_0(x) &= 1, \\ L_1(x) &= -x + 1, \\ -xL_n(x) &= (n+1)L_{n+1}(x) - (2n+1)L_n(x) + nL_{n-1}(x), \quad n = 1, 2, \dots. \end{aligned}$$

We see from Lemma 6.13 that the Laguerre polynomials $\{L_n(x)\}$ form an orthogonal set with respect to the exponential distribution; however, they do not fulfill our normalization (the orthogonal polynomials should be monic). As is easily seen from (6.7), the leading term of $L_n(x)$ is $(-x)^n/n!$ so that the orthogonal polynomials associated with the exponential distribution are given by

$$\begin{aligned} \tilde{L}_n(x) &= (-1)^n n! \, L_n(x) \\ &= (-1)^n e^x \frac{d^n}{dx^n}(x^n e^{-x}), \qquad n = 0, 1, 2, \dots. \end{aligned} \tag{6.9}$$

Proposition 6.16. *The orthogonal polynomials* $\{\tilde{L}_n(x)\}$ *associated with the exponential distribution obey the following recurrence relation:*

$$\begin{aligned} \tilde{L}_0(x) &= 1, \\ \tilde{L}_1(x) &= x - 1, \\ x\tilde{L}_n(x) &= \tilde{L}_{n+1}(x) + (2n+1)\tilde{L}_n(x) + n^2 \tilde{L}_{n-1}(x), \quad n = 1, 2, \dots. \end{aligned} \tag{6.10}$$

In particular, the Jacobi coefficients $(\{\omega_n\}, \{\alpha_n\})$ *of the exponential distribution are given by*

$$\omega_n = n^2, \quad \alpha_n = 2n - 1, \qquad n = 1, 2, \dots.$$

Proof. The recurrence relation (6.10) follows immediately from (6.9) and Lemma 6.15. The rest is then apparent. □

We readily obtain the following:

Theorem 6.17. *Let* $(\Gamma, \{\Psi_n\}, B^+, B^-)$ *be the interacting Fock space associated with Jacobi sequence* $\{\omega_n = n^2\}$. *Then, for* $m = 1, 2, \dots$ *we have*

$$\langle \Psi_0, (B^+ + B^- + 2N + 1)^m \Psi_0 \rangle = \int_0^{+\infty} x^m e^{-x} dx, \tag{6.11}$$

where N *is the number operator defined by* $N\Psi_n = n\Psi_n$.

Let us go back to Theorem 6.10 and consider the case of $p = 1$. Then (6.4) becomes

$$\lim_{\substack{d,v\to\infty \\ 2d/v\to 1}} \left\langle \left(\frac{A_{d,v}}{\sqrt{d(v-d)}} \right)^m \right\rangle_o = \langle \Psi_0, (B^+ + B^- + 2N)^m \Psi_0 \rangle, \tag{6.12}$$

for $m = 1, 2, \ldots$. Then, comparing (6.11) and (6.12), we obtain the following:

Proposition 6.18. *Let $A_{v,d}$ be the adjacency matrix of the Johnson graph $J(v,d)$. Then we have*

$$\lim_{\substack{d,v\to\infty \\ 2d/v\to 1}} \left\langle \left(\frac{A_{v,d}}{\sqrt{d(v-d)}} \right)^m \right\rangle_o = \int_{-1}^{+\infty} x^m \mathrm{e}^{-(x+1)} dx, \quad m = 1, 2, \ldots.$$

6.4 Geometric Distribution and Meixner Polynomials

Having obtained the explicit form of the probability distribution specified in Theorem 6.10 for $p = 1$, we shall discuss in this section the case of $0 < p < 1$.

Before going into the main argument we mention some well-known results on the Pascal distribution and the Meixner polynomials.

Definition 6.19. For $0 < c < 1$ and $\beta > 0$, the *negative binomial distribution* or the *Pascal distribution* with parameters β, c is defined by

$$(1-c)^\beta \sum_{k=0}^{\infty} \frac{c^k}{k!} \frac{\Gamma(\beta+k)}{\Gamma(\beta)} \delta_k. \tag{6.13}$$

If $\beta = 1$, then (6.13) becomes

$$(1-c) \sum_{k=0}^{\infty} c^k \delta_k,$$

which is called the *geometric distribution* with parameter c.

Definition 6.20. Let $\beta > 0$ and $0 < c < 1$. The polynomials $\{M_n(x; \beta, c) = M_n(x)\}$ uniquely specified by

$$\begin{aligned} M_0(x) &= 1, \\ M_1(x) &= \left(1 - \frac{1}{c}\right) x + \beta, \\ (1-c) x M_n(x) &= -c M_{n+1}(x) + \{(1+c)n + c\beta\} M_n(x) \\ &\quad - n(n+\beta-1) M_{n-1}(x), \quad n = 1, 2, \ldots, \end{aligned} \tag{6.14}$$

are called the *Meixner polynomials* (of the first kind) with parameters β, c.

Proposition 6.21. *The Meixner polynomials $\{M_n(x;\beta,c)\}$ form an orthogonal set with respect to the negative binomial distribution with parameters β, c. More precisely,*

$$(1-c)^\beta \sum_{k=0}^\infty M_n(k;\beta,c)M_m(k,\beta,c)\,\frac{c^k}{k!}\,\frac{\Gamma(\beta+k)}{\Gamma(\beta)} = 0, \quad m \neq n,$$

$$(1-c)^\beta \sum_{k=0}^\infty \{M_n(k;\beta,c)\}^2\,\frac{c^k}{k!}\,\frac{\Gamma(\beta+k)}{\Gamma(\beta)} = \frac{n!}{c^n}\,\frac{\Gamma(\beta+n)}{\Gamma(\beta)}.$$

The generating function of the Meixner polynomials is given by

$$\sum_{n=0}^\infty M_n(x;\beta,c)\,\frac{z^n}{n!} = \left(1-\frac{z}{c}\right)^x (1-z)^{-x-\beta}. \tag{6.15}$$

It follows from (6.15) that

$$M_n(x;\beta,c) = (-1)^n n! \sum_{k=0}^n \binom{x}{k}\binom{-x-\beta}{n-k} c^{-k}.$$

Hence the leading coefficient of $M_n(x;\beta,c)$ is

$$(-1)^n n! \sum_{k=0}^n \frac{(-1)^{n-k}}{k!(n-k)!} c^{-k} = (-1)^n \left(\frac{1-c}{c}\right)^n.$$

Then

$$\tilde{M}_n(x) = (-1)^n \left(\frac{c}{1-c}\right)^n M_n(x)$$

becomes a monic polynomial. Thus we have the following:

Proposition 6.22. *The orthogonal polynomials $\{\tilde{M}_n(x)\}$ associated with the Pascal distribution* (6.13) *obey the recurrence relation:*

$$\begin{aligned}
\tilde{M}_0(x) &= 1, \\
\tilde{M}_1(x) &= x - \frac{\beta c}{1-c}, \\
x\tilde{M}_n(x) &= \tilde{M}_{n+1}(x) + \frac{(1+c)n+\beta c}{1-c}\tilde{M}_n(x) \\
&\quad + \frac{cn(n+\beta-1)}{(1-c)^2}\tilde{M}_{n-1}(x), \qquad n = 1, 2, \dots.
\end{aligned} \tag{6.16}$$

The Jacobi coefficient $(\{\tilde{\omega}_n\}, \{\tilde{\alpha}_n\})$ of the Pascal distribution (6.13) are thus given by

$$\tilde{\omega}_n = \frac{cn(n+\beta-1)}{(1-c)^2}, \quad \tilde{\alpha}_n = \frac{(1+c)(n-1)+\beta c}{1-c}, \qquad n = 1, 2, \dots .$$

Compare the Jacobi coefficient in Theorem 6.10 with $(\{\tilde{\omega}_n\}, \{\tilde{\alpha}_n\})$. Letting $\beta = 1$ and considering their affine transformations, we set

$$\omega_n = \left(\frac{1-c}{\sqrt{c}}\right)^2 \tilde{\omega}_n, \quad \alpha_n = \frac{1-c}{\sqrt{c}}\left(\tilde{\alpha}_n - \frac{c}{1-c}\right). \tag{6.17}$$

Then we have

$$\omega_n = n^2, \quad \alpha_n = \frac{1+c}{\sqrt{c}}(n-1), \qquad n = 1, 2, \dots . \tag{6.18}$$

Now we set

$$c = \frac{p}{2-p}.$$

Note that $0 < c < 1$ if and only if $0 < p < 1$. Then we have

$$\frac{1-c}{\sqrt{c}} = \frac{2(1-p)}{\sqrt{p(2-p)}}, \quad \frac{c}{1-c} = \frac{p}{2(1-p)}, \quad \frac{1+c}{\sqrt{c}} = \frac{2}{\sqrt{p(2-p)}}.$$

Consequently, we see that the Jacobi coefficient in Theorem 6.10 coincides with $(\{\omega_n\}, \{\alpha_n\})$ in (6.18). They are an affine transformation of the Jacobi coefficient associated with the geometric distribution as in (6.17).

Proposition 6.23. *The right-hand side of* (6.4) *is given by*

$$\left\langle \Psi_0, \left(B^+ + B^- + \frac{2N}{\sqrt{p(2-p)}}\right)^m \Psi_0 \right\rangle = \sum_{k=0}^{\infty} x_k^m \frac{2(1-p)}{2-p}\left(\frac{p}{2-p}\right)^k, \tag{6.19}$$

where

$$x_k = \frac{2(1-p)}{\sqrt{p(2-p)}}\left(k - \frac{p}{2(1-p)}\right). \tag{6.20}$$

Proof. The effect of affine transformation of a probability to its Jacobi coefficient was established in Proposition 1.49. The right-hand side of (6.19) is concerned with the affinely transformed geometric distribution according to (6.17). □

Summing up Propositions 6.18 and 6.23, we obtain the asymptotic spectral distribution for Johnson graphs.

Theorem 6.24 (CLT for Johnson graphs). *Let $A_{v,d}$ be the adjacency matrix of the Johnson graph $J(v,d)$ and let $0 < p \le 1$. Then there exists a unique probability measure μ satisfying*

$$\lim_{\substack{d,v\to\infty \\ 2d/v\to p}} \left\langle \left(\frac{A_{d,v}}{\sqrt{d(v-d)}}\right)^m \right\rangle_o = \int_{-\infty}^{+\infty} x^m \mu(dx), \qquad m = 1, 2, \dots .$$

The explicit form of μ is given as follows:

(1) *For* $p = 1$,

$$\mu(dx) = \begin{cases} e^{-(x+1)}dx, & x \in [-1, +\infty), \\ 0, & \textit{otherwise}. \end{cases}$$

(2) *For* $0 < p < 1$,

$$\mu = \sum_{k=0}^{\infty} \frac{2(1-p)}{2-p} \left(\frac{p}{2-p}\right)^k \delta_{x_k},$$

where x_k *is defined in* (6.20).

6.5 Asymptotic Spectral Distributions in the Deformed Vacuum States

Lemma 6.25. *A Johnson graph admits a quadratic embedding.*

Proof. Let $J(v, d)$ be a Johnson graph. Consider v dimensional Hilbert space $\mathbb{C}^v$ and fix a complete orthonormal basis $\{e_1, e_2, \dots, e_v\}$. Recall that a vertex $x \in V$ of the Johnson graph is a subset of $\{1, 2, \dots, v\}$ consisting of d elements. We set

$$f(x) = \frac{1}{\sqrt{2}} \sum_{i \in x} e_i, \qquad x \in V.$$

Then, for $x, y \in V$ we have

$$\|f(x) - f(y)\|^2 = \frac{1}{2} \sum_{i \in x \ominus y} \|e_i\|^2 = \frac{1}{2} |x \ominus y| = \partial(x, y),$$

which means that $f : V \to \mathbb{C}^v$ is a quadratic embedding. □

Remark 6.26. The map $f : V \to \mathbb{C}^v$ introduced in the above proof satisfies

$$\|f(x)\|^2 = \frac{1}{2} \sum_{i \in x} \|e_i\|^2 = \frac{|x|}{2} = \frac{d}{2}.$$

Hence $\{f(x)\,;\, x \in V\}$ lies on a sphere of radius $\sqrt{d/2}$.

Proposition 6.27. *For* $0 \le q \le 1$, *the kernel* $(q^{\partial(x,y)})$ *is positive definite on a Johnson graph.*

Proof. This is a consequence of Bożejko's quadratic embedding test (Proposition 2.14). □

It then follows from Theorem 3.27 that $Q = (q^{\partial(x,y)})$ gives rise to a deformed vacuum state for $0 \le q \le 1$, which is denoted by $\langle \cdot \rangle_q$. By using the general formula (3.30), the mean and the variance of the adjacency matrix $A_{v,d}$ in $\langle \cdot \rangle_q$ are easily obtained:

$$\langle A_{v,d}\rangle_q = qd(v-d),$$
$$\Sigma_q^2(A_{v,d}) = d(v-d)\{1+q(v-2)-q^2(v-1)\}.$$

We shall investigate the asymptotic behaviour of

$$\frac{A_{v,d}-\langle A_{v,d}\rangle_q}{\Sigma_q(A_{v,d})}, \tag{6.21}$$

where the limit is taken keeping a suitable balance of v, d, q.

Following the general argument in Sect. 3.4, we set

$$\begin{aligned}\bar{\omega}_n(v,d,q) &= \frac{n^2(d-n+1)(v-d-n+1)}{d(v-d)\{1+q(v-2)-q^2(v-1)\}}\\ &= \frac{n^2\left(\dfrac{d}{v}-\dfrac{n-1}{v}\right)\left(1-\dfrac{d}{v}-\dfrac{n-1}{v}\right)}{\dfrac{d}{v}\left(1-\dfrac{d}{v}\right)\{1+q(v-2)-q^2(v-1)\}},\end{aligned}$$

$$\begin{aligned}\bar{\alpha}_n(v,d,q) &= \frac{(n-1)(v-2(n-1))-qd(v-d)}{\sqrt{d(v-d)\{1+q(v-2)-q^2(v-1)\}}}\\ &= \frac{(n-1)\left(1-\dfrac{2(n-1)}{v}\right)-qd\left(1-\dfrac{d}{v}\right)}{\sqrt{\dfrac{d}{v}\left(1-\dfrac{d}{v}\right)\{1+q(v-2)-q^2(v-1)\}}},\end{aligned}$$

$$\begin{aligned}c_n(v,d,q) &= q^n\sqrt{\binom{d}{n}\binom{v-d}{n}}\\ &= \frac{(qv)^n}{n!}\sqrt{\frac{d}{v}\left(\frac{d}{v}-\frac{1}{v}\right)\cdots\left(\frac{d}{v}-\frac{n-1}{v}\right)}\\ &\quad\times\sqrt{\left(1-\frac{d}{v}\right)\left(1-\frac{d+1}{v}\right)\cdots\left(1-\frac{d+n-1}{v}\right)}.\end{aligned}$$

In order to obtain reasonable limits of the above three sequences, we need to take the limits as

$$v\to\infty, \quad d\to\infty, \quad \frac{2d}{v}\to p, \quad qd\to r, \tag{6.22}$$

where $0 < p \le 1$ and $r \ge 0$ are constants. Note also that, under (6.22),

$$q\to 0, \qquad qv\to\frac{2r}{p}. \tag{6.23}$$

We then easily obtain

$$\omega_n = \lim_{(6.22)} \bar{\omega}_n(v, d, q) = \frac{n^2 p}{p + 2r}, \tag{6.24}$$

$$\alpha_n = \lim_{(6.22)} \bar{\alpha}_n(v, d, q) = \frac{2(n-1) - r(2-p)}{\sqrt{(2-p)(p+2r)}}, \tag{6.25}$$

$$c_n = \lim_{(6.22)} c_n(v, d, q) = \frac{r^n}{n!} \left(\frac{2}{p} - 1 \right)^{n/2}. \tag{6.26}$$

We see immediately that the interacting Fock space describing the limit is, up to constant factors, the same as the one in Theorem 6.10. For simplicity of notations, we use $0 < s \le 1$ defined by

$$\frac{1}{s^2} = \frac{2}{p} - 1 = \frac{1}{c},$$

instead of p. Then (6.22) and (6.23) become

$$\begin{gathered} d \to \infty, \quad v \to \infty, \quad q \to 0, \\ \frac{2d}{v} \to \frac{2s^2}{1+s^2}, \quad qd \to r, \quad qv \to \frac{r(1+s^2)}{s^2}. \end{gathered} \tag{6.27}$$

Moreover, (6.24)–(6.26) become

$$\omega_n = \frac{s^2 n^2}{s^2(1+r) + r}, \tag{6.28}$$

$$\alpha_n = \frac{(1+s^2)(n-1) - r}{\sqrt{s^2(1+r) + r}}, \tag{6.29}$$

$$c_n = \frac{1}{n!} \left(\frac{r}{s} \right)^n. \tag{6.30}$$

In order to describe the limit (6.21), it is more convenient to use the Jacobi sequence $\{\omega_n = n^2\}$ rather than (6.28). Let $(\Gamma, \{\Psi_n\}, B^+, B^-)$ be the interacting Fock space associated with the Jacobi sequence $\{\omega_n = n^2\}$ and N the number operator as usual. Using the sequence in (6.30), we define a coherent vector

$$\Upsilon_{r/s} = \sum_{n=0}^{\infty} \frac{1}{n!} \left(\frac{r}{s} \right)^n \Psi_n, \tag{6.31}$$

which defines a deformed vacuum state.

With these notations we apply Theorem 3.29 to assert the following:

Theorem 6.28 (QCLT for Johnson graphs in the deformed vacuum state). *Let $A_{v,d}$ be the adjacency matrix of the Johnson graph $J(v,d)$ regarded* *as a random variable with respect to the deformed vacuum state* $\langle\cdot\rangle_q$. *Let* $0 < s \le 1$ *and* $r \ge 0$ *be constant numbers. Then, for the quantum components of* $A_{v,d}$ *we have*

$$\lim_{(6.27)} \frac{A^{\pm}_{v,d}}{\Sigma(A_{v,d})} = \frac{sB^{\pm}}{\sqrt{s^2(1+r)+r}},$$
$$\lim_{(6.27)} \frac{A^{\circ}_{v,d} - \langle A_{v,d}\rangle_q}{\Sigma(A_{v,d})} = \frac{(1+s^2)N - r}{\sqrt{s^2(1+r)+r}}, \tag{6.32}$$

where the right-hand sides are random variables in the interacting Fock space $(\Gamma, \{\Psi_n\}, B^+, B^-)$ *associated with a Jacobi sequence* $\{\omega_n = n^2\}$ *with respect to the coherent state* (6.31). *In particular, for* $m = 1, 2, \dots,$

$$\lim_{(6.27)} \left\langle \left(\frac{A_{v,d} - \langle A_{v,d}\rangle_q}{\Sigma_q(A_{v,d})} \right)^m \right\rangle_q$$
$$= \left\langle \left(\frac{s(B^+ + B^- + (s+s^{-1})N - r/s)}{\sqrt{s^2(r+1)+r}} \right)^m \right\rangle_{\Upsilon_{r/s}}. \tag{6.33}$$

Our goal is to obtain a probability measure μ satisfying

$$(6.33) = \int_{-\infty}^{+\infty} x^m \mu(dx), \qquad m = 1, 2, \dots.$$

The explicit computation is somewhat cumbersome; however, the idea is totally clear. We make much of the relevant combinatorial structure of the creation, annihilation and number operators. For notational simplicity, we do not give an explicit form of μ itself but its affine transformation.

Theorem 6.29 (CLT for Johnson graphs in the deformed vacuum state). *Let* $(\Gamma, \{\Psi_n\}, B^+, B^-)$ *be the interacting Fock space associated with the Jacobi sequence* $\{\omega_n = n^2\}$. *For* $0 < s \le 1$ *and* $r \ge 0$ *there exists a unique probability measure* $\nu_{r,s}$ *satisfying*

$$\left\langle \left(B^+ + B^- + \left(s + \frac{1}{s} \right) N + s \right)^m \right\rangle_{\Upsilon_{r/s}} = \int_{-\infty}^{+\infty} x^m \, \nu_{r,s}(dx), \tag{6.34}$$

for $m = 1, 2, \dots$. *The explicit form of* $\nu_{r,s}$ *is as follows:*

(1) *For* $s = 1$,

$$\nu_{r,1}(dx) = \sum_{k=0}^{\infty} e^{-r} \frac{r^k}{k!} \rho_k(x)dx, \quad \rho_k(x) = \begin{cases} \dfrac{x^k}{k!} e^{-x}, & x \ge 0 \\ 0, & x < 0. \end{cases} \tag{6.35}$$

(2) *For* $0 < s < 1$,

$$\nu_{r,s} = \sum_{k=0}^{\infty} \mathrm{e}^{-r} \frac{r^k}{k!} \left\{ \sum_{n=0}^{\infty} \binom{k+n}{n} (1-s^2)^{k+1} s^{2n} \delta_{(s^{-1}-s)(k+n)} \right\}. \tag{6.36}$$

Proof. We give an outline. Let $G(m,r,s)$ denote the left-hand side of (6.34) and consider its expansion:

$$G(m,r,s) = \mathrm{e}^{-r} \sum_{k=0}^{\infty} \frac{r^k}{k!} b_{m,k}(s), \tag{6.37}$$

where $m = 0, 1, 2, \dots$, $r \geq 0$ and $0 < s \leq 1$. It suffices to find a probability measure whose mth moment is $b_{m,k}(s)$. The exponential generating function of $b_{m,k}(s)$ with respect to m will give the Laplace transform of the desired probability measure. To obtain the exponential generating function it is helpful to express $b_{m,k}(s)$ by means of a generating function in s of some sequence. This plan will be realized by introducing a certain combinatorial device as follows:

Step 1. The double sequence $\{b_{m,k}(s)\}$ satisfies the recurrence relation:

$$\begin{aligned} b_{0,k}(s) &= 1, \\ b_{m+1,k}(s) &= s(k+1) b_{m,k+1}(s) + (s^{-1} - s) k b_{m,k}(s), \end{aligned} \tag{6.38}$$

where m, k run over $\{0, 1, 2, \dots\}$ and $0 < s \leq 1$. In fact, set

$$B^\circ = (s + s^{-1}) N + s$$

for simplicity. From the definition of $\Upsilon_{r/s}$ we have

$$G(m,r,s) = \sum_{n=0}^{\infty} \frac{1}{n!} \left(\frac{r}{s}\right)^n K_{m,n},$$

where

$$K_{m,n} = \sum_{\epsilon_1, \dots, \epsilon_m \in \{+,-,\circ\}} \langle B^{\epsilon_1} \cdots B^{\epsilon_m} \Psi_0, \Psi_n \rangle.$$

In view of combinatorics among $\{B^+, B^-, B^\circ\}$, we obtain

$$\begin{aligned} K_{0,0} &= 1, \\ K_{m+1,n} &= (n+1) K_{m,n+1} + \big((s + s^{-1}) n + s\big) K_{m,n} + n K_{m,n-1}, \\ &\qquad n \in \{0, 1, \dots, m+1\}, \quad m = 0, 1, 2, \dots. \end{aligned}$$

This yields a functional equation satisfied by $F(m,r,s) = \mathrm{e}^r G(m,r,s)$:

$$\begin{aligned} F(0,r,s) &= \mathrm{e}^r, \\ F(m+1,r,s) &= rs F_{rr}(m,r,s) + (rs^{-1} + s - rs) F_r(m,r,s), \\ &\qquad m = 0, 1, 2, \dots. \end{aligned}$$

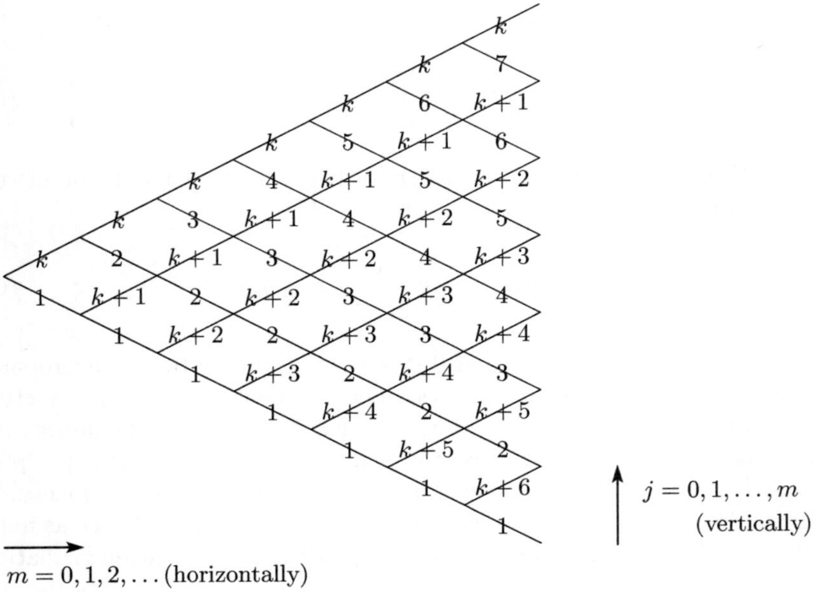

Fig. 6.2. Pascal-like triangle with edge multiplicities

We note from (6.37) that $F(m, r, s)$ is the exponential generating function for $b_{m,k}(s)$ with respect to k. Then, expanding the above equation, we obtain (6.38). For $s = 1$, (6.38) is easily solved. The solution is

$$b_{m,k}(1) = \frac{(k+m)!}{k!}.$$

Step 2. To find an object producing (6.38), we consider the Pascal-like triangle equipped with edge multiplicities as in Fig. 6.2, where $k = 0, 1, 2, \ldots$ is a fixed constant. In a strict wording, such a graph is called a *network*. Let u be a path starting at the root vertex $(0, 0)$ and terminating at an arbitrary vertex (m, j). The product w_u of all edge multiplicities along u is called the *weight* of the path u. Let $d_{m,j}(k)$ denote the sum of the weights of all paths connecting $(0, 0)$ with (m, j), i.e.,

$$d_{m,j}(k) = \sum_u w_u, \tag{6.39}$$

where u runs over all paths connecting the vertices $(0, 0)$ and (m, j). With this notation we readily have

$$\begin{aligned} &d_{m,0}(k) = 1, \\ &d_{m,m}(k) = k^m, \qquad m = 0, 1, 2, \ldots, \\ &d_{m,j}(k) = (j+1)d_{m-1,j}(k) + (k+m-j)d_{m-1,j-1}(k), \\ &\qquad\qquad\qquad\qquad j \in \{1, \ldots, m-1\}, \quad m = 2, 3, \ldots. \end{aligned} \tag{6.40}$$

On the other hand, we claim

$$b_{m,k}(s) = \sum_{j=0}^{m} d_{m,j}(k) s^{m-2j}, \quad m, k \in \{0, 1, 2, \dots\}, \quad 0 < s \le 1. \qquad (6.41)$$

Indeed, the right-hand side satisfies the recurrence relation (6.38) by observing the structure of our Pascal-like triangle.

Step 3. Consider the generating function of $d_{m,j}(k)$ and the exponential generating function of $b_{m,k}(s)$ defined, respectively, by

$$v_k(y, t) = \sum_{m=0}^{\infty} \frac{y^m}{m!} \sum_{j=0}^{m} d_{m,j}(k)\, t^j,$$
$$u_k(x, s) = \sum_{m=0}^{\infty} \frac{b_{m,k}(s)}{m!} x^m.$$

We see from (6.41) that

$$u_k(x, s) = v_k(sx, s^{-2}).$$

Combining this with a partial differential equation satisfied by v_k derived from (6.40), we come to a partial differential equation for $u = u_k$ as

$$2(k + s^2)u + (x + s^2 x - 2s)u_x + (s - s^3)u_s = 0,$$
$$u(0, s) = 1, \qquad 0 < s < 1, \quad -1 < x < 1.$$

This can be solved by the standard method of characteristic curve. The ordinary equations involved therein can be all solved by integration. As a result, we have

$$u_k(x, s) = \frac{1 - s^2}{1 - s^2 e^{(s^{-1}-s)x}} \left(e^{(s^{-1}-s)x} \frac{1 - s^2}{1 - s^2 e^{(s^{-1}-s)x}} \right)^k,$$
$$0 < s < 1, \quad -1 < x < 1.$$

Finally, let $\rho_{k,s}$ be the probability measure of which the Laplace transform is $u_k(x, s)$. Since

$$\int_0^{+\infty} e^{tx} \rho_{k,s}(dt) = \sum_{m=0}^{\infty} \frac{b_{m,k}(s)}{m!} x^m,$$

we see that $\rho_{k,s}$ is the desired probability measure. The proof is completed by checking the Laplace transforms of (6.35) and (6.36). □

Remark 6.30. The probability measures (6.35) and (6.36) are the compound Poisson distributions of gamma and Pascal (with parameter $k + 1, s^2$) distributions, respectively. The random variables that obey such distributions are easily constructed. Let $X_0, X_1, \dots$ and $Y_0, Y_1, \dots$ be two sequences of independent random variables, where X_i obeys the exponential distribution with parameter 1 and Y_i the geometric distribution $\sum_{n=0}^{\infty}(1-s^2)s^{2n}\delta_{(s^{-1}-s)n}$. Let M be a random variable that is independent of X_i's and Y_i's, and obeys the Poisson distribution with parameter r. Then we have

$$\nu_{r,1} \sim X_0 + X_1 + \cdots + X_M,$$
$$\nu_{r,s} \sim Y_0 + (Y_1 + s^{-1} - s) + \cdots + (Y_M + s^{-1} - s).$$

It is noted that $\nu_{r,s}$ converges to $\nu_{r,1}$ weakly as $s \to 1-0$.

Remark 6.31. The quantity $d_{m,j}(k)$ defined in (6.39) is called the *combinatorial dimension function* of the Pascal-like triangle shown in Fig. 6.2. We shall meet such a function more closely in Sect. 12.2 for the Jack graph.

6.6 Odd Graphs

There are some similarities between the Johnson graphs and odd graphs.

Definition 6.32. Let $k \geq 2$ be an integer and set $S = \{1, 2, \dots, 2k-1\}$. The pair

$$V = \{x \subset S\,;\, |x| = k-1\}, \quad E = \{\{x,y\}\,;\, x, y \in V,\, x \cap y = \varnothing\}$$

is called the *odd graph* and is denoted by O_k.

Obviously, O_k is a regular graph of degree k. Note that O_2 is a triangle and O_3 is the Petersen graph (Fig. 2.1).

As in the case of the Johnson graph, the distance between two vertices of an odd graph is characterized by the cardinality of their intersection. Set

$$I_n = \begin{cases} k - 1 - \dfrac{n}{2}, & \text{if } n \text{ is even,} \\ \dfrac{n-1}{2}, & \text{if } n \text{ is odd,} \end{cases} \tag{6.42}$$

where $n = 0, 1, \dots, k-1$. Note that I gives a bijection from $\{0, 1, \dots, k-1\}$ onto itself.

Proposition 6.33. *For a pair of vertices x, y of the odd graph O_k, we have*

$$|x \cap y| = I_n \iff \partial(x,y) = n.$$

Proof. We use a counting argument as in the case of Johnson graphs though the situation is a bit more complicated. Set

$$E_n = \{(x,y) \in V \times V\,;\, |x \cap y| = I_n\},$$
$$F_n = \{(x,y) \in V \times V\,;\, \partial(x,y) = n\}.$$

The assertion is equivalent that $E_n = F_n$ holds for any n. The proof is performed by appealing to induction on n. □

Corollary 6.34. diam $(O_k) = k - 1$.

Corollary 6.35. *The odd graph O_k is distance-transitive and therefore distance-regular.*

Proof. A bijection $\pi : S \to S$ induces a bijection $\tilde{\pi} : V \to V$ in a natural manner. Since $|x \cap y|$ is kept invariant under $\tilde{\pi}$, we see that $\tilde{\pi}$ becomes an automorphism of the graph O_k. Now let $x, y, x', y' \in V$ be such that $\partial(x, y) = \partial(x', y') = n$. By Proposition 6.33 we may set

$$x = \{\alpha_1, \dots, \alpha_I, \beta_1, \dots, \beta_J\}, \quad y = \{\alpha_1, \dots, \alpha_I, \gamma_1, \dots, \gamma_J\},$$

where $I = I_n$, $I + J = k - 1$, $\{\beta_1, \dots, \beta_J\} \cap \{\gamma_1, \dots, \gamma_J\} = \emptyset$. Similarly,

$$x' = \{\alpha'_1, \dots, \alpha'_I, \beta'_1, \dots, \beta'_J\}, \quad y' = \{\alpha'_1, \dots, \alpha'_I, \gamma'_1, \dots, \gamma'_J\}.$$

Take a bijection $\pi : S \to S$ satisfying

$$\pi(\alpha_i) = \alpha'_i, \quad \pi(\beta_i) = \beta'_i, \quad \pi(\gamma_i) = \gamma'_i.$$

Then the automorphism $\tilde{\pi}$ satisfies $\tilde{\pi}(x) = x'$ and $\tilde{\pi}(y) = y'$ and hence O_k is distance-transitive. □

Having observed that the odd graphs $\{O_k\}$ form a family of growing distance-regular graphs, we shall investigate the asymptotic spectral distribution of the adjacency matrix A_k as $k \to \infty$ by applying quantum probabilistic techniques (Theorem 3.21). Our first task is to compute the intersection numbers $p^h_{ij} = p^h_{ij}(k)$ of O_k required in Theorem 3.21.

Proposition 6.36. (1) *For $1 \le n \le k - 1$,*

$$p^n_{1,n-1} = \begin{cases} \dfrac{n}{2}, & \textit{if } n \textit{ is even}, \\ \dfrac{n+1}{2}, & \textit{if } n \textit{ is odd}. \end{cases}$$

(2) *For $0 \le n \le k - 2$,*

$$p^n_{1,n+1} = \begin{cases} k - \dfrac{n}{2}, & \textit{if } n \textit{ is even}, \\ k - \dfrac{n+1}{2}, & \textit{if } n \textit{ is odd}. \end{cases}$$

(3) *For $0 \le n \le k - 1$,*

$$p^n_{1,n} = \begin{cases} 0, & \textit{if } 1 \le n \le k-2, \\ \dfrac{k+1}{2}, & \textit{if } n = k-1 \textit{ and } k \textit{ is odd}, \\ \dfrac{k}{2}, & \textit{if } n = k-1 \textit{ and } k \textit{ is even}. \end{cases}$$

Proof. Just a routine application of Proposition 6.33. We shall prove only (1) for an even n. Let n be an even number such that $1 \le n \le k-1$. Without loss of generality we set

$$o = \{1, 2, \dots, k-1\}, \qquad x = \{1, 2, \dots, I_n, k, k+1, 2k - I_n - 2\}.$$

Then $|o \cap x| = I_n$ so that $\partial(o, x) = n$. Let us find a general form of y such that $\partial(x, y) = 1$ and $\partial(o, y) = n - 1$. In order that $\partial(x, y) = 1$ we have by definition

$$y \subset \{I_n + 1, \dots, k-1\} \cup \{2k - I_n - 1, \dots, 2k - 1\}. \tag{6.43}$$

Since $|y| = k - 1$ and $|\{I_n + 1, \dots, k-1\} \cup \{2k - I_n - 1, \dots, 2k - 1\}| = k$, the vertex y is obtained by deleting one element from the right-hand side of (6.43). There are two cases:

Case 1. If y is obtained by deleting one element in $\{I_n + 1, \dots, k-1\}$, we have

$$|o \cap y| = k - I_n - 2 = \frac{n}{2} - 1 = \frac{(n-1) - 1}{2} = I_{n-1},$$

where we used the assumption of n being even. Hence $\partial(o, y) = n - 1$.

Case 2. If y is obtained by deleting one element in $\{2k - I_n - 1, \dots, 2k-1\}$, we have

$$|o \cap y| = k - I_n - 1 = \frac{n}{2} \neq I_{n-1},$$

which means that $\partial(o, y) \neq n - 1$.

Consequently, a vertex y satisfying $\partial(x, y) = 1$ and $\partial(o, y) = n - 1$ is obtained only in Case 1 and the number of such y's is

$$(k-1) - (I_n + 1) + 1 = k - I_n - 1 = \frac{n}{2}.$$

This proves that $p^n_{1,n-1} = n/2$ for an even n. □

We are now in a position to state the quantum central limit theorem for the odd graphs $\{O_k\}$.

Theorem 6.37 (QCLT for odd graphs). *Let A_k be the adjacency matrix of the odd graph O_k and A_k^ϵ its quantum components, $\epsilon \in \{+, -, \circ\}$. Let $\Gamma_{\{\omega_n\}} = (\Gamma, \{\Psi_n\}, B^+, B^-)$ be the interacting Fock space associated with a Jacobi sequence defined by*

$$\{\omega_n\} = \{1, 1, 2, 2, 3, 3, 4, 4, \dots\}. \tag{6.44}$$

It then holds that

$$\lim_{k \to \infty} \frac{A_k^{\pm}}{\sqrt{k}} = B^{\pm}, \qquad \lim_{k \to \infty} \frac{A_k^{\circ}}{\sqrt{k}} = 0, \tag{6.45}$$

in the sense of stochastic convergence.

Proof. Let $\{p_{ij}^h(k)\}$ denote the intersection numbers of O_k, $k \geq 2$. In view of Theorem 3.21 we need only to find the limits

$$\omega_n = \lim_{k\to\infty} \frac{p_{1,n-1}^n(k)p_{1,n}^{n-1}(k)}{p_{11}^0(k)}, \qquad \alpha_n = \lim_{k\to\infty} \frac{p_{1,n-1}^{n-1}(k)}{\sqrt{p_{11}^0(k)}},$$

which are computed with the help of Proposition 6.36. In fact, for an odd n,

$$\omega_n = \lim_{k\to\infty} \frac{1}{k}\frac{n+1}{2}\left(k - \frac{n-1}{2}\right) = \frac{n+1}{2},$$

and for an even n,

$$\omega_n = \lim_{k\to\infty} \frac{1}{k}\frac{n}{2}\left(k - \frac{n}{2}\right) = \frac{n}{2}.$$

Thus, we obtain (6.44). Similarly, $\{\alpha_n \equiv 0\}$ is proved. □

We now give an intermediate answer to our main question for the odd graphs. (The complete answer will be given in Theorem 6.39.)

Proposition 6.38. *Let A_k be the adjacency matrix of the odd graph O_k. Then there exists a unique probability measure $\mu \in \mathfrak{P}_{\mathrm{fm}}(\mathbb{R})$ such that*

$$\lim_{k\to\infty} \left\langle \left(\frac{A_k}{\sqrt{k}}\right)^m \right\rangle_o = \int_{-\infty}^{+\infty} x^m \mu(dx), \qquad m = 1, 2, \dots. \tag{6.46}$$

The Jacobi coefficient of μ is given by

$$\{\omega_n\} = \{1, 1, 2, 2, 3, 3, 4, 4, \dots\}, \qquad \{\alpha_n \equiv 0\}.$$

In particular, μ is symmetric.

Proof. We maintain the notation in Theorem 6.37. Taking $A_k = A_k^+ + A_k^- + A_k^\circ$ into account, we see from Theorem 6.37 that

$$\lim_{k\to\infty} \left\langle \delta_o, \left(\frac{A_k}{\sqrt{k}}\right)^m \delta_o \right\rangle = \langle \Psi_0, (B^+ + B^-)^m \Psi_0 \rangle, \qquad m = 1, 2, \dots.$$

Since $\{\omega_n\}$ satisfies Carleman's condition in Theorem 1.66, there exists a unique Borel probability measure μ on $\mathbb{R}$ such that

$$\langle \Psi_0, (B^+ + B^-)^m \Psi_0 \rangle = \int_{-\infty}^{+\infty} x^m \mu(dx), \qquad m = 1, 2, \dots.$$

Thus, μ in (6.46) is uniquely determined. That μ is symmetric follows from $\{\alpha_n \equiv 0\}$ and the uniqueness (Proposition 1.48). □

We shall obtain an explicit description of the Borel probability measure μ in Proposition 6.38, where the Jacobi coefficient $(\{\omega_n\},\{\alpha_n\})$ is already obtained. Since μ is uniquely determined by its moment sequence, Theorem 1.97 ensures that the Stieltjes transform of μ admits a convergent continued fraction expansion:

$$G(z) = \int_{-\infty}^{+\infty} \frac{\mu(dx)}{z-x} = \frac{1}{z} - \frac{1}{z} - \frac{1}{z} - \frac{2}{z} - \frac{2}{z} - \frac{3}{z} - \frac{3}{z} - \frac{4}{z} - \frac{4}{z} - \cdots, \tag{6.47}$$

where $z \in \{\mathrm{Im}\, z \neq 0\}$.

Let us compute the continued fraction (6.47). For $n = 1, 2, \dots$ we define a linear fractional transformation:

$$\sigma_n(w) = \frac{n}{z-w}.$$

Then the $2n$th approximant is obtained by

$$G_{2n}(z) = \frac{1}{z} - \overbrace{\frac{1}{z} - \frac{1}{z} - \frac{2}{z} - \frac{2}{z} - \cdots - \frac{n}{z} - \frac{n}{z}}^{2n \text{ terms}} = \sigma_1 \sigma_1^2 \cdots \sigma_n^2(0). \tag{6.48}$$

On the other hand, using

$$\sigma_n^2(w) = \frac{n}{z - \sigma_n(w)} = \frac{n}{z} + \frac{n^2/z}{z^2 - n - zw}, \tag{6.49}$$

we obtain

$$\begin{aligned} &\sigma_1^2 \cdots \sigma_n^2(0) \\ &= \frac{1}{z}\left\{1 + \frac{1^2}{z^2-3} - \frac{2^2}{z^2-5} - \frac{3^2}{z^2-7} - \cdots - \frac{(n-1)^2}{z^2-(2n-1)} - \frac{n^2}{z^2-n}\right\}. \end{aligned}$$

Inserting this into (6.48), we have

$$\begin{aligned} &G_{2n}(z) \\ &= \frac{z}{z^2-1} - \frac{1^2}{z^2-3} - \frac{2^2}{z^2-5} - \frac{3^2}{z^2-7} - \cdots - \frac{(n-1)^2}{z^2-(2n-1)} - \frac{n^2}{z^2-n}. \end{aligned}$$

Since the continued fraction in (6.47) converges in $\{\mathrm{Im}\, z \neq 0\}$, so does

$$\begin{aligned} G(z) &= \lim_{n\to\infty} G_{2n}(z) \\ &= \frac{z}{z^2-1} - \frac{1^2}{z^2-3} - \frac{2^2}{z^2-5} - \frac{3^2}{z^2-7} - \cdots - \frac{(n-1)^2}{z^2-(2n-1)} - \cdots. \end{aligned} \tag{6.50}$$

On the other hand, we know from Proposition 6.16 the Stieltjes transform of the exponential distribution:

$$\int_0^{+\infty} \frac{\mathrm{e}^{-x}}{z-x}\,dx = \frac{1}{z-1}-\frac{1^2}{z-3}-\frac{2^2}{z-5}-\frac{3^2}{z-7}-\cdots, \tag{6.51}$$

which converges in $\mathbb{C} \setminus [0, +\infty)$. Comparing (6.50) and (6.51), we obtain

$$G(z) = \int_0^{+\infty} \frac{z\mathrm{e}^{-x}}{z^2-x}\,dx.$$

Then, replacing x with x^2, we have

$$\begin{aligned} G(z) &= \int_0^{+\infty} \frac{2xz\mathrm{e}^{-x^2}}{z^2-x^2}\,dx \\ &= \int_0^{+\infty} \frac{x\mathrm{e}^{-x^2}}{z-x}\,dx + \int_{-\infty}^{0} \frac{-x\mathrm{e}^{-x^2}}{z-x}\,dx = \int_{-\infty}^{+\infty} \frac{|x|\mathrm{e}^{-x^2}}{z-x}\,dx. \end{aligned} \tag{6.52}$$

Thus, the probability measure μ in (6.47) is given by

$$\mu(dx) = |x|\mathrm{e}^{-x^2}dx, \tag{6.53}$$

which is referred to as the *two-sided Rayleigh distribution.* Using the above explicit form, we rephrase Proposition 6.38 as follows:

Theorem 6.39 (CLT for odd graphs). *For the adjacency matrix A_k of the odd graph O_k we have*

$$\lim_{k\to\infty} \left\langle \left(\frac{A_k}{\sqrt{k}}\right)^m \right\rangle_o = \int_{-\infty}^{+\infty} x^m |x| \mathrm{e}^{-x^2}dx, \qquad m = 1, 2, \dots. \tag{6.54}$$

Proposition 6.40. *For the two-sided Rayleigh distribution* (6.53) *the moments of odd orders vanish and those of even orders are given by*

$$\int_{-\infty}^{+\infty} x^{2m}|x|\mathrm{e}^{-x^2}dx = m!, \qquad m = 0, 1, 2, \dots.$$

Proof. Elementary calculus. □

Exercises

6.1. The Laguerre polynomials are limit cases of the Meixner polynomials. Show that

$$\lim_{c\to 1} M_n\left(\frac{x}{1-c}; 1, c\right) = n! L_n(x).$$

6.2. Let $\mu \in \mathfrak{P}_{\mathrm{fm}}(\mathbb{R})$ and $\{M_m\}$ be its moment sequence. Assume that μ is supported by $[0, +\infty)$ and the solution to a determinate moment problem.

Show that there exists a unique probability distribution $\tilde{\mu} \in \mathfrak{P}_{\mathrm{fm}}(\mathbb{R})$ such that

$$\int_{-\infty}^{+\infty} x^{2m}\tilde{\mu}(dx) = M_m, \qquad \int_{-\infty}^{+\infty} x^{2m+1}\tilde{\mu}(dx) = 0, \qquad m = 0, 1, 2, \dots.$$

Show that if $\mu(dx) = \rho(x)dx$ then $\tilde{\mu}(dx) = |x|\rho(x^2)dx$.

6.3. Consider $J(v, d)$. A vertex x is a subset of $\{1, 2, \dots, v\}$ having cardinality d. We associate a sequence $(\xi_1, \xi_2, \dots, \xi_v)$ by

$$\xi_i = \begin{cases} 1 & \text{if } i \in x, \\ 0 & \text{if } i \notin x. \end{cases}$$

Thus, we obtain an injective map $x \mapsto (\xi_1, \xi_2, \dots, \xi_v) \in \{0, 1\}^v$. On the other hand, with each $(\xi_1, \xi_2, \dots, \xi_v) \in \{0, 1\}^v$ we associate a v-step walk in the first quadrant of $\mathbb{Z}^2$ (or the Pascal triangle) starting from $(0, 0)$ by indicating a direction $(1, 0)$ or $(0, 1)$ according as $\xi = 1$ or $\xi = 0$. Then every walk associated with a vertex of the Johnson graph $J(v, d)$ starts from $(0, 0)$ and reaches $(v, d - v)$. Let $P(v, w)$ be the set of shortest walk connecting $(0, 0)$ and (v, w) in $\mathbb{Z}^2$. Show that the correspondence $x \mapsto (\xi_1, \xi_2, \dots, \xi_v)$ induces a bijection between V and $P(v, d - v)$. Moreover, prove that

$$\partial(x, y) = \frac{1}{2}|\{i\,;\ \xi_i \neq \eta_i\}|.$$

6.4. Prove that $I : n \mapsto I_n$ defined in (6.42) is a bijection from $\{0, 1, \dots, k-1\}$ onto itself. Then prove that the inverse map is given by

$$I^{-1}(i) = \min\{2(k - 1 - i),\ 2i + 1\}.$$

6.5. Complete the proof of Proposition 6.33.

6.6 (q-analogue of Johnson graph). Let $v, d \geq 1$ be a pair of integers satisfying $0 \leq 2d \leq v$, and consider a v-dimensional vector space X over a finite field F_q of order $q \geq 2$. The *q-analogue of Johnson graph* $J_q(v, d)$ is a graph (V, E), where

$$V = \{x \subset X\,;\ d\text{-dimensional subspace}\},$$
$$E = \{\{x, y\}\,;\ x, y \in V,\ \dim(x \cap y) = d - 1\}.$$

(1) Show that $\partial(x, y) = d - \dim(x \cap y)$ holds for $x, y \in V$.
(2) Show that $J_q(v, d)$ is a distance-transitive graph.

Notes

Detailed structure of the Johnson graph is found in Bannai–Ito [17] and Brouwer–Cohen–Neumaier [50]. The spectrum of the Johnson graph $J(v,d)$ for $2d \le v$ is given by

$$\begin{aligned} \lambda_j &= d(v-d) - j(v-j+1), \\ w_j &= \frac{v-2j+1}{v-j+1}\binom{v}{j}, \end{aligned} \qquad j = 0, 1, \dots, d, \tag{6.55}$$

see, e.g., Bannai–Ito [17, Sect. 3.2] and Brouwer–Cohen–Neumaier [50, Sect. 9.1]. The Johnson graph appears in classical probability models too. For example, the simple random walk on the Johnson graph is the classical model of Bernoulli–Laplace, which imitates diffusion of sparse gases.

There are slightly different definitions of the Laguerre polynomials. Our definition owes to Szegö [202] and Chihara [56]. Schoutens [186] provides a compact treatment of various orthogonal polynomials including the Meixner ones in Sect. 6.4.

Asymptotic spectral distributions of the Johnson graphs in the vacuum state were first obtained by Hora [102] by using their spectral data (6.55). By a similar method the case of deformed vacuum states was discussed by Hora [104] again. Quantum central limit theorems developed in Sects. 6.2–6.4 are due to Hashimoto–Hora–Obata [96]. The method of quantum decomposition is much more transparent, in fact, some technical questions in Hora [102,104] are fully resolved. The result on asymptotic spectral distributions in the deformed vacuum states (Sect. 6.5) is due to Hora [105,107].

Several properties of the odd graphs are mentioned by Biggs [29,30]. The result in Sect. 6.6 is due to Igarashi–Obata [115]. Formula (6.51) is also derived as a particular case of the continued fraction expansion for the quotient of hypergeometric functions, see Wall [222, Chap. XVIII (92.7)].

The term 'two-sided Rayleigh distribution' for the probability measure μ in (6.47) is tentative. The term 'Rayleigh distribution' is found, for instance, in Papoulis [175, Sect. 4.3]. The orthogonal polynomials associated with the two-sided Rayleigh distribution are called *generalized Hermite polynomials with parameter* 1/2 by Chihara [56, Sect. 5.2], see also Szegö [202, Problems and Exercises 25].

It is noticeable that the quadratic Jacobi sequence $\omega_n = n^2$ emerges from growing Johnson graphs. Hashimoto [95] studied growing binary trees and got another example of a quadratic Jacobi sequence $\omega_n = 2(2n-1)(n+1)$. His problem is somehow beyond our framework but suggests some interesting future direction.

7

Regular Graphs

We have observed in the previous chapters that a distance-regular graph possesses a significant property from the viewpoint of quantum decomposition, namely, $\Gamma(\mathcal{G})$ is invariant under the actions of quantum components. For asymptotics of spectral distribution, however, the invariance is too much to require and *asymptotic invariance* is sufficient. This property will be formulated for a growing regular graph.

7.1 Integer Lattices

In order to illustrate asymptotic invariance of $\Gamma(\mathcal{G})$, let us consider the integer lattice $\mathbb{Z}^N$. Each $x \in \mathbb{Z}^N$ is expressible in the form

$$x = (\xi_1, \xi_2, \dots, \xi_N) = \sum_{i=1}^{N} \xi_i e_i, \qquad \xi_i \in \mathbb{Z},$$

where $e_1, \dots, e_N$ are the standard basis. Taking $o = (0, 0, \dots, 0)$ to be the origin, we introduce the stratification

$$\mathbb{Z}^N = \bigcup_{n=0}^{\infty} V_n, \qquad V_n = \{x \in \mathbb{Z}^N \,;\, \partial(x, o) = n\},$$

where

$$\partial(x, o) = \sum_{i=1}^{N} |\xi_i|, \qquad x = (\xi_1, \xi_2, \dots, \xi_N).$$

Recall that $\Gamma(\mathbb{Z}^N)$ is the subspace of $\ell^2(\mathbb{Z}^N)$ spanned by

$$\Phi_n = |V_n|^{-1/2} \sum_{x \in V_n} \delta_x, \qquad n = 0, 1, 2, \dots .$$

The adjacency matrix $A = A_N$ admits the quantum decomposition:

A. Hora and N. Obata: Regular Graphs. In: A. Hora and N. Obata, Quantum Probability and Spectral Analysis of Graphs, Theoretical and Mathematical Physics, 175–203 (2007)
DOI 10.1007/3-540-48863-4_7

$$A = A^+ + A^-.$$

Here we note that $A^\circ = 0$ since there is no edge lying in a stratum V_n. The actions of the quantum components A^ϵ on $\Gamma(\mathcal{G})$ are readily known from Theorem 2.26 as follows:

$$A^+\Phi_n = |V_n|^{-1/2} \sum_{y \in V_{n+1}} \omega_-(y)\, \delta_y, \tag{7.1}$$

$$A^-\Phi_n = |V_n|^{-1/2} \sum_{y \in V_{n-1}} \omega_+(y)\, \delta_y. \tag{7.2}$$

Since $\omega_\pm(y)$ is not necessarily constant on each V_n, $\Gamma(\mathcal{G})$ is not invariant under the quantum components.

We shall observe that $\Gamma(\mathbb{Z}^N)$ is asymptotically invariant. The precise estimate being given in the next section, here we only outline our idea. We first consider (7.1). For a large N, most generic vertices in V_{n+1} have the form

$$y = \pm e_{i_1} \pm e_{i_2} \pm \cdots \pm e_{i_{n+1}}, \quad 1 \le i_1 < i_2 < \cdots < i_{n+1} \le N, \tag{7.3}$$

namely, in the right-hand side each e_i appears with multiplicity at most one. Note that

$$|V_{n+1}| = \binom{N}{n+1} 2^{n+1} + O(N^n) = \frac{(2N)^{n+1}}{(n+1)!} + O(N^n), \tag{7.4}$$

where the principal term corresponds to the number of generic vertices. For a generic $y \in V_{n+1}$ we have $\omega_-(y) = n+1$, since a vertex $x \in V_n$ which is adjacent to y is obtained by subtracting one e_{i_k} from the right-hand side of (7.3). Thus, (7.1) becomes

$$\begin{aligned} |V_n|^{1/2} A^+\Phi_n &= \sum_{y \in V_{n+1}} (n+1)\delta_y + O(N^{n/2}) \\ &= (n+1)|V_{n+1}|^{1/2}\Phi_{n+1} + O(N^{n/2}) \end{aligned}$$

(the O-terms express vectors whose norms are of the indicated order). Once again using (7.4), we obtain

$$A^+\Phi_n = \sqrt{2N}\,\sqrt{n+1}\,\Phi_{n+1} + O(1).$$

Similarly,

$$A^-\Phi_n = \sqrt{2N}\,\sqrt{n}\,\Phi_{n-1} + O(N^{-1/2}).$$

Since the variance of $A = A_N$ in the vacuum state being $2N$, taking the normalization into account, we obtain the proper expressions:

$$\frac{A^+}{\sqrt{2N}}\Phi_n = \sqrt{n+1}\,\Phi_{n+1} + O(N^{-1/2}), \tag{7.5}$$

$$\frac{A^-}{\sqrt{2N}}\Phi_n = \sqrt{n}\,\Phi_{n-1} + O(N^{-1}). \tag{7.6}$$

This means that $\Gamma(\mathbb{Z}^N)$ is asymptotically invariant under the (normalized) quantum components. It is immediately seen from (7.5) and (7.6), at a formal level at least, that the actions of normalized quantum components in the limit coincide with those of the annihilation and creation operators of the Boson Fock space $\Gamma_{\mathrm{Boson}} = (\Gamma, \{\Psi_n\}, B^+, B^-)$, i.e.,

$$\lim_{N\to\infty} \frac{A_N^{\pm}}{\sqrt{2N}} = B^{\pm}.$$

Hence, still at a formal level, we obtain

$$\lim_{N\to\infty} \left\langle \left(\frac{A_N}{\sqrt{2N}} \right)^m \right\rangle_o = \langle \Psi_0, (B^+ + B^-)^m \Psi_0 \rangle, \quad m = 1, 2, \dots.$$

As is well known, the right-hand side is the mth moment of the standard Gaussian distribution. Consequently,

$$\lim_{N\to\infty} \left\langle \left(\frac{A_N}{\sqrt{2N}} \right)^m \right\rangle_o = \frac{1}{\sqrt{2\pi}} \int_{-\infty}^{+\infty} x^m \mathrm{e}^{-x^2/2} dx, \quad m = 1, 2, \dots.$$

In other words, the asymptotic spectral distribution of the adjacency matrix of the integer lattice in the vacuum state is the standard Gaussian distribution.

From the next section on, we shall formulate the asymptotic invariance for a general growing regular graph and prove the quantum central limit theorem.

7.2 Growing Regular Graphs

We prepare some notations. Let $\mathcal{G} = (V, E)$ be an arbitrary graph with a fixed origin $o \in V$. As usual, for $n = 0, 1, 2, \dots$ we define

$$V_n = \{x \in V \,;\, \partial(o, x) = n\}$$

and for $x \in V$ and $\epsilon \in \{+, -, \circ\}$,

$$\omega_\epsilon(x) = |\{y \in V \,;\, y \sim x,\ \partial(o, y) = \partial(o, x) + \epsilon\}|.$$

Statistics of $\omega_\epsilon(x)$ will play a crucial role. Whenever $V_n \neq \varnothing$, we define

$$M(\omega_\epsilon|V_n) = \frac{1}{|V_n|} \sum_{x \in V_n} \omega_\epsilon(x),$$

$$\Sigma^2(\omega_\epsilon|V_n) = \frac{1}{|V_n|} \sum_{x \in V_n} \left\{\omega_\epsilon(x) - M(\omega_\epsilon|V_n)\right\}^2,$$

$$L(\omega_\epsilon|V_n) = \max\{\omega_\epsilon(x) \,;\, x \in V_n\}.$$

Namely, $M(\omega_\epsilon|V_n)$ is the mean value of $\omega_\epsilon(x)$ when x runs over V_n, and $\Sigma^2(\omega_\epsilon|V_n)$ its variance. Both $\Sigma^2(\omega_\epsilon|V_n)$ and $L(\omega_\epsilon|V_n)$ indicate fluctuation of $\omega_\epsilon(x)$. The next result is easy to see.

Lemma 7.1. *If $\mathcal{G} = (V, E)$ is a regular graph with degree κ, we have*

$$M(\omega_+|V_n) + M(\omega_-|V_n) + M(\omega_\circ|V_n) = \kappa,$$
$$\Sigma(\omega_+|V_n) \le \Sigma(\omega_-|V_n) + \Sigma(\omega_\circ|V_n).$$

Now consider a growing regular graph $\mathcal{G}^{(\nu)} = (V^{(\nu)}, E^{(\nu)})$, that is, a family of regular graphs with parameter ν running over an infinite directed set. The degree of $\mathcal{G}^{(\nu)}$ is denoted by $\kappa(\nu)$. Each graph $\mathcal{G}^{(\nu)}$ is given an origin $o_\nu \in V^{(\nu)}$ and thereby the stratification

$$V^{(\nu)} = \bigcup_{n=0}^{\infty} V_n^{(\nu)}, \qquad V_n^{(\nu)} = \{y \in V^{(\nu)};\, \partial(o, y) = n\}. \tag{7.7}$$

Note that $V_n^{(\nu)} = \emptyset$ may occur. Let $\Gamma(\mathcal{G}^{(\nu)})$ denote the subspace of $\ell^2(V^{(\nu)})$ spanned by the unit vectors defined by

$$\Phi_n^{(\nu)} = |V_n^{(\nu)}|^{-1/2} \sum_{x \in V_n^{(\nu)}} \delta_x, \qquad n = 0, 1, 2, \dots. \tag{7.8}$$

According to the stratification (7.7) the adjacency matrix A_ν of $\mathcal{G}^{(\nu)}$ admits the quantum decomposition:

$$A_\nu = A_\nu^+ + A_\nu^- + A_\nu^\circ. \tag{7.9}$$

We do not assume that $\Gamma(\mathcal{G}^{(\nu)})$ is invariant under the actions of quantum components A_ν^ϵ, instead we shall formulate asymptotic invariance in terms of how the graph grows. Namely, we consider the following three conditions:

(A1) $\lim_\nu \kappa(\nu) = \infty$.
(A2) For each $n = 1, 2, \dots$ there exists a limit

$$\omega_n = \lim_\nu M(\omega_-|V_n^{(\nu)}) < \infty. \tag{7.10}$$

Moreover,

$$\lim_\nu \Sigma^2(\omega_-|V_n^{(\nu)}) = 0, \tag{7.11}$$
$$W_n \equiv \sup_\nu L(\omega_-|V_n^{(\nu)}) < \infty. \tag{7.12}$$

(A3) For each $n = 0, 1, 2, \dots$ there exists a limit

$$\alpha_{n+1} = \lim_\nu M\left(\frac{\omega_\circ}{\sqrt{\kappa(\nu)}} \,\middle|\, V_n^{(\nu)} \right) = \lim_\nu \frac{M(\omega_\circ|V_n^{(\nu)})}{\sqrt{\kappa(\nu)}} < \infty. \tag{7.13}$$

Moreover,

$$\lim_{\nu} \Sigma^2 \left(\frac{\omega_\circ}{\sqrt{\kappa(\nu)}} \,\middle|\, V_n^{(\nu)} \right) = \lim_{\nu} \frac{\Sigma^2(\omega_\circ|V_n^{(\nu)})}{\kappa(\nu)} = 0, \tag{7.14}$$

$$\sup_{\nu} \frac{L(\omega_\circ|V_n^{(\nu)})}{\sqrt{\kappa(\nu)}} < \infty. \tag{7.15}$$

Remark 7.2. Condition (A2) for $n = 1$ and (A3) for $n = 0$ are automatically satisfied. Also note that $\omega_1 = 1$ and $\alpha_1 = 0$.

Remark 7.3. If $\mathcal{G}^{(\nu)}$ happens to be a finite graph, $M(\omega_\epsilon|V_n^{(\nu)})$ is defined only up to a certain n. This causes, however, no difficulty for defining ω_n and α_n for all n (see the proof of Proposition 7.4).

The meaning of (A1) is clear. Condition (A2) means that, in each stratum most of the vertices have the same number of downward edges, and as the graph grows the fluctuation of that number tends to zero. Condition (A3) is for edges lying in each stratum. The number of such edges may increase as the graph grows, but the growth rate is bounded by $\kappa(\nu)^{1/2}$. We roughly see from conditions (A1), (7.12) and (7.15) that, for a 'generic' vertex $x \in V_n^{(\nu)}$,

$$\omega_+(x) = O(\kappa(\nu)), \qquad \omega_\circ(x) = O(\kappa(\nu)^{1/2}), \qquad \omega_-(x) = O(1),$$

as the graph grows.

Proposition 7.4. *Let $\mathcal{G}^{(\nu)} = (V^{(\nu)}, E^{(\nu)})$ be a growing regular graph satisfying conditions (A1)–(A3). Then, $(\{\omega_n\}, \{\alpha_n\})$ defined therein is a Jacobi coefficient of infinite type.*

Proof. It follows from (A1) that there exists ν_1 such that $V_1^{(\nu)} \neq \varnothing$ for all $\nu > \nu_1$. Take $x \in V_1^{(\nu)}$ and consider the obvious equality:

$$\frac{\omega_+(x)}{\kappa} + \frac{\omega_-(x)}{\kappa} + \frac{\omega_\circ(x)}{\kappa} = 1.$$

By (7.12) and (7.15) there exists $\nu_2 > \nu_1$ such that the first term is positive for all $\nu > \nu_2$, namely, $V_2^{(\nu)} \neq \varnothing$. By induction, we can find $\nu_1 < \nu_2 < \cdots < \nu_n < \cdots$ such that $V_n^{(\nu)} \neq \varnothing$ for all $\nu > \nu_n$, $n = 1, 2, \dots$. Then, for any $x \in V_n^{(\nu)}$ we have $\omega_-(x) \geq 1$, hence $M(\omega_-|V_n^{(\nu)}) \geq 1$. Consequently, $\omega_n \geq 1$ for all n. □

Some part of conditions (A1)–(A3) are rephrased in a slightly different form.

Proposition 7.5. *In conditions (A1)–(A3), we may replace* (7.10) *and* (7.11) *with a single condition: for each $n = 1, 2, \dots$ there exists a constant number ω_n independent of ν such that*

$$\lim_{\nu} \frac{|\{x \in V_n^{(\nu)}\,;\, \omega_-(x) = \omega_n\}|}{|V_n^{(\nu)}|} = 1. \tag{7.16}$$

Proof. Throughout the proof $n = 1, 2, \dots$ is fixed arbitrarily. We first prove that (7.16) implies (7.10) and (7.11). Divide $V_n^{(\nu)}$ into two parts

$$U_{\mathrm{reg}}^{(\nu)} = \{x \in V_n^{(\nu)}\,;\, \omega_-(x) = \omega_n\}, \qquad U_{\mathrm{sing}}^{(\nu)} = \{x \in V_n^{(\nu)}\,;\, \omega_-(x) \neq \omega_n\},$$

where the index n is omitted for simplicity. The average of $\omega_-(x)$ is given by

$$\begin{aligned} M(\omega_-|V_n^{(\nu)}) &= \frac{1}{|V_n^{(\nu)}|}\left\{\sum_{x\in U_{\mathrm{reg}}^{(\nu)}} \omega_-(x) + \sum_{x\in U_{\mathrm{sing}}^{(\nu)}} \omega_-(x)\right\} \\ &= \frac{|U_{\mathrm{reg}}^{(\nu)}|}{|V_n^{(\nu)}|}\,\omega_n + \frac{1}{|V_n^{(\nu)}|}\sum_{x\in U_{\mathrm{sing}}^{(\nu)}} \omega_-(x). \end{aligned}$$

By condition (7.12) we have $\omega_-(x) \le W_n$ for all $x \in V_n^{(\nu)}$ and ν, so that

$$\begin{aligned} |M(\omega_-|V_n^{(\nu)}) - \omega_n| &\le \left(1 - \frac{|U_{\mathrm{reg}}^{(\nu)}|}{|V_n^{(\nu)}|}\right)\omega_n + \frac{|U_{\mathrm{sing}}^{(\nu)}|}{|V_n^{(\nu)}|}\,W_n \\ &\le \frac{|U_{\mathrm{sing}}^{(\nu)}|}{|V_n^{(\nu)}|}\,(\omega_n + W_n). \end{aligned}$$

Here we note from (7.16) that

$$\lim_{\nu} \frac{|U_{\mathrm{sing}}^{(\nu)}|}{|V_n^{(\nu)}|} = 0. \tag{7.17}$$

Then we obtain

$$\lim_{\nu} M(\omega_-|V_n^{(\nu)}) = \omega_n, \tag{7.18}$$

which proves (7.10). We next consider the variance. By Minkowski's inequality, we obtain

$$\begin{aligned}\Sigma(\omega_-|V_n^{(\nu)}) &= \left\{\frac{1}{|V_n^{(\nu)}|}\sum_{x\in V_n^{(\nu)}}(\omega_-(x)-M(\omega_-|V_n^{(\nu)}))^2\right\}^{1/2}\\ &\le \left\{\frac{1}{|V_n^{(\nu)}|}\sum_{x\in V_n^{(\nu)}}(\omega_-(x)-\omega_n)^2\right\}^{1/2}\\ &\quad+\left\{\frac{1}{|V_n^{(\nu)}|}\sum_{x\in V_n^{(\nu)}}(\omega_n-M(\omega_-|V_n^{(\nu)}))^2\right\}^{1/2}\\ &= \left\{\frac{1}{|V_n^{(\nu)}|}\sum_{x\in U_{\mathrm{sing}}^{(\nu)}}(\omega_-(x)-\omega_n)^2\right\}^{1/2} + \big|\omega_n - M(\omega_-|V_n^{(\nu)})\big|.\end{aligned}$$

Since $|\omega_-(x)-\omega_n| \le \omega_-(x)+\omega_n \le W_n+\omega_n$ for $x \in V_n^{(\nu)}$, we have

$$\Sigma(\omega_-|V_n^{(\nu)}) \le \left(\frac{|U_{\mathrm{sing}}^{(\nu)}|}{|V_n^{(\nu)}|}\right)^{1/2}(W_n+\omega_n) + \big|\omega_n - M(\omega_-|V_n^{(\nu)})\big|$$

and hence (7.11) follows by (7.17) and (7.18).

We next show that (7.16) is derived from (7.10) and (7.11). By (7.10), for any $\epsilon > 0$ there exists ν_0 such that

$$|M(\omega_-|V_n^{(\nu)}) - \omega_n| < \epsilon, \quad \nu \ge \nu_0.$$

If $x \in V_n^{(\nu)}$ satisfies $|\omega_-(x)-\omega_n| \ge 2\epsilon$, we have

$$\big|\omega_-(x) - M(\omega_-|V_n^{(\nu)})\big| \ge |\omega_-(x)-\omega_n| - |\omega_n - M(\omega_-|V_n^{(\nu)})| \ge \epsilon.$$

Hence

$$\frac{|\{x\in V_n^{(\nu)}\,;\,|\omega_-(x)-\omega_n|\ge 2\epsilon\}|}{|V_n^{(\nu)}|} \le \frac{|\{x\in V_n^{(\nu)}\,;\,|\omega_-(x)-M(\omega_-|V_n^{(\nu)})|\ge \epsilon\}|}{|V_n^{(\nu)}|}.$$

By Chebyshev's inequality and (7.11) we have

$$\frac{|\{x\in V_n^{(\nu)}\,;\,|\omega_-(x)-\omega_n|\ge 2\epsilon\}|}{|V_n^{(\nu)}|} \le \frac{\Sigma^2(\omega_-|V_n^{(\nu)})}{\epsilon^2} \to 0, \qquad \nu\to\infty. \tag{7.19}$$

We prove that ω_n is an integer. Suppose otherwise; then, since $\omega_-(x)$ is always an integer, we can choose a sufficiently small $\epsilon > 0$ such that

$$V_n^{(\nu)} = \{x \in V_n^{(\nu)}\,;\,|\omega_-(x)-\omega_n| \ge 2\epsilon\}.$$

But this contradicts (7.19) and hence ω_n is an integer. Since $\omega_-(x)$ and ω_n are all integers, we may choose a sufficiently small $\epsilon > 0$ such that

$$\frac{|\{x \in V_n^{(\nu)} \,;\, \omega_-(x) \neq \omega_n\}|}{|V_n^{(\nu)}|} = \frac{|\{x \in V_n^{(\nu)} \,;\, |\omega_-(x) - \omega_n| \geq 2\epsilon\}|}{|V_n^{(\nu)}|}. \tag{7.20}$$

As is shown in (7.19), the right-hand side of (7.20) tends to 0 as $\nu \to \infty$. Therefore

$$\lim_{\nu} \frac{|\{x \in V_n^{(\nu)} \,;\, \omega_-(x) \neq \omega_n\}|}{|V_n^{(\nu)}|} = 0$$

and (7.16) follows. □

During the above proof we have established the following:

Proposition 7.6. *Let $\mathcal{G}^{(\nu)} = (V^{(\nu)}, E^{(\nu)})$ be a growing regular graph satisfying conditions (A1)–(A3). Then, the Jacobi sequence $\{\omega_n\}$ defined therein consists of positive integers.*

7.3 Quantum Central Limit Theorems

First we claim the key result.

Theorem 7.7. *Let $\mathcal{G}^{(\nu)} = (V^{(\nu)}, E^{(\nu)})$ be a growing regular graph satisfying conditions (A1)–(A3) and A_ν its adjacency matrix. Let $(\Gamma, \{\Psi_n\}, B^+, B^-)$ be the interacting Fock space associated with $\{\omega_n\}$ and B° the diagonal operator defined by $\{\alpha_n\}$, where $\{\omega_n\}$ and $\{\alpha_n\}$ are given in conditions (A1)–(A3). Then we have*

$$\lim_{\nu} \left\langle \Phi_j^{(\nu)}, \frac{A_\nu^{\epsilon_m}}{\sqrt{\kappa(\nu)}} \cdots \frac{A_\nu^{\epsilon_1}}{\sqrt{\kappa(\nu)}} \Phi_n^{(\nu)} \right\rangle = \langle \Psi_j, B^{\epsilon_m} \cdots B^{\epsilon_1} \Psi_n \rangle, \tag{7.21}$$

for any $\epsilon_1, \dots, \epsilon_m \in \{+, -, \circ\}$, $m = 1, 2, \dots$, and $j, n = 0, 1, 2, \dots$.

Before going into the proof, we shall give an estimate of $|V_n^{(\nu)}|$.

Lemma 7.8. *Let $\mathcal{G} = (V, E)$ be a regular graph with degree κ. Fix an origin $o \in V$ and consider the stratification $V = \bigcup_n V_n$. Then, for any $n = 1, 2, \dots$ with $V_n \neq \emptyset$ we have*

$$|V_n| = \kappa^n \prod_{j=1}^{n} M(\omega_-|V_j)^{-1} \prod_{j=0}^{n-1} \left(1 - \frac{M(\omega_-|V_j)}{\kappa} - \frac{M(\omega_\circ|V_j)}{\kappa}\right). \tag{7.22}$$

Proof. With the help of the matching identity (Lemma 2.22) we have

$$\begin{aligned}
\kappa|V_{n-1}| &= \sum_{x \in V_{n-1}} \{\omega_+(x) + \omega_-(x) + \omega_\circ(x)\} \\
&= \sum_{y \in V_n} \omega_-(y) + \sum_{x \in V_{n-1}} \omega_-(x) + \sum_{x \in V_{n-1}} \omega_\circ(x) \\
&= M(\omega_-|V_n)|V_n| + M(\omega_-|V_{n-1})|V_{n-1}| + M(\omega_\circ|V_{n-1})|V_{n-1}|.
\end{aligned}$$

Hence,

$$|V_n| = \kappa M(\omega_-|V_n)^{-1}|V_{n-1}| \left(1 - \frac{M(\omega_-|V_{n-1})}{\kappa} - \frac{M(\omega_\circ|V_{n-1})}{\kappa}\right). \quad (7.23)$$

Noting that $V_n \neq \varnothing$ implies $V_{n-1} \neq \varnothing$, ..., $V_1 \neq \varnothing$, we obtain (7.22) by repeated application of (7.23). □

Proposition 7.9. *If a growing regular graph $\mathcal{G}^{(\nu)} = (V^{(\nu)}, E^{(\nu)})$ satisfies conditions (A1)–(A3), we have*

$$\lim_{\nu} \frac{|V_n^{(\nu)}|}{\kappa(\nu)^n} = \frac{1}{\omega_n \cdots \omega_1}, \qquad n = 1, 2, \dots. \quad (7.24)$$

Proof. By Lemma 7.8 we have

$$\frac{|V_n^{(\nu)}|}{\kappa(\nu)^n} = \prod_{j=1}^{n} M(\omega_-|V_j^{(\nu)})^{-1} \prod_{j=0}^{n-1} \left(1 - \frac{M(\omega_-|V_j^{(\nu)})}{\kappa(\nu)} - \frac{M(\omega_\circ|V_j^{(\nu)})}{\kappa(\nu)}\right).$$

The first product converges to $(\omega_n \cdots \omega_1)^{-1}$ by (A2) and the second one to 1 by (A3) so that (7.24) follows. □

The explicit actions of the quantum components A^ϵ are given in (2.31)–(2.33). Inserting the mean values $M(\omega_\epsilon|V_n)$ therein, we obtain

$$\begin{aligned} A^+\Phi_n &= M(\omega_-|V_{n+1}) \left(\frac{|V_{n+1}|}{|V_n|}\right)^{1/2} \Phi_{n+1} \\ &\quad + \frac{1}{\sqrt{|V_n|}} \sum_{y \in V_{n+1}} (\omega_-(y) - M(\omega_-|V_{n+1}))\delta_y, \qquad (7.25) \\ A^-\Phi_n &= M(\omega_+|V_{n-1}) \left(\frac{|V_{n-1}|}{|V_n|}\right)^{1/2} \Phi_{n-1} \\ &\quad + \frac{1}{\sqrt{|V_n|}} \sum_{y \in V_{n-1}} (\omega_+(y) - M(\omega_+|V_{n-1}))\delta_y, \qquad (7.26) \\ A^\circ\Phi_n &= M(\omega_\circ|V_n)\Phi_n \\ &\quad + \frac{1}{\sqrt{|V_n|}} \sum_{y \in V_n} (\omega_\circ(y) - M(\omega_\circ|V_n))\delta_y, \qquad (7.27) \end{aligned}$$

for $n = 0, 1, 2, \dots$, understanding that $A^-\Phi_0 = 0$ for the second formula. It is convenient to unify the above three formulae. We set

$$\gamma_n^+ = M(\omega_-|V_n)\left(\frac{|V_n|}{\kappa|V_{n-1}|}\right)^{1/2}, \quad n = 1, 2, \dots, \tag{7.28}$$

$$\gamma_n^- = M(\omega_+|V_n)\left(\frac{|V_n|}{\kappa|V_{n+1}|}\right)^{1/2}, \quad n = 0, 1, 2, \dots, \tag{7.29}$$

$$\gamma_n^\circ = \frac{M(\omega_\circ|V_n)}{\sqrt{\kappa}}, \quad n = 0, 1, 2, \dots, \tag{7.30}$$

and

$$S_n^+ = \frac{1}{\sqrt{\kappa|V_{n-1}|}} \sum_{y \in V_n} (\omega_-(y) - M(\omega_-|V_n))\delta_y, \quad n = 1, 2, \dots, \tag{7.31}$$

$$S_n^- = \frac{1}{\sqrt{\kappa|V_{n+1}|}} \sum_{y \in V_n} (\omega_+(y) - M(\omega_+|V_n))\delta_y, \quad n = 0, 1, 2, \dots, \tag{7.32}$$

$$S_n^\circ = \frac{1}{\sqrt{\kappa|V_n|}} \sum_{y \in V_n} (\omega_\circ(y) - M(\omega_\circ|V_n))\delta_y, \quad n = 0, 1, 2, \dots. \tag{7.33}$$

In fact,

$$S_0^- = S_0^\circ = 0.$$

For convenience we set

$$S_0^+ = \gamma_{-1}^- \Phi_{-1} = S_{-1}^- = 0.$$

With these notations (7.25)–(7.27) are unified:

$$\frac{A^\epsilon}{\sqrt{\kappa}} \Phi_n = \gamma_{n+\epsilon}^\epsilon \Phi_{n+\epsilon} + S_{n+\epsilon}^\epsilon, \quad \epsilon \in \{+, -, \circ\}, \quad n = 0, 1, 2, \dots. \tag{7.34}$$

Then its repeated action is expressible in a concise form:

$$\begin{aligned} &\frac{A^{\epsilon_m}}{\sqrt{\kappa}} \cdots \frac{A^{\epsilon_1}}{\sqrt{\kappa}} \Phi_n \\ &\quad = \gamma_{n+\epsilon_1}^{\epsilon_1} \gamma_{n+\epsilon_1+\epsilon_2}^{\epsilon_2} \cdots \gamma_{n+\epsilon_1+\dots+\epsilon_m}^{\epsilon_m} \Phi_{n+\epsilon_1+\dots+\epsilon_m} \\ &\qquad + \sum_{k=1}^{m} \underbrace{\gamma_{n+\epsilon_1}^{\epsilon_1} \cdots \gamma_{n+\epsilon_1+\dots+\epsilon_{k-1}}^{\epsilon_{k-1}}}_{(k-1) \text{ times}} \underbrace{\frac{A^{\epsilon_m}}{\sqrt{\kappa}} \cdots \frac{A^{\epsilon_{k+1}}}{\sqrt{\kappa}}}_{(m-k) \text{ times}} S_{n+\epsilon_1+\dots+\epsilon_k}^{\epsilon_k}. \end{aligned} \tag{7.35}$$

By observing the up–down actions of A^ϵ we see immediately that

$$\frac{A^{\epsilon_m}}{\sqrt{\kappa}} \cdots \frac{A^{\epsilon_1}}{\sqrt{\kappa}} \Phi_n = 0$$

unless

$$n+\epsilon_1 \ge 0, \quad n+\epsilon_1+\epsilon_2 \ge 0, \dots, \quad n+\epsilon_1+\epsilon_2+\dots+\epsilon_m \ge 0. \tag{7.36}$$

We need to estimate the error term of (7.35). For $n, q = 1, 2, \dots$ we set

$$M^-_{n,q} = \max\left\{\prod_{j=1}^{q} L(\omega_-|V_{k_j})\,;\, 1 \le k_1, k_2, \dots, k_q \le n\right\}, \tag{7.37}$$

$$M^-_{n,0} = 1.$$

Similarly, taking condition (A3) in mind, we set

$$M^\circ_{n,q} = \max\left\{\prod_{j=1}^{q} \frac{L(\omega_\circ|V_{k_j})}{\sqrt{\kappa}}\,;\, 1 \le k_1, k_2, \dots, k_q \le n\right\}, \tag{7.38}$$

$$M^\circ_{n,0} = 1.$$

Lemma 7.10. *Let $n = 0, 1, 2, \dots$, $m = 1, 2, \dots$ and $\epsilon_1, \dots, \epsilon_m \in \{+, -, \circ\}$. Denote by p, q and r the numbers of $+$, $-$ and $\circ$ in $\{\epsilon_1, \dots, \epsilon_m\}$, respectively. Then, whenever $n + p - q \ge 0$, we have*

$$\left|\left\langle \Phi_{n+p-q}, \frac{A^{\epsilon_m}}{\sqrt{\kappa}} \cdots \frac{A^{\epsilon_1}}{\sqrt{\kappa}} S^+_n \right\rangle\right|$$
$$\le \Sigma(\omega_-|V_n) M^-_{n+p,q} M^\circ_{n+p,r} \frac{\kappa^{p+\frac{r-m-1}{2}}|V_n|}{\sqrt{|V_{n+p-q}||V_{n-1}|}}, \tag{7.39}$$

$$\left|\left\langle \Phi_{n+p-q}, \frac{A^{\epsilon_m}}{\sqrt{\kappa}} \cdots \frac{A^{\epsilon_1}}{\sqrt{\kappa}} S^-_n \right\rangle\right|$$
$$\le \{\Sigma(\omega_-|V_n) + \Sigma(\omega_\circ|V_n)\} M^-_{n+p,q} M^\circ_{n+p,r} \frac{\kappa^{p+\frac{r-m-1}{2}}|V_n|}{\sqrt{|V_{n+p-q}||V_{n+1}|}}, \tag{7.40}$$

$$\left|\left\langle \Phi_{n+p-q}, \frac{A^{\epsilon_m}}{\sqrt{\kappa}} \cdots \frac{A^{\epsilon_1}}{\sqrt{\kappa}} S^\circ_n \right\rangle\right|$$
$$\le \Sigma(\omega_\circ|V_n) M^-_{n+p,q} M^\circ_{n+p,r} \frac{\kappa^{p+\frac{r-m-1}{2}}\sqrt{|V_n|}}{\sqrt{|V_{n+p-q}|}}. \tag{7.41}$$

Proof. We only prove (7.39) as the rest is similar. Since the left-hand side of (7.39) vanishes unless (7.36) is satisfied, it is sufficient to prove it under the condition (7.36). Using the explicit expression (7.31) we obtain

$$\frac{A^{\epsilon_m}}{\sqrt{\kappa}} \cdots \frac{A^{\epsilon_1}}{\sqrt{\kappa}} S^+_n$$
$$= \frac{1}{(\kappa|V_{n-1}|)^{1/2}} \sum_{y \in V_n} (\omega_-(y) - M(\omega_-|V_n)) \frac{A^{\epsilon_m}}{\sqrt{\kappa}} \cdots \frac{A^{\epsilon_1}}{\sqrt{\kappa}} \delta_y$$
$$= \frac{\kappa^{-m/2}}{(\kappa|V_{n-1}|)^{1/2}} \sum_{y \in V_n} (\omega_-(y) - M(\omega_-|V_n)) A^{\epsilon_m} \cdots A^{\epsilon_1} \delta_y. \tag{7.42}$$

Here we introduce a new notation. For $y, z \in V$ and $\epsilon \in \{+, -, \circ\}$ we write $y \overset{\epsilon}{\to} z$ if $z \sim y$ and $\partial(z, o) = \partial(y, o) + \epsilon$. For $y, z \in V$ we put

$$w(y; \epsilon_1, \dots, \epsilon_m; z) = |\{(z_1, \dots, z_{m-1}) \in V^{m-1} ; y \overset{\epsilon_1}{\to} z_1 \overset{\epsilon_2}{\to} z_2 \cdots \overset{\epsilon_{m-1}}{\to} z_{m-1} \overset{\epsilon_m}{\to} z\}|.$$

This counts the number of walks from y to z along edges with directions $\epsilon_1, \dots, \epsilon_m$. Then (7.42) becomes

$$\frac{A^{\epsilon_m}}{\sqrt{\kappa}} \cdots \frac{A^{\epsilon_1}}{\sqrt{\kappa}} S_n^+ = \frac{\kappa^{-m/2}}{(\kappa|V_{n-1}|)^{1/2}} \sum_{y \in V_n} \sum_{z \in V_{n+p-q}} (\omega_-(y) - M(\omega_-|V_n)) w(y; \epsilon_1, \dots, \epsilon_m; z) \delta_z.$$

Therefore,

$$\left\langle \Phi_{n+p-q}, \frac{A^{\epsilon_m}}{\sqrt{\kappa}} \cdots \frac{A^{\epsilon_1}}{\sqrt{\kappa}} S_n^+ \right\rangle = \frac{1}{|V_{n+p-q}|^{1/2}} \frac{\kappa^{-m/2}}{(\kappa|V_{n-1}|)^{1/2}} \times \sum_{y \in V_n} \sum_{z \in V_{n+p-q}} (\omega_-(y) - M(\omega_-|V_n)) w(y; \epsilon_1, \dots, \epsilon_m; z). \tag{7.43}$$

For a fixed $y \in V_n$,

$$\sum_{z \in V_{n+p-q}} w(y; \epsilon_1, \dots, \epsilon_m; z) \tag{7.44}$$

coincides with the number of walks from y to a certain vertex in V_{n+p-q} along m edges with directions $\epsilon_1, \dots, \epsilon_m$ in order. Consider an intermediate vertex $\xi \in V_k$ in such a walk. The number of edges from ξ with $-$ direction is bounded by $L(\omega_-|V_k)$, with $\circ$ direction by $L(\omega_\circ|V_k)$, and with $+$ direction by κ. Given $(\epsilon_1, \dots, \epsilon_m)$, $+$, $-$ and $\circ$ directions appear p, q and r times, respectively, and the intermediate vertex ξ lies in $V_0 \cup V_1 \cup \dots \cup V_{n+p}$. Hence by (7.37) and (7.38) we obtain

$$\sum_{z \in V_{n+p-q}} w(y; \epsilon_1, \dots, \epsilon_m; z) \le \kappa^{p+\frac{r}{2}} M_{n+p,q}^- M_{n+p,r}^\circ, \tag{7.45}$$

where the right-hand side is independent of $y \in V_n$. Combining (7.43) and (7.45), we come to

$$\begin{aligned}
&\left| \left\langle \Phi_{n+p-q}, \frac{A^{\epsilon_m}}{\sqrt{\kappa}} \cdots \frac{A^{\epsilon_1}}{\sqrt{\kappa}} S_n^+ \right\rangle \right| \\
&\le \frac{\kappa^{p+\frac{r}{2}} M^-_{n+p,q} M^\circ_{n+p,r}}{|V_{n+p-q}|^{1/2}} \frac{\kappa^{-m/2}}{(\kappa |V_{n-1}|)^{1/2}} \sum_{y \in V_n} |\omega_-(y) - M(\omega_-|V_n)| \\
&\le \frac{\kappa^{p+\frac{r}{2}-\frac{m}{2}-\frac{1}{2}} M^-_{n+p,q} M^\circ_{n+p,r}}{|V_{n+p-q}|^{1/2} |V_{n-1}|^{1/2}} \left(\sum_{y \in V_n} |\omega_-(y) - M(\omega_-|V_n)|^2 \right)^{1/2} |V_n|^{1/2} \\
&= \Sigma(\omega_-|V_n) M^-_{n+p,q} M^\circ_{n+p,r} \frac{\kappa^{p+\frac{r-m-1}{2}} |V_n|}{|V_{n+p-q}|^{1/2} |V_{n-1}|^{1/2}},
\end{aligned}$$

which proves (7.39). □

Proof of Theorem 7.7. Let $\mathcal{G}_\nu = (V^{(\nu)}, E^{(\nu)})$ be a growing regular graph as stated therein. Given $\epsilon_1, \dots, \epsilon_m \in \{+, -, \circ\}$, $m = 1, 2, \dots$, and $n, j = 0, 1, 2, \dots$ we consider

$$\left\langle \Phi_j^{(\nu)}, \frac{A_\nu^{\epsilon_m}}{\sqrt{\kappa(\nu)}} \cdots \frac{A_\nu^{\epsilon_1}}{\sqrt{\kappa(\nu)}} \Phi_n^{(\nu)} \right\rangle. \tag{7.46}$$

Let p, q, r be the numbers of $+, -, \circ$ appearing in $\{\epsilon_1, \dots, \epsilon_m\}$, respectively. From the up–down action of A^ϵ we see easily that (7.46) vanishes unless (7.36) and $j = n + p - q$ hold. On the other hand, for the same $\epsilon_1, \dots, \epsilon_m$, it follows by the definition of an interacting Fock space $\Gamma_{\{\omega_n\}} = (\Gamma, \{\Psi_n\}, B^+, B^-)$ and the diagonal operator B° that

$$\langle \Psi_j, B^{\epsilon_m} \cdots B^{\epsilon_1} \Psi_n \rangle = 0.$$

Namely, (7.21) is true when (7.36) or $j = n + p - q$ is not fulfilled.

Next we consider the case where both (7.36) and $j = n+p-q$ are fulfilled. Using (7.35), we obtain

$$\begin{aligned}
&\left\langle \Phi_j^{(\nu)}, \frac{A_\nu^{\epsilon_m}}{\sqrt{\kappa(\nu)}} \cdots \frac{A_\nu^{\epsilon_1}}{\sqrt{\kappa(\nu)}} \Phi_n^{(\nu)} \right\rangle \\
&\quad = \gamma^{\epsilon_1}_{n+\epsilon_1} \gamma^{\epsilon_2}_{n+\epsilon_1+\epsilon_2} \cdots \gamma^{\epsilon_m}_{n+\epsilon_1+\dots+\epsilon_m} \\
&\quad\quad + \sum_{k=1}^m \gamma^{\epsilon_1}_{n+\epsilon_1} \cdots \gamma^{\epsilon_{k-1}}_{n+\epsilon_1+\dots+\epsilon_{k-1}} \\
&\quad\quad \times \left\langle \Phi_j^{(\nu)}, \frac{A_\nu^{\epsilon_m}}{\sqrt{\kappa(\nu)}} \cdots \frac{A_\nu^{\epsilon_{k+1}}}{\sqrt{\kappa(\nu)}} S^{\epsilon_k}_{n+\epsilon_1+\dots+\epsilon_k} \right\rangle.
\end{aligned} \tag{7.47}$$

Note that the coefficient γ_n^ϵ depends on ν. The explicit expressions of γ_n^ϵ being given in (7.28)–(7.30), with the help of Proposition 7.9 and conditions (A1)–(A3) we come to

$$\lim_{\nu} \gamma_n^+ = \lim_{\nu} M(\omega_-|V_n) \frac{1}{\sqrt{\omega_n}} = \sqrt{\omega_n}\,, \tag{7.48}$$

$$\lim_{\nu} \gamma_n^- = \lim_{\nu} \{\kappa - M(\omega_-|V_n) - M(\omega_\circ|V_n)\} \frac{\sqrt{\omega_{n+1}}}{\kappa} = \sqrt{\omega_{n+1}}, \tag{7.49}$$

$$\lim_{\nu} \gamma_n^\circ = \alpha_{n+1}. \tag{7.50}$$

Then by the definition of B^ϵ, we obtain

$$\lim_{\nu} \gamma_{n+\epsilon_1}^{\epsilon_1} \gamma_{n+\epsilon_1+\epsilon_2}^{\epsilon_2} \cdots \gamma_{n+\epsilon_1+\cdots+\epsilon_m}^{\epsilon_m} = \langle \Psi_j, B^{\epsilon_m} \cdots B^{\epsilon_1} \Psi_n \rangle.$$

Thus, to our goal it is sufficient to show that the second term of (7.47) vanishes in the limit. Since it is a finite sum, we need only to show that

$$\lim_{\nu} \left\langle \Phi_j^{(\nu)}, \frac{A^{\epsilon_m}}{\sqrt{\kappa(\nu)}} \cdots \frac{A^{\epsilon_{k+1}}}{\sqrt{\kappa(\nu)}} S_{n+\epsilon_1+\cdots+\epsilon_k}^{\epsilon_k} \right\rangle = 0. \tag{7.51}$$

For this it is sufficient to show that the right-hand sides of (7.39)–(7.41) in Lemma 7.10 vanish in the limit, i.e.,

$$\lim_{\nu} \Sigma(\omega_-|V_n) M_{n+p,q}^- M_{n+p,r}^\circ \frac{\kappa^{p+\frac{r-m-1}{2}} |V_n|}{\sqrt{|V_{n+p-q}||V_{n-1}|}} = 0, \tag{7.52}$$

$$\lim_{\nu} \{\Sigma(\omega_-|V_n) + \Sigma(\omega_\circ|V_n)\} M_{n+p,q}^- M_{n+p,r}^\circ \frac{\kappa^{p+\frac{r-m-1}{2}} |V_n|}{\sqrt{|V_{n+p-q}||V_{n+1}|}} = 0, \tag{7.53}$$

$$\lim_{\nu} \Sigma(\omega_\circ|V_n) M_{n+p,q}^- M_{n+p,r}^\circ \frac{\kappa^{p+\frac{r-m-1}{2}} \sqrt{|V_n|}}{\sqrt{|V_{n+p-q}|}} = 0, \tag{7.54}$$

where the suffix ν is omitted for simple notation. We see from (7.37), (7.38) and conditions (A1)–(A3) that $M_{n+p,q}^- M_{n+p,r}^\circ$ converges to a finite limit. On the other hand, by Proposition 7.9

$$\frac{\kappa^{p+\frac{r-m-1}{2}} |V_n|}{\sqrt{|V_{n+p-q}||V_{n-1}|}} = O(1),$$

$$\frac{\kappa^{p+\frac{r-m-1}{2}} |V_n|}{\sqrt{|V_{n+p-q}||V_{n+1}|}} = O(\kappa^{-1}),$$

$$\frac{\kappa^{p+\frac{r-m-1}{2}} \sqrt{|V_n|}}{\sqrt{|V_{n+p-q}|}} = O(\kappa^{-1/2}).$$

Then, (7.52)–(7.54) follows by (A2) and (A3). The proof is now complete. □

Theorem 7.11 (QCLT for regular graphs). *Let $\mathcal{G}^{(\nu)} = (V^{(\nu)}, E^{(\nu)})$ be a growing regular graph satisfying conditions (A1)–(A3) and A_ν its adjacency* *matrix. Let $(\Gamma, \{\Psi_n\}, B^+, B^-)$ be the interacting Fock space associated with $\{\omega_n\}$ and B° the diagonal operator associated with $\{\alpha_n\}$, where $\{\omega_n\}$ and $\{\alpha_n\}$ are given in conditions (A1)–(A3). Then we have*

$$\lim_{\nu} \frac{A_\nu^\epsilon}{\sqrt{\kappa(\nu)}} = B^\epsilon, \qquad \epsilon \in \{+, -, \circ\},$$

in the sense of stochastic convergence with respect to the vacuum states.

Proof. This is a particular case of Theorem 7.7. We need only to take $j = n = 0$ in (7.21). □

7.4 Deformed Vacuum States

We keep the same notations as in the previous section. The deformed vacuum state is defined by

$$\langle a \rangle_q = \langle Q\delta_o, a\delta_o \rangle, \qquad a \in \mathcal{A}(\mathcal{G}),$$

where

$$Q\delta_o = \sum_{x \in V} q^{\partial(x,o)} \delta_x = \sum_{n=0}^{\infty} q^n |V_n|^{1/2} \Phi_n. \tag{7.55}$$

Lemma 7.12. *The mean and variance of the adjacency matrix A in the deformed vacuum state are given by*

$$\langle A \rangle_q = q\kappa, \tag{7.56}$$

$$\Sigma_q^2(A) = \kappa(1-q)(1+q+qM(\omega_\circ|V_1)). \tag{7.57}$$

Proof. The proof is similar to that of Lemma 3.25. Noting that $|V_1| = \kappa$, we have

$$A\delta_o = \sum_{y \in V_1} \delta_y,$$

$$A^2\delta_o = \sum_{y \in V_1} A\delta_y = \sum_{z \in V_2} \omega_-(z)\delta_z + \sum_{y \in V_1} \omega_\circ(y)\delta_y + \kappa\delta_o.$$

Then, (7.56) is immediate. Similarly, we obtain

$$\begin{aligned}\langle A^2 \rangle_q &= \langle Q\delta_o, A^2\delta_o \rangle \\ &= q^2 \sum_{z \in V_2} \omega_-(z) + q \sum_{z \in V_1} \omega_\circ(z) + \kappa \\ &= q^2 |V_2| M(\omega_-|V_2) + q\kappa M(\omega_\circ|V_1) + \kappa. \end{aligned} \tag{7.58}$$

On the other hand, from the obvious relation $\kappa = \omega_-(x) + \omega_+(x) + \omega_\circ(x)$ and the matching identity it follows that

$$\kappa|V_1| = M(\omega_-|V_2)|V_2| + M(\omega_-|V_1)|V_1| + M(\omega_\circ|V_1)|V_1|,$$

see the proof of Lemma 7.8. Then,

$$\begin{aligned} M(\omega_-|V_2)|V_2| &= \kappa^2 - \kappa M(\omega_-|V_1) - \kappa M(\omega_\circ|V_1) \\ &= \kappa^2 - \kappa - \kappa M(\omega_\circ|V_1). \end{aligned} \tag{7.59}$$

Inserting (7.59) into (7.58), we obtain

$$\langle A^2\rangle_q = q^2(\kappa^2 - \kappa - \kappa M(\omega_\circ|V_1)) + q\kappa M(\omega_\circ|V_1) + \kappa. \tag{7.60}$$

Finally, in view of (7.56) and (7.60) we come to

$$\begin{aligned} \Sigma_q^2(A) &= \langle A^2\rangle_q - \langle A\rangle_q^2 \\ &= q^2(\kappa^2 - \kappa - \kappa M(\omega_\circ|V_1)) + q\kappa M(\omega_\circ|V_1) + \kappa - (\kappa q)^2 \\ &= (1-q^2)\kappa - (q^2 - q)\kappa M(\omega_\circ|V_1) \\ &= \kappa(1-q)(1+q+qM(\omega_\circ|V_1)), \end{aligned}$$

as desired. □

Lemma 7.12 is valid for any $q \in \mathbb{R}$ and the validity is independent of the positivity of the deformed vacuum state. However, for normalization we need to assume at least that $\Sigma_q^2(A) > 0$, or equivalently,

$$-\frac{1}{1+M(\omega_\circ|V_1)} < q < 1. \tag{7.61}$$

Under this condition the normalized adjacency matrix is given by

$$\frac{A - \langle A\rangle_q}{\Sigma_q(A)}.$$

Combining the quantum decomposition $A = A^+ + A^- + A^\circ$, we come to

$$\frac{A - \langle A\rangle_q}{\Sigma_q(A)} = \frac{\tilde{A}^+}{\Sigma_q(A)} + \frac{\tilde{A}^-}{\Sigma_q(A)} + \frac{\tilde{A}^\circ}{\Sigma_q(A)}, \tag{7.62}$$

where

$$\tilde{A}^\pm = A^\pm, \qquad \tilde{A}^\circ = A^\circ - \langle A\rangle_q.$$

We are interested in the asymptotics of each component in the right-hand side in (7.62) under conditions (A1)–(A3) keeping a certain balance with q. As is shown below, a natural scaling balance between κ and q is given by

$$q = q(\nu) \to 0, \qquad q\sqrt{\kappa} = q(\nu)\sqrt{\kappa(\nu)} \to \gamma, \tag{7.63}$$

where γ is a constant. Moreover, as is easily verified, for any $\gamma > -1/\alpha_2$ we may choose $q = q(\nu)$ satisfying (7.61) and (7.63).

Lemma 7.13. *Let $\mathcal{G}^{(\nu)} = (V^{(\nu)}, E^{(\nu)})$ be a growing regular graph satisfying conditions (A1)–(A3) and A_ν its adjacency matrix. For $\gamma > -1/\alpha_2$ let $q = q(\nu)$ be chosen in such a way that* (7.61) *and* (7.63) *are satisfied. Then*

$$\lim_\nu \frac{\Sigma_q^2(A_\nu)}{\kappa(\nu)} = 1 + \gamma\alpha_2, \tag{7.64}$$

$$\lim_\nu \frac{\langle A_\nu \rangle_q}{\Sigma_q(A_\nu)} = \frac{\gamma}{\sqrt{1+\gamma\alpha_2}}, \tag{7.65}$$

$$\lim_\nu q^n |V_n^{(\nu)}|^{1/2} = \frac{\gamma^n}{\sqrt{\omega_n \cdots \omega_1}}. \tag{7.66}$$

Proof. It follows from (7.57) that

$$\begin{aligned}\frac{\Sigma_q^2(A)}{\kappa} &= (1-q)(1+q+qM(\omega_\circ|V_1)) \\ &= (1-q)\left(1+q+q\sqrt{\kappa}\,\frac{M(\omega_\circ|V_1)}{\sqrt{\kappa}}\right).\end{aligned}$$

Then, (7.64) follows from condition (A3) and (7.63). By (7.56) and (7.64) we have

$$\lim_\nu \frac{\langle A_\nu \rangle_q}{\Sigma_q(A_\nu)} = \lim_\nu \frac{\kappa q}{\sqrt{\kappa(1+\gamma\alpha_2)}} = \frac{\gamma}{\sqrt{1+\gamma\alpha_2}},$$

which proves (7.65). Finally, using Proposition 7.9, we have

$$\lim_\nu q^{2n}|V_n^{(\nu)}| = \lim_\nu (q\sqrt{\kappa})^{2n}\,\frac{|V_n^{(\nu)}|}{\kappa^n} = \frac{\gamma^{2n}}{\omega_n \cdots \omega_1},$$

from which (7.66) follows. □

Theorem 7.14. *Let $\mathcal{G}^{(\nu)} = (V^{(\nu)}, E^{(\nu)})$ be a growing regular graph satisfying conditions (A1)–(A3). For $\gamma > -1/\alpha_2$ let $q = q(\nu)$ be chosen in such a way that* (7.61) *and* (7.63) *are satisfied. Let A_ν be the adjacency matrix and define*

$$\tilde{A}_\nu^\pm = A_\nu^\pm, \qquad \tilde{A}_\nu^\circ = A_\nu^\circ - \langle A_\nu \rangle_q.$$

Let $(\Gamma, \{\Psi_n\}, B^+, B^-)$ be the interacting Fock space associated with $\{\omega_n\}$ and B° the diagonal operator defined by $\{\alpha_n\}$, where $\{\omega_n\}$ and $\{\alpha_n\}$ are given in conditions (A1)–(A3). Define

$$\tilde{B}^\pm = \frac{B^\pm}{\sqrt{1+\gamma\alpha_2}}, \qquad \tilde{B}^\circ = \frac{B^\circ - \gamma}{\sqrt{1+\gamma\alpha_2}}.$$

Then we have

$$\lim_\nu \left\langle \Phi_j^{(\nu)}, \frac{\tilde{A}_\nu^{\epsilon_m}}{\Sigma_q(A_\nu)} \cdots \frac{\tilde{A}_\nu^{\epsilon_1}}{\Sigma_q(A_\nu)} \Phi_n^{(\nu)} \right\rangle = \langle \Psi_j, \tilde{B}^{\epsilon_m} \cdots \tilde{B}^{\epsilon_1} \Psi_n \rangle, \tag{7.67}$$

for any $\epsilon_1, \dots, \epsilon_m \in \{+, -, \circ\}$, $m = 1, 2, \dots,$ and $j, n = 0, 1, 2, \dots$.

Proof. This follows directly from Theorem 7.7. We need only to change constant factors according to (7.64) and (7.65) in Lemma 7.13. □

We are now in a position to discuss the limit in the deformed vacuum state:

$$\lim_{\nu} \left\langle Q\delta_o, \frac{\tilde{A}_\nu^{\epsilon_m}}{\Sigma_q(A_\nu)} \cdots \frac{\tilde{A}_\nu^{\epsilon_1}}{\Sigma_q(A_\nu)} \delta_o \right\rangle. \tag{7.68}$$

Although $Q\delta_o$ is an infinite sum as in (7.55), the actions of the operators $\tilde{A}_\nu^\epsilon$ are local so that (7.68) becomes

$$\lim_{\nu} \sum_{n=0}^{m} q^n |V_n^{(\nu)}|^{1/2} \left\langle \Phi_n^{(\nu)}, \frac{\tilde{A}_\nu^{\epsilon_m}}{\Sigma_q(A)} \cdots \frac{\tilde{A}_\nu^{\epsilon_1}}{\Sigma_q(A)} \Phi_0^{(\nu)} \right\rangle. \tag{7.69}$$

Then, applying Lemma 7.13 and Theorem 7.14 we see that (7.69) becomes

$$\sum_{n=0}^{m} \frac{\gamma^n}{\sqrt{\omega_n \cdots \omega_1}} \langle \Psi_n, \tilde{B}^{\epsilon_m} \cdots \tilde{B}^{\epsilon_1} \Psi_0 \rangle = \langle \Omega_\gamma, \tilde{B}^{\epsilon_m} \cdots \tilde{B}^{\epsilon_1} \Psi_0 \rangle,$$

where Ω_γ is the coherent vector. Summing up,

Theorem 7.15 (QCLT for regular graphs in the deformed vacuum states). *Notations and assumptions being the same as in Theorem 7.14, we have*

$$\lim_{\nu} \frac{\tilde{A}_\nu^\epsilon}{\Sigma_q(A)} = \tilde{B}^\epsilon, \tag{7.70}$$

in the sense of stochastic convergence with respect to the deformed vacuum state $\langle\cdot\rangle_q$ in the left-hand side and the coherent state $\langle\cdot\rangle_\gamma$ in the right-hand side.

Corollary 7.16. *Notations and assumptions being the same as in Theorem 7.14, we have*

$$\lim_{\nu} \left\langle \left(\frac{A_\nu - \langle A_\nu \rangle_q}{\Sigma_q(A_\nu)} \right)^m \right\rangle_q = \left\langle \Omega_\gamma, \left(\frac{B^+ + B^- + B^\circ - \gamma}{\sqrt{1+\gamma\alpha_2}} \right)^m \Psi_0 \right\rangle, \tag{7.71}$$

for any $m = 1, 2, \ldots$.

Furthermore, if the deformed vacuum state $\langle\cdot\rangle_q$ is positive, there exists a probability measure $\mu \in \mathfrak{P}_{\mathrm{fm}}(\mathbb{R})$ such that

$$\lim_{\nu} \left\langle \left(\frac{A_\nu - \langle A_\nu \rangle_q}{\Sigma_q(A_\nu)} \right)^m \right\rangle_q = \int_{-\infty}^{+\infty} x^m \mu(dx), \quad m = 1, 2, \ldots.$$

This μ is the asymptotic spectral distribution in the deformed vacuum state. By virtue of Corollary 7.16 we can find μ from the following relation:

$$\int_{-\infty}^{+\infty} x^m \mu(dx) = \left\langle \Omega_\gamma, \left(\frac{B^+ + B^- + B^\circ - \gamma}{\sqrt{1+\gamma\alpha_2}} \right)^m \Psi_0 \right\rangle, \quad m = 1, 2, \ldots.$$

7.5 Examples and Remarks

Since a distance-regular graph is regular, a large part of the general theory established in Chap. 3 is covered by our results in Sects. 7.3 and 7.4.

Proposition 7.17. *For a growing distance-regular graph $\mathcal{G}^{(\nu)} = (V^{(\nu)}, E^{(\nu)})$ with intersection numbers $\{p_{ij}^k(\nu)\}$, the conditions (A1)–(A3) are reduced to the following:*

(DR1) $\lim_\nu \kappa(\nu) = \infty$.
(DR2) *For each $n = 1, 2, \dots$ there exists a limit*

$$\omega_n = \lim_\nu p_{1,n-1}^n(\nu).$$

(DR3) *For each $n = 0, 1, 2, \dots$ there exists a limit*

$$\alpha_{n+1} = \lim_\nu \frac{p_{1,n}^n(\nu)}{\sqrt{\kappa(\nu)}}.$$

Proof. Since for a distance-regular graph $\omega_\epsilon(x)$ is constant on each V_n, in conditions (A2) and (A3) we can drop the conditions concerning the fluctuation. Therefore (A2) and (A3) are reduced to the existence of the limits

$$\omega_n = \lim_\nu M(\omega_-|V_n^{(\nu)}) = \lim_\nu p_{1,n-1}^n(\nu),$$

$$\alpha_{n+1} = \lim_\nu \frac{M(\omega_\circ|V_n^{(\nu)})}{\sqrt{\kappa(\nu)}} = \lim_\nu \frac{p_{1,n}^n(\nu)}{\sqrt{\kappa(\nu)}},$$

respectively. This completes the proof. □

Remark 7.18. Condition (DR) in Sect. 3.4 does not require $\kappa(\nu) \to \infty$. That is why the definition of ω_n in (3.23) takes a form different from (DR2). Note that condition (DR) is derived from (DR1)–(DR3).

The somewhat formal argument in Sect. 7.1 on $\mathbb{Z}^N$ is now easily justified. Condition (A1) is obvious since N is taken to be the growing parameter. For (A2) we first note that, if $N \geq n$,

$$|\{x \in V_n^{(N)} ; \omega_-(x) = k\}| = \binom{N}{k}\binom{n-1}{n-k} 2^k, \quad k = 1, 2, \dots, n. \tag{7.72}$$

It is then easily seen that

$$|\{x \in V_n^{(N)} ; \omega_-(x) = n\}| = \binom{N}{n} 2^n,$$

$$|V_n^{(N)}| = \sum_{k=1}^{n} \binom{N}{k}\binom{n-1}{n-k} 2^k,$$

and the ratio of which tends to 1 as $N \to \infty$. It then follows from Proposition 7.5 that (7.10), (7.11) in (A2) are satisfied with $\omega_n = n$. Equation (7.12) is apparent by the bound being n. Since $A^\circ = 0$, Condition (A3) is trivially satisfied. Thus, the growing integer lattice $\mathbb{Z}^N$ as $N \to \infty$ satisfies conditions (A1)–(A3). Note also that the adjacency matrix A_N of $\mathbb{Z}^N$ satisfies

$$\langle A_N \rangle_q = 2Nq, \quad \Sigma_q^2(A_N) = 2N(1-q^2).$$

Theorem 7.19 (CLT for growing integer lattices). *Let A_N be the adjacency matrix of the integer lattice $\mathbb{Z}^N$. Then,*

$$\lim_{N\to\infty} \left\langle \left(\frac{A_N}{\sqrt{2N}} \right)^m \right\rangle_o = \frac{1}{\sqrt{2\pi}} \int_{-\infty}^{+\infty} x^m \mathrm{e}^{-x^2/2}\, dx, \quad m = 1, 2, \dots.$$

Moreover, for any $\gamma \in \mathbb{R}$ we have

$$\lim_{\substack{q\sqrt{2N}\to\gamma \\ q\to 0, N\to\infty}} \left\langle \left(\frac{A_N - \langle A_N\rangle_q}{\Sigma_q(A_N)} \right)^m \right\rangle_q = \frac{1}{\sqrt{2\pi}} \int_{-\infty}^{+\infty} x^m \mathrm{e}^{-x^2/2}\, dx, \quad m = 1, 2, \dots.$$

Proof. It is sufficient to prove the second assertion since the first one follows by taking $\gamma = 0$. We have already observed that the limit is described by the Boson Fock space ($\{\omega_n = n\}$) and $B^\circ = 0$ ($\{\alpha_n = 0\}$). Then, applying Corollary 7.16, we obtain

$$\lim_{\substack{q\sqrt{2N}\to\gamma \\ q\to 0, N\to\infty}} \left\langle \left(\frac{A_N - \langle A_N\rangle_q}{\Sigma_q(A_N)} \right)^m \right\rangle_q = \langle \Omega_\gamma, (B^+ + B^- - \gamma)^m \Psi_0 \rangle,$$

for $m = 1, 2, \dots$. It is known (Exercise 5.6) that the right-hand side coincides with

$$\langle \Psi_0, (B^+ + B^-)^m \Psi_0 \rangle = \frac{1}{\sqrt{2\pi}} \int_{-\infty}^{+\infty} x^m \mathrm{e}^{-x^2/2}\, dx,$$

which completes the proof. □

The Coxeter groups provide interesting examples of growing Cayley graphs. Let Σ be a countable infinite set. A function $m : \Sigma \times \Sigma \to \{1, 2, \dots\} \cup \{\infty\}$ is called a *Coxeter matrix* if (i) $m(s,s) = 1$ for all $s \in \Sigma$, and (ii) $m(s,t) = m(t,s) \geq 2$ for $s \neq t$. Let $\Sigma_1 \subset \Sigma_2 \subset \cdots \subset \Sigma$ be an increasing sequence of subsets of Σ such that $|\Sigma_N| = N$ and $\Sigma = \bigcup \Sigma_N$. For each $N \geq 1$ let G_N be the group generated by Σ_N subject only to the relations:

$$(st)^{m(s,t)} = e, \qquad s, t \in \Sigma_N, \tag{7.73}$$

where e stands for the unit. In case of $m(s,t) = \infty$ we understand that st is of infinite order. The pair (G_N, Σ_N) is called a *Coxeter system* of rank N (i.e., $|\Sigma_N| = N$) and G_N is called a *Coxeter group*. It is known that each $s \in \Sigma_N$ has order two, namely, is not reduced to the unit (this is not very trivial).

The corresponding Cayley graph is denoted by the same symbol (G_N, Σ_N). We consider the family of Cayley graphs (G_N, Σ_N), $N = 1, 2, \dots$, as a growing regular graph.

In fact, it is shown that the inclusion $\Sigma_N \to \Sigma_{N+1}$ extends uniquely an injective homomorphism $G_N \to G_{N+1}$. The inductive limit group, denoted simply by G, is called the *infinite Coxeter group* associated with a Coxeter matrix $\{m(s,t)\}$.

By definition each $g \in G_N$, $g \neq e$, admits an expression of the form

$$x = s_1 s_2 \cdots s_r, \qquad s_i \in \Sigma_N.$$

If r is as small as possible, the expression is called a *reduced expression* and the number $r = |x|$ is called the *length* of x. The length function is well defined on G.

Lemma 7.20. *For any $s \in \Sigma$ and $x \in G$ we have*

$$|sx| = |x| \pm 1, \qquad |xs| = |x| \pm 1.$$

Proof. There exists a unique homomorphism (character) $\chi : G \to \{\pm 1\}$ such that

$$\chi(s) = -1, \qquad s \in \Sigma.$$

For any $x \in G$, taking a reduced expression $x = s_1 s_2 \cdots s_r$, $r = |x|$, we have

$$\chi(x) = \chi(s_1)\chi(s_2)\cdots\chi(s_r) = (-1)^{|x|}, \qquad x \in G.$$

Using this formula, we obtain

$$\chi(sx) = (-1)^{|sx|},$$

and

$$\chi(sx) = \chi(s)\chi(x) = (-1)(-1)^{|x|} = (-1)^{|x|+1}.$$

Therefore, $|sx| \equiv |x| + 1 \pmod 2$. This must be compatible with the triangle inequality (Exercise 7.3)

$$|x| - 1 \leq |sx| \leq |x| + 1, \qquad x \in G.$$

Thus $|sx| = |x| \pm 1$ follows. □

Lemma 7.21 (Deletion condition). *Let $g \in G$ be expressed in the form*

$$g = s_1 s_2 \cdots s_m, \qquad s_i \in \Sigma. \tag{7.74}$$

If $|g| < m$, then there exist a pair of indices $1 \leq i < j \leq m$ such that

$$g = s_1 \cdots \check{s}_i \cdots \check{s}_j \cdots s_m,$$

where $\check{s}$ stands for deletion. Therefore, given $g \in G$ of the form (7.74), its reduced expression is obtained by deleting even number of s_i appearing therein.

The proof is omitted. The deletion condition is quite useful in the study of the Coxeter groups. Lemma 7.20 is also an immediate consequence.

Lemma 7.22. *If $s_1, s_2, \dots, s_n \in \Sigma$ are mutually distinct, then $g = s_1 s_2 \cdots s_n$ is a reduced expression.*

Proof. For $n = 1$ the assertion is obvious since Σ is injectively contained in G. Let $n \geq 2$. Suppose that $g = s_1 s_2 \cdots s_n$ is not a reduced expression though $s_1, s_2, \dots, s_n \in \Sigma$ are mutually distinct. Then by Lemma 7.21,

$$s_1 \cdots s_n = s_1 \cdots \check{s}_i \cdots \check{s}_j \cdots s_n$$

and hence

$$s_i = s_{i+1} \cdots s_{j-1} s_j s_{j-1} \cdots s_{i+1}.$$

Since the right-hand side is of length 1, deleting an even number of elements from the right-hand side leads to a reduced expression. The obtained reduced expression should be one of $\{s_{i+1}, \dots, s_j\}$. This contradicts the assumption that $s_1, \dots, s_n$ are mutually distinct. □

We next study the Coxeter group associated with a Coxeter matrix satisfying $m(s,t) \geq 3$ for any pair $s \neq t$. In that case, the Cayley graph has no cycle with length less than six, i.e., contains neither triangle, square, nor pentagon.

Lemma 7.23. *Assume that $m(s,t) \geq 3$ for any pair $s \neq t$. If $s_1, \dots, s_n \in \Sigma$ are mutually distinct and the relation*

$$s_1 \cdots s_n = sx$$

holds for some $s \in \Sigma$ and $x \in G$ of length $n-1$, then $s = s_1$.

Proof. We prove the assertion by induction on n. For $n = 1$ the assertion is obvious. Assume that $s_1 s_2 = sx$ holds where $s_1, s_2 \in \Sigma$ are mutually distinct, $s \in \Sigma$ and $x \in G$ of length 1. From $s = s_1 s_2 x$ we see easily that $s = s_1$ or $s = s_2$ or $s = x$. If $s = x$ happens, we have $s_1 = s_2$ which yields contradiction. If $s = s_2$ happens, $x = s_1$ and $(s_1 s_2)^2 = e$ which is again contradiction. Consequently, $s = s_1$.

Assume that the assertion is valid up to $n - 1$, $n \geq 2$. Since

$$s s_1 \cdots s_n = x \tag{7.75}$$

is of length $n - 1$, deleting two elements from the left-hand side we obtain a reduced expression of x. If these two elements are chosen from $\{s_1, \dots, s_n\}$, say, s_i, s_j $(i < j)$, we come back to

$$s_1 \cdots \check{s}_i \cdots \check{s}_j \cdots s_n = sx = s_1 \cdots s_n,$$

which is a reduced expression by Lemma 7.22. This is contradiction. Hence, to get a reduced expression of x in (7.75), we need to delete s and s_i for some $i = 1, \dots, n$. In that case we come to

$$s_1 \cdots \check{s}_i \cdots s_n = x,$$

and hence

$$ss_1 \cdots s_{i-1} = s_1 \cdots s_i. \tag{7.76}$$

If $1 \le i \le n-1$, by the assumption of induction we have $s = s_1$. Suppose $i = n$, i.e.,

$$ss_1 \cdots s_{n-1} = s_1 \cdots s_n.$$

By a simple argument with the deletion condition we see that $s \in \{s_1, \dots, s_n\}$. If $s = s_j$, $1 \le j \le n-1$, then

$$s_n = s_{n-1} \cdots s_1 s_j s_1 \cdots s_{n-1},$$

which implies that s_n coincides with some of $\{s_1, \dots s_{n-1}\}$. But this contradicts the assumption. Hence $s = s_n$, i.e.,

$$s_n s_1 \cdots s_{n-1} = s_1 \cdots s_n. \tag{7.77}$$

We shall prove that this does not occur. Note first that (7.77) is equivalent to the following:

$$(s_{n-2} \cdots s_1) s_n (s_1 \cdots s_{n-2}) s_{n-1} = s_{n-1} s_n.$$

Since this is of length 2, deleting an even number of elements from the left-hand side, we obtain a reduced expression of length 2, say, tt'. This is the case of $n = 2$ so we know that $t = s_{n-1}$. But this is impossible. □

Consider the Cayley graph (G_N, Σ_N) with $e \in G_N$ being an origin. We consider as usual the stratification

$$G_N = \bigcup_{n=0}^{\infty} V_n^{(N)}.$$

Statistics of $\omega_-(x)$ is of importance. We see from Lemma 7.20 that $\omega_\circ(x) = 0$ for all $x \in G_N$.

Lemma 7.24. *Assume that $m(s,t) \ge 3$ for any pair $s, t \in \Sigma$, $s \ne t$. Then, for any $n = 1, 2, \dots$ we have*

$$\lim_{N \to \infty} \frac{|\{x \in V_n^{(N)}\,;\, \omega_-(x) = 1\}|}{|V_n^{(N)}|} = 1. \tag{7.78}$$

Proof. The assertion is apparent for $n = 1$. We assume that $n \geq 2$. It follows from Lemma 7.23 that $\omega_-(x) = 1$ for any $x \in V_n^{(N)}$ which admits an expression of the form $x = s_1 \cdots s_n$ with mutually distinct $s_1, \dots, s_n \in \Sigma_N$. The number of such x is $N(N-1)\cdots(N-n+1)$ so that

$$|\{x \in V_n^{(N)} \,;\, \omega_-(x) = 1\}| \geq N(N-1)\cdots(N-n+1).$$

By virtue of the obvious inequality $|V_n^{(N)}| \leq N^n$, we have

$$\frac{|\{x \in V_n^{(N)} \,;\, \omega_-(x) = 1\}|}{|V_n^{(N)}|} \geq \frac{N(N-1)\cdots(N-n+1)}{N^n},$$

from which (7.78) follows. □

Lemma 7.25. *Assume that $m(s,t) \geq 3$ for any pair $s, t \in \Sigma$, $s \neq t$. Then $\omega_-(x) \leq 2$ for all $x \in G$.*

The proof is a tedious application of the deletion condition.

Theorem 7.26 (QCLT for Coxeter groups). *Let (G, Σ) be an infinite Coxeter group with a Coxeter matrix $\{m(s,t)\}$ such that $m(s,t) \geq 3$ for any pair $s, t \in \Sigma$, $s \neq t$. Let $\Sigma_1 \subset \Sigma_2 \subset \cdots$ be an increasing sequence of subsets of Σ such that $\bigcup_{N=1}^{\infty} \Sigma_N = \Sigma$, and consider the Cayley graph of the Coxeter group (G_N, Σ_N) and its adjacency matrix A_N. Then $A_N^\circ = 0$ and*

$$\lim_{N\to\infty} \frac{A_N^{\pm}}{\sqrt{|\Sigma_N|}} = B^{\pm},$$

in the sense of stochastic convergence with respect to the vacuum state, where $B^{\pm}$ are the annihilation and creation operators in the free Fock space.

Proof. It is sufficient to show conditions (A1)–(A3). First (A1) is obvious, since the degree of G_N is $|\Sigma_N|$, which tends to the infinity by assumption. Conditions (7.10) and (7.11) in (A2) follow from Lemma 7.24 due to Proposition 7.5. Moreover, the sequence therein is $\omega_n \equiv 1$ so that the limit is described by the free Fock space. Condition (7.12) in (A2) follows from Lemma 7.25 with $W_n = 2$. Finally, (A3) is obvious since $\omega_\circ(x) = 0$ for all $x \in G_N$. Consequently, our assertion is an immediate consequence of Theorem 7.11. □

It is known that the symmetric group $S(N)$ is generated by the successive transpositions

$$\sigma_1 = (12), \quad \sigma_2 = (23), \quad \dots, \quad \sigma_{N-1} = (N-1\,N).$$

We set $\Sigma_N = \{\sigma_1, \sigma_2, \dots, \sigma_{N-1}\}$. Then $(S(N), \Sigma_N)$ becomes a Coxeter group. Note that the Coxeter matrix is given by

$$m(i,j) = \begin{cases} 3, & |i-j| = 1, \\ 2, & |i-j| \geq 2. \end{cases}$$

Therefore, Theorem 7.26 is not applicable. Instead, we have the following:

Theorem 7.27 (QCLT for symmetric groups). *Let A_N be the adjacency matrix of the Cayley graph $(S(N), \Sigma_N)$. Then $A_N^\circ = 0$ and*

$$\lim_{N\to\infty} \frac{A_N^\pm}{\sqrt{N-1}} = B^\pm,$$

in the sense of stochastic convergence with respect to the vacuum state, where $B^\pm$ are the annihilation and creation operators in the Boson Fock space.

Finally, we discuss a growing regular graph which yields a periodic Jacobi sequence.

Theorem 7.28. *Let $a, b, k \ge 1$ be integers. Define $\kappa = abk$ and*

$$\omega_1 = 1, \quad \omega_2 = a, \quad \omega_3 = b, \quad \omega_4 = a, \quad \omega_5 = b, \dots. \tag{7.79}$$

Then there exists a regular graph $\mathcal{G} = (V, E)$ of degree κ which admits a stratification $V = \bigcup_{n=0}^\infty V_n$ such that $\omega_-(x) = \omega_n$ for $x \in V_n$, $n = 1, 2, \dots$.

Proof. We shall construct a regular graph having the desired properties.

Step 1. Let V_0 and V_1 consist of a single vertex o (origin) and of κ vertices, respectively. We draw edges connecting each vertex in V_1 and o. Then o has κ edges.

Step 2. We construct V_2 and edges connecting between V_1 and V_2. The number of vertices in V_2 is determined by counting such edges. Since each $x \in V_1$ must have $\kappa - 1$ edges connecting with vertices in V_2 and each $y \in V_2$ has a edges connecting with vertices in V_1 by request, we have the relation:

$$(\kappa - 1)|V_1| = a|V_2|.$$

Thus,

$$|V_2| = \frac{\kappa(\kappa-1)}{a}, \tag{7.80}$$

which is an integer for $\kappa = abk$. We must prove that the vertices in V_1 and those in V_2 can be connected by edges in such a way that each vertex $y \in V_2$ has a edges and each $x \in V_1$ has $\kappa - 1$ edges. This is possible by looking at

$$|V_1| = \kappa = \frac{\kappa}{a} \times a, \qquad |V_2| = \frac{\kappa(\kappa-1)}{a} = \frac{\kappa}{a} \times (\kappa - 1).$$

We can divide V_1 and V_2 into $\kappa/a = bk$ subsets:

$$V_1 = \bigcup_{i=1}^{bk} V_1^{(i)}, \quad V_2 = \bigcup_{i=1}^{bk} V_2^{(i)} \quad \text{with} \quad |V_1^{(i)}| = a, \quad |V_2^{(i)}| = \kappa - 1.$$

For each i, we draw edges between $V_1^{(i)}$ and $V_2^{(i)}$ in such a way that any pair $x \in V_1^{(i)}$ and $y \in V_2^{(i)}$ is connected. For distinct i, j there is no edge connecting between $V_1^{(i)}$ and $V_2^{(j)}$. In this way, each $x \in V_1$ has κ edges with $\omega_-(x) = 1$ and each $y \in V_2$ has a edges connecting with vertices in V_1.

Step 3. We construct V_3 and edges connecting between V_2 and V_3. The number of vertices in V_3 is determined by the relation:

$$(\kappa - a)|V_2| = b|V_3|.$$

Hence, in view of (7.80) we have

$$|V_3| = \frac{\kappa(\kappa-1)(\kappa-a)}{ab}.$$

Since

$$|V_2| = \frac{\kappa(\kappa-1)}{a} = \frac{\kappa(\kappa-1)}{ab} \times b, \qquad |V_3| = \frac{\kappa(\kappa-1)}{ab} \times (\kappa - a),$$

a similar argument as in Step 2 allows us to draw edges between V_2 and V_3 in such a way that each vertex in V_2 has $\kappa - a$ edges and each vertex in V_3 has b edges. In total each vertex in $y \in V_2$ has κ edges with $\omega_-(y) = a$.

Step 4. This procedure can be applied repeatedly and we obtain a regular graph of degree κ having the desired properties. □

Remark 7.29. In fact, the number of vertices in each stratum is given by

$$|V_0| = 1, \qquad |V_1| = \kappa,$$
$$|V_{2n}| = \frac{\kappa(\kappa-1)}{a}\left(\frac{(\kappa-a)(\kappa-b)}{ab}\right)^{n-1}, \qquad n \geq 1,$$
$$|V_{2n+1}| = \frac{\kappa(\kappa-1)(\kappa-a)}{ab}\left(\frac{(\kappa-a)(\kappa-b)}{ab}\right)^{n-1}, \qquad n \geq 1.$$

Remark 7.30. There are three trivial cases: (i) $\kappa = 1$, (ii) $\kappa = a \geq 2$ and $b = 1$ and (iii) $\kappa = b \geq 2$ and $a = 1$. Except these cases the regular graph constructed in Theorem 7.28 has infinitely many strata.

Let $a, b \geq 1$ be fixed integers and consider $\mathcal{G}_k$ constructed in Theorem 7.28 as a growing regular graph as $k \to \infty$. Conditions (A1)–(A3) are easily verified. Then, as an immediate consequence of Theorem 7.11, the asymptotic spectral distribution in the vacuum state at $o \in V$ is determined by the Jacobi coefficient

$$\{\omega_n\} = \{1, a, b, a, b, a, b, \dots\}, \qquad \{\alpha_n \equiv 0\}. \tag{7.81}$$

Obviously, the corresponding probability measure is unique. After a routine calculation of the continued fraction (Exercise 7.5) and application of the Stieltjes inversion formula, we obtain the following:

Theorem 7.31. *For $a > 0$ and $b > 0$ we define $\rho_{a,b}(x)$ by*

$$\rho_{a,b}(x) = \frac{\sqrt{2(a+b)x^2 - x^4 - (a-b)^2}}{2\pi|x|\{(b-1)x^2 + a - b + 1\}},$$

for

$$-\sqrt{a} - \sqrt{b} \le x \le -|\sqrt{a} - \sqrt{b}|, \qquad |\sqrt{a} - \sqrt{b}| \le x \le \sqrt{a} + \sqrt{b},$$

and $\rho_{a,b}(x) = 0$ otherwise. Then the probability measure μ whose Jacobi coefficient is (7.81) is given as follows:

(i) *If $1 \le b \le a - 1$,*

$$\mu(dx) = \left(1 - \frac{1}{a-b+1}\right)\delta_0(dx) + \rho_{a,b}(x)dx.$$

(ii) *If $b = a$ or $b = a + 1$,*

$$\mu(dx) = \rho_{a,b}(x)dx.$$

(iii) *If $b \ge a + 2$,*

$$\mu(dx) = \frac{1}{2}\left(1 - \frac{a}{(b-1-a)(b-1)}\right)(\delta_\xi + \delta_{-\xi})(dx) + \rho_{a,b}(x)dx,$$

where $\xi = \sqrt{(b-1-a)/(b-1)}$.

Exercises

7.1. Let $s \ge 0$ and $n \ge 1$ be integers. Show that the number of non-negative integral solutions $(x_1, x_2, \dots, x_n)$ to the equation

$$x_1 + x_2 + \cdots + x_n = s$$

is given by $\binom{s+n-1}{s}$.

7.2. Prove the formula (7.72). [Hint: Use Exercise 7.1.]

7.3. Let (G, Σ) be a Coxeter group. Prove the following properties for the length function:

(1) $|x| = |x^{-1}|$ for $x \in G$.
(2) For $x \in G$, $|x| = 1$ if and only if $x \in \Sigma$.
(3) $|x| - |y| \le |xy| \le |x| + |y|$ for $x, y \in G$.
(4) $|x| - 1 \le |sx| \le |x| + 1$ for $x \in G$ and $s \in \Sigma$.

7.4. Let $a_1, \dots, a_m, k \geq 1$ be integers. Define $\kappa = a_1 a_2 \cdots a_m k$ and

$$\omega_1 = 1, \quad \omega_2 = a_1, \quad \dots, \quad \omega_{m+1} = a_m,$$
$$\omega_{jm+i+1} = a_i\,, \qquad j \geq 0, \quad 1 \leq i \leq m.$$

Prove that there exists a regular graph $\mathcal{G} = (V, E)$ of degree κ which admits a stratification $V = \bigcup_{n=0}^{\infty} V_n$ such that $\omega_-(x) = \omega_n$ for $x \in V_n$, $n \geq 1$. [Hint: Modify the proof of Theorem 7.28.]

7.5. Let $a > 0$ and $b > 0$ be constant numbers. Derive the following formula:

$$\frac{1}{z} \, \frac{1}{-} \, \frac{1}{z} \, \frac{a}{-} \, \frac{a}{z} \, \frac{b}{-} \, \frac{b}{z} \, \frac{a}{-} \, \frac{a}{z} \, \frac{b}{-} \, \frac{b}{z} - \cdots$$

$$= \frac{(2b-1)z^2 + a - b - \sqrt{z^4 - 2(a+b)z^2 + (a-b)^2}}{2z\{(b-1)z^2 + (a-b+1)\}}. \tag{7.82}$$

Determine the branch of the analytic square root.

Notes

Conditions (A1)–(A3) are finally formulated in Hora–Obata [112] after some preliminary consideration. A similar formulation was given in Hashimoto–Hora–Obata [96] in case of $A^\circ = 0$. The method of quantum decomposition clarifies the mechanism of how the coherent states emerge in the limit of deformed vacuum states.

The Coxeter groups provide an interesting class of growing regular graphs. For the deletion condition (Lemma 7.21) see, e.g., Humphreys [114, Sect. 5]. The proof of Lemma 7.25 as well as more properties of geodesics in the Cayley graph of the Coxeter group are given in Szwarc [203].

The classical reduction of Theorem 7.26 is that the Wigner semicircle law is the asymptotic spectral distribution obtained from the Coxeter groups with the off-diagonal elements of the Coxeter matrix being ≥ 3. This result was proved by Fendler [78] in a different manner.

There are many different choices of generators of $S(N)$. Let $\mathcal{T}_N$ be the set of transpositions and consider a Cayley graph $(S(N), \mathcal{T}_N)$. It is then easily seen that conditions (A1)–(A3) are satisfied with $\omega_n = n$ and $W_n = n(n+1)/2$. Hence, the quantum components $A_N^{\pm}$ converge to the creation and annihilation operators on the Boson Fock space and the situation is the same as in Theorem 7.27. This point of view for QCLT on the symmetric group will be further developed in Chaps. 10 and 11, see Theorems 11.12 and 11.13 among others. While, if we take $\{(12), (13), \dots, (1N)\}$ to be the set of generators of $S(N)$, the limit is described by the free Fock space and the asymptotic spectral distribution is the Wigner semicircle law, see Biane [23].

A probability measure with periodic Jacobi coefficient was first derived as an asymptotic spectral distribution (Theorems 7.28 and 7.31) in Hora–Obata [111]. Bożejko (2001) introduced a one-parameter deformation of the free product called the r-free convolution, $0 \le r < 1$, by using the conditionally free products of states developed by Bożejko–Leinert–Speicher [43], Bożejko–Speicher [44]. Although the range of the parameter r is different, it is noticeable that the central limit measure with respect to the r-convolution is given as in (7.82) with $a = r$, $b = 1$. In this connection see also Bożejko–Krystek–Wojakowski [41].

8

Comb Graphs and Star Graphs

There are several different notions of independence in quantum probability. In this chapter we study a growing graph whose adjacency matrix is decomposed into a sum of independent random variables.

8.1 Notions of Independence

Consider two classical random variables X, Y defined on a probability space $(\Omega, \mathcal{F}, P)$. If they are independent, by the product formula we obtain

$$\mathbb{E}(XYXXYXY) = \mathbb{E}(X^4Y^3) = \mathbb{E}(X^4)\mathbb{E}(Y^3). \tag{8.1}$$

In general, such a statistical quantity as above is called a *mixed moment* or a *correlation coefficient*. We understand that the independence gives a rule of calculating mixed moments. In quantum probability theory many different rules can be introduced because of non-commutativity of random variables, where, for example, the first equality in (8.1) may be no longer guaranteed. In this section, we shall mention four different notions of independence, which have been up to now considered most fundamental.

Definition 8.1 (Commutative independence). Let $(\mathcal{A}, \varphi)$ be an algebraic probability space. A family $\{\mathcal{A}_\lambda\}$ of $*$-subalgebras of $\mathcal{A}$ is called *commutative independent* or *tensor independent* (with respect to φ) if

$$\varphi(a_1 \cdots a_m), \qquad a_i \in \mathcal{A}_{\lambda_i},$$

is factorized as follows:

(i) when $\lambda_1 \notin \{\lambda_2, \dots, \lambda_m\}$,

$$\varphi(a_1 \cdots a_m) = \varphi(a_1)\varphi(a_2 \cdots a_m);$$

A. Hora and N. Obata: Comb Graphs and Star Graphs. In: A. Hora and N. Obata, Quantum Probability and Spectral Analysis of Graphs, Theoretical and Mathematical Physics, 205–247 (2007)
DOI 10.1007/3-540-48863-4_8

(ii) otherwise, letting r be the smallest number such that $\lambda_1 = \lambda_r$,

$$\varphi(a_1 \cdots a_m) = \varphi(a_2 \cdots a_{r-1}(a_1 a_r) a_{r+1} \cdots a_m).$$

Note that neither $\mathcal{A}_\lambda$ nor $\mathcal{A}$ is assumed to be commutative.

Definition 8.2 (Free independence). Let $(\mathcal{A}, \varphi)$ be an algebraic probability space. A family $\{\mathcal{A}_\lambda\}$ of $*$-subalgebras of $\mathcal{A}$ is called *free independent* (with respect to φ) if

$$\varphi(a_1 \cdots a_m) = 0$$

holds for any $a_i \in \mathcal{A}_{\lambda_i}$ with $\varphi(a_i) = 0$, $i = 1, 2, \dots, m$, and $\lambda_1 \neq \lambda_2 \neq \cdots \neq \lambda_m$ (any two consecutive indices are different).

Definition 8.3 (Boolean independence). Let $(\mathcal{A}, \varphi)$ be an algebraic probability space and $\mathcal{A}_\lambda \subset \mathcal{A}$ a subset which is closed under the algebraic operations and involution (i.e., a $*$-subalgebra which does not necessarily contain the identity $1_\mathcal{A}$ of $\mathcal{A}$). We say that $\{\mathcal{A}_\lambda\}$ is *Boolean independent* (with respect to φ) if

$$\varphi(a_1 \cdots a_m) = \varphi(a_1)\varphi(a_2 \cdots a_m)$$

for any $a_i \in \mathcal{A}_{\lambda_i}$ with $\lambda_1 \neq \lambda_2 \neq \cdots \neq \lambda_m$.

We need notation. Let $(\Lambda, <)$ be a totally ordered set and consider a finite sequence

$$\lambda_1 \neq \lambda_2 \neq \cdots \neq \lambda_p \neq \cdots \neq \lambda_m \tag{8.2}$$

of elements in Λ, $m \geq 2$. Then λ_p is called a *peak* in (8.2) if (i) $1 < p < m$, $\lambda_{p-1} < \lambda_p$ and $\lambda_p > \lambda_{p+1}$; or (ii) $p = 1$ and $\lambda_1 > \lambda_2$; or (iii) $p = m$ and $\lambda_{m-1} < \lambda_m$.

Definition 8.4 (Monotone independence). Let $(\mathcal{A}, \varphi)$ be an algebraic probability space. Let $(\Lambda, <)$ be a totally ordered set and for each $\lambda \in \Lambda$, $\mathcal{A}_\lambda \subset \mathcal{A}$ a subset which is closed under the algebraic operations and involution. We say that $\{\mathcal{A}_\lambda\}$ is *monotone independent* (with respect to φ) if

$$\varphi(a_1 \cdots a_m) = \varphi(a_p)\varphi(a_1 \cdots \breve{a}_p \cdots a_m) \qquad (\breve{a}_p\text{: omission})$$

for any $a_i \in \mathcal{A}_{\lambda_i}$ with λ_p being a peak in $\lambda_1 \neq \lambda_2 \neq \cdots \neq \lambda_m$.

The above definitions indicate only the first step of calculating a mixed moment $\varphi(a_1 \cdots a_m)$. Table 8.1 shows how mixed moments of $a \in \mathcal{A}_1$ and $b \in \mathcal{A}_2$ are factorized when $\{\mathcal{A}_1, \mathcal{A}_2\}$ is commutative, free, Boolean, or monotone independent (for the monotone independence the natural order $1 < 2$ is adopted). In actual computation the following formulae are useful.

Lemma 8.5. *Let $(\mathcal{A}, \varphi)$ be an algebraic probability space. Let $a_i \in \mathcal{A}$ and set $\bar{a}_i = a_i - \varphi(a_i)$, $i = 1, 2, \dots, m$. Then*

$$a_1 \cdots a_m = a_1 \cdots \bar{a}_i \cdots a_m + \varphi(a_i) a_1 \cdots \check{a}_i \cdots a_m \,, \tag{8.3}$$

$$\varphi(a_1 \cdots a_m) = \varphi(\bar{a}_1 \cdots \bar{a}_m) + \sum_{i=1}^{m} \varphi(a_i) \varphi(\underbrace{a_1 \cdots a_{i-1}}_{i-1} \check{a}_i \underbrace{\bar{a}_{i+1} \cdots \bar{a}_m}_{m-i}). \tag{8.4}$$

Proof. Exercise. □

Table 8.1. Illustrating the factorization rules

	Commutative	Free	Boolean	Monotone
$\varphi(aba)$	$\varphi(a^2)\varphi(b)$	$\varphi(a^2)\varphi(b)$	$\varphi(a)^2\varphi(b)$	$\varphi(a^2)\varphi(b)$
$\varphi(bab)$	$\varphi(a)\varphi(b^2)$	$\varphi(a)\varphi(b^2)$	$\varphi(a)\varphi(b)^2$	$\varphi(a)\varphi(b)^2$
$\varphi(abab)$	$\varphi(a^2)\varphi(b^2)$	$\varphi(a)^2\varphi(b^2)$ $+\varphi(a^2)\varphi(b)^2$ $-\varphi(a)^2\varphi(b)^2$	$\varphi(a)^2\varphi(b)^2$	$\varphi(a^2)\varphi(b)^2$

Example 8.6. Let $(\Omega, \mathcal{F}, P)$ be a classical probability space and $(L^{\infty-}(\Omega), \mathbb{E})$ the associated algebraic probability space, see Example 1.4. Let $\{X_\lambda\}$ be a family of real random variables having moments of all orders and denote by $\mathcal{A}_\lambda$ the $*$-subalgebra of $L^{\infty-}(\Omega)$ generated by X_λ. If $\{X_\lambda\}$ is independent (in the sense of the classical probability theory), then $\{\mathcal{A}_\lambda\}$ is commutative independent.

Example 8.7. The *free Fock space* over a Hilbert space $\mathcal{H}$ is the direct sum Hilbert space defined by

$$\Gamma_{\text{free}}(\mathcal{H}) = \bigoplus_{n=0}^{\infty} \mathcal{H}^{\otimes n}, \qquad \mathcal{H}^{\otimes 0} = \mathbb{C}\Omega,$$

where $\mathcal{H}^{\otimes 0}$ is a one-dimensional Hilbert space equipped with a distinguished unit vector Ω called the *vacuum vector*. For $h \in \mathcal{H}$ we define the *left annihilation operator* $l(h) \in \mathcal{B}(\Gamma_{\text{free}}(\mathcal{H}))$ by

$$l(h)\Omega = 0, \qquad l(h)(f_1 \otimes \cdots \otimes f_n) = \langle h, f_1 \rangle f_2 \otimes \cdots \otimes f_n.$$

The adjoint operator, called the *left creation operator*, is characterized by

$$l^*(h)\Omega = h, \qquad l^*(h)(f_1 \otimes \cdots \otimes f_n) = h \otimes f_1 \otimes \cdots \otimes f_n.$$

Taking an orthonormal set $\{e_n\}$ of $\mathcal{H}$, define $\mathcal{A}_n \subset \mathcal{B}(\Gamma_{\mathrm{free}}(\mathcal{H}))$ to be the $*$-subalgebra generated by $\{l(e_n), l^*(e_n)\}$. Then $\{\mathcal{A}_n\}$ is free independent with respect to the vacuum state Ω. The proof is straightforward. The free Fock space introduced in Example 1.32 is a special case of $\Gamma_{\mathrm{free}}(\mathcal{H})$, where $\mathcal{H}$ is one-dimensional.

Theorem 8.8. *For $n = 1, 2, \dots, N$ let $\mathcal{H}_n$ be a Hilbert space with a distinguished unit vector $\Omega_n \in \mathcal{H}_n$ and consider an algebraic probability space:*

$$(\mathcal{B}(\mathcal{H}_1 \otimes \cdots \otimes \mathcal{H}_N), \Omega_1 \otimes \cdots \otimes \Omega_N).$$

Let $P_n \in \mathcal{B}(\mathcal{H}_n)$ be the projection onto the one-dimensional subspace spanned by Ω_n and $\mathcal{A}_n$ the set of operators of the form

$$P_1 \otimes \cdots \otimes P_{n-1} \otimes S_n \otimes P_{n+1} \otimes \cdots \otimes P_N, \qquad S_n \in \mathcal{B}(\mathcal{H}_n). \tag{8.5}$$

Then $\{\mathcal{A}_n\}$ is Boolean independent.

Proof. Note that $\mathcal{A}_n$ is closed under the algebraic operations and involution. ($\mathcal{A}_n$ might contain the identity or might not.) For simplicity we set

$$\mathcal{H} = \mathcal{H}_1 \otimes \cdots \otimes \mathcal{H}_N, \quad \Omega = \Omega_1 \otimes \cdots \otimes \Omega_N.$$

Let $m \geq 2$ and take $n_1 \neq n_2 \neq \cdots \neq n_m$ from $\{1, 2, \dots, N\}$. For $a_i \in \mathcal{A}_{n_i}$ we need to show that

$$\langle \Omega, a_1 \cdots a_m \Omega \rangle = \langle \Omega, a_1 \Omega \rangle \langle \Omega, a_2 \cdots a_m \Omega \rangle. \tag{8.6}$$

We set

$$a_i = P_1 \otimes \cdots \otimes \overset{n_i\text{th}}{S_i} \otimes \cdots \otimes P_N, \qquad S_i \in \mathcal{B}(\mathcal{H}_{n_i}).$$

Noting that $n_1 \neq n_2$, we observe that

$$\begin{aligned} a_2^* a_1^* \Omega &= (P_1 \otimes \cdots \otimes \overset{n_2\text{th}}{S_2^*} \otimes \cdots \otimes P_N)(P_1 \otimes \cdots \otimes \overset{n_1\text{th}}{S_1^*} \otimes \cdots \otimes P_N)\Omega \\ &= \Omega_1 \otimes \cdots \otimes S_2^* \Omega_{n_2} \otimes \cdots \otimes P_{n_1} S_1^* \Omega_{n_1} \otimes \cdots \otimes \Omega_N. \end{aligned}$$

Since $P_{n_1} S_1^* \Omega_{n_1} = \langle \Omega_{n_1}, S_1^* \Omega_{n_1} \rangle \Omega_{n_1} = \langle \Omega, a_1^* \Omega \rangle \Omega_{n_1}$, we have

$$\begin{aligned} a_2^* a_1^* \Omega &= \langle \Omega, a_1^* \Omega \rangle \, \Omega_1 \otimes \cdots \otimes S_2^* \Omega_{n_2} \otimes \cdots \otimes \Omega_{n_1} \otimes \cdots \otimes \Omega_N \\ &= \langle \Omega, a_1^* \Omega \rangle \, a_2^* \Omega. \end{aligned}$$

Hence

$$\begin{aligned} \langle \Omega, a_1 \cdots a_m \Omega \rangle &= \langle a_2^* a_1^* \Omega, a_3 \cdots a_m \Omega \rangle \\ &= \overline{\langle \Omega, a_1^* \Omega \rangle} \, \langle a_2^* \Omega, a_3 \cdots a_m \Omega \rangle \\ &= \langle \Omega, a_1 \Omega \rangle \, \langle \Omega, a_2 a_3 \cdots a_m \Omega \rangle, \end{aligned}$$

which proves (8.6). □

Theorem 8.9. *For $n = 1, 2, \dots, N$ let $\mathcal{H}_n$ be a Hilbert space. Consider an algebraic probability space:*

$$(\mathcal{B}(\mathcal{H}_1 \otimes \cdots \otimes \mathcal{H}_N), \psi \otimes \Omega_2 \cdots \otimes \Omega_N),$$

where ψ is a state on $\mathcal{B}(\mathcal{H}_1)$ and Ω_n a vector state on $\mathcal{B}(\mathcal{H}_n)$ corresponding to a unit vector (denoted by the same symbol) in $\mathcal{H}_n$ for $n = 2, 3, \dots, N$. Let $\mathcal{A}_n$ be the set of operators of the form

$$1_1 \otimes \cdots \otimes 1_{n-1} \otimes S_n \otimes P_{n+1} \otimes \cdots \otimes P_N, \qquad S_n \in \mathcal{B}(\mathcal{H}_n),$$

where 1_n is the identity of $\mathcal{B}(\mathcal{H}_n)$ and $P_n \in \mathcal{B}(\mathcal{H}_n)$ the projection onto the one-dimensional subspace spanned by Ω_n. Then $\{\mathcal{A}_n\}$ is monotone independent. (Here $\{1, 2, \dots, N\}$ is equipped with the usual total order.)

The proof is similar to that of Theorem 8.8 and is omitted.

Remark 8.10. The Boolean independence yields a rather trivial situation when $\mathcal{A}_\lambda$ contains the identity. This remains even if the condition in Definition 8.3 is weakened in such a way that a_i is taken from $\mathcal{A}_{\lambda_i} \setminus \mathbb{C}$. Assume that $\{\mathcal{A}_1, \mathcal{A}_2\}$ is Boolean independent and that $\mathcal{A}_1$ contains the identity, i.e., a $*$-subalgebra. Take $a_i \in \mathcal{A}_i \setminus \mathbb{C}$ and consider $\varphi(a_2^*(a_1 + 1)a_2)$. Applying the independence and then linearity, we obtain

$$\begin{aligned} \varphi(a_2^*(a_1 + 1)a_2) &= \varphi(a_2^*)\varphi(a_1 + 1)\varphi(a_2) \\ &= \varphi(a_2^*)\varphi(a_1)\varphi(a_2) + \varphi(a_2^*)\varphi(a_2). \end{aligned} \tag{8.7}$$

On the other hand, applying first linearity and then independence, we obtain

$$\begin{aligned} \varphi(a_2^*(a_1 + 1)a_2) &= \varphi(a_2^* a_1 a_2) + \varphi(a_2^* a_2) \\ &= \varphi(a_2^*)\varphi(a_1)\varphi(a_2) + \varphi(a_2^* a_2). \end{aligned} \tag{8.8}$$

We then see from (8.7) and (8.8) that

$$\varphi(a_2^* a_2) = \varphi(a_2^*)\varphi(a_2) = |\varphi(a_2)|^2.$$

Similarly, from $\varphi(a_2(a_1 + 1)a_2^*)$ we obtain

$$\varphi(a_2 a_2^*) = |\varphi(a_2)|^2.$$

Consequently, $a_2 \overset{\mathrm{s}}{=} \varphi(a_2)1$ follows by Proposition 1.23. In other words, $\mathcal{A}_2$ is reduced essentially to the $*$-subalgebra $\mathbb{C}1$. A similar situation occurs in the case of monotone independence.

Definition 8.11. Let $(\mathcal{A}, \varphi)$ be an algebraic probability space and $\{a_n\}$ be a sequence of random variables. Let $\mathcal{A}_n^0$ be the linear span of elements of the form

$$a_n^{\epsilon_1} \cdots a_n^{\epsilon_m}, \qquad \epsilon_i \in \{1, *\}, \quad m = 1, 2, \dots,$$

and set $\mathcal{A}_n = \mathcal{A}_n^0 + \mathbb{C}1$, which is the $*$-subalgebra generated by a_n. We say that $\{a_n\}$ is *commutative* or *free independent* if so is $\{\mathcal{A}_n\}$. We say that $\{a_n\}$ is *Boolean* or *monotone independent* if so is $\{\mathcal{A}_n^0\}$.

Remark 8.12. $\mathcal{A}_n^0$ is closed under the algebraic operations and involution. But it cannot be decided by definition whether or not the identity $1_{\mathcal{A}}$ is contained in $\mathcal{A}_n^0$.

8.2 Singleton Condition and Central Limit Theorems

We first formulate a common property possessed by the four notions of independence defined in the previous section. Let Λ be a set and consider a finite sequence

$$\lambda_1, \lambda_2, \dots, \lambda_m, \qquad \lambda_i \in \Lambda. \tag{8.9}$$

We say that λ_s is a *singleton* in (8.9) if λ_s appears only once, i.e., if $\lambda_s \neq \lambda_i$ for all $i \neq s$.

Definition 8.13. Let $(\mathcal{A}, \varphi)$ be an algebraic probability space and for $\lambda \in \Lambda$, $\mathcal{A}_\lambda \subset \mathcal{A}$ a subset closed under the algebraic operations and involution. We say that $\{\mathcal{A}_\lambda\}$ satisfies the *singleton condition* if

$$\varphi(a_1 \cdots a_s \cdots a_m) = 0, \qquad a_i \in \mathcal{A}_{\lambda_i},$$

whenever λ_s is a singleton in $\lambda_1, \dots, \lambda_m$ and $\varphi(a_s) = 0$.

Proposition 8.14. *Let $(\mathcal{A}, \varphi)$ and $\{\mathcal{A}_\lambda\}$ be as in Definitions 8.1–8.4. If $\{\mathcal{A}_\lambda\}$ is commutative, free, Boolean or monotone independent, then $\{\mathcal{A}_\lambda\}$ satisfies the singleton condition.*

Proof. For the case of commutative, Boolean or monotone independence the assertion follows immediately. Consider the case of free independence. It is sufficient to prove that

$$\varphi(a_1 \cdots a_s \cdots a_m) = \varphi(a_s)\varphi(a_1 \cdots \check{a}_s \cdots a_m) \tag{8.10}$$

for $a_i \in \mathcal{A}_{\lambda_i}$ with $\lambda_1 \neq \lambda_2 \neq \cdots \neq \lambda_m$ and λ_s is a singleton. We utilize induction on m. The assertion is trivial for $m = 1$. Assume that the assertion is valid up to $m-1$, $m \geq 2$. We set $\bar{a}_i = a_i - \varphi(a_i)$ and start with

$$\varphi(a_1 \cdots a_m) = \varphi(\bar{a}_1 \cdots \bar{a}_m) + \sum_{i=1}^{m} \varphi(a_i)\varphi(\underbrace{a_1 \cdots a_{i-1}}_{i-1}\, \check{a}_i\, \underbrace{\bar{a}_{i+1} \cdots \bar{a}_m}_{m-i}), \tag{8.11}$$

see Lemma 8.5. By free independence the first term in (8.11) vanishes so that

$$\varphi(a_1 \cdots a_m) = \sum_{i=1}^{m} \varphi(a_i)\varphi(\underbrace{a_1 \cdots a_{i-1}}_{i-1}\, \check{a}_i\, \underbrace{\bar{a}_{i+1} \cdots \bar{a}_m}_{m-i}).$$

Then, applying the assumption of induction we have

$$\varphi(a_1 \cdots a_s \cdots a_m) = \sum_{i=1}^{s-1} \varphi(a_i)\varphi(\bar{a}_s)\varphi(a_1 \cdots \check{a}_i \cdots \check{\bar{a}}_s \cdots \bar{a}_m)$$
$$+ \varphi(a_s)\varphi(a_1 \cdots \check{a}_s \cdots \bar{a}_m)$$
$$+ \sum_{i=s+1}^{m} \varphi(a_s)\varphi(a_i)\varphi(a_1 \cdots \check{a}_s \cdots \check{a}_i \cdots \bar{a}_m). \qquad (8.12)$$

Since $\varphi(\bar{a}_s) = 0$, (8.12) becomes

$$\varphi(a_s)\Big\{ \varphi(a_1 \cdots a_{s-1}\check{a}_s\bar{a}_{s+1} \cdots \bar{a}_m)$$
$$+ \sum_{i=s+1}^{m} \varphi(a_i)\varphi(a_1 \cdots \check{a}_s \cdots a_{i-1}\check{a}_i\bar{a}_{i+1} \cdots \bar{a}_m) \Big\}.$$

Checking $\{\cdots\} = \varphi(a_1 \cdots \check{a}_s \cdots a_m)$ without difficulty, we obtain (8.10). □

For a sequence $\{a_n\}$ of random variables in an algebraic probability space $(\mathcal{A}, \varphi)$ consider its partial sum defined by

$$S_N = \sum_{n=1}^{N} a_n. \qquad (8.13)$$

We are interested in the limit of a mixed moment

$$\varphi(S_N^{\epsilon_1} \cdots S_N^{\epsilon_m}), \qquad \epsilon_1, \dots, \epsilon_m \in \{1, *\}, \qquad (8.14)$$

as $N \to \infty$ under a suitable scaling. Our strategy is simple. Expanding (8.14) into a sum of $\varphi(a_{n_1}^{\epsilon_1} \cdots a_{n_m}^{\epsilon_m})$, we eliminate the terms which do not contribute to the limit. We prepare some notation. Let $\mathfrak{M}(m, N)$ denote the set of maps from $\{1, 2, \dots, m\}$ into $\{1, 2, \dots, N\}$. Then, (8.14) becomes

$$\varphi(S_N^{\epsilon_1} \cdots S_N^{\epsilon_m}) = \sum_{n \in \mathfrak{M}(m,N)} \varphi(a_{n_1}^{\epsilon_1} \cdots a_{n_m}^{\epsilon_m}). \qquad (8.15)$$

We introduce some subclasses of $\mathfrak{M}(m, N)$. Set

$$\mathfrak{M}_s(m, N) = \{n \in \mathfrak{M}(m, N)\,;\ \text{surjective}\},$$
$$\mathfrak{M}_i(m, N) = \{n \in \mathfrak{M}(m, N)\,;\ \text{injective}\},$$
$$\mathfrak{M}_o(m, N) = \{n \in \mathfrak{M}(m, N)\,;\ i < j \Rightarrow n(i) < n(j)\},$$
$$\mathfrak{M}_p(m, N) = \{n \in \mathfrak{M}(m, N)\,;\ |n^{-1}(i)| = 0 \text{ or } 2\}.$$

Definition 8.15. Let $\{a_n\}$ be a sequence of random variables in an algebraic probability space $(\mathcal{A}, \varphi)$. We say that $\{a_n\}$ has *uniformly bounded mixed moments* if for each $m = 1, 2, \dots,$

$$\sup\left\{ |\varphi(a_{n_1}^{\epsilon_1} \cdots a_{n_m}^{\epsilon_m})|\,;\ \begin{matrix} \epsilon_1, \dots, \epsilon_m \in \{1, *\} \\ n_1, \dots, n_m \in \{1, 2, \dots\} \end{matrix} \right\} < \infty. \qquad (8.16)$$

Proposition 8.16. *Let $\{a_n\}$ be a sequence of random variables in an algebraic probability space $(\mathcal{A}, \varphi)$. For each $n = 1, 2, \dots$ let $\mathcal{A}_n^0$ be the linear span of $\{a_n^{\epsilon_1} \cdots a_n^{\epsilon_m}\,;\ \epsilon_i \in \{1, *\},\ m = 1, 2, \dots\}$. Assume the following three conditions:*

(i) $\varphi(a_n) = \varphi(a_n^*) = 0$;
(ii) $\{a_n\}$ *has uniformly bounded mixed moments;*
(iii) $\{\mathcal{A}_n^0\}$ *satisfies the singleton condition.*

Then, for $m = 1, 2, \dots,$

$$\lim_{N\to\infty} \varphi\left(\frac{S_N^{\epsilon_1}}{\sqrt{N}} \cdots \frac{S_N^{\epsilon_{2m-1}}}{\sqrt{N}}\right) = 0 \tag{8.17}$$

$$\lim_{N\to\infty} \varphi\left(\frac{S_N^{\epsilon_1}}{\sqrt{N}} \cdots \frac{S_N^{\epsilon_{2m}}}{\sqrt{N}}\right) = \lim_{N\to\infty} N^{-m} \sum_{n\in\mathfrak{M}_p(2m,N)} \varphi(a_{n_1}^{\epsilon_1} \cdots a_{n_{2m}}^{\epsilon_{2m}}). \tag{8.18}$$

Here we understand that if the right-hand side of (8.18) *exists then the left-hand side exists and they coincide.*

Proof. We divide the right-hand side of (8.15) according to the cardinality of the image of $n \in \mathfrak{M}(m, N)$. Suppose that the image of n consists of p points, $1 \le p \le m$. These points being uniquely enumerated as $\sigma_1 < \sigma_2 < \cdots < \sigma_p$, we obtain $\sigma \in \mathfrak{M}_o(p, N)$. Taking into account the partition:

$$\{1, 2, \dots, m\} = \bigcup_{i=1}^{p} n^{-1}(\sigma_i),$$

we define $\pi \in \mathfrak{M}(m, p)$ by $\pi \restriction_{n^{-1}(\sigma_i)} = i$. Then $\pi \in \mathfrak{M}_s(m, p)$ and $n = \sigma \circ \pi$. This expression being unique, we obtain a one-to-one correspondence between the set of maps $n \in \mathfrak{M}(m, N)$ with the image consisting of p points and $\mathfrak{M}_o(p, N) \times \mathfrak{M}_s(m, p)$. Hence (8.15) becomes

$$\varphi(S_N^{\epsilon_1} \cdots S_N^{\epsilon_m}) = \sum_{p=1}^{m} \sum_{\pi\in\mathfrak{M}_s(m,p)} \sum_{\sigma\in\mathfrak{M}_o(p,N)} \varphi(a_{\sigma\circ\pi(1)}^{\epsilon_1} \cdots a_{\sigma\circ\pi(m)}^{\epsilon_m}). \tag{8.19}$$

By assumption (ii) we take a constant M_m in such a way that

$$|\varphi(a_{n_1}^{\epsilon_1} \cdots a_{n_m}^{\epsilon_m})| \le M_m, \qquad n_1, \dots, n_m \in \{1, 2, \dots\}.$$

Then we have

$$N^{-m/2} \left| \sum_{\pi\in\mathfrak{M}_s(m,p)} \sum_{\sigma\in\mathfrak{M}_o(p,N)} \varphi(a_{\sigma\circ\pi(1)}^{\epsilon_1} \cdots a_{\sigma\circ\pi(m)}^{\epsilon_m}) \right|$$
$$\le M_m N^{-m/2} |\mathfrak{M}_s(m, p)| |\mathfrak{M}_o(p, N)|. \tag{8.20}$$

Since

$$\lim_{N\to\infty} N^{-p}|\mathfrak{M}_o(p,N)| = \lim_{N\to\infty} N^{-p}\binom{N}{p} = \frac{1}{p!},$$

we see from (8.20) that for $p < m/2$,

$$\lim_{N\to\infty} N^{-m/2} \sum_{\pi\in\mathfrak{M}_s(m,p)} \sum_{\sigma\in\mathfrak{M}_o(p,N)} \varphi(a^{\epsilon_1}_{\sigma\circ\pi(1)} \cdots a^{\epsilon_m}_{\sigma\circ\pi(m)}) = 0. \tag{8.21}$$

Consequently, from (8.19) we obtain

$$\lim_{N\to\infty} N^{-m/2}\varphi(S_N^{\epsilon_1}\cdots S_N^{\epsilon_m})$$
$$= \lim_{N\to\infty} \sum_{m/2\le p\le m} \sum_{\pi\in\mathfrak{M}_s(m,p)} \sum_{\sigma\in\mathfrak{M}_o(p,N)} \varphi(a^{\epsilon_1}_{\sigma\circ\pi(1)} \cdots a^{\epsilon_m}_{\sigma\circ\pi(m)}). \tag{8.22}$$

We next consider the singleton condition. Take p with

$$\frac{m}{2} \le p \le m \tag{8.23}$$

and $\pi \in \mathfrak{M}_s(m,p)$. Set

$$S_j = \pi^{-1}(j), \qquad j \in \{1,2,\dots,p\}.$$

If $|S_j| = 1$ occurs for some j, by the singleton condition (iii) and assumption (i) we have

$$\sum_{\sigma\in\mathfrak{M}_o(p,N)} \varphi(a^{\epsilon_1}_{\sigma\circ\pi(1)} \cdots a^{\epsilon_m}_{\sigma\circ\pi(m)}) = 0. \tag{8.24}$$

Suppose that $|S_j| \ge 2$ for all j. Then

$$m = \sum_{j=1}^{p} |S_j| \ge 2p, \tag{8.25}$$

which is possible, due to (8.23), only when $m = 2p$ and $|S_j| = 2$ for all j. Hence, if m is odd, (8.22) equals 0 by (8.24). Thus (8.17) follows. If $m = 2p$ is even,

$$\begin{aligned}
\lim_{N\to\infty} & N^{-m/2}\varphi(S_N^{\epsilon_1}\cdots S_N^{\epsilon_m}) \\
&= \lim_{N\to\infty} N^{-m/2} \sum_{\substack{\pi\in\mathfrak{M}_s(2m,p) \\ \text{2-1 map}}} \sum_{\sigma\in\mathfrak{M}_o(p,N)} \varphi(a^{\epsilon_1}_{\sigma\circ\pi(1)} \cdots a^{\epsilon_m}_{\sigma\circ\pi(m)}) \\
&= \lim_{N\to\infty} N^{-m/2} \sum_{n\in\mathfrak{M}_p(2m,N)} \varphi(a^{\epsilon_1}_{n_1} \cdots a^{\epsilon_{2m}}_{n_{2m}}),
\end{aligned}$$

which proves (8.18). □

Remark 8.17. If $\{a_n\}$ is a sequence of bounded operators on a Hilbert space such that $\sup_n \|a_n\| < \infty$, then $\{a_n\}$ has uniformly bounded mixed moments in a vector state. This criterion is also valid for a C^*-probability space $(\mathcal{A}, \varphi)$.

Remark 8.18. One may consider another scaling such as

$$\lim_{N\to\infty} N^{-\alpha m} \varphi(S_N^{\epsilon_1} \cdots S_N^{\epsilon_m}), \qquad \alpha > 0.$$

However, as is seen during the above proof, $\alpha = 1/2$ is the unique choice for the reasonable limit under condition (ii).

For a sequence of real random variables, Proposition 8.16 becomes slightly simpler.

Proposition 8.19. *Let $\{a_n\}$ be a sequence of random variables in an algebraic probability space $(\mathcal{A}, \varphi)$ satisfying the following conditions:*

(i) a_n *is real, i.e.,* $a_n = a_n^*$;
(ii) $\varphi(a_n) = 0$;
(iii) $\{a_n\}$ *has uniformly bounded mixed moments.*

For each $n = 1, 2, \dots$ let $\mathcal{A}_n^0$ be the linear span of $\{a_n, a_n^2, a_n^3, \dots\}$. If $\{\mathcal{A}_n^0\}$ satisfies the singleton condition, for $m = 1, 2, \dots$ we have

$$\lim_{N\to\infty} \varphi\left(\left(\frac{1}{\sqrt{N}}\sum_{n=1}^{N} a_n\right)^{2m-1}\right) = 0,$$

$$\lim_{N\to\infty} \varphi\left(\left(\frac{1}{\sqrt{N}}\sum_{n=1}^{N} a_n\right)^{2m}\right) = \lim_{N\to\infty} N^{-m} \sum_{n\in\mathfrak{M}_p(2m,N)} \varphi(a_{n_1} \cdots a_{n_{2m}}).$$

We can now derive from Propositions 8.14 and 8.19 the explicit forms of quantum central limit theorems associated with four different notions of independence. For simplicity of the statements we extract the common condition:

(CC) Let $(\mathcal{A}, \varphi)$ be an algebraic probability space and $\{a_n\}$ a sequence of random variables. Assume the following conditions:

(i) a_n is real, i.e., $a_n^* = a_n$;
(ii) a_n is normalized, i.e., $\varphi(a_n) = 0$ and $\varphi(a_n^2) = 1$;
(iii) $\{a_n\}$ has uniformly bounded mixed moments.

For each $n = 1, 2, \dots$ let $\mathcal{A}_n^0$ be the linear span of $\{a_n, a_n^2, a_n^3, \dots\}$ and set $\mathcal{A}_n = \mathcal{A}_n^0 + \mathbb{C}1$. For this discrimination see Remark 8.12.

Theorem 8.20 (Commutative CLT). *Notations and assumptions being as in (CC), if $\{a_n\}$ is commutative independent, then*

$$\lim_{N\to\infty} \varphi\left(\left(\frac{1}{\sqrt{N}}\sum_{n=1}^{N} a_n\right)^{m}\right) = \frac{1}{\sqrt{2\pi}} \int_{-\infty}^{+\infty} x^m e^{-x^2/2} dx, \quad m = 1, 2, \dots,$$

where the probability measure appearing on the right-hand side is the standard Gaussian distribution.

Proof. Since $\{a_n\}$ is commutative independent, so is $\{\mathcal{A}_n\}$ by definition. Then by Proposition 8.14, $\{\mathcal{A}_n\}$ satisfies the singleton condition, hence so is $\{\mathcal{A}_n^0\}$. Applying Proposition 8.19, for the moments of odd orders we have

$$\lim_{N\to\infty} \varphi\biggl(\Bigl(\frac{1}{\sqrt{N}}\sum_{n=1}^{N} a_n\Bigr)^{2m-1}\biggr) = 0 = \frac{1}{\sqrt{2\pi}}\int_{-\infty}^{+\infty} x^{2m-1}\mathrm{e}^{-x^2/2}dx,$$

while for the moments of even degree,

$$\lim_{N\to\infty} \varphi\biggl(\Bigl(\frac{1}{\sqrt{N}}\sum_{n=1}^{N} a_n\Bigr)^{2m}\biggr) = \lim_{N\to\infty} N^{-m} \sum_{n\in\mathfrak{M}_p(2m,N)} \varphi(a_{n_1}\cdots a_{n_{2m}}).$$

Since

$$\varphi(a_{n_1}\cdots a_{n_{2m}}) = \varphi(a_{i_1}^2)\cdots\varphi(a_{i_m}^2) = 1, \qquad n\in\mathfrak{M}_p(2m,N),$$

the right-hand side becomes

$$\lim_{N\to\infty} N^{-m}|\mathfrak{M}_p(2m,N)| = \lim_{N\to\infty} N^{-m}\binom{N}{m}\frac{(2m)!}{2^m} = \frac{(2m)!}{2^m m!},$$

which coincides with the $2m$th moment of the standard Gaussian distribution. Namely,

$$\lim_{N\to\infty} \varphi\biggl(\Bigl(\frac{1}{\sqrt{N}}\sum_{n=1}^{N} a_n\Bigr)^{2m}\biggr) = \frac{1}{\sqrt{2\pi}}\int_{-\infty}^{+\infty} x^{2m}\mathrm{e}^{-x^2/2}dx,$$

which completes the proof. □

Theorem 8.21 (Free CLT). *Notations and assumptions being as in (CC), if $\{a_n\}$ is free independent, we have*

$$\lim_{N\to\infty} \varphi\biggl(\Bigl(\frac{1}{\sqrt{N}}\sum_{n=1}^{N} a_n\Bigr)^{m}\biggr) = \frac{1}{2\pi}\int_{-2}^{+2} x^m\sqrt{4-x^2}\,dx, \qquad m=1,2,\dots,$$

where the probability measure appearing on the right-hand side is the Wigner semicircle law.

Proof. The proof is similar to that of Theorem 8.20. Each $n\in\mathfrak{M}_p(2m,N)$ defines a pair partition of $\{1,2,\dots,2m\}$ by its counter images. We then see that $\varphi(a_{n_1}\cdots a_{n_{2m}}) = 1$ if the pair partition defined by n is non-crossing, and $= 0$ otherwise. Hence, for $m=1,2,\dots,$ we have

$$\begin{aligned}
&\lim_{N\to\infty} \varphi\biggl(\Bigl(\frac{1}{\sqrt{N}}\sum_{n=1}^{N} a_n\Bigr)^{2m}\biggr) \\
&= \lim_{N\to\infty} N^{-m}\binom{N}{m} m!\,|\mathcal{P}_{\mathrm{NCP}}(2m)| = |\mathcal{P}_{\mathrm{NCP}}(2m)| = \frac{(2m)!}{m!(m+1)!},
\end{aligned}$$

which coincides with the Catalan number, i.e., the $2m$th moment of the Wigner semicircle law. □

Theorem 8.22 (Boolean CLT). *Notations and assumptions being as in (CC), if $\{a_n\}$ is Boolean independent, we have*

$$\lim_{N\to\infty} \varphi\bigg(\bigg(\frac{1}{\sqrt{N}}\sum_{n=1}^{N} a_n\bigg)^m\bigg) = \frac{1}{2}\int_{-\infty}^{+\infty} x^m(\delta_{-1}+\delta_{+1})(dx), \quad m=1,2,\dots,$$

where the probability measure appearing on the right-hand side is the Bernoulli distribution.

Proof. The proof is similar to those of Theorems 8.20 and 8.21. For $n \in \mathfrak{M}_p(2m,N)$,

$$\varphi(a_{n_1}\cdots a_{n_{2m}}) = \begin{cases} 1, & n_1=n_2,\ \dots,\ n_{2m-1}=n_{2m}\,, \\ 0, & \text{otherwise.} \end{cases}$$

Hence

$$\lim_{N\to\infty} \varphi\bigg(\bigg(\frac{1}{\sqrt{N}}\sum_{n=1}^{N} a_n\bigg)^{2m}\bigg) = \lim_{N\to\infty} N^{-m}\binom{N}{m} m! = 1.$$

This is the $2m$th moment of the Bernoulli distribution. □

Theorem 8.23 (Monotone CLT). *Notations and assumptions being as in (CC), if $\{a_n\}$ is monotone independent, we have for $m=1,2,\dots$*

$$\lim_{N\to\infty} \varphi\bigg(\bigg(\frac{1}{\sqrt{N}}\sum_{n=1}^{N} a_n\bigg)^m\bigg) = \frac{1}{\pi}\int_{-\sqrt{2}}^{+\sqrt{2}} \frac{x^m}{\sqrt{2-x^2}}\, dx, \tag{8.26}$$

where the probability measure appearing on the right-hand side is the normalized arcsine law.

By Proposition 8.19, for an odd m, the left-hand side of (8.26) is zero as well as the right-hand side. It is then essential to show (8.26) for an even m. The proof requires, however, some technical preparations and will be deferred in Sect. 8.4.

8.3 Integer Lattices and Homogeneous Trees: Revisited

Let G be a discrete group and consider a Cayley graph (G,Σ), see Definition 2.1. The adjacency matrix A is canonically identified with an element in $\mathbb{C}[G]$, i.e.,

$$A = \sum_{g\in\Sigma} g. \tag{8.27}$$

Let us write Σ as a disjoint union:

$$\Sigma = \{s_1,s_2,\dots,s_k\} \cup \{g_1,g_2,\dots,g_l\} \cup \{g_1^{-1},g_2^{-1},\dots,g_l^{-1}\},$$

where s_i is of order 2 and g_i of order greater than 2. Then (8.27) becomes

$$A = \sum_{i=1}^{k} s_i + \sum_{i=1}^{l} (g_i + g_i^{-1}). \tag{8.28}$$

Given a state φ on $\mathbb{C}[G]$, we consider s_i and $g_i + g_i^{-1}$ as real random variables. Generally speaking, for the (asymptotic) spectral distribution of A we need to tackle statistical relations among them. In particular, if (8.28) is a sum of independent random variables, strong tools of quantum probability theory are available.

Proposition 8.24. *Let G be a discrete group which is the direct product of subgroups $\{G_\lambda\}$. Then $\{\mathbb{C}[G_\lambda]\}$ is commutative independent in $(\mathbb{C}[G], \delta_e)$.*

Proof. We check directly the conditions in Definition 8.1. Let us consider

$$\langle \delta_e, a_1 \cdots a_m \delta_e \rangle, \qquad a_k \in \mathbb{C}[G_{\lambda_k}].$$

Using the canonical identification $\mathbb{C}[G_{\lambda_k}] \cong C_0(G_{\lambda_k})$, we may write

$$a_k = \sum_{g_k \in G_{\lambda_k}} a_k(g_k) g_k.$$

Then

$$\begin{aligned} \langle \delta_e, a_1 \cdots a_m \delta_e \rangle &= \sum_{\substack{g_1 \in G_{\lambda_1} \\ \cdots \\ g_m \in G_{\lambda_m}}} a_1(g_1) \cdots a_m(g_m) \langle \delta_e, \delta_{g_1 \cdots g_m} \rangle \\ &= \sum_{g_1 \cdots g_m = e} a_1(g_1) \cdots a_m(g_m). \end{aligned} \tag{8.29}$$

Suppose that $\lambda_1 \notin \{\lambda_2, \dots, \lambda_m\}$. Then $g_1 \cdots g_m = e$ occurs only when $g_1 = e$. Hence (8.29) becomes

$$\begin{aligned} \langle \delta_e, a_1 \cdots a_m \delta_e \rangle &= \sum_{g_2 \cdots g_m = e} a_1(e) a_2(g_2) \cdots a_m(g_m) \\ &= \langle \delta_e, a_1 \delta_e \rangle \langle \delta_e, a_2 \cdots a_m \delta_e \rangle, \end{aligned}$$

which is the desired factorization property of the commutative independence. Suppose next that $\lambda_1 \in \{\lambda_2, \dots, \lambda_m\}$ and let $r \in \{2, 3, \dots, m\}$ be the smallest number such that $\lambda_1 = \lambda_r$. We need to show that

$$\langle \delta_e, a_1 \cdots a_m \delta_e \rangle = \langle \delta_e, a_2 \cdots (a_1 a_r) \cdots a_m \delta_e \rangle.$$

But this is obvious because G_λ and G_μ commute whenever $\lambda \neq \mu$. □

The N-dimensional integer lattice is the Cayley graph of the additive group $\mathbb{Z}^N$ equipped with the canonical generators:

$$e_{\pm k} = \pm e_k = (0, \dots, \pm 1, \dots, 0), \qquad k = 1, 2, \dots, N,$$

where ± 1 sits at the kth position. The adjacency matrix A_N becomes

$$A_N = \sum_{k=1}^{N} (e_k + e_{-k}). \tag{8.30}$$

Since $\mathbb{Z}^N$ is the N-fold direct product of $\mathbb{Z}$, by Proposition 8.24 the right-hand side of (8.30) is a sum of commutative independent random variables. By normalization we obtain

$$\frac{A_N}{\sqrt{2N}} = \frac{1}{\sqrt{N}} \sum_{k=1}^{N} \frac{e_k + e_{-k}}{\sqrt{2}},$$

to which we apply the commutative central limit theorem (Theorem 8.20). This is an alternative method of proving (the first half of) Theorem 7.19. We thus come to the following:

Theorem 8.25 (CLT for integer lattices). *Let A_N denote the adjacency matrix of the integer lattice $\mathbb{Z}^N$. Then,*

$$\lim_{N \to \infty} \left\langle \delta_o, \left(\frac{A_N}{\sqrt{2N}} \right)^m \delta_o \right\rangle = \frac{1}{\sqrt{2\pi}} \int_{-\infty}^{+\infty} x^m e^{-x^2/2} dx, \tag{8.31}$$

for $m = 1, 2, \dots,$ where o is the unit element of $\mathbb{Z}^N$.

Before going into the discussion on homogeneous trees, we consider the following:

Proposition 8.26. *Let G be a discrete group which is the free product of subgroups $\{G_\lambda\}$. Then $\{\mathbb{C}[G_\lambda]\}$ is free independent in $(\mathbb{C}[G], \delta_e)$.*

Proof. We maintain the same notations as used in the proof of Proposition 8.24. We assume that

$$\langle \delta_e, a_k \delta_e \rangle = 0, \qquad k = 1, 2, \dots, m.$$

Then

$$a_k = \sum_{g_k \in G_{\lambda_k} \setminus \{e\}} a_k(g_k) g_k$$

and hence (8.29) reads

$$\langle \delta_e, a_1 \cdots a_m \delta_e \rangle = \sum_{\substack{g_1 \cdots g_m = e \\ g_1 \neq e, \dots, g_m \neq e}} a_1(g_1) \cdots a_m(g_m).$$

Since G is the free product of $\{G_\lambda\}$, the constraint for the sum is never satisfied so that

$$\langle \delta_e, a_1 \cdots a_m \delta_e \rangle = 0,$$

which completes the proof. □

Let F_N be the free group on N generators, say, $g_1, \dots, g_N$. Then F_N is the free product of subgroups $\langle g_k \rangle \cong \mathbb{Z}$ generated by g_k. Equipped with $\Sigma = \{g_1, g_1^{-1}, \dots, g_N, g_N^{-1}\}$, the free group becomes a Cayley graph, which is isomorphic to the homogeneous tree of degree $2N$. The adjacency matrix A_{2N} is given by

$$A_{2N} = \sum_{k=1}^{N} (g_k + g_k^{-1})$$

and taking the normalization we obtain

$$\frac{A_{2N}}{\sqrt{2N}} = \frac{1}{\sqrt{N}} \sum_{k=1}^{N} \frac{g_k + g_k^{-1}}{\sqrt{2}},$$

where the right-hand side is a sum of free independent random variables. Then, applying the free central limit theorem (Theorem 8.21) we come to the following:

Theorem 8.27 (CLT for homogeneous trees). *Let A_{2N} be the adjacency matrix of the Cayley graph of the free group F_N, i.e., of the homogeneous tree of degree $2N$. Then,*

$$\lim_{N\to\infty} \left\langle \delta_e, \left(\frac{A_{2N}}{\sqrt{2N}} \right)^m \delta_e \right\rangle = \frac{1}{2\pi} \int_{-2}^{+2} x^m \sqrt{4 - x^2}\, dx, \tag{8.32}$$

for $m = 1, 2, \dots$.

Theorem 8.25 deals with the direct product of the infinite cyclic group $\mathbb{Z}$, while Theorem 8.27 with the free product of it. These results remain valid when $\mathbb{Z}$ is replaced with an arbitrary finite cyclic group.

Having introduced four concepts of independence in Sect. 8.1, we naturally wonder the counterparts for the monotone independence and Boolean independence. The rest of this chapter will be devoted to this topic.

8.4 Monotone Trees and Monotone Central Limit Theorem

We do not go directly into the proof of the monotone central limit theorem (Theorem 8.23). Instead we start with a prototype which is interesting in itself from the viewpoint of spectral analysis of a graph.

Fix an integer $N \geq 1$. For $n = 0, 1, 2, \ldots, N$ let $V_n^{(N)}$ denote the set of increasing natural numbers

$$x = (i_1, i_2, \ldots, i_n), \quad 1 \leq i_1 < i_2 < \cdots < i_n \leq N,$$

where $V_0^{(N)} = \{\varnothing\}$. Set

$$V^{(N)} = \bigcup_{n=0}^{N} V_n^{(N)}. \tag{8.33}$$

By definition $x = (i_1, i_2, \ldots, i_m)$ and $y = (j_1, j_2, \ldots, j_n)$ in $V^{(N)}$ are connected by an edge, i.e., $\{x, y\} \in E^{(N)}$ if

$$(i_2, \ldots, i_m) = (j_1, j_2, \ldots, j_n) \quad \text{or} \quad (i_1, i_2, \ldots, i_m) = (j_2, \ldots, j_n).$$

Thus $\mathcal{M}_N = (V^{(N)}, E^{(N)})$ becomes a graph which is called a *monotone tree.* By construction there is no cycle so that $\mathcal{M}_N$ is a finite tree. Note that (8.33) is the stratification of $\mathcal{M}_N$ with respect to the origin $\varnothing$.

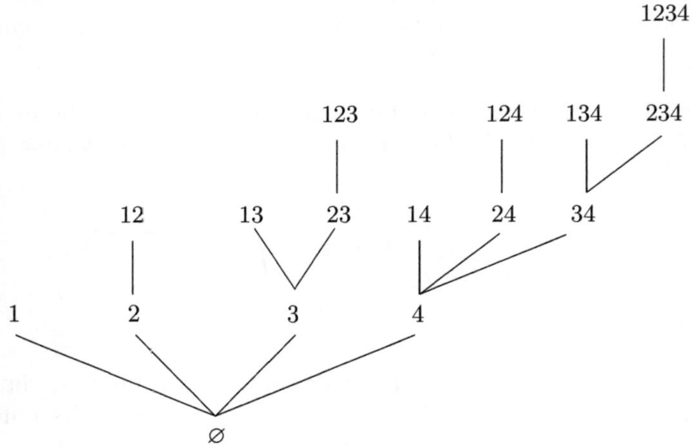

Fig. 8.1. Monotone tree $\mathcal{M}_4$

Let A_N be the adjacency matrix of $\mathcal{M}_N$ which acts on the Hilbert space $\ell^2(V^{(N)})$ in a usual manner. For $x = (i_1, \ldots, i_n) \in V^{(N)}$ we write $\delta_x = \delta_{i_1, \ldots, i_n}$. For $k = 1, 2, \ldots, N$ we define $b_k^{\pm} \in \mathcal{B}(\ell^2(V^{(N)}))$ by

$$b_k^+ \delta_x = \begin{cases} \delta_k, & \text{if } x = \varnothing, \\ \delta_{k,i_1,\dots,i_n}, & \text{if } x = (i_1, \dots, i_n) \text{ with } i_1 > k, \\ 0, & \text{otherwise,} \end{cases} \tag{8.34}$$

$$b_k^- \delta_x = \begin{cases} \delta_\varnothing, & \text{if } x = (i_1) \text{ with } i_1 = k, \\ \delta_{i_2,\dots,i_n}, & \text{if } x = (i_1, i_2, \dots, i_n) \text{ with } i_1 = k, \\ 0, & \text{otherwise.} \end{cases} \tag{8.35}$$

Lemma 8.28. *For $k = 1, 2, \dots, N$, $b_k^\pm$ are mutually adjoint. Moreover, it holds that*

$$A_N = \sum_{k=1}^N (b_k^+ + b_k^-). \tag{8.36}$$

Proof. Straightforward. □

The Hilbert space $\ell^2(V^{(N)})$ equipped with the pair of operators $\{b_k^+\}, \{b_k^-\}$ is called a *monotone Fock space*. We call b_k^+ and b_k^- the *monotone creation operator* and *monotone annihilation operator*, respectively. The unit vector $\delta_\varnothing$ is called the *vacuum vector*.

Proposition 8.29. *Let $(\ell^2(V^{(N)}), \{b_k^+\}, \{b_k^-\})$ be a monotone Fock space. For each $k = 1, 2, \dots, N$, the distribution of the real random variable $b_k^+ + b_k^-$ in the vacuum state is the Bernoulli distribution $(\delta_{-1} + \delta_{+1})/2$.*

Proof. As is seen easily by definition, the actions of $b_k^\pm$ are given by

$$b_k^+ : \delta_\varnothing \to \delta_k \to 0, \qquad b_k^- : \delta_k \to \delta_\varnothing \to 0. \tag{8.37}$$

Then we have

$$\begin{aligned} \langle \delta_\varnothing, (b_k^+ + b_k^-)^m \delta_\varnothing \rangle &= \sum_{\epsilon_1,\dots,\epsilon_m \in \{\pm\}} \langle \delta_\varnothing, b_k^{\epsilon_m} \cdots b_k^{\epsilon_1} \delta_\varnothing \rangle \\ &= \begin{cases} 1, & \text{if } m \text{ is even,} \\ 0, & \text{otherwise,} \end{cases} \end{aligned}$$

which coincides, apparently, with the $2m$th moment of the Bernoulli distribution $(\delta_{-1} + \delta_{+1})/2$. □

In other words, $b_k^+ + b_k^-$ is an algebraic realization of the Bernoulli random variable X such that $P(X = +1) = P(X = -1) = 1/2$, that is, the coin toss (normalized so as to have mean 0 and variance 1). We are naturally interested in

$$\lim_{N \to \infty} \frac{1}{\sqrt{N}} \sum_{k=1}^N (b_k^+ + b_k^-), \tag{8.38}$$

as an analogue of the celebrated de Moivre–Laplace theorem. If $\{b_k^+ + b_k^-\}$ were independent in the sense of classical probability (i.e., commutative independent), the limit would obey the standard Gaussian distribution. In fact, they are not so and we have the following:

Theorem 8.30 (Monotone de Moivre–Laplace theorem). *Let b_k^+ and b_k^- be respectively the monotone creation and annihilation operators as above. Then, for $m = 1, 2, \ldots$ we have*

$$\lim_{N\to\infty} \left\langle \delta_\varnothing, \left(\frac{1}{\sqrt{N}} \sum_{k=1}^{N} (b_k^+ + b_k^-) \right)^m \delta_\varnothing \right\rangle = \frac{1}{\pi} \int_{-\sqrt{2}}^{+\sqrt{2}} \frac{x^m}{\sqrt{2 - x^2}}\, dx, \tag{8.39}$$

where the probability measure appearing on the right-hand side is the normalized arcsine law.

Proof. Let A_N be the adjacency matrix of $\mathcal{M}_N$. We see from Lemma 8.28 that

$$\left\langle \delta_\varnothing, \left(\frac{1}{\sqrt{N}} \sum_{k=1}^{N} (b_k^+ + b_k^-) \right)^m \delta_\varnothing \right\rangle = N^{-m/2} \langle \delta_\varnothing, A_N^m \delta_\varnothing \rangle. \tag{8.40}$$

Define

$$D(N, m) = \langle \delta_\varnothing, A_N^m \delta_\varnothing \rangle,$$

which coincides with the number of m-step walks in the monotone tree $\mathcal{M}_N$ which start at $\varnothing$ and terminate at itself. Obviously, $D(N, m) = 0$ for an odd m. On the other hand, the right-hand side of (8.39) is also zero for an odd m. Thus, both sides being zero, (8.39) holds for an odd m. We need to prove that (8.39) holds for an even m, i.e.,

$$\lim_{N\to\infty} \frac{D(N, 2m)}{N^m} = \frac{1}{\pi} \int_{-\sqrt{2}}^{+\sqrt{2}} \frac{x^{2m}}{\sqrt{2 - x^2}}\, dx, \qquad m = 1, 2, \ldots. \tag{8.41}$$

The proof of (8.41) will be given after some lemmas. □

Let us formulate notation once again. We readily know that

$$D(N, 2m) = \langle \delta_\varnothing, A_N^{2m} \delta_\varnothing \rangle. \tag{8.42}$$

Set $D(0, 0) = 1$ and $D(0, 2m) = 0$ for $m = 1, 2, \ldots$.

Lemma 8.31. $\{D(N, 2m)\}$ *obeys the following recurrence relation:*

$$D(N, 0) = 1, \quad N \geq 1, \tag{8.43}$$

$$D(N, 2m) = \sum_{n=1}^{m} D(N, 2m - 2n) \sum_{k=1}^{N} D(k - 1, 2n - 2), \quad N \geq 1,\ m \geq 1. \tag{8.44}$$

Proof. (8.43) is obvious. For $N \geq 1$, $m \geq 1$ we shall prove (8.44). Expanding (8.42) in terms of $b_k^{\pm}$, we obtain

$$D(N, m) = \sum_{\substack{k_1, \dots, k_{2m} \\ \epsilon_1, \dots, \epsilon_{2m} \in \{\pm\}}} \langle \delta_\varnothing, b_{k_{2m}}^{\epsilon_{2m}} \cdots b_{k_1}^{\epsilon_1} \delta_\varnothing \rangle, \tag{8.45}$$

where $k_1, \dots, k_{2m}$ run over $\{1, 2, \dots, N\}$. Due to the actions of $b_k^{\pm}$ the inner product on the right-hand side is equal to 1 if

$$b_{k_{2m}}^{\epsilon_{2m}} \cdots b_{k_1}^{\epsilon_1} \delta_\varnothing = \delta_\varnothing \tag{8.46}$$

and zero otherwise. Observing the up–down actions of $b_k^{\pm}$, we see that (8.46) occurs only when $(\epsilon_1, \dots, \epsilon_{2m})$ forms a Catalan path, i.e.,

$$\begin{aligned} &\epsilon_1 + \cdots + \epsilon_k \geq 0, \quad k = 1, 2, \dots, 2m - 1, \\ &\epsilon_1 + \cdots + \epsilon_{2m} = 0. \end{aligned}$$

Let $\mathcal{C}_m$ be the set of Catalan paths of length $2m$. Thus, (8.45) becomes

$$D(N, 2m) = \sum_{\mathcal{C}_m} \sum_{k_1, \dots, k_{2m}} \langle \delta_\varnothing, b_{k_{2m}}^{\epsilon_{2m}} \cdots b_{k_1}^{\epsilon_1} \delta_\varnothing \rangle, \tag{8.47}$$

where the first sum is taken for $(\epsilon_1, \dots, \epsilon_{2m}) \in \mathcal{C}_m$ and the second for $k_1, \dots, k_{2m}$ running over $\{1, 2, \dots, N\}$.

We shall divide the first sum of (8.47). For $n = 1, 2, \dots, m$ let $\mathcal{C}_m^n$ be the set of Catalan paths $(\epsilon_1, \dots, \epsilon_{2m}) \in \mathcal{C}_m$ such that

$$\begin{aligned} &\epsilon_1 + \cdots + \epsilon_k > 0, \quad k = 1, 2, \dots, 2n - 1, \\ &\epsilon_1 + \cdots + \epsilon_{2n} = 0, \end{aligned}$$

see Fig. 8.2. Then (8.47) becomes

$$D(N, 2m) = \sum_{n=1}^{m} \sum_{\mathcal{C}_m^n} \sum_{k_1, \dots, k_{2m}} \langle \delta_\varnothing, b_{k_{2m}}^{\epsilon_{2m}} \cdots b_{k_1}^{\epsilon_1} \delta_\varnothing \rangle. \tag{8.48}$$

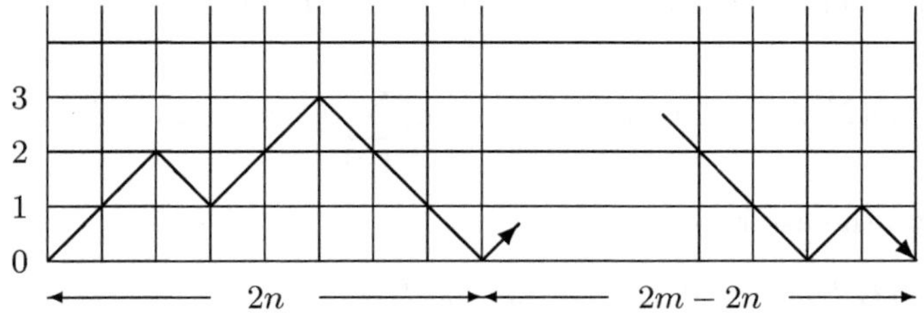

Fig. 8.2. $(\epsilon_1, \dots, \epsilon_{2m}) \in \mathcal{C}_m^n$

For $(\epsilon_1, \dots, \epsilon_{2m}) \in \mathcal{C}_m^n$ we have $b_{k_{2n}}^{\epsilon_{2n}} \cdots b_{k_1}^{\epsilon_1} \delta_\varnothing = \delta_\varnothing$ or $= 0$, so that

$$\langle \delta_\varnothing, b_{k_{2m}}^{\epsilon_{2m}} \cdots b_{k_1}^{\epsilon_1} \delta_\varnothing \rangle = \langle \delta_\varnothing, b_{k_{2m}}^{\epsilon_{2m}} \cdots b_{k_{2n+1}}^{\epsilon_{2n+1}} \delta_\varnothing \rangle \langle \delta_\varnothing, b_{k_{2n}}^{\epsilon_{2n}} \cdots b_{k_1}^{\epsilon_1} \delta_\varnothing \rangle.$$

Moreover, $\langle \delta_\varnothing, b_{k_{2n}}^{\epsilon_{2n}} \cdots b_{k_1}^{\epsilon_1} \delta_\varnothing \rangle$ can be non-zero (in fact, $= 1$) only when $\epsilon_1 = +$, $\epsilon_{2n} = -$, $k_1 = k_{2n} \equiv k$ with $1 \le k \le N$. In that case,

$$\langle \delta_\varnothing, b_{k_{2n}}^{\epsilon_{2n}} \cdots b_{k_1}^{\epsilon_1} \delta_\varnothing \rangle = \langle \delta_\varnothing, b_k^- b_{k_{2n-1}}^{\epsilon_{2n-1}} \cdots b_{k_2}^{\epsilon_2} b_k^+ \delta_\varnothing \rangle = \langle \delta_k, b_{k_{2n-1}}^{\epsilon_{2n-1}} \cdots b_{k_2}^{\epsilon_2} \delta_k \rangle.$$

Then (8.48) becomes

$$\begin{aligned} D(N, 2m) = \sum_{n=1}^{m} \Bigg\{ \sum_{\mathcal{C}_{m-n}} \sum_{k_{2n+1}, \dots, k_{2m}} \langle \delta_\varnothing, b_{k_{2m}}^{\epsilon_{2m}} \cdots b_{k_{2n+1}}^{\epsilon_{2n+1}} \delta_\varnothing \rangle \\ \times \sum_{k=1}^{N} \sum_{\mathcal{C}_{n-1}} \sum_{k_2, \dots, k_{2n-1}} \langle \delta_k, b_{k_{2n-1}}^{\epsilon_{2n-1}} \cdots b_{k_2}^{\epsilon_2} \delta_k \rangle \Bigg\}. \end{aligned} \tag{8.49}$$

As for the first sum in (8.49), by the definition of $D(N, 2m-2n)$ we have

$$\sum_{\mathcal{C}_{m-n}} \sum_{k_{2n+1}, \dots, k_{2m}} \langle \delta_\varnothing, b_{k_{2m}}^{\epsilon_{2m}} \cdots b_{k_{2n+1}}^{\epsilon_{2n+1}} \delta_\varnothing \rangle = D(N, 2m-2n). \tag{8.50}$$

On the other hand, for the second sum in (8.49), we note that

$$\sum_{\mathcal{C}_{n-1}} \sum_{k_2, \dots, k_{2n-1}} \langle \delta_k, b_{k_{2n-1}}^{\epsilon_{2n-1}} \cdots b_{k_2}^{\epsilon_2} \delta_k \rangle, \quad 1 \le k \le N, \tag{8.51}$$

is the number of walks of length $2n-2$ that start at the vertex (k), terminate at itself and do not pass through $\varnothing$. For $k \ge 2$, by the self-similar structure of the monotone tree, there is a one-to-one correspondence between the set of such walks and the set of walks of length $2n-2$ in $\mathcal{M}_{k-1}$ that start at $\varnothing$ and terminate at itself. Therefore, for $2 \le k \le N$ we have

$$\sum_{\mathcal{C}_{n-1}} \sum_{k_2, \dots, k_{2n-1}} \langle \delta_k, b_{k_{2n-1}}^{\epsilon_{2n-1}} \cdots b_{k_2}^{\epsilon_2} \delta_k \rangle = D(k-1, 2n-2). \tag{8.52}$$

For $k = 1$, since (8.51) is 1 if $n = 1$ and is 0 otherwise, (8.52) is valid also for $k = 1$. Consequently, inserting (8.50) and (8.52) into (8.49), we obtain

$$D(N, 2m) = \sum_{n=1}^{m} D(N, 2m-2n) \sum_{k=1}^{N} D(k-1, 2n-2),$$

which proves the assertion. □

Lemma 8.32. *For $m = 0, 1, 2, \dots$ the limit*

$$D_m = \lim_{N \to \infty} \frac{D(N, 2m)}{N^m}$$

exists and $\{D_m\}$ *obeys the following recurrence relation:*

$$D_0 = 1, \quad D_m = \sum_{n=1}^{m} \frac{1}{n} D_{m-n} D_{n-1}, \quad m = 1, 2, \ldots. \tag{8.53}$$

Proof. By induction on m. First for $m = 0$, it follows from (8.43) that

$$D_0 = \lim_{N\to\infty} D(N, 0) = 1.$$

Next assume that $m \geq 1$ and $D_0, D_1, \ldots, D_{m-1}$ exist. By (8.44) we obtain

$$\frac{D(N, 2m)}{N^m} = \sum_{n=1}^{m} \frac{D(N, 2m-2n)}{N^{m-n}} \frac{1}{N^n} \sum_{k=1}^{N} D(k-1, 2n-2). \tag{8.54}$$

For the first term, using the assumption of induction, we have

$$\lim_{N\to\infty} \frac{D(N, 2m-2n)}{N^{m-n}} = D_{m-n}, \qquad 1 \leq n \leq m. \tag{8.55}$$

We need to consider

$$\begin{aligned} &\lim_{N\to\infty} \frac{1}{N^n} \sum_{k=1}^{N} D(k-1, 2n-2) \\ &= \lim_{N\to\infty} \frac{1}{N^n} \sum_{k=1}^{N-1} D(k, 2n-2), \qquad 1 \leq n \leq m. \end{aligned} \tag{8.56}$$

Once again by assumption of induction,

$$D_{n-1} = \lim_{k\to\infty} \frac{D(k, 2n-2)}{k^{n-1}}, \qquad 1 \leq n \leq m,$$

so that

$$\lim_{N\to\infty} \left(\sum_{k=1}^{N-1} k^{n-1} \right)^{-1} \sum_{k=1}^{N-1} k^{n-1} \frac{D(k, 2n-2)}{k^{n-1}} = D_{n-1}. \tag{8.57}$$

This is due to the elementary fact that if a sequence $\{a_n\}$ converges to α, then the mean sequence $\{(a_1 + \cdots + a_n)/n\}$ also converges to α. On the other hand, note that

$$\lim_{N\to\infty} \frac{1}{N^n} \sum_{k=1}^{N-1} k^{n-1} = \lim_{N\to\infty} \frac{1}{N} \sum_{k=1}^{N-1} \left(\frac{k}{N} \right)^{n-1} = \int_0^1 x^{n-1} dx = \frac{1}{n}.$$

Then, (8.56) becomes

$$\lim_{N\to\infty} \frac{1}{N^n} \sum_{k=1}^{N} D(k-1, 2n-2) = \frac{1}{n} D_{n-1}. \tag{8.58}$$

We see from (8.55) and (8.58) that the limit on the right-hand side of (8.54) exists and (8.53) holds. □

Lemma 8.33. *Let D_m be defined as in Lemma 8.32. Then,*

$$D_m = \binom{-1/2}{m}(-2)^m = \frac{(2m)!}{2^m m! m!}, \qquad m = 0, 1, 2, \ldots .$$

Proof. Define the generating function $g(z)$ of $\{D_m\}$ by

$$g(z) = \sum_{m=0}^{\infty} D_m z^m. \tag{8.59}$$

Comparing the recurrence relation satisfied by the Catalan numbers $\{C_m\}$, which is given by

$$C_0 = 1, \qquad C_m = \sum_{n=1}^{m} C_{m-n} C_{n-1}, \qquad m \geq 1,$$

and (8.53), we see that $g(z)$ admits a positive radius of convergence. Applying Lemma 8.32 to (8.59), we obtain

$$g(z) - 1 = g(z) \sum_{n=1}^{\infty} \frac{D_{n-1}}{n} z^n.$$

By differentiation we easily obtain the differential equation

$$g'(z) = g(z)^3, \qquad g(0) = 1,$$

whose solution is given by

$$g(z) = \frac{1}{\sqrt{1-2z}} = \sum_{m=0}^{\infty} \binom{-1/2}{m} (-2z)^m. \tag{8.60}$$

The assertion follows then immediately by comparing (8.59) and (8.60). □

Proof of Theorem 8.30 (continuation). It follows from Lemmas 8.32 and 8.33 that

$$\lim_{N\to\infty} \frac{D(N, 2m)}{N^m} = D_m = \frac{(2m)!}{2^m m! m!}. \tag{8.61}$$

On the other hand, elementary calculus shows that (8.61) coincides with the $2m$th moment of the normalized arcsine law. This completes the proof of (8.41). □

We now go into the proof of the monotone central limit theorem (Theorem 8.23). Consider a non-crossing pair partition $\vartheta \in \mathcal{P}_{\mathrm{NCP}}(2m)$ given as in

$$\vartheta = \{\{l_1, r_1\}, \ldots, \{l_m, r_m\}\}, \quad l_1 < l_2 < \cdots < l_m, \quad l_i < r_i.$$

Let $\tilde{\mathfrak{M}}^o_\vartheta(2m,N)$ be the set of maps $n \in \mathfrak{M}(2m,N)$ such that

(i) $n(l_i) = n(r_i)$ for all $i = 1, 2, \dots$;
(ii) $n(l_i) < n(l_j)$ if $[l_i, r_i] \subset [l_j, r_j]$ and $[l_i, r_i] \neq [l_j, r_j]$.

Note that

$$\bigcup_{\vartheta \in \mathcal{P}_{\mathrm{NCP}}(2m)} \tilde{\mathfrak{M}}^o_\vartheta(2m,N) \tag{8.62}$$

is a disjoint union. In fact, for $n \in \tilde{\mathfrak{M}}^o_\vartheta(2m,N)$ take $1 \le i_1 < i_2 < \dots < i_{2k} \le 2m$ such that

$$n(i_1) = n(i_2) = \dots = n(i_{2k}) \neq n(j), \quad j \notin \{i_1, i_2, \dots, i_{2k}\}.$$

By condition (i),

$$\{i_1, i_2, \dots, i_{2k}\} = \bigcup_{s=1}^{k} \{l_{i_s}, r_{i_s}\}. \tag{8.63}$$

Among the subsets appearing on the right-hand side the proper inclusion relation is not allowed due to condition (ii). Moreover, since the right-hand side of (8.63) gives rise to a non-crossing pair partition, we see that

$$\{\{l_{i_1}, r_{i_1}\}, \dots, \{l_{i_k}, r_{i_k}\}\} = \{\{i_1, i_2\}, \dots, \{i_{2k-1}, i_{2k}\}\}.$$

In this way, $n \in \tilde{\mathfrak{M}}^o_\vartheta(2m,N)$ determines a non-crossing pair partition uniquely.

Lemma 8.34. *For $N \ge 1$ and $m \ge 1$ it holds that*

$$D(N,2m) = \sum_{\vartheta \in \mathcal{P}_{\mathrm{NCP}}(2m)} |\tilde{\mathfrak{M}}^o_\vartheta(2m,N)|.$$

Proof. It is sufficient to construct a one-to-one correspondence between (8.62) and the set of $2m$-step walks in the monotone tree $\mathcal{M}_N$ starting and terminating at $\varnothing$. Such a walk is represented by means of the monotone annihilation and creation operators as follows:

$$b_{k_{2m}}^{\epsilon_{2m}} \cdots b_{k_1}^{\epsilon_1} \delta_\varnothing, \tag{8.64}$$

where $k_1, \dots, k_{2m} \in \{1, 2, \dots, N\}$ and $\epsilon_1, \dots, \epsilon_{2m} \in \{\pm\}$ are uniquely determined. More precisely, (8.64) expresses such a walk if and only if (8.64) is reduced to $\delta_\varnothing$. Since $\mathcal{M}_N$ is a tree, $(\epsilon_1, \epsilon_2, \dots, \epsilon_{2m})$ is a Catalan path, i.e., belongs to $\mathcal{C}_m$. Through the canonical correspondence between $\mathcal{C}_m$ and $\mathcal{P}_{\mathrm{NCP}}(2m)$, a non-crossing pair partition $\vartheta \in \mathcal{P}_{\mathrm{NCP}}(2m)$ is associated with (8.64). Then, in view of (8.64), we may define a map $n \in \tilde{\mathfrak{M}}^o_\vartheta(2m,N)$ by

$$n(i) = k_i, \qquad i \in \{1, 2, \dots, 2m\}.$$

This correspondence satisfies the desired property. □

Proof of Theorem 8.23. It is sufficient to prove that

$$\lim_{N\to\infty} \varphi\Big(\Big(\frac{1}{\sqrt{N}}\sum_{n=1}^{N} a_n\Big)^{2m}\Big) = \frac{1}{\pi}\int_{-\sqrt{2}}^{+\sqrt{2}} \frac{x^{2m}}{\sqrt{2-x^2}}\,dx, \tag{8.65}$$

for $m = 1, 2, \ldots$. We readily know by Proposition 8.19 that

$$\begin{aligned}&\lim_{N\to\infty} \varphi\Big(\Big(\frac{1}{\sqrt{N}}\sum_{n=1}^{N} a_n\Big)^{2m}\Big)\\ &= \lim_{N\to\infty} N^{-m} \sum_{n\in\mathfrak{M}_p(2m,N)} \varphi(a_{n_1}\cdots a_{n_{2m}}).\end{aligned} \tag{8.66}$$

With each $n \in \mathfrak{M}_p(2m, N)$ we associate a pair partition $\vartheta_n \in \mathcal{P}_{\mathrm{P}}(2m)$ in a natural manner. Note that $\varphi(a_{n_1}\cdots a_{n_{2m}}) = 0$ unless ϑ_n is non-crossing. In fact, $k \equiv \max\{n_1, \ldots, n_{2m}\}$ appears twice therein. If they are not close to each other, by monotone independence

$$\varphi(a_{n_1}\cdots a_{n_{2m}}) = \varphi(a_k)\varphi(a_{n_1}\cdots \breve{a}_k \cdots a_{n_{2m}}) = 0.$$

Hence $\varphi(a_{n_1}\cdots a_{n_{2m}}) \neq 0$ implies that a_k appears as a_k^2. In that case, applying monotone independence again, we have

$$\varphi(a_{n_1}\cdots a_{n_{2m}}) = \varphi(a_k^2)\varphi(a_{n_1}\cdots \breve{a}_k\breve{a}_k \cdots a_{n_{2m}}) = \varphi(a_{n_1}\cdots \breve{a}_k\breve{a}_k \cdots a_{n_{2m}}).$$

Repeating this argument, we see that $\varphi(a_{n_1}\cdots a_{n_{2m}}) \neq 0$ implies that ϑ_n is non-crossing. Set

$$\vartheta_n = \{\{l_1, r_1\}, \ldots \{l_m, r_m\}\}, \quad l_1 < l_2 < \cdots < l_m, \quad l_i < r_i.$$

In the above argument, we have seen that $\varphi(a_{n_1}\cdots a_{n_{2m}}) \neq 0$ implies the following properties:

(i) $n(l_i) = n(r_i)$ for all $i = 1, 2, \ldots$;
(i′) $n(l_1), \ldots, n(l_m)$ are mutually distinct;
(ii′) $n(l_i) > n(l_j)$ if $[l_i, r_i] \subset [l_j, r_j]$ and $[l_i, r_i] \neq [l_j, r_j]$.

Let $\mathfrak{M}^*_\vartheta(2m, N)$ denote the set of $n \in \mathfrak{M}(2m, N)$ satisfying the above conditions. We know readily that

$$\varphi(a_{n_1}\cdots a_{n_{2m}}) = \begin{cases} 1, & \text{if } n \in \mathfrak{M}^*_\vartheta(2m, N),\\ 0, & \text{otherwise.}\end{cases}$$

Thus, (8.66) becomes

$$\lim_{N\to\infty} \varphi\Big(\Big(\frac{1}{\sqrt{N}}\sum_{n=1}^{N} a_n\Big)^{2m}\Big)$$
$$= \lim_{N\to\infty} N^{-m} \sum_{\vartheta\in\mathcal{P}_{\rm NCP}(2m)} \sum_{n\in\mathfrak{M}^*_\vartheta(2m,N)} \varphi(a_{n_1}\cdots a_{n_{2m}})$$
$$= \lim_{N\to\infty} N^{-m} \sum_{\vartheta\in\mathcal{P}_{\rm NCP}(2m)} |\mathfrak{M}^*_\vartheta(2m,N)|. \tag{8.67}$$

Note that there is a one-to-one correspondence between $\tilde{\mathfrak{M}}^o_\vartheta(2m,N)$ and the one satisfying condition (ii′) instead of (ii). Then, in view of $|\tilde{\mathfrak{M}}^o_\vartheta(2m,N)| - |\mathfrak{M}^*_\vartheta(2m,N)| = O(N^{m-1})$, we see that (8.67) becomes

$$= \lim_{N\to\infty} N^{-m} \sum_{\vartheta\in\mathcal{P}_{\rm NCP}(2m)} |\tilde{\mathfrak{M}}^o_\vartheta(2m,N)|.$$

Now applying Lemma 8.34, we obtain

$$\lim_{N\to\infty} \varphi\Big(\Big(\frac{1}{\sqrt{N}}\sum_{n=1}^{N} a_n\Big)^{2m}\Big) = \lim_{N\to\infty} \frac{D(N,2m)}{N^m}, \tag{8.68}$$

which coincides with

$$= \frac{(2m)!}{2^m m! m!} = \frac{1}{\pi}\int_{-\sqrt{2}}^{+\sqrt{2}} \frac{x^{2m}}{\sqrt{2-x^2}}\, dx$$

by Lemmas 8.32 and 8.33. □

8.5 Comb Product

Let $\mathcal{G}^{(1)} = (V^{(1)}, E^{(1)})$ and $\mathcal{G}^{(2)} = (V^{(2)}, E^{(2)})$ be two graphs and assume that the second graph is given a distinguished vertex $o \in V^{(2)}$. Consider the Cartesian product

$$V = V^{(1)} \times V^{(2)}$$

and set

$$E = \big\{\{(x,y),(x',y')\}\,;\,(x,y),(x',y') \in V,\ (x,y)\sim(x',y')\big\},$$

where $(x,y)\sim(x',y')$ means that one of the following conditions is satisfied:

(i) $x \sim x'$ and $y = y' = o$;
(ii) $x = x'$ and $y \sim y'$.

Then $\mathcal{G} = (V,E)$ becomes a graph (which is locally finite and connected whenever so are both $\mathcal{G}^{(1)}$ and $\mathcal{G}^{(2)}$). We call $\mathcal{G}$ the *comb product* of $\mathcal{G}^{(1)}$ and $\mathcal{G}^{(2)}$ with a contact vertex $o \in V^{(2)}$ and write

$$\mathcal{G} = \mathcal{G}^{(1)} \rhd_o \mathcal{G}^{(2)}.$$

In that case, $\mathcal{G}^{(1)}$ and $\mathcal{G}^{(2)}$ are sometimes called a *backbone* and a *finger*, respectively. In fact, the comb product $\mathcal{G}$ is obtained by grafting a copy of $\mathcal{G}^{(2)}$ at the vertex o into each vertex of $\mathcal{G}^{(1)}$, see Fig. 8.3.

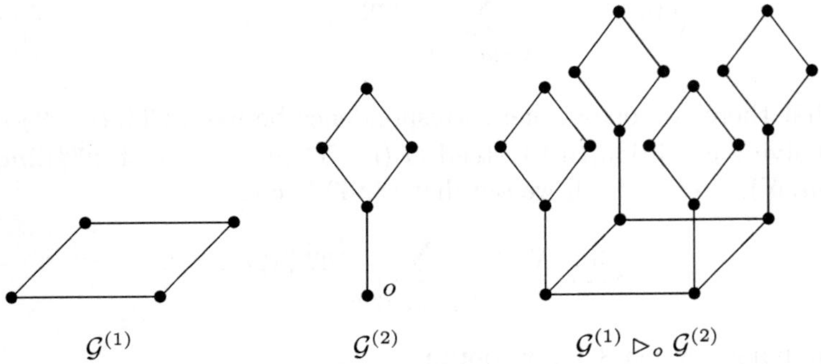

Fig. 8.3. Comb product

Let $A^{(i)}$ be the adjacency matrix of $\mathcal{G}^{(i)}$. The adjacency matrix of the comb product $\mathcal{G}^{(1)} \rhd_o \mathcal{G}^{(2)}$ is denoted by $A^{(1)} \rhd_o A^{(2)}$.

Lemma 8.35. *The matrix elements of $A^{(1)} \rhd_o A^{(2)}$ are given by*

$$(A^{(1)} \rhd_o A^{(2)})_{(x,y),(x',y')} = A^{(1)}_{xx'}\delta_{yo}\delta_{y'o} + \delta_{xx'}A^{(2)}_{yy'}, \tag{8.69}$$

where $x, x' \in V^{(1)}$ and $y, y' \in V^{(2)}$.

Proof. A simple computation shows that

$$A^{(1)}_{xx'}\delta_{yo}\delta_{y'o} + \delta_{xx'}A^{(2)}_{yy'} = \begin{cases} 1, & \text{if } x \sim x' \text{ and } y = y' = o, \\ 1, & \text{if } x = x' \text{ and } y \sim y', \\ 0, & \text{otherwise.} \end{cases}$$

Then, in view of conditions (i) and (ii) above, we see that the right-hand side of (8.69) takes value 1 if and only if $(x, y) \sim (x', y')$. In other words, the right-hand side of (8.69) coincides with the matrix element of the adjacency matrix of the comb product $\mathcal{G}^{(1)} \rhd_o \mathcal{G}^{(2)}$. □

Obviously, the comb product is not commutative; still it is associative.

Lemma 8.36. *For $i = 1, 2, 3$ let $\mathcal{G}^{(i)}$ be a graph and assume that a distinguished vertex o_i of $\mathcal{G}^{(i)}$ is chosen for $i = 2, 3$. Then we have*

$$(\mathcal{G}^{(1)} \rhd_{o_2} \mathcal{G}^{(2)}) \rhd_{o_3} \mathcal{G}^{(3)} = \mathcal{G}^{(1)} \rhd_{(o_2,o_3)} (\mathcal{G}^{(2)} \rhd_{o_3} \mathcal{G}^{(3)}).$$

Proof. Straightforward from (8.69). □

Whenever there is no danger of confusion, we omit the suffix o and write $\mathcal{G}^{(1)} \rhd \mathcal{G}^{(2)}$ and $A^{(1)} \rhd A^{(2)}$ for brevity. The adjacency matrix $A^{(1)} \rhd A^{(2)}$ acts on $\ell^2(V_1 \times V_2) \cong \ell^2(V_1) \otimes \ell^2(V_2)$. Let us examine this action in detail.

Lemma 8.37. *Notations being as above,*

$$A^{(1)} \rhd A^{(2)} = A^{(1)} \otimes P^{(2)} + 1 \otimes A^{(2)}, \tag{8.70}$$

where $P^{(2)}$ is the projection from $\ell^2(V_2)$ onto the one-dimensional subspace spanned by δ_o, i.e., defined by

$$P^{(2)}\psi(y) = \langle \delta_o, \psi \rangle \delta_o(y) = \delta_{yo}\psi(o), \quad \psi \in \ell^2(V_2), \quad y \in V^{(2)}. \tag{8.71}$$

Proof. Let $\phi \in \ell^2(V_1)$ and $\psi \in \ell^2(V_2)$. Then, in view of (8.69) we have

$$\begin{aligned}
&(A^{(1)} \rhd A^{(2)})(\phi \otimes \psi)(x,y) \\
&\quad = \sum_{x',y'} \left(A^{(1)}_{xx'} \delta_{yo}\delta_{y'o} + \delta_{xx'} A^{(2)}_{yy'} \right) \phi(x')\psi(y') \\
&\quad = \sum_{x'} A^{(1)}_{xx'} \delta_{yo}\phi(x')\psi(o) + \sum_{y'} A^{(2)}_{yy'} \phi(x)\psi(y') \\
&\quad = (A^{(1)}\phi)(x)\delta_{yo}\psi(o) + \phi(x)(A^{(2)}\psi)(y).
\end{aligned} \tag{8.72}$$

By (8.71) the first term becomes

$$(A^{(1)}\phi)(x)\delta_{yo}\psi(o) = (A^{(1)}\phi)(x)(P^{(2)}\psi)(y) = (A^{(1)}\phi \otimes P^{(2)}\psi)(x,y).$$

On the other hand, the second term in (8.72) becomes

$$\phi(x)(A^{(2)}\psi)(y) = (\phi \otimes A^{(2)}\psi)(x,y).$$

Then the assertion follows immediately. □

Proposition 8.38. *For $i = 1, 2, \dots, n$ let $\mathcal{G}^{(i)} = (V^{(i)}, E^{(i)})$ be a graph and assume that a distinguished vertex $o_i \in V^{(i)}$ is chosen for $i = 2, \dots, n$. Then the adjacency matrix of the comb product $\mathcal{G}^{(1)} \rhd_{o_2} \mathcal{G}^{(2)} \rhd_{o_3} \cdots \rhd_{o_n} \mathcal{G}^{(n)}$ admits a decomposition of the form*

$$\begin{aligned}
&A^{(1)} \rhd A^{(2)} \rhd \cdots \rhd A^{(n)} \\
&\quad = \sum_{i=1}^{n} \overbrace{1^{(1)} \otimes \cdots \otimes 1^{(i-1)}}^{i-1} \otimes A^{(i)} \otimes \overbrace{P^{(i+1)} \otimes \cdots \otimes P^{(n)}}^{n-i},
\end{aligned} \tag{8.73}$$

where $1^{(i)}$ is the identity operator on $\ell^2(V^{(i)})$ and $P^{(i)}$ the projection from $\ell^2(V^{(i)})$ onto the one-dimensional subspace spanned by δ_{o_i}.

Proof. By repeated application of Lemmas 8.36 and 8.37. □

Combining Theorem 8.9 and Proposition 8.38, we obtain the following interesting result.

Theorem 8.39. *Notations and assumptions being the same as in Proposition 8.38, consider a state of the form*

$$\Omega_n = \psi \otimes \delta_{o_2} \otimes \cdots \otimes \delta_{o_n},$$

where ψ is an arbitrary state on $\mathcal{B}(\ell^2(V^{(1)}))$. Then, the right-hand side of (8.73) is a sum of monotone independent random variables with respect to Ω_n.

As a consequence of the monotone central limit theorem (Theorem 8.23), we prove the following:

Theorem 8.40 (CLT for comb powers). *Let $\mathcal{G} = (V, E)$ be a graph with a distinguished vertex $o \in V$. Consider the n-fold comb power and its adjacency matrix:*

$$\mathcal{G}^{\rhd n} = \overbrace{\mathcal{G} \rhd_o \mathcal{G} \rhd_o \cdots \rhd_o \mathcal{G}}^{n \text{ times}}, \quad A^{\rhd n} = \overbrace{A \rhd_o A \rhd_o \cdots \rhd_o A}^{n \text{ times}}.$$

Then it holds that

$$\lim_{n\to\infty} \left\langle \Omega_n, \left(\frac{A^{\rhd n}}{\sqrt{n\kappa(o)}} \right)^m \Omega_n \right\rangle = \frac{1}{\pi} \int_{-\sqrt{2}}^{+\sqrt{2}} \frac{x^m}{\sqrt{2 - x^2}}\, dx, \quad m = 1, 2, \dots,$$

where $\Omega_n = \delta_o \otimes \cdots \otimes \delta_o$ (n times) and $\kappa(o)$ is the degree of the distinguished vertex $o \in V$.

Proof. Applying the decomposition (8.73) to the present case of a comb power, we have

$$A^{\rhd n} = \sum_{i=1}^{n} \overbrace{1 \otimes \cdots \otimes 1}^{i-1} \otimes A \otimes \overbrace{P \otimes \cdots \otimes P}^{n-i} \equiv \sum_{i=1}^{n} X_i,$$

where P is the projection from $\ell^2(V)$ onto the one-dimensional subspace spanned by δ_o. We know from Theorem 8.39 that $\{X_i\}$ is monotone independent. Obviously, $\{X_i\}$ satisfies conditions (i) and (iii) in (CC). Moreover, since $\langle \delta_o, A\delta_o \rangle = 0$,

$$\langle \Omega_n, X_i \Omega_n \rangle = \prod_{j=1}^{i-1} \langle \delta_o, \delta_o \rangle \times \langle \delta_o, A\delta_o \rangle \times \prod_{j=i+1}^{n} \langle \delta_o, P\delta_o \rangle = 0.$$

Similarly,

$$\begin{aligned}\langle \Omega_n, X_i^2 \Omega_n \rangle &= \prod_{j=1}^{i-1} \langle \delta_o, \delta_o \rangle \times \langle \delta_o, A^2 \delta_o \rangle \times \prod_{j=i+1}^{n} \langle \delta_o, P^2 \delta_o \rangle \\ &= \langle \delta_o, A^2 \delta_o \rangle = \kappa(o),\end{aligned}$$

which is the number of two-step walks from o to itself. Hence $\{X_i/\sqrt{\kappa(o)}\}$ satisfies all the conditions in Theorem 8.23 and our assertion is a direct consequence of it. □

8.6 Comb Lattices

As a simple example, we shall study the spectral property of the adjacency matrix of the two-dimensional *comb lattice* $\mathbb{Z} \triangleright_0 \mathbb{Z}$, see Fig. 8.4.

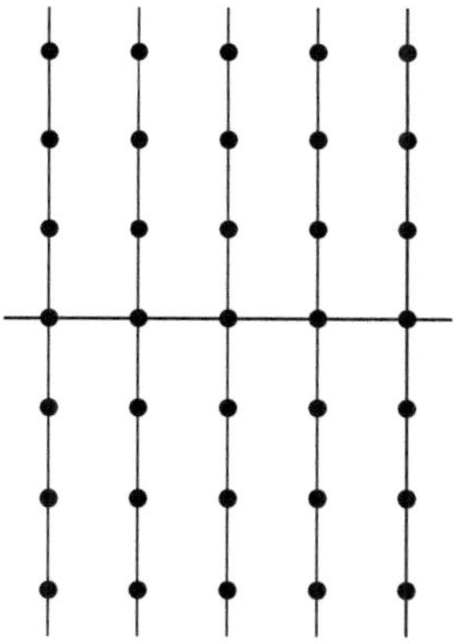

Fig. 8.4. Comb lattice $\mathbb{Z} \triangleright \mathbb{Z}$

Let us start with the one-dimensional integer lattice $\mathbb{Z}$, where two vertices $i, j \in \mathbb{Z}$ are adjacent by definition if $|i - j| = 1$. The adjacency matrix of $\mathbb{Z}$ is denoted by A. Taking $0 \in \mathbb{Z}$ to be the origin, we introduce a stratification:

$$\mathbb{Z} = \bigcup_{n=0}^{\infty} V_n\,, \qquad V_0 = \{0\}, \quad V_n = \{\pm n\}, \quad n \geq 1, \tag{8.74}$$

and associated unit vectors in $\ell^2(\mathbb{Z})$:

$$\Phi_0 = \delta_0, \qquad \Phi_n = \frac{1}{\sqrt{2}}(\delta_n + \delta_{-n}), \quad n \geq 1.$$

Then $\{\Phi_n\}$ is an orthonormal set in $\ell^2(\mathbb{Z})$. Let $\Gamma(\mathbb{Z})$ be the linear space spanned by $\{\Phi_n\}$. As is easily verified, $\Gamma(\mathbb{Z})$ is invariant under the action of A. In fact, we have

$$\begin{cases} A\Phi_0 = \sqrt{2}\,\Phi_1, \\ A\Phi_1 = \Phi_2 + \sqrt{2}\,\Phi_0, \\ A\Phi_m = \Phi_{m+1} + \Phi_{m-1}, \quad m = 2, 3, \dots . \end{cases} \tag{8.75}$$

Let $A = A^+ + A^-$ be the quantum decomposition. Then, (8.75) means that

$$(\Gamma(\mathbb{Z}), \{\Phi_n\}, A^+ \restriction_{\Gamma(\mathbb{Z})}, A^- \restriction_{\Gamma(\mathbb{Z})})$$

is an interacting Fock space associated with a Jacobi sequence given by

$$\omega_1 = 2, \ \omega_2 = \omega_3 = \dots = 1.$$

By a routine calculation we come to the following:

Proposition 8.41. *Let A be the adjacency matrix of the one-dimensional integer lattice $\mathbb{Z}$. The spectral distribution of A in the vacuum state δ_0 is the arcsine law with variance 2, i.e.,*

$$\langle \delta_0, A^m \delta_0 \rangle = \frac{1}{\pi} \int_{-2}^{+2} \frac{x^m}{\sqrt{4 - x^2}}\, dx, \qquad m = 1, 2, \dots . \tag{8.76}$$

Note that A is a bounded operator on $\ell^2(\mathbb{Z})$. Since $\Gamma(\mathbb{Z})$ is invariant under A, so is the orthogonal complement $\Gamma(\mathbb{Z})^\perp$. We set

$$\Psi_n = \frac{1}{\sqrt{2}}(\delta_{n+1} - \delta_{-n-1}), \qquad n = 0, 1, 2, \dots .$$

Then $\{\Psi_n\}$ forms an orthonormal basis of $\Gamma(\mathbb{Z})^\perp$. Let $\Gamma_-(\mathbb{Z})$ be the linear space spanned by $\{\Psi_n\}$. We see by direct computation that

$$A\Psi_0 = \Psi_1\,, \qquad A\Psi_n = \Psi_{n+1} + \Psi_{n-1}, \quad n = 1, 2, \dots . \tag{8.77}$$

In other words,

$$(\Gamma_-(\mathbb{Z}), \{\Psi_n\}, A^+ \restriction_{\Gamma_-(\mathbb{Z})}, A^- \restriction_{\Gamma_-(\mathbb{Z})})$$

is an interacting Fock space associated with a Jacobi sequence $\{\omega_n \equiv 1\}$, i.e., the free Fock space. Consequently,

Proposition 8.42. *Let A be the adjacency matrix of the one-dimensional integer lattice $\mathbb{Z}$. The spectral distribution of A in the vector state corresponding to the state vector*

$$\Psi_0 = \frac{1}{\sqrt{2}}\,(\delta_{+1} - \delta_{-1})$$

is the Wigner semicircle law, namely,

$$\langle \Psi_0, A^m \Psi_0 \rangle = \frac{1}{2\pi} \int_{-2}^{+2} x^m \sqrt{4 - x^2}\, dx, \qquad m = 1, 2, \dots . \tag{8.78}$$

We are now in a position to discuss the two-dimensional comb lattice $\mathbb{Z} \rhd_0 \mathbb{Z}$. The adjacency matrix is denoted by

$$A^{(2)} = A \rhd_0 A$$

for simplicity. Then, $A^{(2)}$ acts on $\ell^2(\mathbb{Z} \times \mathbb{Z}) \cong \ell^2(\mathbb{Z}) \otimes \ell^2(\mathbb{Z})$ and, as is already shown in Lemma 8.37, we have

$$A^{(2)} = A \otimes P + 1 \otimes A, \tag{8.79}$$

where P is the projection from $\ell^2(\mathbb{Z})$ onto the one-dimensional subspace spanned by $\delta_0 = \Phi_0$. On the other hand, $\ell^2(\mathbb{Z})$ admits an orthogonal decomposition:

$$\ell^2(\mathbb{Z}) = \overline{\Gamma_+(\mathbb{Z})} \oplus \overline{\Gamma_-(\mathbb{Z})},$$

where we set $\Gamma_+(\mathbb{Z}) = \Gamma(\mathbb{Z})$ for notational convenience. Both $\Gamma_\pm(\mathbb{Z})$ are invariant under the action of A as was shown in the first half of this section, and also under the action of P. Consequently, we have

$$\ell^2(\mathbb{Z}) \otimes \ell^2(\mathbb{Z}) = \bigoplus_{\epsilon, \epsilon' \in \{\pm\}} \overline{\Gamma_\epsilon(\mathbb{Z})} \otimes \overline{\Gamma_{\epsilon'}(\mathbb{Z})} \tag{8.80}$$

and each of the four orthogonal components is invariant under the action of $A^{(2)}$. Note that the state vectors Φ_0, Ψ_0 considered in the case of one-dimensional integer lattice yield state vectors

$$\begin{aligned}
\Phi_0 \otimes \Phi_0 &\in \Gamma_+(\mathbb{Z}) \otimes \Gamma_+(\mathbb{Z}), \\
\Psi_0 \otimes \Phi_0 &\in \Gamma_-(\mathbb{Z}) \otimes \Gamma_+(\mathbb{Z}), \\
\Phi_0 \otimes \Psi_0 &\in \Gamma_+(\mathbb{Z}) \otimes \Gamma_-(\mathbb{Z}), \\
\Psi_0 \otimes \Psi_0 &\in \Gamma_-(\mathbb{Z}) \otimes \Gamma_-(\mathbb{Z}).
\end{aligned}$$

We are interested in the spectral distribution of $A^{(2)}$ in the vector state corresponding to the above state vectors.

We need the following result whose proof is omitted.

Theorem 8.43 (Muraki's formula). *For $k = 1, 2, \dots, n$ let a_k be a real random variable in an algebraic probability space $(\mathcal{A}, \varphi)$. Let μ_k be the distribution of a_k and assume that μ_k is the solution of a determinate moment problem. If $\{a_k\}$ is monotone independent, it holds that*

$$H_{a_1 + \cdots + a_n}(z) = H_{a_1}(H_{a_2}(\cdots H_{a_n}(z) \cdots)), \tag{8.81}$$

where $H_a(z)$ is the reciprocal Stieltjes transform of the distribution μ_a of $a = a^* \in \mathcal{A}$, i.e.,

$$H_a(z) = \frac{1}{G_a(z)}, \quad G_a(z) = \int_{-\infty}^{+\infty} \frac{\mu_a(dx)}{z - x}, \quad \mathrm{Im}\,(z) > 0.$$

Theorem 8.44. *The spectral distribution of the adjacency matrix $A^{(2)}$ in the vacuum state $\Phi_0 \otimes \Phi_0$ is the arcsine law with variance 4 (see Fig. 8.5), i.e.,*

$$\langle \Phi_0 \otimes \Phi_0, (A^{(2)})^m \Phi_0 \otimes \Phi_0 \rangle = \frac{1}{\pi} \int_{-2\sqrt{2}}^{+2\sqrt{2}} \frac{x^m}{\sqrt{8-x^2}}\, dx, \qquad m = 1, 2, \dots .$$

Proof. Let ν_1 and ν_2 be the distribution of $A \otimes P$ and $1 \otimes A$ in the vacuum state $\Phi_0 \otimes \Phi_0$, respectively. As is easy to see, ν_1 and ν_2 coincide with the distribution of A in Φ_0, which is by Proposition 8.41 the arcsine law:

$$\nu_1(dx) = \nu_2(dx) = \frac{dx}{\pi\sqrt{4-x^2}}, \qquad |x| < 2.$$

Note that $\nu_1 = \nu_2$ is the solution of a determinate moment problem because it has a compact support. The reciprocal Stieltjes transform of $\nu_1 = \nu_2$ is known:

$$H_{\nu_1}(z) = H_{\nu_2}(z) = \sqrt{z^2 - 4}.$$

Let μ be the spectral distribution of $A^{(2)} = A \otimes P + 1 \otimes A$ in the vacuum state $\Phi_0 \otimes \Phi_0$. Since $\{A \otimes P, 1 \otimes A\}$ is monotone independent with respect to $\Phi_0 \otimes \Phi_0$, by Muraki's formula (Theorem 8.43) we obtain

$$H_\mu(z) = H_{\nu_1}(H_{\nu_2}(z)) = \sqrt{H_{\nu_2}(z)^2 - 4} = \sqrt{z^2 - 8},$$

that is,

$$G_\mu(z) = \frac{1}{\sqrt{z^2 - 8}},$$

from which we easily see that μ is the arcsine law with variance 4. □

Theorem 8.45. *The spectral distribution of $A^{(2)}$ in $\Psi_0 \otimes \Phi_0$ is absolutely continuous with respect to the Lebesgue measure and the density function $\rho(x)$ is given by*

$$\rho(x) = \begin{cases} \dfrac{1}{2\pi}\left(\sqrt{8-x^2} - \sqrt{4-x^2}\right), & |x| \le 2, \\ \dfrac{1}{2\pi}\sqrt{8-x^2}, & 2 \le |x| \le 2\sqrt{2}, \\ 0, & \textit{otherwise}. \end{cases}$$

(See Fig. 8.5)

Proof. We first see from Theorem 8.9 that

$$A^{(2)} = A \otimes P + 1 \otimes A$$

is a sum of random variables which are monotone independent with respect to $\Psi_0 \otimes \Phi_0$ as well. Hence our argument is similar to the proof of Theorem 8.44. Let

ν_1 and ν_2 denote the distributions of $A \otimes P$ and $1 \otimes A$ in $\Psi_0 \otimes \Phi_0$, respectively. We already know from Proposition 8.42 that ν_1 is the Wigner semicircle law and from Proposition 8.41 that ν_2 is the arcsine law with variance 2. Their reciprocal Stieltjes transforms are

$$H_{\nu_1}(z) = \frac{2}{z - \sqrt{z^2 - 4}}, \qquad H_{\nu_2}(z) = \sqrt{z^2 - 4}.$$

Let μ denote the distribution of $A^{(2)}$ in $\Psi_0 \otimes \Phi_0$. Applying Muraki's formula, we obtain

$$H_\mu(z) = H_{\nu_1}(H_{\nu_2}(z)) = \frac{2}{\sqrt{z^2 - 4} - \sqrt{z^2 - 8}}.$$

Hence the Stieltjes transform of μ is given by

$$G_\mu(z) = \frac{1}{H_\mu(z)} = \frac{\sqrt{z^2 - 4} - \sqrt{z^2 - 8}}{2}.$$

Then the density function is obtained by elementary calculus with the help of the Stieltjes inversion formula. □

Theorem 8.46. *The spectral distributions of $A^{(2)}$ in $\Phi_0 \otimes \Psi_0$ and $\Psi_0 \otimes \Psi_0$ are the Wigner semicircle law (see Fig. 8.5).*

Proof. Taking (8.79) into account, we have

$$(A^{(2)})^m = \sum_{k=0}^{m} A^k \otimes X(m-k, k),$$

where $X(m-k, k)$ is the sum of $A^{m-k}P^k$ and its all possible permutations. Then,

$$\langle \Phi_0 \otimes \Psi_0, (A^{(2)})^m \Phi_0 \otimes \Psi_0 \rangle = \sum_{k=0}^{m} \langle \Phi_0, A^k \Phi_0 \rangle \langle \Psi_0, X(m-k, k) \Psi_0 \rangle. \qquad (8.82)$$

Since $\Psi_0 \in \Gamma(\mathbb{Z})^\perp$ and $\Gamma(\mathbb{Z})^\perp$ is invariant under A and P, $X(m-k,k)\Psi_0 \in \Gamma(\mathbb{Z})^\perp$. Moreover, since P acts on $\Gamma(\mathbb{Z})^\perp$ as a zero operator, if $X(m-k,k)$ contains P, that is, if $1 \le k \le m$, we have $X(m-k,k)\Psi_0 = 0$. Thus, (8.82) becomes

$$\langle \Phi_0 \otimes \Psi_0, (A^{(2)})^m \Phi_0 \otimes \Psi_0 \rangle = \langle \Psi_0, X(m, 0) \Psi_0 \rangle = \langle \Psi_0, A^m \Psi_0 \rangle.$$

Namely, the distribution of $A^{(2)}$ in $\Phi_0 \otimes \Psi_0$ is the same as that of A in Ψ_0, which is the Wigner semicircle law as is shown in Proposition 8.42. The distribution of $A^{(2)}$ in $\Psi_0 \otimes \Psi_0$ is obtained in a similar fashion. □

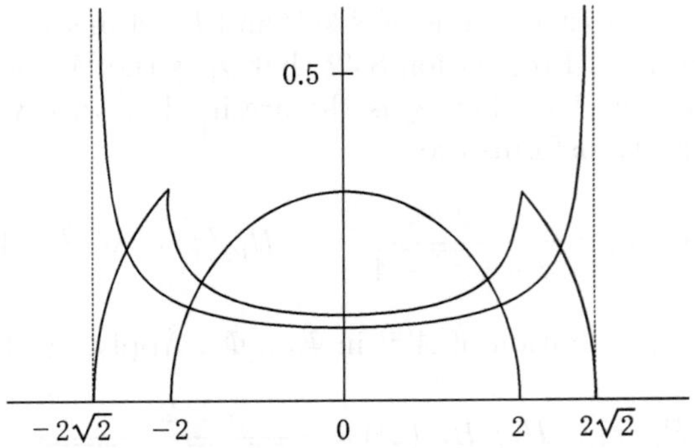

Fig. 8.5. Spectral distributions of the adjacency matrix of the comb lattice

8.7 Star Product

The star product is a very simple idea. Consider two graphs $\mathcal{G}^{(1)} = (V^{(1)}, E^{(1)})$ and $\mathcal{G}^{(2)} = (V^{(2)}, E^{(2)})$ and assume that a distinguished vertex $o_i \in V^{(i)}$ is chosen for each $i = 1, 2$. The *star product* of $\mathcal{G}^{(1)}$ and $\mathcal{G}^{(2)}$ with contact vertices o_1 and o_2 is obtained by uniting them at the contact vertices, see Fig. 8.6. Such a star product is denoted by

$$\mathcal{G}^{(1)}{}_{o_1}\!\star_{o_2} \mathcal{G}^{(2)} \quad \text{or simply} \quad \mathcal{G}^{(1)} \star \mathcal{G}^{(2)}.$$

Accordingly, the adjacency matrix is denoted by

$$A^{(1)}{}_{o_1}\!\star_{o_2} A^{(2)} \quad \text{or simply} \quad A^{(1)} \star A^{(2)}.$$

For our purpose the following formal definition will be useful. Set

$$V = \{(x, o_2)\,;\, x \in V^{(1)}\} \cup \{(o_1, y)\,;\, y \in V^{(2)}\} \subset V^{(1)} \times V^{(2)}$$

and

$$E = \big\{\{(x,y),(x',y')\}\,;\,(x,y),(x',y') \in V,\ (x,y) \sim (x',y')\big\},$$

where $(x, y) \sim (x', y')$ means that one of the following conditions is satisfied:

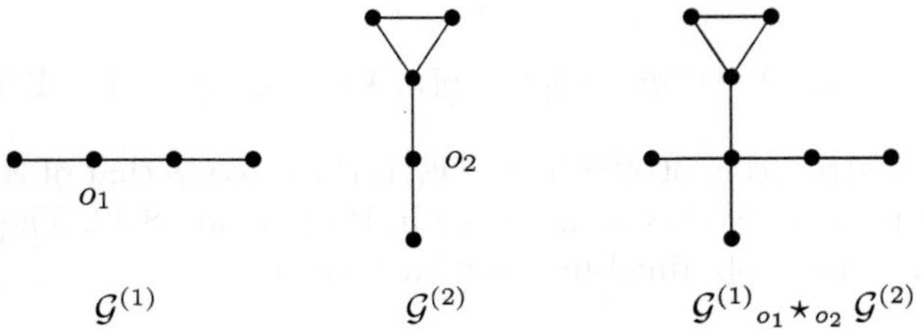

Fig. 8.6. Star product

(i) $x \sim x'$ and $y = y' = o_2$;
(ii) $x = x' = o_1$ and $y \sim y'$.

Then $\mathcal{G} = (V, E)$ becomes a graph which is nothing but the star product introduced above.

Lemma 8.47. *The matrix elements of* $A^{(1)}{}_{o_1}\star_{o_2} A^{(2)}$ *are given by*

$$(A^{(1)}{}_{o_1}\star_{o_2} A^{(2)})_{(x,y),(x',y')} = A^{(1)}_{xx'}\delta_{yo_2}\delta_{y'o_2} + \delta_{xo_1}\delta_{x'o_1}A^{(2)}_{yy'} \tag{8.83}$$

for $(x, y), (x', y') \in V$.

The verification is straightforward. The right-hand side of (8.83) is defined for all $(x, y), (x', y') \in V^{(1)} \times V^{(2)}$ and becomes a symmetric matrix whose entries take values in $\{0, 1\}$ and diagonal ones vanish. Therefore, it is the adjacency matrix of a graph which is not necessarily connected. Our star product is the connected component containing (o_1, o_2).

Lemma 8.48. *The star product is commutative and associative, i.e.,*

$$\mathcal{G}^{(1)}{}_{o_1}\star_{o_2} \mathcal{G}^{(2)} = \mathcal{G}^{(2)}{}_{o_2}\star_{o_1} \mathcal{G}^{(1)},$$
$$(\mathcal{G}^{(1)}{}_{o_1}\star_{o_2} \mathcal{G}^{(2)})_{(o_1,o_2)}\star_{o_3} \mathcal{G}^{(3)} = \mathcal{G}^{(1)}{}_{o_1}\star_{(o_2,o_3)} (\mathcal{G}^{(2)}{}_{o_2}\star_{o_3} \mathcal{G}^{(3)}).$$

We now examine the action of $A^{(1)}{}_{o_1}\star_{o_2} A^{(2)}$ on the Hilbert space $\ell^2(V) \subset \ell^2(V_1 \times V_2) \cong \ell^2(V_1) \otimes \ell^2(V_2)$. By Lemma 8.47 we may regard $A^{(1)}{}_{o_1}\star_{o_2} A^{(2)}$ as an operator acting on $\ell^2(V_1) \otimes \ell^2(V_2)$.

Lemma 8.49. *Regarded as an operator on* $\ell^2(V_1) \otimes \ell^2(V_2)$,

$$A^{(1)}{}_{o_1}\star_{o_2} A^{(2)} = A^{(1)} \otimes P^{(2)} + P^{(1)} \otimes A^{(2)}, \tag{8.84}$$

where $P^{(i)}$ *is the projection from* $\ell^2(V_i)$ *onto the one-dimensional subspace spanned by* δ_{o_i}. *Moreover,* $\ell^2(V)$ *is invariant under the action of* $A^{(1)}{}_{o_1}\star_{o_2} A^{(2)}$.

Proof. For simplicity we write $A^{(1)} \star A^{(2)} = A^{(1)}{}_{o_1}\star_{o_2} A^{(2)}$. By definition and Lemma 8.47,

$$\begin{aligned}\langle \delta_{(x,y)}, A^{(1)} \star A^{(2)} \delta_{(x',y')}\rangle &= (A^{(1)} \star A^{(2)})_{(x,y),(x',y')} \\ &= A^{(1)}_{xx'}\delta_{yo_2}\delta_{y'o_2} + \delta_{xo_1}\delta_{x'o_1}A^{(2)}_{yy'}.\end{aligned} \tag{8.85}$$

On the other hand, since

$$P^{(1)}\delta_{x'} = \langle \delta_{o_1}, \delta_{x'}\rangle\delta_{o_1} = \delta_{x'o_1}\delta_{o_1}$$

we have

$$\langle \delta_x, P^{(1)}\delta_{x'}\rangle = \delta_{x'o_1}\langle \delta_x, \delta_{o_1}\rangle = \delta_{x'o_1}\delta_{xo_1}.$$

Hence (8.85) becomes

$$\begin{aligned}&\langle \delta_{(x,y)}, A^{(1)} \star A^{(2)} \delta_{(x',y')} \rangle \\ &= \langle \delta_x, A^{(1)} \delta_{x'} \rangle \langle \delta_y, P^{(2)} \delta_{y'} \rangle + \langle \delta_x, P^{(1)} \delta_{x'} \rangle \langle \delta_y, A^{(2)} \delta_{y'} \rangle \\ &= \langle \delta_x \otimes \delta_y, (A^{(1)} \otimes P^{(2)})(\delta_{x'} \otimes \delta_{y'}) \rangle + \langle \delta_x \otimes \delta_y, (P^{(1)} \otimes A^{(2)})(\delta_{x'} \otimes \delta_{y'}) \rangle,\end{aligned}$$

which proves (8.84). Then, since $\ell^2(V)$ is spanned by

$$\{\delta_x \otimes \delta_{o_2} \,;\, x \in V^{(1)}\} \cup \{\delta_{o_1} \otimes \delta_y \,;\, y \in V^{(2)}\},$$

it is invariant under the action of $A^{(1)} \star A^{(2)}$. □

Proposition 8.50. *For $i = 1, 2, \dots, n$ let $\mathcal{G}^{(i)} = (V^{(i)}, E^{(i)})$ be a graph with a distinguished vertex $o_i \in V^{(i)}$. Then the adjacency matrix of the star product $\mathcal{G}^{(1)} \star \mathcal{G}^{(2)} \star \cdots \star \mathcal{G}^{(n)}$ admits a decomposition of the form*

$$\begin{aligned}&A^{(1)} \star A^{(2)} \star \cdots \star A^{(n)} \\ &\quad = \sum_{i=1}^{n} \overbrace{P^{(1)} \otimes \cdots \otimes P^{(i-1)}}^{i-1} \otimes A^{(i)} \otimes \overbrace{P^{(i+1)} \otimes \cdots \otimes P^{(n)}}^{n-i}, \qquad (8.86)\end{aligned}$$

where $P^{(i)}$ is the projection from $\ell^2(V^{(i)})$ onto the one-dimensional subspace spanned by δ_{o_i}.

Proof. By repeated application of Lemma 8.49. □

Combining Theorem 8.8 and Proposition 8.50, we obtain the following interesting result.

Theorem 8.51. *Notations and assumptions being the same as in Proposition 8.50, consider a state of the form*

$$\Omega_n = \delta_{o_1} \otimes \delta_{o_2} \otimes \cdots \otimes \delta_{o_n}.$$

Then, the right-hand side of (8.86) *is a sum of Boolean independent random variables with respect to Ω_n.*

We focus on the star powers. Let $\mathcal{G} = (V, E)$ be a graph equipped with a fixed origin $o \in V$. For $N \geq 1$ the *N-fold star power* is defined by

$$\mathcal{G}^{\star N} = \overbrace{\mathcal{G} \star \mathcal{G} \star \cdots \star \mathcal{G}}^{N \text{ times}}.$$

The initial graph $\mathcal{G}$ is naturally considered as a subgraph of $\mathcal{G}^{\star N}$, which we call a *leaf*. Let A be the adjacency matrix of $\mathcal{G}$ and $A^{\star N}$ that of $\mathcal{G}^{\star N}$. The vertex set of $\mathcal{G}^{\star N}$ is denoted by $V^{(N)}$, which is a subset of the Cartesian product V^N. Taking $(o, o, \dots, o)$ to be the origin of $\mathcal{G}^{\star N}$, we introduce the stratification and the quantum decomposition:

$$V^{(N)} = \bigcup_{n=1}^{\infty} V_n^{(N)}, \qquad A^{\star N} = (A^{\star N})^+ + (A^{\star N})^- + (A^{\star N})^\circ.$$

For simplicity of notation we write $o = (o, o, \dots, o)$.

Lemma 8.52. *Notations being as above, assume that $\Gamma(\mathcal{G})$ is invariant under the quantum components A^ϵ of A, $\epsilon \in \{+,-,\circ\}$, and let $(\{\omega_n\},\{\alpha_n\})$ be the associated Jacobi coefficient. Then, $\Gamma(\mathcal{G}^{\star N})$ is also invariant under $(A^{\star N})^\epsilon$, $\epsilon \in \{+,-,\circ\}$, and the associated Jacobi coefficient is given by*

$$(\{N\omega_1,\omega_2,\omega_3,\dots\},\ \{\alpha_1,\alpha_2,\alpha_3,\dots\}). \tag{8.87}$$

Proof. As usual, for $\epsilon \in \{+,-,\circ\}$ we set

$$\omega_\epsilon(x) = |\{y \in V\,;\, y \sim x,\ \partial(o,y) = \partial(o,x)+\epsilon\}|, \quad x \in V,$$
$$\omega_\epsilon^{(N)}(x') = |\{y' \in V^{(N)}\,;\, y' \sim x',\ \partial(o,y') = \partial(o,x')+\epsilon\}|, \quad x' \in V^{(N)}.$$

Since $\Gamma(\mathcal{G})$ is invariant under the actions of the quantum components A^ϵ, we see that

$$\omega_n = \frac{|V_n|}{|V_{n-1}|}\,\omega_-(y)^2, \quad y \in V_n, \tag{8.88}$$
$$\alpha_n = \omega_\circ(y), \quad y \in V_{n-1}, \tag{8.89}$$

where ω_n and α_n are defined independently of the choice of y (see Proposition 2.27). On the other hand, we see from construction of the star graph that for $n = 1,2,\dots,$

$$\omega_-^{(N)}(y') = \omega_-(y), \qquad y' \in V_n^{(N)}, \quad y \in V_n, \tag{8.90}$$
$$\omega_\circ^{(N)}(y') = \omega_\circ(y), \qquad y' \in V_{n-1}^{(N)}, \quad y \in V_{n-1}, \tag{8.91}$$

It follows easily (see also Proposition 2.27) that $\Gamma(\mathcal{G}^{\star N})$ is invariant under the quantum components $(A^{\star N})^\epsilon$, $\epsilon \in \{+,-,\circ\}$.

Let $(\{\omega_n^{(N)}\},\{\alpha_n^{(N)}\})$ be the associated Jacobi coefficient. Then, by (8.88) and (8.90) we see that for $n = 2,3,\dots,$

$$\omega_n^{(N)} = \frac{|V_n^{(N)}|}{|V_{n-1}^{(N)}|}\,\omega_-^{(N)}(y')^2 = \frac{N|V_n|}{N|V_{n-1}|}\,\omega_-(y)^2 = \omega_n,$$

where $y' \in V_n^{(N)}$ and $y \in V_n$. And for $n = 1$,

$$\omega_1^{(N)} = \frac{|V_1^{(N)}|}{|V_0^{(N)}|}\,\omega_-^{(N)}(y')^2 = \frac{N|V_1|}{|V_0|}\,\omega_-(y)^2 = N\omega_1,$$

where $y' \in V_1^{(N)}$ and $y \in V_1$. Similarly, by (8.89) and (8.91) we obtain for $n = 1,2,\dots,$

$$\alpha_n^{(N)} = \omega_\circ^{(N)}(y') = \omega_\circ(y) = \alpha_n,$$

where $y' \in V_{n-1}^{(N)}$ and $y \in V_{n-1}$. This completes the proof. □

Theorem 8.53 (CLT for star powers). *Let $\mathcal{G} = (V, E)$ be a graph equipped with a distinguished vertex $o \in V$. Let A be the adjacency matrix of $\mathcal{G}$ and assume that $\Gamma(\mathcal{G})$ is invariant under the quantum components of A. Then, for $m = 1, 2, \dots,$*

$$\lim_{N\to\infty} \left\langle \Omega_N, \left(\frac{A^{\star N}}{\sqrt{N\kappa(o)}}\right)^m \Omega_N \right\rangle = \frac{1}{2}\int_{-\infty}^{+\infty} x^m (\delta_{+1} + \delta_{-1})(dx), \tag{8.92}$$

where $\Omega_N = \delta_o \otimes \cdots \otimes \delta_o$ (N times) and $\kappa(o)$ is the degree of $o \in V$.

Proof. Since $A^{\star N}$ is decomposed into a sum of Boolean independent random variables, by Theorem 8.51, the assertion is a direct consequence of the Boolean central limit theorem (Theorem 8.22).

We here give a more direct proof. Let $(\{\omega_n\}, \{\alpha_n\})$ be the Jacobi coefficient derived from A. It then follows from Lemma 8.52 that the Jacobi coefficient of $A^{\star N}$ is given by

$$\{N\omega_1, \omega_2, \dots\}, \qquad \{\alpha_1, \alpha_2, \dots\}.$$

Note that

$$\langle \Omega_N, A^{\star N}\Omega_N\rangle = 0, \qquad \langle \Omega_N, (A^{\star N})^2\Omega_N\rangle = N\kappa(o).$$

Then $A^{\star N}/\sqrt{N\kappa(o)}$ becomes a normalized random variable, whose spectral distribution in the vacuum state Ω_N is determined by the Jacobi coefficient:

$$\left\{1, \frac{\omega_2}{N\omega_1}, \frac{\omega_3}{N\omega_1}, \dots\right\}, \qquad \left\{\frac{\alpha_1}{\sqrt{N\omega_1}}, \frac{\alpha_2}{\sqrt{N\omega_1}}, \dots\right\}, \tag{8.93}$$

where we used $\omega_1 = \kappa(o)$. Hence, letting $N \to \infty$ we see that (8.93) converge to

$$\{1, 0, 0, \dots\}, \qquad \{0, 0, \dots\},$$

which is the Jacobi coefficient of the Bernoulli distribution $(\delta_{+1} + \delta_{-1})/2$. Since the mth moment is expressed in terms of the first m terms of the Jacobi coefficient (e.g., by the Accardi–Bożejko formula), (8.92) follows. □

Here is a concrete example. Let $\mathcal{G} = (V, E)$ be the half line of integers, i.e., $V = \{0, 1, 2, \dots\}$ and $i \sim j$ if and only if $|i - j| = 1$. The adjacency matrix is denoted by A. The N-fold star power $\mathcal{G}^{\star N}$ is called a *star lattice*, see Fig. 8.7.

Since $(\ell^2(V), \{\delta_n\}, A^+, A^-)$ is a free Fock space, the Jacobi coefficient of the spectral distribution of A in the vacuum δ_0 (in fact, the Wigner semicircle law) is given by $(\{\omega_n \equiv 1\}, \{\alpha_n \equiv 0\})$. We see by Lemma 8.52 that the Jacobi coefficient of the spectral distribution μ_N of $A^{\star N}$ is given by

$$\{\omega_1 = N,\ \omega_2 = \omega_3 = \cdots = 1\}, \quad \{\alpha_n \equiv 0\}.$$

Therefore, μ_N is the Kesten measure with parameter $N, 1$. For the explicit form see Sect. 4.1.

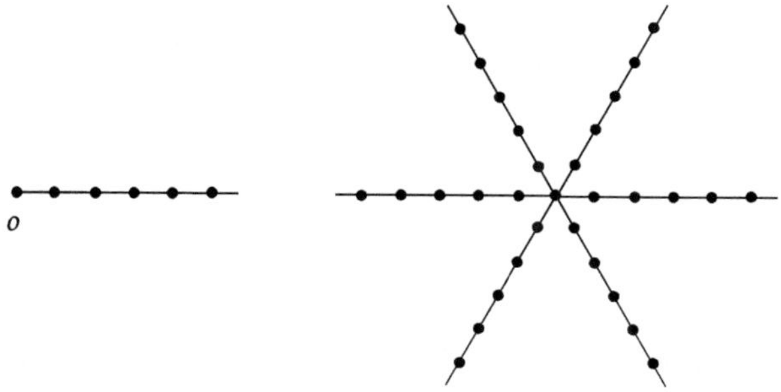

Fig. 8.7. Star lattice

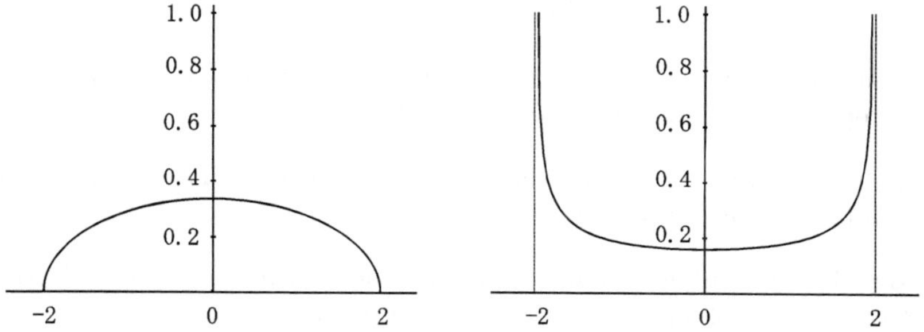

Fig. 8.8. Star lattices: $N = 1$ (semicircle law), $N = 2$ (arcsine law)

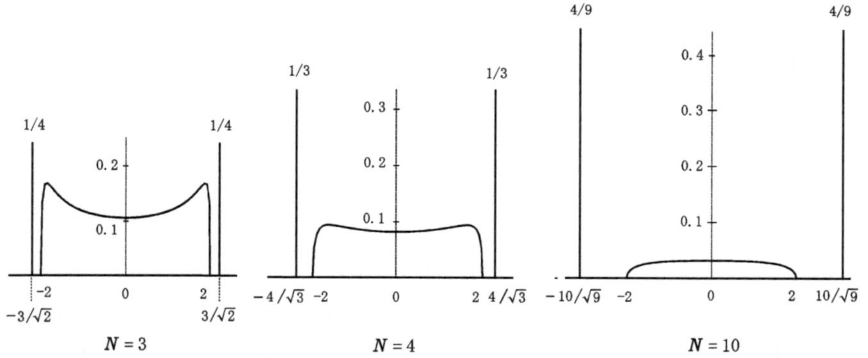

Fig. 8.9. Spectral distributions of star lattices: $\mathbb{Z}_+^{\star N}$

In fact, μ_1 is the Wigner semicircle law and μ_2 the arcsine law with variance 2 (see Fig. 8.8).

For $N \geq 3$ we have (see Fig. 8.9)

$$\mu_N(dx) = \rho_N(x)dx + \frac{N-2}{2N-2}\left(\delta_{-N/\sqrt{N-1}} + \delta_{N/\sqrt{N-1}}\right)(dx), \qquad (8.94)$$

where

$$\rho_N(x) = \frac{1}{2\pi}\frac{N\sqrt{4-x^2}}{N^2-(N-1)x^2}, \qquad -2 \le x \le 2. \tag{8.95}$$

Exercises

8.1. Prove Lemma 8.5.

8.2. Let $\mathcal{H}$ be a Hilbert space and $\{e_i\} \subset \mathcal{H}$ an orthonormal set. Let $l(e_i)$ and $l^*(e_i)$ be the left annihilation operator and the left creation operator acting on the free Fock space $\Gamma_{\text{free}}(\mathcal{H})$. Prove that

$$l(e_i)l^*(e_j) = \delta_{ij}1.$$

8.3. Using Theorem 8.25, prove that (8.31) remains valid if δ_o on the left-hand side is replaced with an arbitrary δ_x, $x \in \mathbb{Z}^N$.

8.4. Using Theorem 8.27, prove that (8.32) remains valid if δ_e on the left-hand side is replaced with an arbitrary δ_x, $x \in F_N$.

8.5. Let A be the adjacency matrix of the one-dimensional integer lattice $\mathbb{Z}$. Take $\alpha, \beta \in \mathbb{C}$ satisfying $|\alpha|^2 + |\beta|^2 = 1$ and define

$$\omega = \omega_{\alpha,\beta} = \alpha\Phi_0 + \beta\Psi_0 = \alpha\delta_0 + \frac{\beta}{\sqrt{2}}(\delta_{+1} - \delta_{-1}).$$

Show that

$$\langle \omega, A^m\omega\rangle = \int_{-2}^{+2} x^m \rho_{\alpha,\beta}(x)dx, \qquad m = 1, 2, \dots,$$

where

$$\rho_{\alpha,\beta}(x) = \frac{|\alpha|^2}{\pi\sqrt{4-x^2}} + \frac{|\beta|^2}{2\pi}\sqrt{4-x^2}.$$

8.6. Let $\mathcal{M}_N$ be the monotone tree equipped with the stratification with respect to $\varnothing$. Let $A_N = A_N^+ + A_N^-$ be the quantum decomposition of the adjacency matrix. Prove that $\Gamma(\mathcal{M}_N)$ is invariant under A_N^+ but not under A_N^-.

8.7. Let $\mathcal{G}$ be a graph with a distinguished vertex o as is indicated in Fig. 8.10. Consider the N-fold star power $\mathcal{G}^{\star N}$. Let μ_N be the spectral distribution of the adjacency matrix in the vacuum state δ_o and $G_N(z)$ the Stieltjes transform. Verify the following:

(1) For $\mathcal{L}_2$,

$$G_N(z) = \frac{z}{z^2 - N}, \qquad \mu_N = \frac{1}{2}(\delta_{+\sqrt{N}} + \delta_{-\sqrt{N}}).$$

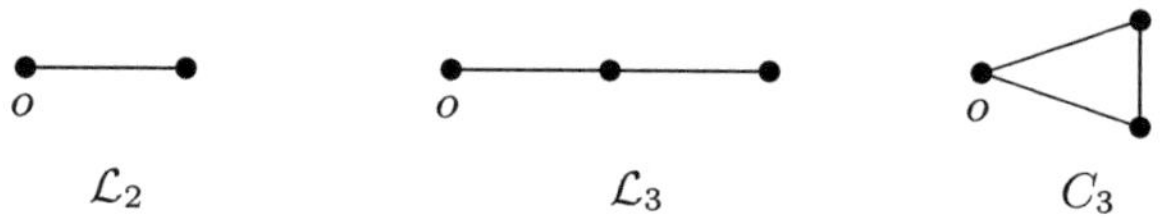

Fig. 8.10. Consider the star powers

(2) For $\mathcal{L}_3$,

$$G_N(z) = \frac{z^2 - 1}{z^3 - (N+1)z},$$

$$\mu_N = \frac{1}{N+1}\,\delta_0 + \frac{N}{2(N+1)}\left(\delta_{+\sqrt{N+1}} + \delta_{-\sqrt{N+1}}\right).$$

(3) For C_3,

$$G_N(z) = \frac{z-1}{z^2 - z - 2N}, \quad \mu_N = p_- \delta_{\alpha_+} + p_+ \delta_{\alpha_-},$$

where

$$\alpha_\pm = \frac{1 \pm \sqrt{1+8N}}{2}, \qquad p_\pm = \frac{1}{2}\left(1 \pm \frac{1}{\sqrt{1+8N}}\right).$$

8.8. Let $\rho_N(x)$ be as in (8.95). Show that for $3 \le N \le 6$, $\rho_N(x)$ has two local maxima at

$$x = \pm\sqrt{\frac{8-(N-4)^2}{N-1}}$$

and for $N \ge 7$ it has just one at $x = 0$.

Notes

The notion of commutative (or tensor in some literatures) independence is abstracted from the usual independence in classical probability theory. It is noted, however, that $\mathcal{A}$ is not assumed to be commutative. Such an algebraic formulation traces back to Cockroft–Hudson [59], Cushen–Hudson [61], Giri–von Waldenfels [86] in the 1970s.

The notion of free independence (freeness) was introduced and investigated by Voiculescu [213, 214] in the 1980s, see also Avitzour [15] and Bożejko [35] for some relevant study. The free probability theory has grown up to a research field with wide spectrum, see e.g., the monographs Hiai–Petz [99], Speicher [197] and Voiculescu–Dykema–Nica [216].

The notion of Boolean independence traces back to the regular free product of Bożejko [36]. Connections with convolution product and combinatorics were studied by Speicher–Woroudi [199]. For the tensor representation of Boolean independent random variables, see Lenczewski [143]. For the relevant stochastic calculus, see Ben Ghorbal–Schürmann [21].

The notion of monotone independence was discovered by Lu [149] and Muraki [164] in different contexts, and was studied extensively by Muraki [165, 167] as the third paradigm after classical and free probability theories. For the monotone central limit theorem, see also Liebscher [147]. For the tensor representation of monotone independent random variables, see also Franz [81]. Further relevant topics are discussed by Proskurin [179], Proskurin–Iksanov [180] and Wysoczański [227].

There are some attempts to axiomatize and unify these notions of independence. Schürman [188] initiated classification of independence in terms of axiomatic properties of products of algebraic probability spaces. Speicher [196] introduced the notion of *universal product* and characterized tensor, free and Boolean products. This classification was achieved within the categorical framework by Ben Ghorbal–Schürmann [20]. Muraki [166] introduced the notion of *quasi-universal product* and characterized the monotone independence as well, see also Franz [79–81], Lenczewski [143–146] and Muraki [168]. Further notions of independence have been studied by Accardi–Hashimoto–Obata [5], Bożejko–Speicher [44], among others.

For two real random variables a_1, a_2 in an algebraic probability space $(\mathcal{A}, \varphi)$ we are interested in the distribution μ of $a = a_1 + a_2$. For $i = 1, 2$ let $\mathcal{A}_i$ be the $*$-algebra generated by a_i and μ_i the distribution of a_i. Depending on whether $\{\mathcal{A}_1, \mathcal{A}_2\}$ is commutative, free, Boolean or monotone independent, we call μ the commutative, free, Boolean or monotone convolution of μ_1 and μ_2. Since the convolution product is linear with respect to the cumulants of the distributions, the so-called *moment-cumulant formula* is important, see e.g., Lehner [139–141]. The discussion is interesting also from the viewpoint of combinatorics. The commutative convolution coincides with the usual convolution in the classical probability theory. For the free convolution see Hiai–Petz [99], Voiculescu–Dykema–Nica [216], for the Boolean convolution see Bożejko–Wysoczański [49], Privault [178], Speicher–Woroudi [199], Stoica [201], and for the monotone convolution see Muraki [167]. The Fermion convolution of Oravecz [174] is a variant of the Boolean convolution. For interpolation or deformation of these convolutions, see Bożejko–Krystek–Wojakowski [41], Bożejko–Wysoczański [48, 49], Krystek–Yoshida [136, 137] and Yoshida [228].

Definition 8.54 (Bożejko–Wysoczański [48, 49]). Let $\mu \in \mathfrak{P}_{\mathrm{fm}}(\mathbb{R})$ be a probability measure which is the solution of a determinate moment problem. Let $(\{\omega_n\}, \{\alpha_n\})$ be the Jacobi coefficient of μ. For $t > 0$, the probability measure whose Jacobi coefficient is given by

$$(\{t\omega_1, \omega_2, \omega_3, \dots\}, \{t\alpha_1, \alpha_2, \alpha_3, \dots\})$$

is called the *t-transform* of μ and is denoted by $U_t\mu$.

This is related to Theorem 8.53. Let μ and μ_N be the spectral distributions of A and $A^{\star N}$, respectively. Then we have $\mu_N = U_N\mu$ since $\alpha_1 = 0$ in our situation. Moreover, it was pointed out by Bożejko–Wysoczański [49] that $U_N\mu$ is the Boolean convolution power of μ, see also Speicher–Woroudi [199].

Comb graphs, in particular, comb lattices, have been recently studied in physical literatures in connection with Bose–Einstein condensation (BEC), see Baldi–Burioni–Cassi [19], Burioni et al. [51–53]. For mathematical treatment of BEC, see also Matsui [156]. As is immediately seen from (8.94), for $N \geq 3$ there is a spectral gap:

$$\gamma = \frac{N}{\sqrt{N-1}} - 2 = \frac{(\sqrt{N-1}-1)^2}{\sqrt{N-1}} > 0.$$

A finite volume approximation of $\mathcal{G}^{\star N}$ gives a discrete measure which approaches to μ_N by taking the infinite volume limit. It is also interesting to study asymptotic behaviour of the discrete spectrum between $[2, 2+\gamma]$ along this approximation for the *hidden spectrum* of Burioni–Cassi–Vezzani [53].

9

The Symmetric Group and Young Diagrams

In the rest of this book we shall discuss asymptotic analysis for adjacency matrices and representations of the symmetric groups from the viewpoint of quantum probability theory. The purpose of this chapter is to assemble basic notions and tools in representation theory of the symmetric groups. Most materials are standard though the analytic description of Young diagrams, which is essential for the study of asymptotic behaviour of a representation of $S(n)$ as $n \to \infty$, would be less familiar to the readers. This fantastic idea is due to Vershik and Kerov.

9.1 Young Diagrams

Definition 9.1. A *Young diagram* of size $n \geq 0$ is a non-increasing sequence of integers:

$$\lambda_1 \geq \lambda_2 \geq \cdots \geq \lambda_j \geq \cdots \geq 0$$

such that

$$\sum_{j=1}^{\infty} \lambda_j = n.$$

In that case we write $\lambda = (\lambda_1 \geq \lambda_2 \geq \cdots)$. The size of λ is denoted by $|\lambda|$. The set of Young diagrams of size n is denoted by $\mathbb{Y}_n$.

Graphical representation of a Young diagram is convenient. With a Young diagram λ of size $|\lambda| = n$ we associate an array of n boxes, where the number of boxes of the jth row is λ_j. A Young diagram of size 0 is an empty diagram $\varnothing$. There are three different ways of drawing a Young diagram as is shown in Fig. 9.1. For a Young diagram $\lambda = (\lambda_1 \geq \lambda_2 \geq \cdots)$ we define

$$\mathrm{col}\,(\lambda) = \lambda_1\,, \quad \mathrm{row}\,(\lambda) = \max\{j\,;\, \lambda_j > 0\}.$$

These are the numbers of columns and rows of λ, though not necessarily standard notations.

A. Hora and N. Obata: The Symmetric Group and Young Diagrams. In: A. Hora and N. Obata, Quantum Probability and Spectral Analysis of Graphs, Theoretical and Mathematical Physics, 249–270 (2007)
DOI 10.1007/3-540-48863-4_9

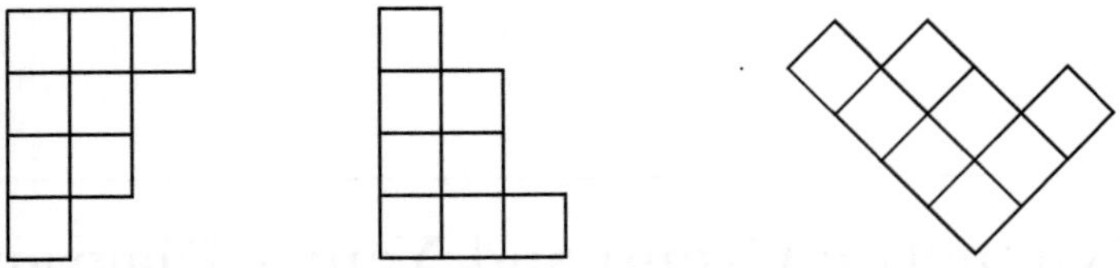

Fig. 9.1. Young diagram $(1^1 2^2 3^1)$

There is an alternative notation for a Young diagram. For $\lambda \in \mathbb{Y}_n$ let $m_j(\lambda)$ denote the number of rows of length j (or j-rows, for short). A Young diagram being uniquely specified by $\{m_1(\lambda), m_2(\lambda), \dots\}$, we also write

$$\lambda = (1^{m_1(\lambda)} 2^{m_2(\lambda)} \cdots j^{m_j(\lambda)} \cdots).$$

For example, Fig. 9.1 shows a Young diagram

$$\lambda = (3 \geq 2 \geq 2 \geq 1 \geq 0 \geq \cdots) = (1^1 2^2 3^1).$$

For this Young diagram we have $|\lambda| = 8$, $\mathrm{col}\,(\lambda) = 3$ and $\mathrm{row}\,(\lambda) = 4$.

For an integer $n \geq 0$ let $S(n)$ denote the symmetric group of degree n consisting of permutations of n letters, say, $\{1, 2, \dots, n\}$. By definition $S(0) = \{e\}$. In representation theory of the symmetric group, a crucial role is played by the Young diagrams because there exists a one-to-one correspondence among $\mathbb{Y}_n$, the conjugacy classes of $S(n)$, and the equivalence classes of irreducible representations of $S(n)$.

Every $g \in S(n)$ is uniquely decomposed into a product of disjoint cycles. Let m_j be the number of cycles of length j (or j-cycle) appearing in the product. The *cycle type* of g is a Young diagram defined by

$$\rho = \rho(g) = (1^{m_1} 2^{m_2} \cdots j^{m_j} \cdots),$$

where the size is $n = \sum_{j=1}^{n} j m_j$.

Proposition 9.2. *Two elements in $S(n)$ belong to the same conjugacy class if and only if their cycle types coincide. Moreover, the map $\rho : S(n) \to \mathbb{Y}_n$ induces a one-to-one correspondence between the conjugacy classes of $S(n)$ and $\mathbb{Y}_n$.*

For a Young diagram $\rho \in \mathbb{Y}_n$ we denote by C_ρ the corresponding conjugacy class in $S(n)$.

Proposition 9.3. *It holds that*

$$|C_\rho| = \frac{n!}{z_\rho}, \qquad z_\rho = \prod_{j=1}^{n} j^{m_j(\rho)} m_j(\rho)!\,. \tag{9.1}$$

Let $\mathcal{T}_n$ be the set of all transpositions in $S(n)$. Since every $g \in S(n)$ is expressed as a product of transpositions, $(S(n), \mathcal{T}_n)$ becomes a Cayley graph. Namely, the vertex set is $S(n)$ itself and two vertices $g, h \in S(n)$ are joined by an edge if and only if gh^{-1} is a transposition. Set

$$l(g) = \partial(g, e), \qquad g \in S(n),$$

which is the distance between g and the unit element e in the Cayley graph, i.e., the minimal number of transpositions needed to express g. Let $c(g)$ denote the number of cycles (including trivial cycles, i.e., cycles of length 1) in $S(n)$. Then we have the following:

Proposition 9.4. *$l(g) = n - c(g)$ for $g \in S(n)$.*

The number of *inversions* of $g \in S(n)$ is defined by

$$\mathrm{inv}(g) = |\{i < j\, ;\, g(i) > g(j)\}|,$$

see Fig. 9.2 for examples. The sign of g is denoted by $\mathrm{sgn}(g)$. Then

$$\mathrm{sgn}(g) = (-1)^{l(g)} = (-1)^{\mathrm{inv}(g)}. \tag{9.2}$$

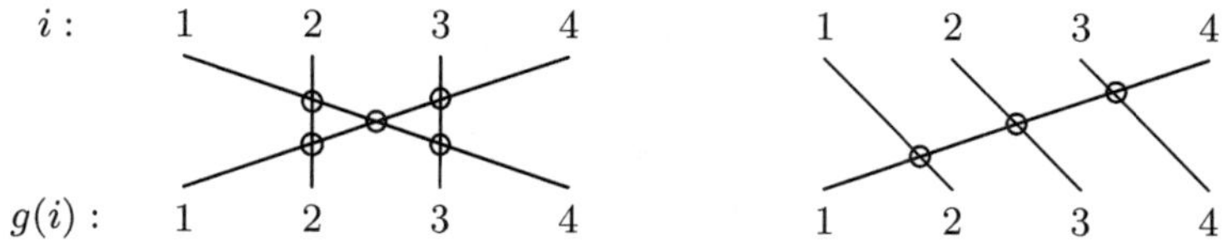

Fig. 9.2. Inversion in $S(4)$: $\mathrm{inv}(14) = 5$ and $\mathrm{inv}(1\ 2\ 3\ 4) = 3$

Since l is a class (or central) function, i.e., is constant on each conjugacy class, the stratification of $S(n)$ by means of l is coarser than the partition into conjugacy classes.

Proposition 9.5. *Let $\rho \in \mathbb{Y}_n$. Then for $g \in C_\rho$ it holds that*

$$l(g) = n - \sum_{j=1}^{n} m_j(\rho) = \sum_{j=2}^{n} (j-1) m_j(\rho).$$

Definition 9.6. Let $\lambda \in \mathbb{Y}_n$. A *Young tableau* of λ-shape is an array of the letters $\{1, 2, \ldots, n\}$ obtained by putting one by one into the boxes of λ. A Young tableau is said to be *standard* if the letters are in an increasing order along every row and column. Let $\mathrm{Tab}(\lambda)$ and $\mathrm{STab}(\lambda)$ denote the sets of Young tableau of λ-shape and of standard ones, respectively.

In the above definition, the Young diagram should be placed in the English manner. The increasing order along a row (resp. column) should be taken from the left (resp. top) to the right (resp. bottom).

In a Young diagram λ the box at the cross of the ith row and the jth column is said to have *indices* (i, j) or called the (i, j)-*box* for short. The north-west corner (displayed in the English manner) has by definition indices $(1, 1)$. For a Young tableau T, the letter in the (i, j)-box is denoted by $T(i, j)$. If T is standard, we have by definition

$$T(1,1) = 1, \quad T(i,1) < T(i,2) < \cdots, \quad T(1,j) < T(2,j) < \cdots.$$

For $\lambda \in \mathbb{Y}_n$, let f^λ be the number of standard Young tableau of λ-shape, i.e.,

$$f^\lambda = |\mathrm{STab}(\lambda)|.$$

Let b be a box contained in a Young diagram λ. We write $b \in \lambda$ for simplicity. Let (i, j) be the indices of b, i.e., b is the (i, j)-box. The *hook* of b is the set of boxes in λ having indices (i, j') with $j' \geq j$ or (i', j) with $i' \geq i$, see Fig. 9.3. The number of boxes in the hook of b is denoted by $h_\lambda(b)$ and is called the *hook length* of b. The hook length is essential for evaluation of f^λ.

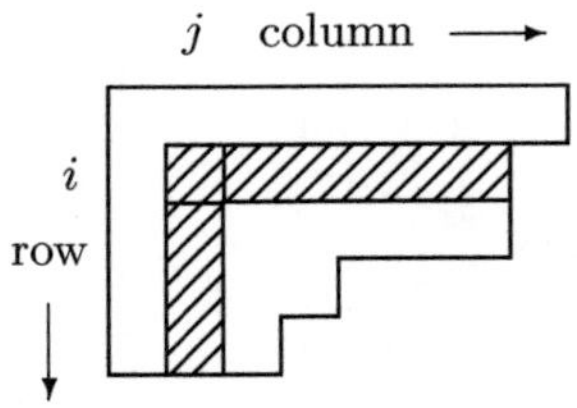

Fig. 9.3. Hook of the (i, j)-box

Theorem 9.7 (Hook formula I). *It holds that*

$$f^\lambda = \frac{n!}{\prod_{b \in \lambda} h_\lambda(b)}, \qquad \lambda \in \mathbb{Y}_n. \tag{9.3}$$

Theorem 9.8 (Hook formula II). *For $\lambda \in \mathbb{Y}_n$ let l_i be the hook length of the ith box in the first column, i.e., $l_i = \lambda_i + \mathrm{row}\,(\lambda) - i$. Then*

$$f^\lambda = n! \, \frac{\prod_{1 \leq j < k \leq \mathrm{row}\,(\lambda)} (l_j - l_k)}{\prod_{j=1}^{\mathrm{row}\,(\lambda)} l_j!}. \tag{9.4}$$

Before going into the proof of the hook formulae, we need more notation. A box in a Young diagram λ is called a *corner* of λ if the diagram obtained by removing it remains a Young diagram of size $|\lambda| - 1$. The Young diagram in Fig. 9.1 has three corners. When λ has k corners, we enumerate them from the left to the right as $\square_1, \square_2, \ldots, \square_k$. This enumeration can be done in either of the three manners of displaying.

Proof of Theorems 9.7 and 9.8 (outline). Set $p = \mathrm{row}(\lambda)$. For $p = 1$ the proof is direct. Assume that $p \geq 2$. The right-hand side of (9.4) is rewritten by using

$$\frac{\prod_{j=1}^{p} l_j!}{\prod_{1 \leq j < k \leq p}(l_j - l_k)} = \prod_{j=1}^{p-1} \frac{l_j!}{\prod_{k=j+1}^{p}(l_j - l_k)} l_p! \,.$$

Here $l_j!/\prod_{k=j+1}^{p}(l_j - l_k)$ agrees with a product of the hook lengths of the boxes in the jth row. This implies equivalence of (9.3) and (9.4).

The hook formula is verified by induction on n. Check (9.4) directly for small n. For example, $f^{(1^1 2^1)} = 2$. For a given $\Lambda \in \mathbb{Y}_n$, take the corners $\square_1, \ldots, \square_q$ of Λ and set $\lambda^{(i)} = \Lambda \setminus \square_i$ (removing the box $\square_i$ from Λ) for $i = 1, \ldots, q$. By the definition of a standard tableau, we have

$$f^{\Lambda} = \sum_{i=1}^{q} f^{\lambda^{(i)}}, \qquad f^{\lambda^{(i)}} \in \mathbb{Y}_{n-1}.$$

We know by the assumption of induction that (9.4) holds for $\lambda = \lambda^{(i)}$. Then, by careful computation one can verify that (9.4) holds for $\lambda = \Lambda$. $\square$

9.2 Irreducible Representations of the Symmetric Group

Let $\Delta(x_1, \ldots, x_n)$ be the difference product defined by

$$\Delta(x_1, \ldots, x_n) = \prod_{1 \leq i < j \leq n} (x_i - x_j) = \det \begin{pmatrix} x_1^{n-1} & \ldots & x_1^2 & x_1 & 1 \\ x_2^{n-1} & \ldots & x_2^2 & x_2 & 1 \\ \vdots & & \vdots & \vdots & \vdots \\ x_n^{n-1} & \ldots & x_n^2 & x_n & 1 \end{pmatrix},$$

which agrees with Vandermonde's determinant up to the sign $(-1)^{n(n-1)/2}$. Let $\lambda = (\lambda_1 \geq \lambda_2 \geq \cdots) \in \mathbb{Y}_n$ be a Young diagram. With each $T \in \mathrm{Tab}(\lambda)$ we assign a polynomial $\Delta(T)$ as follows. If λ is a 1-column diagram, denoting by $i_1, i_2, \ldots$ the letters in the boxes of T from the top to the bottom, we set

$$\Delta(T) = \Delta(x_{i_1}, x_{i_2}, \ldots).$$

For a general $T \in \mathrm{Tab}(\lambda)$ with a general $\lambda \in \mathbb{Y}_n$, denoting by T_j the jth column of T, we define

$$\Delta(T) = \prod_{j=1}^{\mathrm{col}(\lambda)} \Delta(T_j).$$

We call $\Delta(T)$ the *Specht polynomial* associated with $T \in \mathrm{Tab}(\lambda)$. It is checked that $\Delta(T)$ is a homogeneous polynomial in $x_1, \dots, x_n$ with degree

$$n(\lambda) = \sum_{j=1}^{\mathrm{col}(\lambda)} \frac{1}{2} \lambda'_j (\lambda'_j - 1), \tag{9.5}$$

where λ' denotes the transposed diagram of λ.

The symmetric group $S(n)$ acts on the Young tableaux in a natural manner, i.e., for $g \in S(n)$,

$$(gT)(i,j) = g(T(i,j)), \qquad T \in \mathrm{Tab}(\lambda), \quad \lambda \in \mathbb{Y}_n.$$

On the other hand, $S(n)$ acts on a polynomial $F(x_1, \dots, x_n)$ by coordinate permutation:

$$(gF)(x_1, \dots, x_n) = F(x_{g(1)}, \dots, x_{g(n)}).$$

We see that

$$\Delta(gT) = g\Delta(T), \qquad T \in \mathrm{Tab}(\lambda), \quad \lambda \in \mathbb{Y}_n.$$

Hence the linear space S_λ spanned by $\{\Delta(T)\,;\, T \in \mathrm{Tab}(\lambda)\}$ is $S(n)$-invariant. In other words, (U_λ, S_λ) becomes a representation of $S(n)$. We call S_λ the *Specht module* associated with $\lambda \in \mathbb{Y}_n$.

Theorem 9.9. *The set $\{\Delta(T)\,;\, T \in \mathrm{STab}(\lambda)\}$ forms a basis of S_λ. In particular, $f^\lambda = \dim S_\lambda$.*

For simplicity we set $\dim \lambda = \dim S_\lambda$. An element of $S(n-1)$ is regarded as a permutation in $S(n)$ which fixes the letter n. We thus consider $S(n-1)$ as a subgroup of $S(n)$. Let us investigate $S(n-1)$-invariant subspaces of S_Λ for $\Lambda \in \mathbb{Y}_n$. Let $\square_1, \square_2, \dots, \square_q$ denote the corners of Λ as before. A standard tableau of Λ-shape contains the letter n at one of its corners. For each $i = 1, \dots, q$ define $\lambda^{(i)} \in \mathbb{Y}_{n-1}$ by

$$\lambda^{(i)} = \Lambda \setminus \square_i.$$

Let A_i be the sets of Specht polynomials $\Delta(T)$ with $T \in \mathrm{STab}(\Lambda)$ such that the ith corner $\square_i$ of T is filled with the letter n. The linear space spanned by A_i is denoted by V_i. Note that $V_i \cap V_j = \{0\}$ if $i \neq j$. Setting

$$\tilde{V}_i = V_i + \cdots + V_q,$$

we have a descending sequence of subspaces

$$S_\Lambda = \tilde{V}_1 \supsetneq \tilde{V}_2 \supsetneq \cdots \supsetneq \tilde{V}_q = V_q.$$

Any element of $\tilde{V}_i = V_i \oplus \tilde{V}_{i+1}$ is of the form

$$\sum_{\Delta(T)\in A_i} \alpha_T \Delta(T) + v, \quad \alpha_T \in \mathbb{C}, \quad v \in \tilde{V}_{i+1}.$$

Let $\bar{T} \in \mathrm{STab}(\lambda^{(i)})$ denote the tableau obtained by removing the corner $\square_i$ (filled with n) of T. We get a linear surjection ϕ_i from $\tilde{V}_i$ onto $S_{\lambda^{(i)}}$ by

$$\phi_i\Big(\sum_{\Delta(T)\in A_i} \alpha_T \Delta(T) + v \Big) = \sum_{\Delta(T)\in A_i} \alpha_T \Delta(\bar{T}).$$

Lemma 9.10. *Let $i = 1, \ldots, q$.*

(1) *$\tilde{V}_i$ is $S(n-1)$-invariant.*
(2) *ϕ_i intertwines the actions of $S(n-1)$, namely,*

$$\phi_i \circ U_\Lambda(g) = U_{\lambda^{(i)}}(g) \circ \phi_i, \qquad g \in S(n-1).$$

Since $\ker \phi_i = \tilde{V}_{i+1}$, we see from Lemma 9.10 with the G-homomorphism theorem that

$$\tilde{V}_i/\tilde{V}_{i+1} \simeq_{S(n-1)} S_{\lambda^{(i)}}, \qquad i = 1, \ldots, q.$$

By using complete reducibility first for $\tilde{V}_{q-1}$, then for $\tilde{V}_{q-2}$ and so on, we have a sequence of subspaces $W_1, \ldots, W_q$ of S_Λ such that

$$S_\Lambda = W_1 \oplus \cdots \oplus W_q$$

and

$$W_q = V_q \simeq_{S(n-1)} S_{\lambda^{(q)}}, \quad W_i \simeq_{S(n-1)} S_{\lambda^{(i)}}, \quad i = 1, \ldots, q-1. \tag{9.6}$$

If a Young diagram λ is obtained from another Λ by removing a box, which necessarily lies at the corner, we use the notation $\lambda \nearrow \Lambda$. Then (9.6) implies

$$\mathrm{Res}^{S(n)}_{S(n-1)} U_\Lambda \simeq \bigoplus_{\lambda:\lambda\nearrow\Lambda} U_\lambda, \qquad \Lambda \in \mathbb{Y}_n, \tag{9.7}$$

where Res means the restriction as usual.

Theorem 9.11. *For $\lambda \in \mathbb{Y}_n$ let $U_\lambda = (U_\lambda, S_\lambda)$ be the representation of $S(n)$ defined above.*

(1) *U_λ is an irreducible representation of $S(n)$.*
(2) *U_λ and U_μ are not equivalent if $\lambda \neq \mu$.*
(3) *$\{U_\lambda\,;\, \lambda \in \mathbb{Y}_n\}$ is a system of complete representatives of the equivalence classes of irreducible representations of $S(n)$.*

Proof. We shall prove (1) and (2) by induction. It is directly checked from their actions that $U_{(2^1)}$ and $U_{(1^2)}$ are not equivalent as irreducible representations (with dimension 1) of $S(2)$. Assume that the assertions (1) and (2) are valid for $S(n-1)$ with $n \geq 3$.

Let $\lambda \in \mathbb{Y}_n$. We show irreducibility of U_λ. The corners of λ are denoted by $\square_1, \ldots, \square_q$. The decomposition of S_λ is given as in (9.6). The assumption of induction implies that $W_1, \ldots, W_q$ are the only $S(n-1)$-invariant irreducible subspaces of S_λ. Let W be a non-trivial $S(n)$-invariant subspace of S_λ. Then W is decomposed into a direct sum of $S(n-1)$-invariant irreducible subspaces. Here each summand of W coincides with one of W_i's. Let $W^\perp$ be the orthogonal complement of W with respect to an $S(n)$-invariant inner product in S_λ. Since $W^\perp$ is also $S(n)$-invariant, $W^\perp$ is a direct sum of some of W_i's unless it is non-trivial. Since $S_\lambda = W \oplus W^\perp$, we have either $W \supset W_q$ or $W^\perp \supset W_q$. Note that $W_q = V_q$ contains a Specht polynomial $\Delta(T)$ such that the corner $\square_q$ of T has letter n. Letting $S(n)$ act on this polynomial $\Delta(T)$, we get all the Specht polynomials of shape λ. Hence $W \supset W_q$ implies $W = S_\lambda$. Similarly, $W^\perp \supset W_q$ implies $W^\perp = S_\lambda$, which is impossible since W is non-trivial.

Let $\lambda, \mu \in \mathbb{Y}_n$ and $\lambda \neq \mu$. We see from (9.7) that

$$S_\lambda \simeq_{S(n-1)} \bigoplus_{\nu:\nu \nearrow \lambda} S_\nu, \quad S_\mu \simeq_{S(n-1)} \bigoplus_{\nu:\nu \nearrow \mu} S_\nu.$$

If $n \geq 3$, $\lambda \neq \mu$ implies $\{\nu \in \mathbb{Y}_{n-1}\,;\, \nu \nearrow \lambda\} \neq \{\nu \in \mathbb{Y}_{n-1}\,;\, \nu \nearrow \mu\}$. Then, since U_λ and U_μ have different decompositions when restricted onto $S(n-1)$, they are not equivalent as representations of $S(n-1)$, much less of $S(n)$.

(3) It is known from the general representation theory for finite groups that the number of non-equivalent irreducible representations coincides with that of the conjugacy classes. For $S(n)$ the number agrees with $|\mathbb{Y}_n|$, whence $\{U_\lambda\,;\, \lambda \in \mathbb{Y}_n\}$ exhausts the irreducible representations of $S(n)$ up to equivalence. □

Let $\mathbb{Y}$ be the set of all Young diagrams:

$$\mathbb{Y} = \bigcup_{n=0}^{\infty} \mathbb{Y}_n,$$

where $\mathbb{Y}_0$ consists of the empty diagram $\varnothing$. According to the size of Young diagrams, $\mathbb{Y}$ is stratified, where $\mathbb{Y}_n$ forms the nth stratum (or the nth level). We define λ and μ to be adjacent if either $\lambda \nearrow \mu$ or $\mu \nearrow \lambda$. The resulting graph, with $\mathbb{Y}$ being the vertex set, is called the *Young graph.* For simplicity the Young graph is also denoted by $\mathbb{Y}$ (see Fig. 9.4).

Let $\lambda, \mu \in \mathbb{Y}$ with $|\lambda| < |\mu|$. Among paths connecting λ and μ there are shortest ones. A shortest path is of the form

$$\lambda = \nu^{(0)} \nearrow \nu^{(1)} \nearrow \cdots \nearrow \nu^{(i)} \nearrow \cdots \nearrow \nu^{(k)} = \mu$$

and $k = |\mu| - |\lambda|$ coincides with the length of this path. We shall often drop the term 'shortest' when we discuss the Young graph. A (shortest) path starting at $\varnothing$ and terminating at $\lambda \in \mathbb{Y}_n$ describes the history of growth of Young diagrams to λ, and is identified with a standard tableau of λ-shape. Hence the number of paths connecting $\varnothing$ to λ coincides with f^λ.

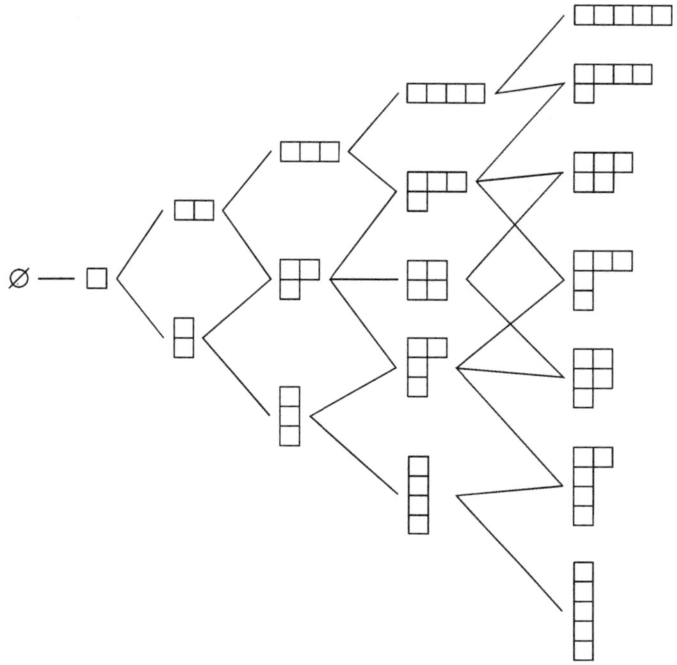

Fig. 9.4. Young graph

Iterating decompositions of S_λ as in (9.6), we assign a unique one-dimensional subspace of S_λ to each path from λ back to $\varnothing$ on the Young graph. The resulting basis, called the *Young basis*, of S_λ is thus parameterized by the set of paths connecting $\varnothing$ to λ. With these notations and the Frobenius reciprocity, we see that (9.7) becomes

$$\mathrm{Ind}_{S(n-1)}^{S(n)} U_\lambda \simeq \bigoplus_{\Lambda:\lambda\nearrow\Lambda} U_\Lambda, \tag{9.8}$$

where the left-hand side is the induced representation. The structure described in (9.7) and (9.8) will be used to investigate spectral properties of the Jucys–Murphy element.

9.3 The Jucys–Murphy Element

Definition 9.12. The element of $J_{n-1} \in \mathbb{C}[S(n)]$ defined by

$$J_{n-1} = (1\ n) + (2\ n) + \cdots + (n-1\ n)$$

is called the *Jucys–Murphy element*. For $\lambda \in \mathbb{Y}_n$, $U_\lambda(J_{n-1})$ is called the *Jucys–Murphy operator* on S_λ.

Recall that $S(n-1)$ is identified with the subgroup of $S(n)$ consisting of permutations which fix the letter n. It is then obvious that J_{n-1} commutes with $S(n-1)$. The spectral properties of the Jucys–Murphy element are of importance.

Let $\Lambda \in \mathbb{Y}_n$ and $\square_1, \dots, \square_q$ its corners with (i_k, j_k) being the indices of $\square_k$. The irreducible decomposition of S_Λ as $S(n-1)$-module is given in (9.6). As in the previous section, let V_k be the linear space spanned by $\Delta(T)$ with $T \in \mathrm{STab}(\Lambda)$ such that the letter n lies in $\square_k$, and set $\tilde{V}_k = V_k + \cdots + V_q$, i.e., $\tilde{V}_k$ is the linear space spanned by $\Delta(T)$ with $T \in \mathrm{STab}(\Lambda)$ such that the letter n lies somewhere in $\square_k, \dots, \square_q$. Then S_Λ is direct sum of V_k's.

Lemma 9.13. *Let $S \in \mathrm{STab}(\Lambda)$. If the letter n lies in a corner $\square_k$ with indices (i_k, j_k), then*

$$U_\Lambda(J_{n-1})\Delta(S) = (j_k - i_k)\Delta(S) + v, \quad v \in \tilde{V}_{k+1}.$$

In general, for a box $b \in \lambda$ with indices (i,j) we call $j-i$ the *content* of b. The contents of the corners $j_1 - i_1, \dots, j_q - i_q$ are distinct.

We need a total order among the standard tableaux of a given shape (see Fig. 9.5).

Definition 9.14. Let Λ be a Young diagram. Let S and T be standard tableaux of Λ-shape. We decide the order $S < T$ if

(i) the letter n lies at a more right box in T than in S;
(ii) otherwise (n lies in the same box), the letter $n-1$ lies at a more right box in T than in S;
(iii) otherwise, repeating the similar procedure.

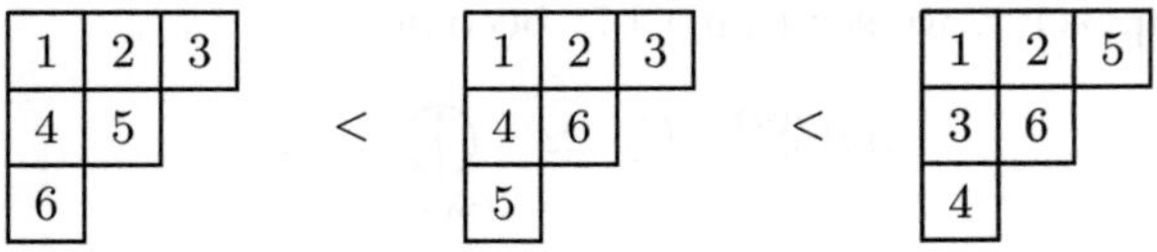

Fig. 9.5. Standard tableaux of $(1^1 2^1 3^1)$-shape

We enumerate the standard tableaux of Λ-shape as $T_1 < T_2 < \cdots$. Then $\{\Delta(T_i)\,;\, i = 1, 2, \dots\}$ becomes a basis of S_Λ. Lemma 9.13 means triangularization of $U_\Lambda(J_{n-1})$:

$$U_\Lambda(J_{n-1})(\Delta(T_1)\,\Delta(T_2)\,\cdots) = (\Delta(T_1)\,\Delta(T_2)\,\cdots)\begin{pmatrix} D_1 & & O \\ & \ddots & \\ * & & D_q \end{pmatrix}, \qquad (9.9)$$

where each block on the diagonal line is a scalar matrix

$$D_k = (j_k - i_k) I_{\dim \lambda^{(k)}}, \qquad k = 1, 2, \ldots, q,$$

of degree $\dim \lambda^{(k)}$ with $\lambda^{(k)} = \Lambda \setminus \Box_k$.

Corollary 9.15. *The spectrum of $U_\Lambda(J_{n-1})$ is given by*

$$\begin{matrix} j_1 - i_1 & j_2 - i_2 & \ldots & j_q - i_q \\ \dim \lambda^{(1)} & \dim \lambda^{(2)} & \ldots & \dim \lambda^{(q)} \end{matrix}.$$

Proposition 9.16. *The Jucys–Murphy operator $U_\Lambda(J_{n-1})$ for $\Lambda \in \mathbb{Y}_n$ is diagonalizable. Each W_k in* (9.6) *is the eigenspace associated with the eigenvalue $j_k - i_k$.*

Proof. Let E_k and E'_k be the eigenspace and the generalized eigenspace of $U_\Lambda(J_{n-1})$ belonging to the eigenvalue $j_k - i_k$, respectively. Clearly, $E'_k \supset E_k \neq \{0\}$. Since J_{n-1} commutes with $S(n-1)$, E_k and E'_k are $S(n-1)$-invariant subspaces of S_Λ. Using representation matrix (9.9) of U_Λ and noting $(j_k - i_k) - (j_l - i_l) \neq 0$ for $k \neq l$, we see that $E'_k \subset \tilde{V}_k$.

Recall the irreducible decomposition (9.6) of S_Λ into W_k's. We see from Lemma 9.13 that $W_q = V_q = \tilde{V}_q \subset E_q$, so that $E_q = E'_q = W_q$. To use induction on k, assume that $E_l = E'_l = W_l$ holds for $l = k+1, \ldots, q$. In the irreducible decomposition

$$\tilde{V}_k = W_k \oplus W_{k+1} \oplus \cdots \oplus W_q$$

into mutually non-equivalent $S(n-1)$-invariant subspaces, we have $W_{k+1} = E_{k+1}, \ldots, W_q = E_q$. Since E_k is a non-trivial $S(n-1)$-invariant subspace of $\tilde{V}_k$, it is a direct sum of some of $W_k, \ldots, W_q$. Now E_k cannot contain $E_{k+1}, \ldots, E_q$ by the definition of an eigenspace. Hence we have $E_k = W_k$. Similarly, we get $E'_k = W_k$. This completes the proof. □

9.4 Analytic Description of a Young Diagram

Let us place a Young diagram in the Russian style as in Fig. 9.6, where the size of each box is adjusted to be $\sqrt{2} \times \sqrt{2}$. The boundary part indicated by a bold line is called the *profile* of a Young diagram. A profile is the graph of a particular continuous function.

Definition 9.17. An $\mathbb{R}$-valued function ω defined on $\mathbb{R}$ is called a *rectangular diagram* if

(i) $\omega(x)$ is continuous and piecewise linear;
(ii) $\omega'(x) = \pm 1$ except finitely many points $x \in \mathbb{R}$;
(iii) $\omega(x) = |x|$ if $|x|$ is large enough.

Let $\mathbb{D}_0$ be the set of rectangular diagrams.

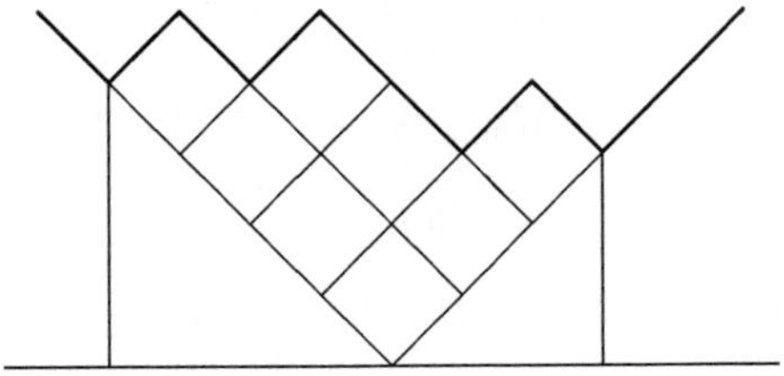

Fig. 9.6. The profile of $\lambda = (1^1 2^2 3^1)$, where $\mathrm{col}(\lambda) = 3$ and $\mathrm{row}(\lambda) = 4$

By taking the profile a Young diagram is identified with a rectangular diagram. In this way we understand the inclusion $\mathbb{Y} \subset \mathbb{D}_0$.

There is a convenient parameterization for a rectangular diagram. With each $\omega \in \mathbb{D}_0$ we associate an interlacing sequence

$$x_1 < y_1 < \cdots < x_{r-1} < y_{r-1} < x_r, \tag{9.10}$$

where x_i and y_j are the horizontal coordinates of a local minimum (valley) and a local maximum (peak), respectively, see Fig. 9.7. The interlacing sequence (9.10) is called the *min–max coordinates* of ω.

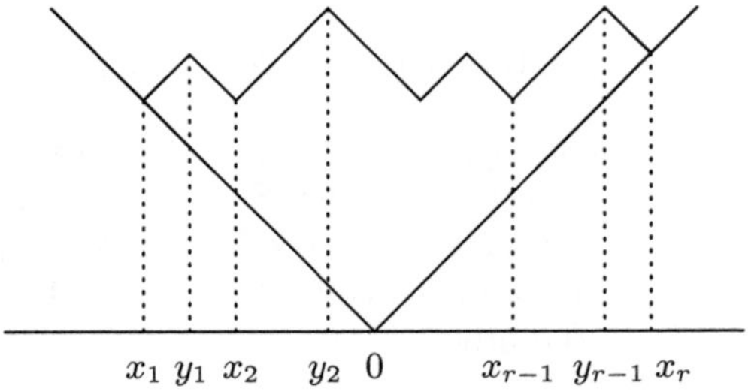

Fig. 9.7. Min–max coordinates

Lemma 9.18. *Let* $\omega \in \mathbb{D}_0$ *be a rectangular diagram with min–max coordinates being given as in* (9.10). *Then,*

$$\sum_{i=1}^{r} x_i = \sum_{i=1}^{r-1} y_i. \tag{9.11}$$

Proof. Verified easily by induction on the number of valleys. □

Lemma 9.19. *A rectangular diagram is a Young diagram if and only if the min–max coordinates are all integers.*

Proof. Obvious. □

With each Young diagram we associate in two ways measures on the real line. In fact, we do it for a rectangular diagram. Let $\lambda \in \mathbb{D}_0$ and

$$x_1 < y_1 < \cdots < x_{r-1} < y_{r-1} < x_r \tag{9.12}$$

its min–max coordinates. Putting a plus (+) charge at a valley and a minus (−) charge at a peak, we define

$$\tau_\lambda = \sum_{i=1}^{r} \delta_{x_i} - \sum_{i=1}^{r-1} \delta_{y_i} .$$

This becomes a signed probability measure on $\mathbb{R}$ and is called the *Rayleigh measure* of λ. It is obvious that $\lambda \mapsto \tau_\lambda$ is injective.

Proposition 9.20. *Let $\lambda \in \mathbb{D}_0$. The kth moment of τ_λ is given by*

$$M_k(\tau_\lambda) = \int_{-\infty}^{+\infty} x^k \left(\frac{\lambda(x) - |x|}{2} \right)'' dx, \quad k = 1, 2, \dots, \tag{9.13}$$

where the derivative is taken in the distribution sense.

Proof. We only need to note that

$$\tau_\lambda = \left(\frac{\lambda(x) - |x|}{2} \right)'' + \delta_0, \tag{9.14}$$

which is straightforward by definition. □

In particular, $M_0(\tau_\lambda) = 1$, $M_1(\tau_\lambda) = 0$, and $M_2(\tau_\lambda)$ agrees with the area confined by the profile $\lambda(x)$ and $|x|$. If $\lambda \in \mathbb{Y}_n$, then $M_2(\tau_\lambda) = 2n$.

The second measure associated to a Young diagram is defined as follows. Let $\lambda \in \mathbb{D}_0$ be given with min–max coordinates as in (9.12). Consider the rational function having the zeros and the poles respectively at the max and min coordinates:

$$\frac{(z - y_1) \cdots (z - y_{r-1})}{(z - x_1) \cdots (z - x_r)}, \qquad z \in \mathbb{C}. \tag{9.15}$$

We then see a partial fraction expansion

$$\frac{(z - y_1) \cdots (z - y_{r-1})}{(z - x_1) \cdots (z - x_r)} = \frac{\mu_1}{z - x_1} + \cdots + \frac{\mu_r}{z - x_r}, \tag{9.16}$$

where

$$\mu_i = \frac{(x_i - y_1) \cdots (x_i - y_{r-1})}{(x_i - x_1) \cdots (x_i - x_{i-1})(x_i - x_{i+1}) \cdots (x_i - x_r)}.$$

Note that $\mu_i > 0$ for all i since x_i and y_i are interlacing. Moreover, multiplying z on the both sides of (9.16) and letting $z \to \infty$, we see that

$$\sum_{i=1}^{r} \mu_i = 1.$$

Therefore,

$$\mathfrak{m}_\lambda = \sum_{i=1}^{r} \mu_i \delta_{x_i} \tag{9.17}$$

becomes a probability measure on $\mathbb{R}$, which is called the *transition measure* of λ. In fact, (9.16) is the Stieltjes transform of $\mathfrak{m}_\lambda$. One may expect from the term 'transition measure' a relation to a Markov chain. This point will be clarified in Sect. 9.6.

As for relation between the Rayleigh measures τ_λ and the transition measures $\mathfrak{m}_\lambda$, we prove the following:

Proposition 9.21. *Let $\lambda \in \mathbb{D}_0$. The moments of $\mathfrak{m}_\lambda$ and of τ_λ are related as*

$$\sum_{n=0}^{\infty} \frac{M_n(\mathfrak{m}_\lambda)}{z^n} = \exp\left\{\sum_{n=1}^{\infty} \frac{M_n(\tau_\lambda)}{n} \frac{1}{z^n}\right\}, \tag{9.18}$$

where the series in both sides converge for large $|z|$.

Proof. Let $x_1 < y_1 < \cdots < y_{r-1} < x_r$ be the min–max coordinates of $\lambda \in \mathbb{D}_0$. We first note that

$$\sum_{n=0}^{\infty} \frac{M_n(\mathfrak{m}_\lambda)}{z^n} = \frac{(1-y_1/z)\cdots(1-y_{r-1}/z)}{(1-x_1/z)\cdots(1-x_r/z)}.$$

Then the right-hand side becomes

$$\begin{aligned}
&= \exp\Big\{\log\Big(1-\frac{y_1}{z}\Big) + \cdots + \log\Big(1-\frac{y_{r-1}}{z}\Big) \\
&\qquad - \log\Big(1-\frac{x_1}{z}\Big) - \cdots - \log\Big(1-\frac{x_r}{z}\Big)\Big\} \\
&= \exp\left\{\sum_{k=1}^{\infty} \frac{1}{k}\left(\sum_{i=1}^{r} x_i^k - \sum_{i=1}^{r-1} y_i^k\right)\frac{1}{z^k}\right\} \\
&= \exp\left\{\sum_{k=1}^{\infty} \frac{M_k(\tau_\lambda)}{k} \frac{1}{z^k}\right\},
\end{aligned}$$

which shows (9.18). □

It follows from (9.18) that $\{M_n(\mathfrak{m}_\lambda)\}$ and $\{M_n(\tau_\lambda)\}$ are expressed by polynomials of each other. In particular, we have

$$M_1(\mathfrak{m}_\lambda) = M_1(\tau_\lambda) = 0,$$
$$M_2(\mathfrak{m}_\lambda) = \frac{1}{2}(M_1(\tau_\lambda)^2 + M_2(\tau_\lambda)).$$

For a Young diagram $\lambda \in \mathbb{Y}$ we have $M_2(\mathfrak{m}_\lambda) = |\lambda|$.

Proposition 9.22. *The map $\lambda \mapsto \mathfrak{m}_\lambda$ is a bijection between the rectangular diagrams $\mathbb{D}_0$ and the probabilities on $\mathbb{R}$ with finite supports and mean zero.*

Proof. We already know that the transition measure $\mathfrak{m}_\lambda$ has the desired properties. Conversely, suppose that a probability measure μ is given as in the right-hand side of (9.17) with $x_1 < \cdots < x_r$ and mean zero. We set

$$\frac{\mu_1}{z - x_1} + \cdots + \frac{\mu_r}{z - x_r} = \frac{f(z)}{(z - x_1)\cdots(z - x_r)},$$

where $f(z)$ is a polynomial in z with degree $r - 1$. Since μ is a probability measure, $f(z)$ is monic. Moreover, the sign of $f(x_i)$ varies alternatively as i runs through $1, \dots, r$. Hence $f(z)$ has a unique root in each interval (x_i, x_{i+1}), which is denoted by y_i. We thus obtain the expression of (9.16). Following the computation in the proof of Proposition 9.21, we have

$$\sum_{i=1}^{r} x_i - \sum_{i=1}^{r-1} y_i = M_1(\mu) = 0.$$

Hence the interlacing sequence $x_1 < y_1 < \cdots < y_{r-1} < x_r$ determines $\lambda \in \mathbb{D}_0$. Clearly, this gives an inverse map of $\lambda \mapsto \mathfrak{m}_\lambda$. □

9.5 A Basic Trace Formula

For $\lambda \in \mathbb{Y}$ we set

$$\chi^\lambda = \operatorname{tr} U_\lambda, \quad \tilde{\chi}^\lambda = \frac{1}{\dim \lambda} \chi^\lambda.$$

These are the irreducible characters corresponding to $\lambda \in \mathbb{Y}$ and the normalized one, respectively. Let $\mathbb{E}_n : \mathbb{C}[S(n+1)] \to \mathbb{C}[S(n)]$ be the linear map uniquely specified by

$$\mathbb{E}_n(g) = \begin{cases} g, & \text{if } g \in S(n), \\ 0, & \text{if } g \in S(n+1) \setminus S(n). \end{cases}$$

The map $\mathbb{E}_n$ is often called the *conditional expectation* and is the dual of the natural extension of the injection $S(n) \to S(n+1)$. Let $\mathcal{Z}(\mathbb{C}[S(n)])$ denote the centre of $\mathbb{C}[S(n)]$. Since $J_n \in \mathbb{C}[S(n+1)]$ commutes with $S(n)$, $\mathbb{E}_n J_n^k$ belongs to $\mathcal{Z}(\mathbb{C}[S(n)])$ for all $k = 1, 2, \dots$.

Theorem 9.23 (Basic trace formula). *For* $\lambda \in \mathbb{Y}_n$, $n = 1, 2, \dots,$ *we have*

$$\tilde{\chi}^\lambda(\mathbb{E}_n J_n^k) = M_k(\mathfrak{m}_\lambda), \quad k = 1, 2, \dots. \tag{9.19}$$

Before going into the proof of Theorem 9.23, we prepare a general result. Let G be a finite group and H its subgroup. Consider the conditional expectation $\mathbb{E} : \mathbb{C}[G] \to \mathbb{C}[H]$. We have

$$(\mathbb{E}b)a = \mathbb{E}(ba), \quad b \in \mathbb{C}[G], \quad a \in \mathbb{C}[H].$$

Let $\hat{G}$ and $\hat{H}$ denote the equivalence classes of irreducible representations of G and H, respectively. We choose an irreducible representation (ρ_β, W_β) [resp. (π_α, V_α)] from $\beta \in \hat{G}$ [resp. $\alpha \in \hat{H}$]. We have algebra isomorphisms

$$\mathbb{C}[H] \simeq \bigoplus_{\alpha \in \hat{H}} \mathrm{End}(V_\alpha) \simeq \bigoplus_{\alpha \in \hat{H}} V_\alpha \otimes V_\alpha^*, \tag{9.20}$$

$$\mathbb{C}[G] \simeq \bigoplus_{\beta \in \hat{G}} \mathrm{End}(W_\beta) \simeq \bigoplus_{\beta \in \hat{G}} W_\beta \otimes W_\beta^*. \tag{9.21}$$

The minimal central projection e_α in $\mathbb{C}[H]$ associated with $\alpha \in \hat{H}$ corresponds to the identity $I_{V_\alpha} \in \mathrm{End}(V_\alpha)$ in (9.20). Let L_H and L_G denote the left regular representations of H and G, respectively. Then, for $a \in \mathbb{C}[H]$,

$$L_H(ae_\alpha) \sim \bigoplus_{\gamma \in \hat{H}} \pi_\gamma(a)\pi_\gamma(e_\alpha) \otimes I_{V_\gamma^*} \sim \pi_\alpha(a) \otimes I_{V_\alpha^*} \oplus \bigoplus_{\gamma \neq \alpha} 0,$$

where $\sim$ denotes similarity between operators. Here we used

$$\pi_\gamma(e_\alpha) = \delta_{\alpha,\gamma} I_{V_\alpha},$$

which is readily seen from (9.20). Taking traces of both sides, we have

$$\dim \alpha \mathrm{tr}\, \pi_\alpha(a) = \mathrm{tr}\, L_H(ae_\alpha) = |H|(ae_\alpha)_e,$$

where $(\cdot)_e$ means the e-component, i.e., the coefficient of the unit element. Putting $a = \mathbb{E}b$ for $b \in \mathbb{C}[G]$, we have

$$\begin{aligned} \dim \alpha \mathrm{tr} \pi_\alpha(\mathbb{E}b) &= |H|((\mathbb{E}b)e_\alpha)_e = |H|(\mathbb{E}(be_\alpha))_e = |H|(be_\alpha)_e \\ &= \frac{1}{[G:H]} \mathrm{tr}\, L_G(be_\alpha) = \frac{1}{[G:H]} \sum_{\beta \in \hat{G}} \dim \beta \mathrm{tr}\, \rho_\beta(be_\alpha), \end{aligned}$$

where the last equality followed from (9.21). We thus come to the following:

Lemma 9.24. *For* $b \in \mathbb{C}[G]$ *it holds that*

$$\mathrm{tr}\, \pi_\alpha(\mathbb{E}b) = \frac{1}{[G:H] \dim \alpha} \sum_{\beta \in \hat{G}} \dim \beta \mathrm{tr}\, (\rho_\beta(b)\rho_\beta(e_\alpha)),$$

where $\rho_\beta(e_\alpha)$ *is the projection onto the* α*-component of* W_β, *i.e., the sum of images of elements in* $\mathrm{Hom}_H(V_\alpha, W_\beta)$.

We now go back to the symmetric groups. Applying Lemma 9.24 to the case where $G = S(n+1)$, $H = S(n)$ and $b = J_n^k$, we get

$$\tilde{\chi}^\lambda(\mathbb{E}_n J_n^k) = \frac{1}{(n+1)\dim^2\lambda} \sum_{\Lambda\in\mathbb{Y}_{n+1}} \dim\Lambda \operatorname{tr}(U_\Lambda(J_n^k)U_\Lambda(e_\lambda)) \tag{9.22}$$

for $\lambda \in \mathbb{Y}_n$ and $k = 1, 2, \ldots$.

We then employ spectral properties of the Jucys–Murphy operator. Let $x_1 < y_1 < \cdots < y_{r-1} < x_r$ be the min–max coordinates of $\lambda \in \mathbb{Y}_n$. The Young diagram obtained by putting one box at the sth valley of λ is denoted by $\Lambda^{(s)} \in \mathbb{Y}_{n+1}$, where $s = 1, 2, \ldots, r$.

Lemma 9.25. *Notations and assumptions being as above, it holds that*

$$\tilde{\chi}^\lambda(\mathbb{E}_n J_n^k) = \sum_{s=1}^{r} x_s^k \frac{\dim\Lambda^{(s)}}{(n+1)\dim\lambda}. \tag{9.23}$$

Proof. Since $U_\Lambda(e_\lambda)$ in (9.22) is the projection onto the λ-component in S_Λ, we see from (9.6) that it is non-trivial if and only if $\lambda \nearrow \Lambda$. Moreover, if $\lambda \nearrow \Lambda$, it follows from Proposition 9.16 that $U_\Lambda(J_n)$ acts on the λ-component as scalar multiplication by the content of $\Lambda \setminus \lambda$. This content coincides with the min coordinate of λ caused by removing the box from Λ (see Fig. 9.8). If $\Lambda = \Lambda^{(s)}$, $s \in \{1, \ldots, r\}$, then this content is equal to the sth min coordinate x_s of λ. We thus get from (9.22)

$$\begin{aligned}\tilde{\chi}^\lambda(\mathbb{E}_n J_n^k) &= \frac{1}{(n+1)\dim^2\lambda} \sum_{s=1}^{r} \dim\Lambda^{(s)} x_s^k \operatorname{tr}(U_{\Lambda^{(s)}}(e_\lambda)) \\ &= \sum_{s=1}^{r} x_s^k \frac{\dim\Lambda^{(s)}}{(n+1)\dim\lambda},\end{aligned}$$

which completes the proof. □

Lemma 9.26. *Notations and assumptions being as above,*

$$\frac{\dim\Lambda^{(s)}}{(n+1)\dim\lambda} = \mathfrak{m}_\lambda(\{x_s\}), \quad s = 1, \ldots, r. \tag{9.24}$$

Proof. We shall rewrite the left-hand side of (9.24) in terms of min–max coordinates by using the hook formula (Theorems 9.7 and 9.8) which gives the dimensions of irreducible representations of symmetric groups. Let $\Lambda^{(s)}$ be produced by putting a box at the sth valley of $\lambda \in \mathbb{Y}_n$ and the box has indices (i, j) in $\Lambda^{(s)}$ (see Fig. 9.8). We then have

$$\frac{\dim\Lambda^{(s)}}{(n+1)\dim\lambda} = \frac{\prod_{b\in\lambda} h_\lambda(b)}{\prod_{b\in\Lambda^{(s)}} h_{\Lambda^{(s)}}(b)}. \tag{9.25}$$

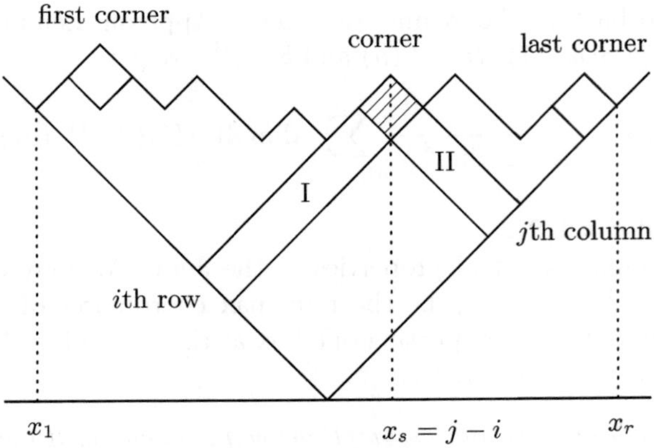

Fig. 9.8. Content of a corner and min coordinate

It suffices to consider hooks at the boxes in zones I and II in Fig. 9.8, since other hook lengths coincide in λ and $\Lambda^{(s)}$. Let $h_\lambda(k,l)$ denote the hook length of the (k,l)-box in λ. We have

$$\begin{aligned}
&h_{\Lambda^{(s)}}(i,1) = -x_1 - i + j = x_s - x_1, \\
&h_{\Lambda^{(s)}}(i,2) = h_\lambda(i,1), \\
&\quad \cdots \\
&h_{\Lambda^{(s)}}(i, y_1 - x_1) = h_\lambda(i, y_1 - x_1 - 1), \\
&h_\lambda(i, y_1 - x_1) = -x_1 - i + j - (y_1 - x_1) = x_s - y_1, \\
&h_{\Lambda^{(s)}}(i, y_1 - x_1 + 1) = -x_1 - (x_2 - y_1) - i + j - (y_1 - x_1) = x_s - x_2,
\end{aligned}$$

and so on. This yields

$$\frac{\prod_{b\in\text{zone I}} h_\lambda(b)}{\prod_{b\in\text{zone I}} h_{\Lambda^{(s)}}(b)} = \frac{x_s - y_1}{x_s - x_1} \cdots \frac{x_s - y_{s-1}}{x_s - x_{s-1}}.$$

If we flip the Young diagram and consider, instead of λ, the one which has the min–max coordinates $-x_r < -y_{r-1} < \cdots < -y_1 < -x_1$, we can apply the case of zone I to evaluate the case of zone II for λ and $\Lambda^{(s)}$. Thus

$$\begin{aligned}
\frac{\prod_{b\in\text{zone II}} h_\lambda(b)}{\prod_{b\in\text{zone II}} h_{\Lambda^{(s)}}(b)} &= \frac{-x_s - (-y_{r-1})}{-x_s - (-x_r)} \cdots \frac{-x_s - (-y_s)}{-x_s - (-x_{s+1})} \\
&= \frac{x_s - y_s}{x_s - x_{s+1}} \cdots \frac{x_s - y_{r-1}}{x_s - x_r}.
\end{aligned}$$

Taking the definition of transition measure $\mathfrak{m}_\lambda$ into account, we get (9.24). □

Proof of Theorem 9.23. Immediate from Lemmas 9.25 and 9.26. □

9.6 Plancherel Measures

A probability measure on the set of Young diagrams gives rise to a random measure on $\mathbb{R}$ through the transition measures or Rayleigh measures. A probability measure on the path space of the Young graph yields a random growth process of Young diagrams. In this section we introduce the most fundamental one in both senses, the Plancherel measure.

A function φ on $\mathbb{Y}$ is said to be *harmonic* if it satisfies

$$\varphi(\lambda) = \sum_{\Lambda:\lambda\nearrow\Lambda} \varphi(\Lambda).$$

An example of a positive normalized harmonic function is given by

$$\varphi(\lambda) = \frac{\dim\lambda}{|\lambda|!}, \qquad \lambda \in \mathbb{Y}. \tag{9.26}$$

Here normalization means $\varphi(\varnothing) = 1$.

The set of all infinite paths on $\mathbb{Y}$ starting at $\varnothing$ is denoted by $\mathfrak{T}$. An element of $\mathfrak{T}$ is expressed as

$$t = (\varnothing = t(0) \nearrow t(1) \nearrow \cdots \nearrow t(n) \nearrow \cdots),$$

where $t(n) \in \mathbb{Y}_n$ is the Young diagram sitting at the nth stratum along $t \in \mathfrak{T}$. Therefore, $\mathfrak{T}$ is identified with the set of infinite standard tableaux. With the relative topology induced from the product space $\prod_n \mathbb{Y}_n$, $\mathfrak{T}$ is compact and is equipped with the topological Borel σ-field. Given a finite path

$$u = (\varnothing \nearrow \lambda^{(1)} \nearrow \cdots \nearrow \lambda^{(n)} = \lambda), \quad \lambda \in \mathbb{Y}_n,$$

we define a cylindrical subset of $\mathfrak{T}$ by

$$C_u = \{t \in \mathfrak{T};\, t(j) = \lambda^{(j)},\, j = 0, 1, \ldots, n\}. \tag{9.27}$$

The σ-field of $\mathfrak{T}$ is generated by the cylindrical subsets. For C_u given in (9.27) we set

$$\mathfrak{P}(C_u) = \frac{\dim\lambda}{n!} = \varphi(\lambda), \tag{9.28}$$

where φ is the harmonic function defined in (9.26). It is shown that $\mathfrak{P}$ consistently determines a finitely additive measure on the cylindrical subsets of $\mathfrak{T}$ and that $\mathfrak{P}$ is extended to a σ-additive measure on $\mathfrak{T}$ by virtue of compactness of $\mathfrak{T}$ and the Hopf extension theorem. Obviously, $\mathfrak{P}$ is a probability measure, which we call the *Plancherel measure* on the path space $\mathfrak{T}$. The (marginal) distribution of $\mathfrak{P}$ on the nth stratum is called the *Plancherel measure* on $\mathbb{Y}_n$ and is denoted by $\mathfrak{P}_n$.

The number of the finite paths starting at $\varnothing$ and terminating at λ agrees with f^λ, hence with $\dim\lambda$ by Theorem 9.9. We have

$$\mathfrak{P}_n(\lambda) = \mathfrak{P}(\{t \in \mathfrak{T}\,;\, t(n) = \lambda\}) = \frac{\dim^2\lambda}{n!}, \qquad \lambda \in \mathbb{Y}_n. \tag{9.29}$$

Since (9.29) corresponds to the irreducible decomposition of the regular representation of $S(n)$, the term 'Plancherel measure' is reasonable from the viewpoint of Fourier analysis. As is seen from (9.28), $\mathfrak{P}(C_u)$ depends only on the terminating vertex of a finite path u. Such a measure on $\mathfrak{T}$ is said to be *central.*

Let us consider $X_n(t) = t(n)$ as a random variable on probability space $(\mathfrak{T}, \mathfrak{P})$. Then $\{X_n\,;\, n = 0, 1, 2, \dots\}$ becomes a Markov chain. In fact,

$$\mathfrak{P}\big(t(n+1) = \Lambda \,\big|\, t(1) = \lambda^{(1)}, \dots, t(n-1) = \lambda^{(n-1)}, t(n) = \lambda\big)$$
$$= \begin{cases} \dfrac{\dim\Lambda}{(n+1)\dim\lambda}, & \text{if } \lambda \nearrow \Lambda, \\ 0, & \text{otherwise.} \end{cases} \tag{9.30}$$

The Markov chain on $\mathbb{Y}$ with transition probability (9.30) is called the *Plancherel growth process.* We now understand from (9.24) and (9.30) the meaning of the transition measure $\mathfrak{m}_\lambda$ due to Kerov.

The Plancherel growth process induces transition of the Rayleigh measures as follows. Recall that a Young diagram $\lambda \in \mathbb{Y}_n$ is identified with a Rayleigh measure τ_λ in such a way that

$$\tau_\lambda = \sum_{i=1}^{r} \delta_{x_i} - \sum_{i=1}^{r-1} \delta_{y_i}, \qquad \lambda = (x_1 < y_1 < \dots < y_{r-1} < x_r),$$

whence τ_λ represents a configuration in $\mathbb{Z}$ such that $+1$ at x_i, -1 at y_i and 0 elsewhere (see Fig. 9.9).

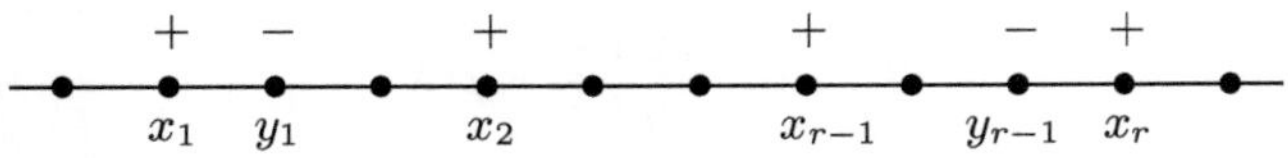

Fig. 9.9. Configuration corresponding to τ_λ

Recall that $\mathfrak{m}_\lambda$ is a probability measure on $\{x_1, x_2, \dots, x_r\}$. For the next step, one site x_j is chosen according to $\mathfrak{m}_\lambda$. Subtract the value $+2$ from the cite x_j to be shared with the two neighbouring sites. Thus we obtain a new configuration which corresponds again to a Rayleigh measure, hence to a Young diagram. Figure 9.10 illustrates the transition from λ to λ'.

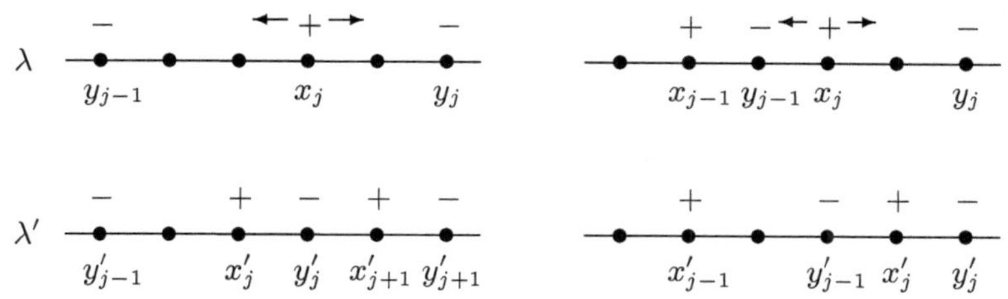

Fig. 9.10. Plancherel growth process

Exercises

9.1. Show that $l(g) = n - c(g)$ for $g \in S(n)$. [Proposition 9.4]

9.2. Show that $l(g) \equiv \mathrm{inv}(g) \pmod 2$ for $g \in S(n)$. [(9.2)]

9.3. Show that

$$\sum_{j=1}^{\mathrm{col}(\lambda)} \lambda'_j(\lambda'_j - 1) = \sum_{i=1}^{n} (i-1)\lambda_i \,, \lambda \in \mathbb{Y}_n.$$

9.4. Let G be a finite group acting on a finite dimensional complex vector space V. Assume that V is decomposed into a direct sum $V = \bigoplus_i W_i$, where each W_i is non-trivial ($V \neq 0$), G-invariant and irreducible, and mutually non-equivalent. Show that

(1) a non-trivial G-invariant irreducible subspace of V coincides with one of W_i's.
(2) every G-invariant subspace of V is a direct sum of some W_i's.

[Hint: Schur's lemma]

9.5. Find out the Pascal triangle embedded in the Young graph. Which diagrams are its vertices?

9.6. Prove Lemma 9.18.

9.7. Show that the sequence $y_1 < \cdots < y_{r-1}$ determined by (9.16) and (9.17) satisfies the interlacing property: $x_1 < y_1 < \cdots < y_{r-1} < x_r$.

9.8. Verify that the finitely additive measure on the cylindrical subsets of $\mathfrak{T}$ defined by (9.28) can be extended to a unique countably additive measure on $\mathfrak{T}$.

9.9. Let G be a finite group and H a subgroup. Let $i : \mathbb{C}[H] \to \mathbb{C}[G]$ be the canonical injective $*$-homomorphism. Show that i^* coincides with the conditional expectation $\mathbb{C}[G] \to \mathbb{C}[H]$.

9.10 (Plancherel formula for $S(n)$). Let f be a function on $S(n)$. Define the Fourier transform $\hat{f}$ by

$$\hat{f}(\lambda) = \sum_{x \in S(n)} f(x) U_\lambda(x), \qquad \lambda \in \mathbb{Y}_n.$$

Show the following identity:

$$\sum_{x \in S(n)} \overline{f_1(x)} f_2(x) = \sum_{\lambda \in \mathbb{Y}_n} \varphi_{\mathrm{tr}}(\hat{f}_1(\lambda)^* \hat{f}_2(\lambda))\, \mathfrak{P}_n(\lambda),$$

where φ_{tr} is the normalized trace on square matrices.

Notes

There exist a wide variety of literatures on symmetric groups and their representations with various flavours. Here we refer only to James [118] and Sagan [182] as standard textbooks. Diaconis [70] suggests us vast scope for applications of symmetric groups.

We mainly followed Terada–Harada [205] in this chapter to state results on irreducible representations of $S(n)$ and spectral properties of the Jucys–Murphy elements. The inductive structure of $\{S(n)\}$ plays an important role. From this point of view a thorough treatment of representation theory of the symmetric group, or more generally of the Coxeter groups is due to Okounkov–Vershik [173].

For the proof of the hook formula (Theorems 9.7 and 9.8) see, e.g., Simon [192].

The detailed arguments given in Sects. 9.2 and 9.3 can be found in Terada–Harada [205, Sect. 3.1] and [205, Sect. 3.2], respectively. The proofs of Theorem 9.9 and Lemmas 9.10 and 9.13 are based on the Garnir relation for the Specht polynomials, see [205, Sect. 3.1].

An analytic description (the Russian form) of Young diagrams is suitable for considering limiting procedures. See Kerov's monograph [126] for various aspects and applications of this method.

The relation in Proposition 9.21 is satisfied also by the complete symmetric functions and the power sums. See Kerov–Olshanski [127] and Ivanov–Olshanski [117] for usefulness of the symmetric function approach in asymptotic representation theory.

Spectral structure of the Jucys–Murphy operators of $S(n)$ is well known. The form of Theorem 9.23 is found in Biane [27].

Potential theory on the Young graph is essential in harmonic analysis on the infinite symmetric group, which is a vivid research field. See Vershik–Kerov [212], Kerov–Olshanski–Vershik [129], its extended version [130] and references cited therein.

10

The Limit Shape of Young Diagrams

This chapter is devoted to the celebrated *limit shape of Young diagrams*, which is the milestone towards the asymptotic representation theory of the symmetric groups. To our goal we shall not take the quickest way but employ a new idea from quantum probability theory. Our approach is largely combinatorial. In other words, we make much of the *moment method* which has been developed in previous chapters from various aspects. It is our hope that this framework will also serve as an accessible introduction to asymptotic representation theory.

10.1 Continuous Diagrams

Having discussed in Sect. 9.4 the analytic description of Young diagrams, we shall prepare in this section the framework for their limiting objects.

Definition 10.1. An $\mathbb{R}$-valued function defined on $\mathbb{R}$ is called a *continuous diagram* if

(i) $|\omega(x_1) - \omega(x_2)| \le |x_1 - x_2|$ for all $x_1, x_2 \in \mathbb{R}$;
(ii) $\omega(x) = |x|$ if $|x|$ is large enough.

The set of continuous diagrams is denoted by $\mathbb{D}$.

By definition we have an obvious inclusion relation

$$\mathbb{Y} \subset \mathbb{D}_0 \subset \mathbb{D},$$

where $\mathbb{Y}$ and $\mathbb{D}_0$ are the sets of Young diagrams and of rectangular diagrams, respectively.

The transition measure and the Rayleigh measure defined for a rectangular diagram are extended to a continuous diagram by approximation. For $K > 0$ we set

$$\mathbb{D}_K = \{\omega \in \mathbb{D}\,;\, \omega(x) = |x|,\ |x| > K\}. \tag{10.1}$$

A. Hora and N. Obata: The Limit Shape of Young Diagrams. In: A. Hora and N. Obata, Quantum Probability and Spectral Analysis of Graphs, Theoretical and Mathematical Physics, 271–296 (2007)
DOI 10.1007/3-540-48863-4_10

Every continuous diagram belongs to $\mathbb{D}_K$ for a sufficiently large K. Given $\omega \in \mathbb{D}_K$, we take an approximating sequence $\lambda_n \in \mathbb{D}_0 \cap \mathbb{D}_K$, which converges to ω in the uniform norm. Consider

$$\frac{\omega(x) - |x|}{2}, \qquad \frac{\lambda_n(x) - |x|}{2}.$$

Both are almost everywhere differentiable. The derivatives have compact supports contained in $[-K, K]$ and their absolute values are bounded by one. Note that for any $f \in C[-K, K]$,

$$\lim_{n\to\infty} \int_{-\infty}^{+\infty} f(x) \left(\frac{\lambda_n(x) - |x|}{2} \right)' dx = \int_{-\infty}^{+\infty} f(x) \left(\frac{\omega(x) - |x|}{2} \right)' dx. \tag{10.2}$$

In fact, (10.2) is valid for a polynomial $f(x)$, as is easily verified by integration by parts. Then, we apply the fact that the polynomials are dense in $C[-K, K]$ with respect to the uniform norm. Now it follows from (9.13) and (10.2) that $\{M_k(\tau_{\lambda_n}) ; n = 1, 2, \dots\}$ is convergent and hence so is $\{M_k(\mathfrak{m}_{\lambda_n}) ; n = 1, 2, \dots\}$. Again using approximation by polynomials, it turns out that

$$\left\{ \int_{-K}^{K} f(x) \mathfrak{m}_{\lambda_n}(dx) \, ; \, n = 1, 2, \dots \right\}$$

converges for all $f \in C[-K, K]$. Set

$$M(f) = \lim_{n\to\infty} \int_{-K}^{K} f(x) \mathfrak{m}_{\lambda_n}(dx), \qquad f \in C[-K, K].$$

As is seen from (10.2), the functional M depends only on ω and is independent of the choice of an approximating sequence $\{\lambda_n\}$. Obviously, M is linear, positive and normalized on $C[-K, K]$ so that by the Riesz–Markov representation theorem (Theorem 1.3) there exists a unique probability measure μ supported by $[-K, K]$ such that

$$M(f) = \int_{-K}^{K} f(x) \mu(dx), \qquad f \in C[-K, K].$$

Clearly, μ is defined independently of K whenever $\mathrm{supp}(\omega(x) - |x|) \subset [-K, K]$. We write $\mu = \mathfrak{m}_\omega$ and call it the *transition measure* of $\omega \in \mathbb{D}$. Thus we have

$$\lim_{n\to\infty} \int_{-\infty}^{+\infty} f(x) \mathfrak{m}_{\lambda_n}(dx) = \int_{-\infty}^{+\infty} f(x) \mathfrak{m}_\omega(dx), \quad f \in C(\mathbb{R}). \tag{10.3}$$

It follows from (9.18) that for $\lambda \in \mathbb{D}_0$ and $z \in \mathbb{C} \setminus \mathrm{supp}(\lambda(x) - |x|)$,

$$\int_{-\infty}^{+\infty} \frac{1}{z - x} \mathfrak{m}_\lambda(dx) = \frac{1}{z} \exp\left\{ \int_{-\infty}^{+\infty} \frac{1}{x - z} \left(\frac{\lambda(x) - |x|}{2} \right)' dx \right\}.$$

Hence from (10.2) and (10.3) we see the following:

Proposition 10.2. *For a continuous diagram $\omega \in \mathbb{D}$ let $\mathfrak{m}_\omega$ be the transition measure. We have*

$$\int_{-\infty}^{+\infty} \frac{\mathfrak{m}_\omega(dx)}{z-x} = \frac{1}{z} \exp\left\{ \int_{-\infty}^{+\infty} \frac{1}{x-z} \Big(\frac{\omega(x)-|x|}{2} \Big)' dx \right\}. \tag{10.4}$$

Moreover, let $\omega \in \mathbb{D}$ such that $\omega'(x)$ is of bounded variation. Then $\frac{1}{2}(\omega(x) - |x|)''$ is a signed measure of finite total variation. Taking (9.14) for a rectangular diagram into account, we may call

$$\tau_\omega = \left(\frac{\omega(x)-|x|}{2} \right)'' + \delta_0 \tag{10.5}$$

the *Rayleigh measure* of ω. In this case, the characterizing relation (10.4) can be translated into

$$\log \left(\int_{-\infty}^{+\infty} \frac{1}{1-\zeta x} \mathfrak{m}_\omega(dx) \right) = \int_{-\infty}^{+\infty} \left(\log \frac{1}{1-\zeta x} \right) \tau_\omega(dx), \tag{10.6}$$

where $\zeta = 1/z$ has a sufficiently small absolute value.

Example 10.3 (triangular diagram). Let $a > 0$. A triangular diagram (Fig. 10.1) is defined by

$$\omega(x) = \begin{cases} a, & |x| \le a, \\ |x|, & |x| \ge a. \end{cases}$$

The transition measure $\mathfrak{m}_\omega$ is the arcsine law with the density

$$\frac{1}{\pi\sqrt{a^2-x^2}}, \qquad |x| < a,$$

while the Rayleigh measure is given by

$$\tau_\omega = \frac{\delta_{-a} + \delta_a}{2}.$$

The verification is left to the reader [Exercise 10.2].

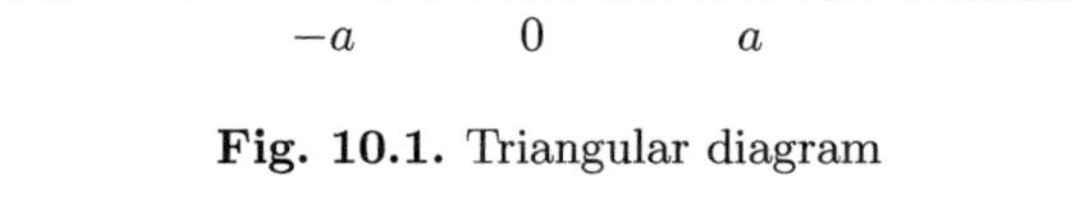

Fig. 10.1. Triangular diagram

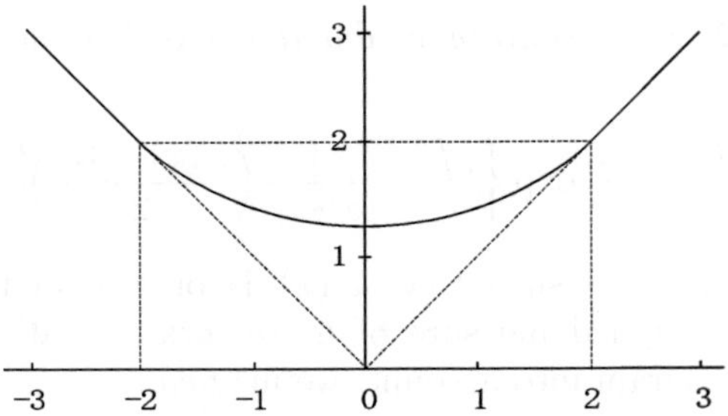

Fig. 10.2. $\Omega(x)$: The limit shape of Young diagrams

Example 10.4 (Limit shape of Young diagrams). Consider the continuous diagram $\Omega \in \mathbb{D}$ defined by

$$\Omega(x) = \begin{cases} \dfrac{2}{\pi}\left(x \arcsin \dfrac{x}{2} + \sqrt{4 - x^2}\right), & |x| \le 2, \\ |x|, & |x| > 2. \end{cases} \tag{10.7}$$

This is known to be the limit shape of Young diagrams and the relevant discussion will be given later (see Fig. 10.2).

Here we compute its Rayleigh measure τ_Ω and its transition measure $\mathfrak{m}_\Omega$. We see from (9.13) that

$$M_k(\tau_\Omega) = \begin{cases} \dbinom{2p}{p}, & \text{if } k \text{ is even and } k = 2p, \\ 0, & \text{if } k \text{ is odd,} \end{cases}$$

and from a general binomial expansion that

$$\sum_{p=0}^{\infty} \binom{2p}{p} \frac{1}{z^{2p+1}} = \frac{1}{\sqrt{z^2 - 4}}, \qquad |z| > 2.$$

Hence

$$\int_{-\infty}^{+\infty} \frac{\tau_\Omega(dx)}{z - x} = \frac{1}{\sqrt{z^2 - 4}}. \tag{10.8}$$

Integrating termwise, we have

$$\frac{1}{z} \exp\left\{ \sum_{p=1}^{\infty} \binom{2p}{p} \frac{1}{2p} z^{-2p} \right\} = \frac{z - \sqrt{z^2 - 4}}{2}, \qquad |z| > 2.$$

Hence (9.18) yields

$$\int_{-\infty}^{+\infty} \frac{\mathfrak{m}_\Omega(dx)}{z - x} = \frac{z - \sqrt{z^2 - 4}}{2}. \tag{10.9}$$

Here in (10.8) and (10.9) the branch is taken so that $\sqrt{z^2-4}>0$ for $z>2$ and $\sqrt{z^2-4}<0$ for $z<-2$ hold. Taking boundary values on the real axis, we see that the transition and Rayleigh measures of Ω are given by the Wigner semicircle law and the arcsine law, which have the densities

$$\frac{d\mathfrak{m}_\Omega}{dx} = \frac{1}{2\pi}\sqrt{4-x^2}\,, \qquad |x| \le 2, \tag{10.10}$$

$$\frac{d\tau_\Omega}{dx} = \frac{1}{\pi\sqrt{4-x^2}}\,, \qquad |x| < 2, \tag{10.11}$$

respectively (see Definitions 1.73 and 4.6).

10.2 The Limit Shape of Young Diagrams

Discovery of the limit shape of Young diagrams by Vershik–Kerov and Logan–Shepp in the 70s brought a breakthrough in asymptotic theory for representations of symmetric groups. This result specifies the shape of the most likely Young diagram with respect to the Plancherel measure $\mathfrak{P}$. In other words, it is a concentration phenomenon observed in irreducible decomposition of the regular representation of $S(n)$ as $n \to \infty$. In the rest of this chapter, we discuss the limit shape in the framework of law of large numbers in probability theory.

Let L_n denote the left regular representation of $S(n)$ on $\ell^2(S(n))$. It is well known that L_n is decomposed into a direct sum of irreducible representations as

$$L_n \simeq \bigoplus_{\lambda \in \mathbb{Y}_n} (\dim\lambda) U_\lambda.$$

Applying the normalized character, we have

$$\delta_e = \tilde{\chi}^{L_n} = \sum_{\lambda\in\mathbb{Y}_n} \frac{\dim^2\lambda}{n!}\tilde{\chi}^\lambda = \sum_{\lambda\in\mathbb{Y}_n} \mathfrak{P}_n(\lambda)\tilde{\chi}^\lambda.$$

Identifying $\lambda \in \mathbb{Y}_n$ as a rectangular diagram in $\mathbb{D}_0$ through its profile, we rescale it by $1/\sqrt{n}$ both horizontally and vertically. Define $\lambda^{\sqrt{n}} \in \mathbb{D}_0$ by

$$\lambda^{\sqrt{n}}(x) = \frac{1}{\sqrt{n}}\lambda(\sqrt{n}\,x).$$

It is easily seen that the area confined by the profile of $\lambda^{\sqrt{n}}$ and $|x|$ is kept constant 2 independently of n. In the limit as $n \to \infty$ the dominant components with respect to the Plancherel measure $\mathfrak{P}$ concentrate at the shape Ω in (10.7). This phenomenon is formulated in the following law of large numbers, of which the proof will be mentioned in Sect. 10.7.

Theorem 10.5 (Strong LLN for Young diagrams). *For $\mathfrak{P}$-almost all paths $t \in \mathfrak{T}$,*

$$\lim_{n\to\infty} \sup_{x\in\mathbb{R}} |t(n)^{\sqrt{n}}(x) - \Omega(x)| = 0,$$

where $t(n) \in \mathbb{Y}_n$ denotes the nth Young diagram along t.

If we look at the irreducible decomposition of the regular representation of $S(n)$ by zooming out in $1/\sqrt{n}$-scaling, almost all that we see are the irreducible components corresponding to those Young diagrams which have shapes quite close to Ω. From the macroscopic point of view, only 'Ω-component' can be observed. In this sense Ω in (10.7) is called the limit shape of Young diagrams.

In Theorem 10.5, the distance between two continuous diagrams is measured by the uniform norm

$$\sup_{x\in\mathbb{R}} |\omega_1(x) - \omega_2(x)|, \qquad \omega_1, \omega_2 \in \mathbb{D}.$$

Let us consider another topology on $\mathbb{D}$. Consider semi-distances on $\mathbb{D}$

$$|M_k(\mathfrak{m}_{\omega_1}) - M_k(\mathfrak{m}_{\omega_2})|, \qquad k = 1, 2, \ldots, \tag{10.12}$$

determined by the moments of transition measures and

$$\left|\int_{-\infty}^{\infty} x^k \left(\frac{\omega_1(x) - |x|}{2}\right)' dx - \int_{-\infty}^{\infty} x^k \left(\frac{\omega_2(x) - |x|}{2}\right)' dx\right|, \qquad k = 1, 2, \ldots, \tag{10.13}$$

for $\omega_1, \omega_2 \in \mathbb{D}$.

Lemma 10.6. *The topologies on $\mathbb{D}$ induced by the semi-distances* (10.12) *and* (10.13) *coincide.*

Proof. It follows from (10.4) that the moments of the transition measure $\mathfrak{m}_\omega$ and

$$\left\{\int_{-\infty}^{\infty} x^k \left(\frac{\omega(x) - |x|}{2}\right)' dx\right\}$$

are mutually expressible in polynomials. Hence the assertion follows. □

The topology on $\mathbb{D}$ induced by the semi-distances (10.12), or equivalently by (10.13), is called the *moment topology* on $\mathbb{D}$.

Lemma 10.7. *Let $K > 0$ and $\mathbb{D}_K$ be as in* (10.1). *The moment topology and the uniform topology coincide on $\mathbb{D}_K$.*

Proof. It is obvious that the moment topology is weaker on $\mathbb{D}_K$. We show the converse. We have

$$\sup_{x\in\mathbb{R}} |\omega_1(x) - \omega_2(x)| \le 2\int_{-K}^{K} \left|\left(\frac{\omega_1(x) - |x|}{2}\right)' - \left(\frac{\omega_2(x) - |x|}{2}\right)'\right| dx$$

for $\omega_1, \omega_2 \in \mathbb{D}_K$. The L^1-norm in the right-hand side is approximated by canonical pairings with continuous functions on $[-K, K]$, while a continuous function is uniformly approximated by polynomials on $[-K, K]$. This completes the proof of the assertion. □

Remark 10.8. The moment topology and the uniform topology are not comparable on $\mathbb{D}$.

Actually, the method of limit theorems in algebraic probability we have developed in the preceding chapters gives insight and finds approach to the limit shape problem. The Jucys–Murphy element, regarded as a random variable in a suitable algebraic probability space determined by a character, plays a key role through the basic trace formula (9.19). In the sequel, we develop analysis for moments of the Jucys–Murphy element and resulting limit theorem, arriving then at the result of the limit shape. We begin with rough analysis which yields the limit shape as weak law of large numbers (Theorem 10.18). Then we proceed to finer argument to obtain the formulation as strong law of large numbers (Theorem 10.24).

10.3 The Modified Young Graph

Analysis for moments of Jucys–Murphy elements J_n forms a core of our approach to the problems around the limit shape of Young diagrams. An essential property of J_n is that it commutes with $S(n)$ and hence $\mathbb{E}_n J_n^k$ belongs to the centre of $\mathbb{C}[S(n)]$.

We start with a general finite group G. Let $\mathbb{C}[G]$ denote the group $*$-algebra and $\mathcal{Z}(\mathbb{C}[G])$ its centre. For a conjugacy class C of G we define $A_C \in \mathbb{C}[G]$ by

$$A_C = \sum_{g \in C} g.$$

For simplicity, the same symbol A_C stands for the operator on $\ell^2(G)$ defined through the left regular representation of G. We call A_C the *adjacency matrix* associated with C.

Lemma 10.9. *The set $\{A_C\}$, where C runs over the conjugacy classes of G, is a linear basis of $\mathcal{Z}(\mathbb{C}[G])$.*

Let $\{\delta_x \,;\, x \in G\}$ be the canonical orthonormal basis of $\ell^2(G)$. Since

$$(A_C)_{xy} = \langle \delta_x, A_C \delta_y \rangle = \sum_{g \in C} \langle \delta_x, g\delta_y \rangle = \sum_{g \in C} \langle \delta_x, \delta_{gy} \rangle, \qquad x, y \in G,$$

we have

$$(A_C)_{xy} = \begin{cases} 1, & xy^{-1} \in C, \\ 0, & \text{otherwise.} \end{cases}$$

The *intersection numbers* $p^{\gamma}_{\alpha\beta}$ are defined by

$$A_{C_\alpha} A_{C_\beta} = \sum_{\gamma} p^{\gamma}_{\alpha\beta} A_{C_\gamma}. \tag{10.14}$$

As is easily verified, taking $x \in C_\gamma$ arbitrarily, we see that

$$\begin{aligned} p^{\gamma}_{\alpha\beta} &= |\{(y,z) \in C_\alpha \times C_\beta\,;\, x = yz\}| \\ &= |\{z \in C_\alpha\,;\, z^{-1}x \in C_\beta\}|. \end{aligned} \tag{10.15}$$

In general, the full description of the intersection numbers $p^{\gamma}_{\alpha\beta}$ is difficult. For the symmetric groups, an important progress was made by Ivanov and Kerov (see also Sect. 11.5).

We now go back to the symmetric group $S(n)$. For a Young diagram $\rho \in \mathbb{Y}_n$ let A_ρ denote the adjacency matrix associated to the conjugacy class determined by ρ. For the sake of combinatorial arguments, we introduce an array of Young diagrams parametrizing conjugacy classes. This structure will also be used for quantum central limit theorem in the next chapter.

We consider the Cayley graph of $S(n)$ with all the transpositions adopted as its generators. Recall that $c(x)$ denotes the number of cycles (including trivial ones) appearing in the cycle decomposition of $x \in S(n)$ (Sect. 9.1). The graph distance between x and the unit e is given by

$$\partial(x,e) = l(x) = n - c(x)$$

(see Proposition 9.4).

The set of Young diagrams without one-box rows is denoted by $\overline{\mathbb{Y}}$. Namely,

$$\overline{\mathbb{Y}} = \{\rho \in \mathbb{Y}\,;\, m_1(\rho) = 0\}.$$

By definition, the empty diagram $\varnothing$ belongs to $\overline{\mathbb{Y}}$. For a given diagram $\rho \in \overline{\mathbb{Y}}$, we take an integer $n \geq 1$ such that $|\rho| \leq n$ and consider $x \in S(n)$ which has the cycle decomposition of type ρ. Then,

$$n - c(x) = l(x) = |\rho| - \mathrm{row}\,(\rho).$$

Thus it is convenient to set

$$l(\rho) = |\rho| - \mathrm{row}\,(\rho), \qquad \rho \in \overline{\mathbb{Y}}. \tag{10.16}$$

We call l the *length function* on $\overline{\mathbb{Y}}$.

Remark 10.10. In many literatures, $l(\lambda)$ denotes the number of rows of Young diagram λ while it is denoted by $\mathrm{row}(\lambda)$ in this book. Our $l(\rho)$ comes from the length function on the Cayley graph.

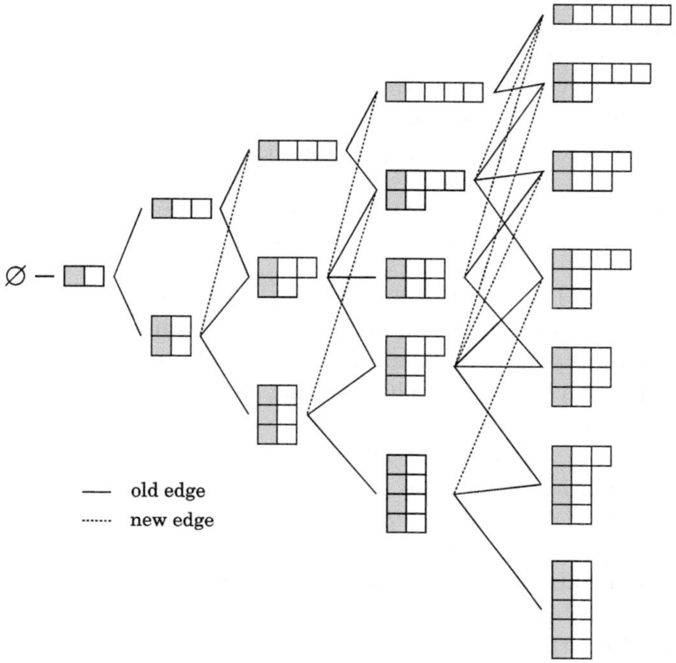

Fig. 10.3. Modified Young graph

This length function $l(\rho)$ brings us a stratified arrangement of the diagrams in $\overline{\mathbb{Y}}$. In other words, we arrange diagrams ρ having the common value $l(\rho)$ in the same stratum. Note that $l(\rho) = l(\rho')$ does not imply $|\rho| = |\rho'|$ contrary to the usual Young graph. In fact, the stratification of $\overline{\mathbb{Y}}$ is obtained from the Young graph $\mathbb{Y}$ in such a way that each $\rho \in \mathbb{Y}$ is glued a copy of the first (longest) column to the most left side of ρ (see Fig. 10.3).

The infinite symmetric group is defined by

$$S(\infty) = \bigcup_{n=0}^{\infty} S(n),$$

namely, is the group of finite permutations of $\{1, 2, \dots\}$. With each $g \in S(\infty)$, being decomposed into a product of disjoint cycles of length ≥ 2, we associate a Young diagram in $\overline{\mathbb{Y}}$, which is denoted by $\operatorname{type}(g)$. Then, $g \mapsto \operatorname{type}(g)$ induces a one-to-one correspondence between the conjugacy classes of $S(\infty)$ and $\overline{\mathbb{Y}}$. By restriction, the conjugacy classes of $S(n)$ are parametrized by $\{\rho \in \overline{\mathbb{Y}}; |\rho| \leq n\}$. A Young diagram in $\mathbb{Y}_n$ is considered as a union of a Young diagram $\rho \in \overline{\mathbb{Y}}$, $|\rho| \leq n$, and one-column one $1^{n-|\rho|}$. The corresponding conjugacy class of $S(n)$ is then denoted by

$$C_{\rho\cup(1^{n-|\rho|})} \quad \text{or} \quad C_{(\rho,1^{n-|\rho|})}.$$

The graph structure of $\overline{\mathbb{Y}}$ is introduced in such a way that $\rho, \sigma \in \overline{\mathbb{Y}}$ are joined by an edge if there exist $n = 1, 2, \ldots$ and $x, (i\,j) \in S(n)$ such that $\rho = \text{type}\,(x)$, $\sigma = \text{type}\,((i\,j)x)$. We refer to the resulting graph as the *modified Young graph*, which is denoted by $\overline{\mathbb{Y}}$.

We study the action of a transposition more in detail.

Lemma 10.11. *Let $x, (i\,j) \in S(n)$ and set $\rho = \text{type}\,(x)$, $\sigma = \text{type}\,((i\,j)x)$. Then, either $l(\sigma) = l(\rho) + 1$ or $l(\rho) - 1$ holds.*

If $l(\sigma) = l(\rho) + 1$ occurs, the action of $(i\,j)$ on x is called *up* and otherwise *down*. These actions are discriminated by graphical consideration. Write down the cycle notation of $x \in S(n)$ and the Young tableau of size n. Delete the rows consisting of a single box. The resulting shape is $\rho = \text{type}\,(x) \in \overline{\mathbb{Y}}$. Mark the boxes occupied by i, j with $\circ, \times$. If i or j lies in the deleted boxes (this happens when i or j is fixed by x), we mark $\circ$ or $\times$ outside ρ. Figure 10.4 exhausts possibilities of location of $\circ, \times$. The left three cases cause the up actions while the right three the down. For example, in the third case therein, the action of $(i\,j)$ makes the two cycles containing i, j join into a single cycle.

As mentioned above, $\overline{\mathbb{Y}}$ is constructed from the Young graph $\mathbb{Y}$ by gluing a copy of the first column. The edges of the Young graph are inherited as those of $\overline{\mathbb{Y}}$. In Fig. 10.3, they are referred to as the old edges for $\overline{\mathbb{Y}}$, while the other edges are said to be new.

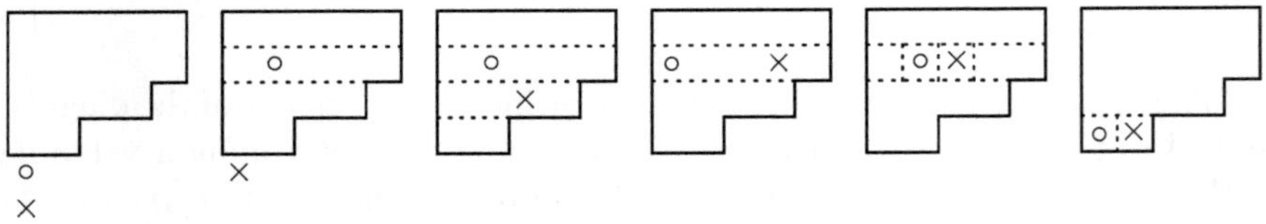

Fig. 10.4. Action of a transposition

10.4 Moments of the Jucys–Murphy Element

We are interested in a power of the Jucys–Murphy element J_n. Let us start with the expansion

$$J_n^k = \sum_{i_1,\ldots,i_k \in \{1,\ldots,n\}} (i_k\,*) \cdots (i_2\,*)(i_1\,*), \tag{10.17}$$

where $*$ stands for the letter $n+1$. A sequence $i_1, i_2, \ldots, i_k$ is called *admissible* if $(i_k\,*) \cdots (i_1\,*) \in S(n)$. To each walk

$$e \to (i_1\,*) \to \cdots \to (i_k\,*) \cdots (i_1\,*) \tag{10.18}$$

in the Cayley graph of $S(n+1)$, we assign a walk

$$\varnothing \to \rho_1 \to \cdots \to \rho_k \tag{10.19}$$

in $\overline{\mathbb{Y}}$ where

$$\rho_j = \text{type}\,((i_j\,*)\cdots(i_1\,*)).$$

Equation (10.19) is called the projection of (10.18) onto $\overline{\mathbb{Y}}$. A k-walk on $\overline{\mathbb{Y}}$ starting at $\varnothing$ is said to be *admissible* if it is the projection of a k-walk on $S(n+1)$ caused by an admissible sequence $i_1, \dots, i_k$. Actually, it suffices to take a sufficiently large n in this definition. Each step $\rho_j \to \rho_{j+1}$ is said to be *up* or *down* according as $l(\rho_{j+1}) = l(\rho_j) + 1$ or $l(\rho_j) - 1$.

For an admissible walk on $\overline{\mathbb{Y}}$, we consider *degree of freedom* or *growth order* at each step. Imagine large n's. Let $x \in S(n+1)$ be decomposed into non-trivial cycles as in Fig. 10.4, where $\{i, *\} = \{\circ, \times\}$ and $* = n+1$. When $(i\,*)$ acts on x as down, we regard it as degree of freedom 0 or growth order 1 because both i and $*$ are chosen among the letters appearing already in the non-trivial cycle of x as in the right three cases. Although the action of $(i\,*)$ in the third case is up, it is degree of freedom 0 or growth order 1 by the same reason with the above cases. In the second case, we can take $* = \circ$ and $i = \times$. Since the choice of i is possible in $n+1-|\text{type}\,(x)|$ ways, we regard this step as degree of freedom 1 or growth order n. By the same reason, the first case is referred to as degree of freedom 1 or growth order n.

For a characterization of the terminating vertices of the admissible walks of length k on $\overline{\mathbb{Y}}$, we need some notation. Set

$$\overline{\mathbb{Y}}_k = \overline{\mathbb{Y}} \cap \mathbb{Y}_k = \{\sigma \in \mathbb{Y}_k\,;\, m_1(\sigma) = 0\}.$$

For $\sigma \in \overline{\mathbb{Y}}$ we denote by σ° the Young diagram obtained from $\sigma \in \overline{\mathbb{Y}}$ by removing all 2-rows and the first column as indicated in Fig. 10.5. For each k, the correspondence $\sigma \mapsto \sigma^\circ$ is an injection from $\overline{\mathbb{Y}}_k$ into $\overline{\mathbb{Y}}$.

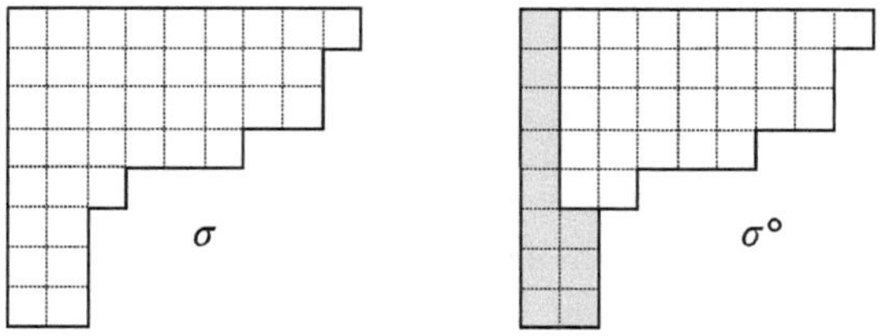

Fig. 10.5. The operation $\sigma \mapsto \sigma^\circ$

Lemma 10.12. *For $\rho \in \overline{\mathbb{Y}}$ and $k = 1, 2, \dots$, the following three conditions are equivalent:*

(i) *there exist an integer $n \geq 1$ and an admissible sequence $i_1, \dots, i_k \in \{1, \dots, n\}$ such that* $\text{type}\,((i_k\,*)\cdots(i_1\,*)) = \rho$*;*

(ii) $l(\rho) \equiv k \pmod 2$ *and* $|\rho| + \mathrm{row}\,(\rho) \leq k$;
(iii) $\rho = \sigma^\circ$ *holds for some* $\sigma \in \overline{\mathbb{Y}}_k$.

Proof. (i) $\Longrightarrow$ (ii). Note first that if condition (i) holds for some n then it holds for any natural number $\geq n$. Let u and d denote the numbers of up and down steps respectively of the k-walk from $\varnothing$ to ρ on $\overline{\mathbb{Y}}$ associated with the path

$$e \to (i_1 \ *) \to \cdots \to (i_k \ *) \cdots (i_1 \ *)$$

on $S(n+1)$. We have $u + d = k$ and $u - d = l(\rho)$ so that

$$u = \frac{k + l(\rho)}{2} \quad \text{and} \quad d = \frac{k - l(\rho)}{2}. \tag{10.20}$$

Hence k and $l(\rho)$ have the same parity. Since a cycle in $(i_k \ *) \cdots (i_1 \ *)$ which does not contain $*$ cannot be produced until a down step occurs, we have

$$\mathrm{row}\,(\rho) \leq d = \frac{k - l(\rho)}{2}.$$

Consequently,

$$k \geq l(\rho) + 2\,\mathrm{row}\,(\rho) = |\rho| + \mathrm{row}\,(\rho),$$

as desired.

(ii) $\Longrightarrow$ (i). We shall construct $(i_k \ *) \cdots (i_1 \ *) \in S(n)$ of ρ-type. Let

$$\rho = (\rho_1 \geq \rho_2 \geq \cdots \geq \rho_{\mathrm{row}(\rho)}).$$

First take distinct $i_1, \ldots, i_{\rho_1} \in \{1, \ldots, n\}$ and $i_{\rho_1 + 1} = i_1$. Then

$$(i_{\rho_1+1} \ *)(i_{\rho_1} \ *) \cdots (i_1 \ *) = (i_1 i_2 \ldots i_{\rho_1}) \in S(n).$$

Next take distinct $j_1, \ldots, j_{\rho_2} \in \{1, \ldots, n\} \setminus \{i_1, \ldots, i_{\rho_1}\}$ and make ρ_2-cycle in $S(n)$ similarly. Repeating this procedure to get distinct row (ρ) cycles in $S(n)$, we count the number of steps as

$$(\rho_1 + 1) + \cdots + (\rho_{\mathrm{row}(\rho)} + 1) = |\rho| + \mathrm{row}\,(\rho) \leq k.$$

As for the remaining $k - (|\rho| + \mathrm{row}\,(\rho))$ steps, which is even, we repeat multiplying $(s \ *)$ by using letter s which never appeared.

(ii) $\Longleftrightarrow$ (iii) is straightforward by the definition of σ°. □

Recall that $\mathbb{E}_n J_n^k$ is central, namely belongs to $\mathcal{Z}(\mathbb{C}[S(n)])$. Then by Lemma 10.12,

$$\mathbb{E}_n J_n^k = \sum_{\sigma \in \overline{\mathbb{Y}}_k} K_{\sigma;n} A_{\sigma^\circ \cup (1^{n - |\sigma^\circ|})}, \tag{10.21}$$

where $K_{\sigma;n}$ is a non-negative integer. In fact, $K_{\sigma;n}$ is the number put on $x \in C_{\sigma^\circ \cup (1^{n-|\sigma^\circ|})}$, $\sigma^\circ = \mathrm{type}\,(x)$, in $\mathbb{E}_n J_n^k$.

Lemma 10.13. *The coefficient in* (10.21) *satisfies asymptotically*

$$K_{\sigma;n} \asymp n^{m_2(\sigma)}, \qquad \sigma \in \overline{\mathbb{Y}}_k,$$

namely the ratio of both the sides converges to a non-zero constant as $n \to \infty$.

Proof. In a k-walk from $\varnothing$ to σ° on $\overline{\mathbb{Y}}$, (10.20) yields the numbers of up and down steps as

$$u = \frac{k + l(\sigma^\circ)}{2}, \qquad d = \frac{k - l(\sigma^\circ)}{2}, \tag{10.22}$$

respectively. Although the up steps have three possibilities in Fig. 10.4, we can reach σ° without passing the third case as is seen from the choice of $i_1, \dots, i_k$ in the proof of Lemma 10.12. The growth order of the first and second cases is n, while it is 1 in the third. Hence the order of $K_{\sigma;n}$ does not change even if we exclude walks with up steps of the third pattern. Then, for an admissible k-walk on $\overline{\mathbb{Y}}$ terminating at σ°, the choice of admissible sequences $i_1, \dots, i_k$ has order n^u where u is in (10.22). The number of k-walks on $\overline{\mathbb{Y}}$ is independent of n. The cardinality of $C_{\sigma^\circ \cup (1^{n-|\sigma^\circ|})}$ is of order $n^{|\sigma^\circ|}$. Hence the order of $K_{\sigma;n}$ has the power

$$u - |\sigma^\circ| = \frac{|\sigma| + l(\sigma^\circ)}{2} - |\sigma^\circ| = m_2(\sigma),$$

which proves the assertion. □

Remark 10.14. Actually, a more explicit expression for $K_{\sigma;n}$, not only the growth order, is obtained by Hora [108] as

$$\mathbb{E}_n J_n^k = \sum_{\sigma \in \overline{\mathbb{Y}}_k} |\mathcal{P}_{\mathrm{NC}(\sigma)}(k)| n^{l(\sigma)} \big(1 + O(n^{-1})\big) \frac{A_{\sigma^\circ \cup (1^{n-|\sigma^\circ|})}}{|C_{\sigma^\circ \cup (1^{n-|\sigma^\circ|})}|}.$$

Here $\mathcal{P}_{\mathrm{NC}(\sigma)}(k)$ denotes the set of non-crossing partitions ϑ of $\{1, \dots, k\}$ of type $\sigma \in \overline{\mathbb{Y}}_k$, where the block lengths of ϑ consist of $\{\sigma_1, \sigma_2, \dots\}$. The above expression will not be used in what follows.

10.5 The Limit Shape as a Weak Law of Large Numbers

Recall that the state on a group $*$-algebra $\mathbb{C}[G]$ determined by normalized positive definite function φ on G is denoted by the same symbol φ for simplicity. We shall use below, for example, the delta function δ_e and normalized irreducible characters as φ. Let J_n denote the Jucys-Murphy element in $\mathbb{C}[S(n+1)]$. We begin with

Proposition 10.15.

$$\delta_e(J_n^k) = \begin{cases} \dfrac{1}{(k/2)+1} \dbinom{k}{k/2} n^{k/2} + O(n^{k/2-1}), & \text{if } k \text{ is even}, \\ 0, & \text{otherwise}. \end{cases}$$

Proof. Let δ_e act on the expansion (10.17). The surviving terms satisfy that the corresponding k-walks start at $\varnothing$ and terminate at $\varnothing$. Such walks do not exist if k is odd. Let $k = 2p$ be even. An admissible k-walk on $\overline{\mathbb{Y}}$ returning at $\varnothing$ has p up steps. Consider a walk on $S(n+1)$ projected onto the k-walk. Once appeared a non-trivial cycle not containing $*$, then, in order for the associated walk on $\overline{\mathbb{Y}}$ to return to $\varnothing$, we need an up step at which $*$ interacts with the non-trivial cycle, namely a step at which a letter, say i, is chosen from the non-trivial cycle and the transposition $(i\,*)$ acts. However, since this up step is of growth order 1, the walks on $S(n+1)$ containing such a step are of order $O(n^{p-1})$ and hence negligible.

We thus have only to take into account the remaining cases, that is, the k-walks on $\overline{\mathbb{Y}}$ moving within the upper boundary in Fig. 10.3. In this kind of k-walk starting at $\varnothing$ and terminating at $\varnothing$, each up step has growth order n. Hence the number of the corresponding walks on $S(n+1)$ is $n^p(1 + O(1/n))$. On the other hand, we already encountered these k-walks on the upper boundary of $\overline{\mathbb{Y}}$ (though it is easy to compute the number directly). They have bijective correspondence to $\mathcal{C}_p$, the Catalan paths of length $2p$. The cardinality was given by the Catalan number $\frac{1}{p+1}\binom{2p}{p}$. □

Proposition 10.16.

$$\delta_e((\mathbb{E}_n J_n^k)^2) - \delta_e(J_n^k)^2 = O(n^{k-1}).$$

Proof. Consider the expansion (10.21). From Lemma 10.13 the e-component of $(\mathbb{E}_n J_n^k)^2$ is

$$\sum_{\sigma\in\overline{\mathbb{Y}}_k} K_{\sigma;n}^2 |C_{\sigma^\circ\cup(1^{n-|\sigma^\circ|})}| = \sum_{\sigma\in\overline{\mathbb{Y}}_k} c_\sigma n^{2m_2(\sigma)} n^{|\sigma^\circ|}(1+O(n^{-1}))$$

$$= \sum_{\sigma\in\overline{\mathbb{Y}}_k} c_\sigma n^{l(\sigma)+m_2(\sigma)}(1+O(n^{-1}))$$

with constant c_σ independent of n. For $\sigma \in \overline{\mathbb{Y}}_k$, $l(\sigma)+m_2(\sigma) = k$ if and only if k is even and $\sigma = (2^{k/2})$. The other σ's satisfy $l(\sigma)+m_2(\sigma) \le k-1$. When k is even and $\sigma = (2^{k/2})$, $\sigma^\circ = \varnothing$ holds and the corresponding term in the left-hand side is the square of the e-component of $\mathbb{E}_n J_n^k$, i.e., $\delta_e(J_n^k)^2$. □

Lemma 10.17. $\tilde{\chi}^\lambda$ *is multiplicative on* $\mathcal{Z}(\mathbb{C}[S(n)])$.

Proof. Easily verified in view of the values on irreducible decomposition

$$\mathbb{C}[S(n)] \simeq \bigoplus_{\lambda\in\mathbb{Y}_n} M(\dim\lambda, \mathbb{C})$$

of an irreducible character. □

Applying the estimate of the mean of $\mathbb{E}_n J_n^k$ (Proposition 10.15) and the estimate of the variance (Proposition 10.16) with (10.10), we have

$$\begin{aligned}
&\delta_e((\mathbb{E}_n J_n^k - n^{k/2} M_k(\mathfrak{m}_\Omega))^2) \\
&= \delta_e(\mathbb{E}_n J_n^k \mathbb{E}_n J_n^k) - 2\delta_e(J_n^k) n^{k/2} M_k(\mathfrak{m}_\Omega) + n^k M_k(\mathfrak{m}_\Omega)^2 \\
&= \delta_e(J_n^k)^2 + O(n^{k-1}) - 2\delta_e(J_n^k) n^{k/2} M_k(\mathfrak{m}_\Omega) + n^k M_k(\mathfrak{m}_\Omega)^2 \\
&= O(n^{k-1}).
\end{aligned}$$

On the other hand, we have from Lemma 10.17 and Theorem 9.23,

$$\begin{aligned}
\delta_e((\mathbb{E}_n J_n^k - n^{k/2} M_k(\mathfrak{m}_\Omega))^2) &= \sum_{\lambda \in \mathbb{Y}_n} \mathfrak{P}_n(\lambda) \tilde{\chi}^\lambda((\mathbb{E}_n J_n^k - n^{k/2} M_k(\mathfrak{m}_\Omega))^2) \\
&= \sum_{\lambda \in \mathbb{Y}_n} \mathfrak{P}_n(\lambda) \{\tilde{\chi}^\lambda(\mathbb{E}_n J_n^k) - n^{k/2} M_k(\mathfrak{m}_\Omega)\}^2 \\
&= \sum_{\lambda \in \mathbb{Y}_n} \mathfrak{P}_n(\lambda) \{M_k(\mathfrak{m}_\lambda) - M_k(\mathfrak{m}_\Omega) n^{k/2}\}^2.
\end{aligned}$$

Hence

$$\sum_{\lambda \in \mathbb{Y}_n} \mathfrak{P}_n(\lambda) \{n^{-k/2} M_k(\mathfrak{m}_\lambda) - M_k(\mathfrak{m}_\Omega)\}^2 = O(n^{-1}). \tag{10.23}$$

Combining (10.23) with the Chebyshev inequality and noting that

$$n^{-k/2} M_k(\mathfrak{m}_\lambda) = M_k(\mathfrak{m}_{\lambda\sqrt{n}}), \qquad \lambda \in \mathbb{Y}_n,$$

and (9.29), we obtain the following:

Theorem 10.18 (Weak LLN for Young diagrams). *For any $\epsilon > 0$ and $k = 1, 2, \ldots$, it holds that*

$$\lim_{n \to \infty} \mathfrak{P}(\{t \in \mathfrak{T};\ |M_k(\mathfrak{m}_{t(n)\sqrt{n}}) - M_k(\mathfrak{m}_\Omega)| \geq \epsilon\}) = 0.$$

10.6 More on Moments of the Jucys–Murphy Element

The estimate of (10.23) being not sufficient for deriving the limit shape of Young diagrams as a strong law of large numbers, we need to estimate the fourth moment of $\mathbb{E}_n J_n^k$, i.e.,

$$\delta_e\left(\left(\frac{\mathbb{E}_n J_n^k}{n^{k/2}} - M_k(\mathfrak{m}_\Omega)\right)^4\right). \tag{10.24}$$

The goal of this section is the following:

Proposition 10.19.

$$\delta_e\left(\left(\frac{\mathbb{E}_n J_n^k}{n^{k/2}} - M_k(\mathfrak{m}_\Omega)\right)^4\right) = O(n^{-2}). \tag{10.25}$$

The proof is divided into several steps. Going back to the expression (10.21), we need to take into account not only the highest term of $K_{\sigma;n}$ but also the second term. It follows from Lemma 10.13 that the order of $K_{\sigma;n}$ is $n^{m_2(\sigma)}$. Define $\alpha_\sigma, \beta_\sigma \in \mathbb{R}$ by

$$K_{\sigma;n} = n^{m_2(\sigma)}(\alpha_\sigma + \beta_\sigma n^{-1} + O(n^{-2}))$$

if $m_2(\sigma) > 0$, and set $\beta_\sigma = 0$ if $m_2(\sigma) = 0$. In view of (10.21), we obtain

$$\delta_e(J_n^k) = K_{(2^{k/2});n} = n^{k/2}(\alpha_{(2^{k/2})} + \beta_{(2^{k/2})} n^{-1} + O(n^{-2})),$$

where we understand that every coefficient is 0 if k is odd. Combining this with Proposition 10.15, we have

$$\alpha_{(2^{k/2})} = M_k(\mathfrak{m}_\Omega). \tag{10.26}$$

For the estimate of (10.24) we consider the expansion

$$\begin{aligned}\delta_e\left(\left(\frac{\mathbb{E}_n J_n^k}{n^{k/2}} - M_k(\mathfrak{m}_\Omega)\right)^4\right) &= M_k(\mathfrak{m}_\Omega)^4 \\ &\quad - 4\delta_e\left(\left(\frac{\mathbb{E}_n J_n^k}{n^{k/2}}\right) M_k(\mathfrak{m}_\Omega)^3\right) \\ &\quad + 6\delta_e\left(\left(\frac{\mathbb{E}_n J_n^k}{n^{k/2}}\right)^2 M_k(\mathfrak{m}_\Omega)^2\right) \\ &\quad - 4\delta_e\left(\left(\frac{\mathbb{E}_n J_n^k}{n^{k/2}}\right)^3 M_k(\mathfrak{m}_\Omega)\right) \\ &\quad + \delta_e\left(\left(\frac{\mathbb{E}_n J_n^k}{n^{k/2}}\right)^4\right).\end{aligned} \tag{10.27}$$

The estimate of each term in the right-hand side will be given in the following Lemmas.

Lemma 10.20.

$$\delta_e\left(\left(\frac{\mathbb{E}_n J_n^k}{n^{k/2}}\right) M_k(\mathfrak{m}_\Omega)^3\right) = \alpha_{(2^{k/2})}^3(\alpha_{(2^{k/2})} + \beta_{(2^{k/2})} n^{-1} + O(n^{-2})).$$

Proof. This is immediately seen from (10.26). □

Lemma 10.21.

$$\delta_e\left(\left(\frac{\mathbb{E}_n J_n^k}{n^{k/2}}\right)^2 M_k(\mathfrak{m}_\Omega)^2\right)$$
$$= \alpha_{(2^{k/2})}^2\left\{\alpha_{(2^{k/2})}^2 + \left(2\alpha_{(2^{k/2})}\beta_{(2^{k/2})} + \sum_{s\geq 3}\frac{\alpha^2_{(2^{(k-s)/2}s^1)}}{s-1}\right)n^{-1} + O(n^{-2})\right\}.$$

Proof. We have

$$\begin{aligned}\delta_e((\mathbb{E}_n J_n^k)^2) &= \sum_{\sigma\in\overline{\mathbb{Y}}_k}\sum_{\tau\in\overline{\mathbb{Y}}_k} K_{\sigma;n}K_{\tau;n}\delta_e(A_{\sigma^\circ\cup(1^{n-|\sigma^\circ|})}A_{\tau^\circ\cup(1^{n-|\tau^\circ|})})\\ &= \sum_{\sigma\in\overline{\mathbb{Y}}_k} K_{\sigma;n}^2|C_{\sigma^\circ\cup(1^{n-|\sigma^\circ|})}|.\end{aligned}$$

Since the order of each term is n to the power

$$2m_2(\sigma) + |\sigma^\circ| = k - (m_3(\sigma) + m_4(\sigma) + \cdots),$$

the terms corresponding to $\sigma = (2^{k/2})$ and $(2^{(k-s)/2}s^1)$ contribute with the order n^k and n^{k-1}, respectively while the other terms are $O(n^{k-2})$. Hence we obtain

$$\delta_e\left(\left(\frac{\mathbb{E}_n J_n^k}{n^{k/2}}\right)^2 M_k(\mathfrak{m}_\Omega)^2\right) = M_k(\mathfrak{m}_\Omega)^2 n^{-k}\sum_{\sigma\in\overline{\mathbb{Y}}_k} K_{\sigma;n}^2|C_{\sigma^\circ\cup(1^{n-|\sigma^\circ|})}| \tag{10.28}$$

with

$$\begin{aligned}&\sum_{\sigma\in\overline{\mathbb{Y}}_k} K_{\sigma;n}^2|C_{\sigma^\circ\cup(1^{n-|\sigma^\circ|})}|\\ &\quad = K^2_{(2^{k/2});n} + \sum_{s\geq 3} K^2_{(2^{(k-s)/2}s^1);n}|C_{(s-1,1^{n-|(s-1)|})}| + O(n^{k-2}).\end{aligned} \tag{10.29}$$

With the help of

$$\begin{aligned}K^2_{(2^{k/2});n} &= (\alpha_{(2^{k/2})}n^{k/2} + \beta_{(2^{k/2})}n^{k/2-1} + \cdots)^2\\ &= \alpha^2_{(2^{k/2})}n^k + 2\alpha_{(2^{k/2})}\beta_{(2^{k/2})}n^{k-1} + O(n^{k-2})\end{aligned}$$

and

$$\begin{aligned}K^2_{(2^{(k-s)/2}s^1);n} &= (\alpha_{(2^{(k-s)/2}s^1)}n^{(k-s)/2} + \beta_{(2^{(k-s)/2}s^1)}n^{(k-s)/2-1} + \cdots)^2\\ &= \alpha^2_{(2^{(k-s)/2}s^1)}n^{k-s} + O(n^{k-s-1}),\end{aligned}$$

(10.29) becomes

$$= \alpha_{(2^{k/2})}^2 n^k + 2\alpha_{(2^{k/2})}\beta_{(2^{k/2})} n^{k-1} + O(n^{k-2})$$
$$+ \sum_{s \geq 3} \left\{ \frac{\alpha_{(2^{(k-s)/2} s^1)}^2}{s-1} n^{k-1} + O(n^{k-2}) \right\}$$
$$= \alpha_{(2^{k/2})}^2 n^k + \left\{ 2\alpha_{(2^{k/2})}\beta_{(2^{k/2})} + \sum_{s \geq 3} \frac{\alpha_{(2^{(k-s)/2} s^1)}^2}{s-1} \right\} n^{k-1} + O(n^{k-2}).$$

Inserting this into (10.28), we complete the proof. □

Lemma 10.22.

$$\delta_e\left(\left(\frac{\mathbb{E}_n J_n^k}{n^{k/2}}\right)^3 M_k(\mathfrak{m}_\Omega)\right)$$
$$= \alpha_{(2^{k/2})}^4 + \alpha_{(2^{k/2})} \left\{ 3\alpha_{(2^{k/2})}^2 \beta_{(2^{k/2})} + 3\sum_{s \geq 3} \alpha_{(2^{k/2})} \frac{\alpha_{(2^{(k-s)/2} s^1)}^2}{s-1} \right\} n^{-1}$$
$$+ O(n^{-2}).$$

Proof. We start with

$$\delta_e((\mathbb{E}_n J_n^k)^3)$$
$$= \sum_{\rho \in \overline{\mathbb{Y}}_k} \sum_{\sigma \in \overline{\mathbb{Y}}_k} \sum_{\tau \in \overline{\mathbb{Y}}_k} K_{\rho;n} K_{\sigma;n} K_{\tau;n}$$
$$\times \delta_e(A_{\rho^\circ \cup (1^{n-|\rho^\circ|})} A_{\sigma^\circ \cup (1^{n-|\sigma^\circ|})} A_{\tau^\circ \cup (1^{n-|\tau^\circ|})}).$$

Here we have

$$\delta_e(A_{\rho^\circ \cup (1^{n-|\rho^\circ|})} A_{\sigma^\circ \cup (1^{n-|\sigma^\circ|})} A_{\tau^\circ \cup (1^{n-|\tau^\circ|})})$$
$$= |\{(z,x) \in C_{\rho^\circ \cup (1^{n-|\rho^\circ|})} \times C_{\sigma^\circ \cup (1^{n-|\sigma^\circ|})} \,;\, zx \in C_{\tau^\circ \cup (1^{n-|\tau^\circ|})}\}|, \qquad (10.30)$$

indeed $zx \in C_{\tau^\circ \cup (1^{n-|\tau^\circ|})}$ if and only if $(zx)^{-1} \in C_{\tau^\circ \cup (1^{n-|\tau^\circ|})}$. Take $x \in C_{\sigma^\circ \cup (1^{n-|\sigma^\circ|})}$ and set $l = \mathrm{row}\,(\rho^\circ)$. Expressing $z \in C_{\rho^\circ \cup (1^{n-|\rho^\circ|})}$ as a product of transpositions

$$(k_{j_l} \ldots k_1) \cdots (i_{j_1} \ldots i_1) = (k_{j_l-1} k_{j_l}) \cdots (k_1 k_2) \cdots (i_{j_1-1} i_{j_1}) \cdots (i_1 i_2)$$

and multiplying it onto x from the left, we count the resulting elements falling into $C_{\tau^\circ \cup (1^{n-|\tau^\circ|})}$. The number of steps is

$$(j_1 - 1) + \cdots + (j_l - 1) = |\rho^\circ| - \mathrm{row}\,(\rho^\circ) = l(\rho^\circ).$$

Assume that there exists an $l(\rho^\circ)$-walk on $\overline{\mathbb{Y}}$ starting at σ° and terminating at τ°. Otherwise the right-hand side of (10.30) clearly vanishes. Let u and d denote the numbers of the up and down steps of this walk respectively. Then we have

$$u + d = l(\rho^\circ), \quad u - d = l(\tau^\circ) - l(\sigma^\circ),$$

hence,

$$u = \frac{l(\rho^\circ) + l(\tau^\circ) - l(\sigma^\circ)}{2}.$$

For each x, the highest case of the growth order of up steps is such that, in decomposition into transpositions of every cycle of ρ°, the first step is of order n^2 and the others of order n. In this highest case, we have as an upper bound of the order $n^{u+\mathrm{row}(\rho^\circ)}$. Furthermore, all up steps take this form only if $\tau^\circ = \rho^\circ \cup \sigma^\circ$. In fact, first in order for $(i_1 i_2)$ to have order n^2, neither i_1 nor i_2 is contained in x. Next i_3 is taken from outside of the 2-row consisting of i_1 and i_2. If i_3 is picked up from x, the up step by $(i_2 i_3)$ is of order 1, not n. Hence i_3 must be taken from outside of x and $(i_1 i_2)$ and the action of $(i_2 i_3)$ produces the 3-row $(i_3 i_2 i_1)$. Continuing this procedure, we come to $\tau^\circ = \rho^\circ \cup \sigma^\circ$. Hence, if $\tau^\circ \neq \rho^\circ \cup \sigma^\circ$, there exists an up step of order 1 or at least one of l actions $(i_1 i_2), \ldots, (k_1 k_2)$ is not an up step of order n^2, and thus we have a smaller upper bound $n^{u+\mathrm{row}(\rho^\circ)-1}$. Since x varies in the way of order $n^{|\sigma^\circ|}$ in $C_{\sigma^\circ \cup (1^{n-|\sigma^\circ|})}$, we get the estimate

$$(10.30) = \begin{cases} O(n^{u+\mathrm{row}(\rho^\circ)+|\sigma^\circ|}), & \tau^\circ = \rho^\circ \cup \sigma^\circ, \\ O(n^{u+\mathrm{row}(\rho^\circ)+|\sigma^\circ|-1}), & \tau^\circ \neq \rho^\circ \cup \sigma^\circ, \end{cases} \tag{10.31}$$

where

$$\begin{aligned} u + \mathrm{row}\,(\rho^\circ) + |\sigma^\circ| &= \frac{1}{2}(l(\rho^\circ) + l(\tau^\circ) - l(\sigma^\circ)) + \mathrm{row}\,(\rho^\circ) + |\sigma^\circ| \\ &= \frac{|\rho| + |\sigma| + |\tau|}{2} - \mathrm{row}\,(\tau) - m_2(\rho) - m_2(\sigma) \\ &= \frac{3}{2}k - \mathrm{row}\,(\tau) - m_2(\rho) - m_2(\sigma). \end{aligned}$$

Therefore the order of

$$K_{\rho;n} K_{\sigma;n} K_{\tau;n} \delta_e(A_{\rho^\circ \cup (1^{n-|\rho^\circ|})} A_{\sigma^\circ \cup (1^{n-|\sigma^\circ|})} A_{\tau^\circ \cup (1^{n-|\tau^\circ|})}) \tag{10.32}$$

is bounded by n to the power

$$\begin{aligned} &m_2(\rho) + m_2(\sigma) + m_2(\tau) + \frac{3}{2}k - \mathrm{row}\,(\tau) - m_2(\rho) - m_2(\sigma) \\ &= \frac{3}{2}k - (m_3(\tau) + m_4(\tau) + \cdots). \end{aligned}$$

Hence

$$\delta_e\Big(\Big(\frac{\mathbb{E}_n J_n^k}{n^{k/2}}\Big)^3 M_k(\mathfrak{m}_\Omega)\Big) = \sum O(n^{-(m_3(\tau)+m_4(\tau)+\cdots)}), \tag{10.33}$$

where the number of the summands and the O-constants are independent of n. We see that terms of order 1 or n^{-1} can appear in (10.33) only if

(a) $\tau = (2^{k/2})$;
(b) $\tau = (2^{(k-s)/2}s^1)$ for $s \geq 3$.

First in case of (a), we have $\tau^\circ = \varnothing$. Then

$$(10.32) = \begin{cases} K_{\rho;n}^2 K_{(2^{k/2});n} |C_{\rho^\circ \cup (1^{n-|\rho^\circ|})}|, & \rho^\circ = \sigma^\circ, \\ 0, & \text{otherwise.} \end{cases}$$

This is estimated as in (10.29). In particular, it is $O(n^{3k/2-2})$ unless $\rho^\circ = \varnothing$ nor $\rho^\circ = ((s-1)^1)$. Note that $K_{(2^{k/2});n} \asymp n^{k/2}$ by Lemma 10.13. Secondly in case of (b), $\tau^\circ = ((s-1)^1)$ and the O-term in (10.33) is $O(n^{-1})$. In view of (10.31), we know that, if $\tau^\circ \neq \rho^\circ \cup \sigma^\circ$,

$$(10.32) = O(n^{3k/2-2})$$

holds. We can thus assume that $\tau^\circ = \rho^\circ \cup \sigma^\circ$. Then either ρ° or σ° is $\varnothing$ and the other is τ°.

Consequently, contributions to (10.33) up to $O(n^{-1})$ are given by the terms such that $\rho^\circ, \sigma^\circ, \tau^\circ$ are all $\varnothing$ or one of them is $\varnothing$ and the others coincide with $((s-1)^1)$. Hence

$$\begin{aligned}
&\delta_e\left(\left(\frac{\mathbb{E}_n J_n^k}{n^{k/2}}\right)^3 M_k(\mathfrak{m}_\Omega)\right) \\
&= M_k(\mathfrak{m}_\Omega) n^{-3k/2} \\
&\quad \times \left\{K_{(2^{k/2});n}^3 + 3\sum_{s\geq 3} K_{(2^{k/2});n} K_{(2^{(k-s)/2}s^1);n}^2 |C_{(s-1,1^{n-(s-1)})}|\right\} \\
&\quad + O(n^{-2}) \\
&= M_k(\mathfrak{m}_\Omega) n^{-3k/2}\left\{\alpha_{(2^{k/2})}^3 n^{3k/2} + 3\alpha_{(2^{k/2})}^2 \beta_{(2^{k/2})} n^{3k/2-1} + O(n^{3k/2-2})\right. \\
&\quad \left. + 3\sum_{s\geq 3} \alpha_{(2^{k/2})} \frac{\alpha_{(2^{(k-s)/2}s^1)}^2}{s-1} n^{3k/2-1} + O(n^{3k/2-2})\right\} + O(n^{-2}) \\
&= \alpha_{(2^{k/2})}\left\{\alpha_{(2^{k/2})}^3 + \left(3\alpha_{(2^{k/2})}^2 \beta_{(2^{k/2})} + 3\sum_{s\geq 3} \alpha_{(2^{k/2})} \frac{\alpha_{(2^{(k-s)/2}s^1)}^2}{s-1}\right) n^{-1}\right\} \\
&\quad + O(n^{-2}).
\end{aligned}$$

This completes the proof. □

Let us pick up an estimate for intersection numbers obtained in the above proof. We set for simplicity

$$p^{\gamma}_{\alpha\beta}(n) = p^{\gamma\cup(1^{n-|\gamma|})}_{\alpha\cup(1^{n-|\alpha|})\,\beta\cup(1^{n-|\beta|})}, \qquad \alpha,\beta,\gamma \in \mathbb{Y}, |\alpha|, |\beta|, |\gamma| \le n. \tag{10.34}$$

It follows from (10.30) and (10.31) that

$$\begin{aligned}
&\delta_e(A_{\alpha\cup(1^{n-|\alpha|})}A_{\beta\cup(1^{n-|\beta|})}A_{\gamma\cup(1^{n-|\gamma|})}) \\
&\qquad = \begin{cases} O(n^{(l(\alpha)+l(\gamma)-l(\beta))/2+\mathrm{row}(\alpha)+|\beta|}), & \gamma = \alpha\cup\beta, \\ O(n^{(l(\alpha)+l(\gamma)-l(\beta))/2+\mathrm{row}(\alpha)+|\beta|-1}), & \gamma \neq \alpha\cup\beta. \end{cases}
\end{aligned}$$

On the other hand, the left-hand side is

$$\sum_{\pi} p^{\pi}_{\alpha\beta}(n)\delta_e(A_{\pi\cup(1^{n-|\pi|})}A_{\gamma\cup(1^{n-|\gamma|})}) = p^{\gamma}_{\alpha\beta}(n)|C_{\gamma\cup(1^{n-|\gamma|})}|.$$

Hence we have

$$\begin{aligned}
p^{\gamma}_{\alpha\beta}(n) &= O(n^{(l(\alpha)+l(\gamma)-l(\beta))/2+\mathrm{row}(\alpha)+|\beta|-|\gamma|}) \\
&= O(n^{(|\alpha|+\mathrm{row}(\alpha)+|\beta|+\mathrm{row}(\beta)-|\gamma|-\mathrm{row}(\gamma))/2}),
\end{aligned} \tag{10.35}$$

in which the orders of the both sides coincide if and only if $\gamma = \alpha\cup\beta$.

Lemma 10.23.

$$\begin{aligned}
&\delta_e\left(\left(\frac{\mathbb{E}_n J_n^k}{n^{k/2}}\right)^4\right) \\
&\quad = \alpha^4_{(2^{k/2})} + \left\{4\alpha^3_{(2^{k/2})}\beta_{(2^{k/2})} + 6\sum_{s\ge 3}\alpha^2_{(2^{k/2})}\frac{\alpha^2_{(2^{(k-s)/2}s^1)}}{s-1}\right\}n^{-1} + O(n^{-2}).
\end{aligned}$$

Proof. We begin with

$$\begin{aligned}
&\delta_e((\mathbb{E}_n J_n^k)^4) \\
&\quad = \sum_{\pi\in\overline{\mathbb{Y}}_k}\sum_{\rho\in\overline{\mathbb{Y}}_k}\sum_{\sigma\in\overline{\mathbb{Y}}_k}\sum_{\tau\in\overline{\mathbb{Y}}_k} K_{\pi;n}K_{\rho;n}K_{\sigma;n}K_{\tau;n} \\
&\qquad \times \delta_e(A_{\pi^\circ\cup(1^{n-|\pi^\circ|})}A_{\rho^\circ\cup(1^{n-|\rho^\circ|})}A_{\sigma^\circ\cup(1^{n-|\sigma^\circ|})}A_{\tau^\circ\cup(1^{n-|\tau^\circ|})}) \\
&\quad = \sum_{\pi\in\overline{\mathbb{Y}}_k}\sum_{\rho\in\overline{\mathbb{Y}}_k}\sum_{\sigma\in\overline{\mathbb{Y}}_k}\sum_{\tau\in\overline{\mathbb{Y}}_k} K_{\pi;n}K_{\rho;n}K_{\sigma;n}K_{\tau;n} \\
&\qquad \times \sum_{\kappa} p^{\kappa}_{\pi^\circ\rho^\circ}(n)p^{\kappa}_{\sigma^\circ\tau^\circ}(n)|C_\kappa \cup (1^{n-|\kappa|})|.
\end{aligned} \tag{10.36}$$

By (10.35) we have

$$p^{\kappa}_{\pi^\circ\rho^\circ}(n) = O(n^{(|\pi^\circ|+\mathrm{row}(\pi^\circ)+|\rho^\circ|+\mathrm{row}(\rho^\circ)-|\kappa|-\mathrm{row}(\kappa))/2}) \tag{10.37}$$

$$p^{\kappa}_{\sigma^\circ\tau^\circ}(n) = O(n^{(|\sigma^\circ|+\mathrm{row}(\sigma^\circ)+|\tau^\circ|+\mathrm{row}(\tau^\circ)-|\kappa|-\mathrm{row}(\kappa))/2}). \tag{10.38}$$

Moreover, in (10.37) and (10.38), the orders of the both hand sides can coincide only if $\pi^\circ \cup \rho^\circ = \kappa$ and $\sigma^\circ \cup \tau^\circ = \kappa$ respectively. Inserting (10.37) and (10.38) into (10.36) and noting Lemma 10.13, we have

$$\delta_e((\mathbb{E}_n J_n^k)^4) = \sum O(n^{2k-\mathrm{row}\,(\kappa)}).$$

Hence the terms with row $(\kappa) \geq 2$ are negligible. If row $(\kappa) = 1$, i.e., κ is a one-row diagram, then the above order is at most n^{2k-1}. Hence lack of coincidence of the both sides' orders in either (10.37) or (10.38) yields the upper bound $O(n^{2k-2})$ and makes the term negligible. We can then assume that $\pi^\circ \cup \rho^\circ = \kappa$ and $\sigma^\circ \cup \tau^\circ = \kappa$ hold, namely, that $\pi^\circ = \varnothing$ or $\rho^\circ = \varnothing$ holds and $\sigma^\circ = \varnothing$ or $\tau^\circ = \varnothing$ holds. As a result, in order to obtain an estimate for $\delta_e((\mathbb{E}_n J_n^k)^4)$ up to the order of n^{-1}, we need only to consider the following two cases:

(a) $\kappa = \varnothing$;
(b) one of $\{\pi^\circ, \rho^\circ\}$ is $\varnothing$ while the other is one row, and the same also for $\{\sigma^\circ, \tau^\circ\}$.

In case of (a), $\pi = \rho$ and $\sigma = \tau$. These terms contribute to $\delta_e((\mathbb{E}_n J_n^k)^4)$ as

$$\sum_{\pi\in\overline{\mathbb{Y}}_k}\sum_{\sigma\in\overline{\mathbb{Y}}_k} K^2_{\pi;n}K^2_{\sigma;n}\delta_e(A^2_{\pi^\circ\cup(1^{n-|\pi^\circ|})})\delta_e(A^2_{\sigma^\circ\cup(1^{n-|\sigma^\circ|})})$$
$$= \sum_{\pi\in\overline{\mathbb{Y}}_k}\sum_{\sigma\in\overline{\mathbb{Y}}_k} K^2_{\pi;n}K^2_{\sigma;n}|C_{\pi^\circ\cup(1^{n-|\pi^\circ|})}||C_{\sigma^\circ\cup(1^{n-|\sigma^\circ|})}|.$$

The order of a summand is n to the power

$$2k - (m_3(\pi) + m_4(\pi) + \cdots) - (m_3(\sigma) + m_4(\sigma) + \cdots).$$

We need to consider two cases:

(a1) $\pi = \sigma = (2^{k/2})$;
(a2) one of $\{\pi, \sigma\}$ is $(2^{k/2})$ while the other is $(2^{(k-s)/2}s^1)$.

We then have

$$K^4_{(2^{k/2});n} + 2\sum_{s\geq 3} K^2_{(2^{k/2});n}K^2_{(2^{(k-s)/2}s^1);n}\frac{1}{s-1}n(n-1)\cdots(n-s+2)$$
$$= \alpha^4_{(2^{k/2})}n^{2k} + 4\alpha^3_{(2^{k/2})}\beta_{(2^{k/2})}n^{2k-1} + O(n^{2k-2})$$
$$+ 2\Big\{\sum_{s\geq 3}\alpha^2_{(2^{k/2})}\alpha^2_{(2^{(k-s)/2}s^1)}\frac{n^{2k-1}}{s-1} + O(n^{2k-2})\Big\}$$
$$= \alpha^4_{(2^{k/2})}n^{2k} + \Big\{4\alpha^3_{(2^{k/2})}\beta_{(2^{k/2})} + 2\sum_{s\geq 3}\alpha^2_{(2^{k/2})}\alpha^2_{(2^{(k-s)/2}s^1)}\frac{1}{s-1}\Big\}n^{2k-1}$$
$$+ O(n^{2k-2}).$$

In case of (b), there are four cases where $\varnothing$'s are placed. We then have

$$\begin{aligned}
&4\sum_{s\geq 3}\sum_{t\geq 3} K^2_{(2^{k/2});n} K_{(2^{(k-s)/2}s^1);n} K_{(2^{(k-t)/2}t^1);n} \\
&\qquad \times \delta_e(A_{(s-1,1^{n-(s-1)})} A_{(t-1,1^{n-(t-1)})}) \\
&\quad = 4\sum_{s\geq 3} K^2_{(2^{k/2});n} K^2_{(2^{(k-s)/2}s^1);n} |C_{(s-1,1^{n-(s-1)})}| \\
&\quad = 4\sum_{s\geq 3} \alpha^2_{(2^{k/2})} n^k \alpha^2_{(2^{(k-s)/2}s^1)} n^{k-s} \frac{1}{s-1} n^{s-1}(1+O(n^{-1})) \\
&\quad = 4\Big\{\sum_{s\geq 3} \alpha^2_{(2^{k/2})} \alpha^2_{(2^{(k-s)/2}s^1)} \frac{1}{s-1}\Big\} n^{2k-1} + O(n^{2k-2}).
\end{aligned}$$

Combining cases (a) and (b), we get

$$\begin{aligned}
&\delta_e((\mathbb{E}_n J_n^k)^4) \\
&= \alpha^4_{(2^{k/2})} n^{2k} + \Big\{4\alpha^3_{(2^{k/2})}\beta_{(2^{k/2})} + 6\sum_{s\geq 3} \alpha^2_{(2^{k/2})} \alpha^2_{(2^{(k-s)/2}s^1)} \frac{1}{s-1}\Big\} n^{2k-1} \\
&\quad + O(n^{2k-2}).
\end{aligned}$$

This completes the proof. □

Proof of Proposition 10.19. The right-hand side of (10.27) being estimated by Lemmas 10.20–10.23, we come to the result easily. □

Proposition 10.19 is a key step to the proof of the strong law of large numbers for Young diagrams (Theorem 10.5).

10.7 The Limit Shape as a Strong Law of Large Numbers

Theorem 10.24 (Strong LLN for Young diagrams). *For $\mathfrak{P}$-almost all paths $t \in \mathfrak{T}$,*

$$\lim_{n\to\infty} |M_k(\mathfrak{m}_{t(n)\sqrt{n}}) - M_k(\mathfrak{m}_\Omega)| = 0, \qquad k = 1, 2, \ldots. \tag{10.39}$$

Proof. Recall the decomposition of the regular character

$$\delta_e = \sum_{\lambda\in\mathbb{Y}_n} \mathfrak{P}_n(\lambda)\tilde{\chi}^\lambda.$$

Then, it follows by Proposition 10.19 that

$$\begin{aligned}
O(n^{-2}) &= \sum_{\lambda\in\mathbb{Y}_n} \mathfrak{P}_n(\lambda)\tilde{\chi}^\lambda\Big(\Big(\frac{\mathbb{E}_n J_n^k}{n^{k/2}} - M_k(\mathfrak{m}_\Omega)\Big)^4\Big) \\
&= \sum_{\lambda\in\mathbb{Y}_n} \mathfrak{P}_n(\lambda)\big\{\tilde{\chi}^\lambda(\mathbb{E}_n J_n^k)n^{-k/2} - M_k(\mathfrak{m}_\Omega)\big\}^4 \\
&= \sum_{\lambda\in\mathbb{Y}_n} \mathfrak{P}_n(\lambda)\big\{M_k(\mathfrak{m}_\lambda)n^{-k/2} - M_k(\mathfrak{m}_\Omega)\big\}^4 \\
&= \sum_{\lambda\in\mathbb{Y}_n} \mathfrak{P}_n(\lambda)\big(M_k(\mathfrak{m}_{\lambda^{\sqrt{n}}}) - M_k(\mathfrak{m}_\Omega)\big)^4 \\
&= \int_{\mathfrak{T}} \big(M_k(\mathfrak{m}_{t(n)^{\sqrt{n}}}) - M_k(\mathfrak{m}_\Omega)\big)^4 \mathfrak{P}(dt) \\
&\geq \epsilon^4\, \mathfrak{P}\big(\{t\in\mathfrak{T};\, |M_k(\mathfrak{m}_{t(n)^{\sqrt{n}}}) - M_k(\mathfrak{m}_\Omega)| \geq \epsilon\}\big)
\end{aligned}$$

for any $\epsilon > 0$. Hence we get

$$\sum_{n=1}^{\infty} \mathfrak{P}\big(\{t\in\mathfrak{T};\, |M_k(\mathfrak{m}_{t(n)^{\sqrt{n}}}) - M_k(\mathfrak{m}_\Omega)| \geq \epsilon\}\big) < \infty.$$

Then, applying the Borel–Cantelli lemma, we obtain

$$\mathfrak{P}\big(\{t\in\mathfrak{T};\, |M_k(\mathfrak{m}_{t(n)^{\sqrt{n}}}) - M_k(\mathfrak{m}_\Omega)| \geq \epsilon \text{ for infinitely many } n\}\big) = 0.$$

Let $\epsilon_1 > \epsilon_2 > \cdots \to 0$ and set

$$\mathfrak{T}_0 = \bigcap_{k=1}^{\infty}\bigcap_{j=1}^{\infty} \{t\in\mathfrak{T};\, |M_k(\mathfrak{m}_{t(n)^{\sqrt{n}}}) - M_k(\mathfrak{m}_\Omega)| < \epsilon_j \text{ with finite exception}\}.$$

Then we have $\mathfrak{P}(\mathfrak{T}_0) = 1$. Since (10.39) holds for all $t = (t(n)) \in \mathfrak{T}_0$, the assertion follows. □

Remark 10.25. Theorem 10.24 is a strong law of large numbers for random Young diagrams taken from the Plancherel ensemble with respect to the moment topology on $\mathbb{D}$, while Theorem 10.5 is that with respect to the uniform topology. As is noted in Lemma 10.7, both topologies coincide when they are restricted on $\mathbb{D}_K$ for some constant $K > 0$. We can therefore replace the topology in Theorem 10.24 by the one in Theorem 10.5 if it is possible for us to reduce arguments onto compact interval $[-K, K]$. The following properties suffice for this reduction:

$$\lim_{n\to\infty} \mathfrak{P}_n\big(\{\lambda\in\mathbb{Y}_n \,|\, \mathrm{row}(\lambda) > c_1\sqrt{n} \text{ or } \mathrm{col}(\lambda) > c_1\sqrt{n}\}\big) = 0, \tag{10.40}$$

$$\sum_{n=0}^{\infty} \mathfrak{P}_n\big(\{\lambda\in\mathbb{Y}_n \,|\, \mathrm{row}(\lambda) > c_2\sqrt{n} \text{ or } \mathrm{col}(\lambda) > c_2\sqrt{n}\}\big) < \infty, \tag{10.41}$$

where c_1, c_2 are some positive constants. It is easy that (10.40) and (10.41) along with Theorem 10.24 imply weak and strong laws of large numbers with respect to the uniform topology respectively. Indeed, we use again the Borel–Cantelli lemma to incorporate (10.41) with Theorem 10.24. Since some efforts in another direction are needed to verify (10.40) and (10.41), we shall not go into details of their proofs. See Notes section of this chapter for bibliographical information.

Exercises

10.1. Deduce the relation (10.6) between the transition measure and the Rayleigh measure.

10.2. Compute the transition measure and the Rayleigh measure of a triangular diagram in Example 10.3.

10.3. For a finite group G the direct product $G \times G$ acts on G from the left transitively by $(g_1, g_2)x = g_1 x g_2^{-1}$. This yields the action on $G \times G$ as

$$(g_1, g_2)(x_1, x_2) = (g_1 x_1 g_2^{-1}, g_1 x_2 g_2^{-1})$$

and decomposition of $G \times G$ into the orbits. Verify a bijective correspondence between the $G \times G$-orbits and the conjugacy classes of G, showing that (x_1, x_2) and (y_1, y_2) belong to the same orbit if and only if $x_1 x_2^{-1}$ and $y_1 y_2^{-1}$ are conjugate.

10.4. In Exercise 10.3, let $G = S(n)$. Show that the orbits $\{\mathcal{O}_C\}$ (C runs over the conjugacy classes of $S(n)$) have structure of an association scheme of Bose–Mesner type (see Sect. 3.1).

10.5. Consider the Cayley graph $(S(N), \mathcal{T})$ where $\mathcal{T}$ is the set of all transpositions. Recall the quantities $\omega_-(x)$, $x \in S(N)$ (Sect. 7.2), $l(x)$, $x \in S(N)$ (Sect. 9.1), $n(\lambda)$, $\lambda \in \mathbb{Y}$ ((9.5)), and type(g), $g \in S(N)$ (Sect. 10.3). Show that

(1) $\omega_-(x) = n(\mathrm{type}(x)')$ holds for $x \in S(N)$;
(2) if $k < N$, $\omega_-(x) \le k(k+1)/2$ holds for $x \in S(N)$ such that $l(x) = k$; moreover the equality is attained if and only if x is an $(N+1)$-cycle.

Notes

Our approach to the limit shape of Young diagrams, which is a so-called *moment method*, is inspired by Biane [25, 26] and Ivanov–Olshanski [117]. It has more algebraic or combinatorial flavour, compared with the original methods due to Vershik–Kerov [211] and Logan–Shepp [148]. Although the estimate we developed in this chapter is rather coarse, it is sufficient to show the strong law of large numbers with respect to the moment topology. On the other hand, it has an advantage of being widely applicable to the cases of irreducible decomposition of other representations of the symmetric groups besides the regular ones.

For the extension of the notions of transition measures and Rayleigh measures to continuous diagrams, see Kerov [123]. Continuous diagrams have other applications besides representation theory, e.g., the limiting configuration of zeros of orthogonal polynomials. See Kerov [126] for details on this topic.

The limit shape problem has an aspect of a variational problem for the integral viewed as a continuous hook formula. For history of this problem, see Vershik–Kerov [211], Logan–Shepp [148] and also Kerov [126]. In these works, they deal with a metric on continuous diagrams naturally introduced by a continuous hook formula, which is closely related to the logarithmic potential, and discuss a large deviation principle with respect to this metric. Comparing the metric with the uniform norm, one reaches the statement of Theorem 10.5. This is the original way to the limit shape. We recognize that a similar procedure appears also in deriving the semicircle law as the maximizer of Voiculescu's free entropy under a given variance since the semicircle law is the transition measure of the limit shape. Note also that Theorem 10.5 gives a lower estimate for the longest row and column of typical Young diagrams.

Upper estimate for the longest row and column was given in Vershik–Kerov [211], see also Kerov [126]. The estimate of (10.40) was already due to Hammersley [90]. Ivanov–Olshanski [117] used this property to deduce the weak law of large numbers for the Plancherel measure. Equation (10.41) requires finer estimate, which is seen in Borodin–Okounkov–Olshanski [33]. Statistical properties, especially fluctuation, of the longest rows and columns of random Young diagrams taken from the Plancherel ensemble have been deeply studied in connection with the similar problem for the largest eigenvalues of random matrices in the Gaussian unitary ensemble. Here we refer to Baik–Deift–Johansson [16], Okounkov [172] and Borodin–Okounkov–Olshanski [33] for reader's convenience of searching these rich fields.

The result in Proposition 10.15 was first pointed out by Biane [23] and opened a door to a field in which free probability and asymptotic representation theory of symmetric groups enjoy rich interplay.

Several examples of the concentration in irreducible decomposition of other representations (than the regular one) of the symmetric group are given by Biane [25]. Similar concentration for the regular representations of a Weyl group is discussed in Hora [109].

11

Central Limit Theorem for the Plancherel Measures of the Symmetric Groups

Having established in the previous chapter the concentration phenomenon at the limit shape for Young diagrams with respect to the Plancherel measure, it is now natural to ask what fluctuation is observed in a small neighbourhood of the limit shape. There have been substantial studies along this line for the Plancherel measure. The present chapter tackles such a central limit problem from the viewpoint of quantum probability theory.

11.1 Kerov's Central Limit Theorem and Fluctuation of Young Diagrams

As was shown in Theorems 10.5, 10.18 and 10.24, the scaled diagram $\lambda^{\sqrt{n}}$ of a Young diagram $\lambda \in \mathbb{Y}_n$ converges to the limit shape Ω namely,

$$\lambda^{\sqrt{n}}(x) - \Omega(x) \longrightarrow 0 \quad \text{as } n \to \infty,$$

with respect to the Plancherel measure. What then can be said about the deviation $\lambda^{\sqrt{n}}(x) - \Omega(x)$ along the vertical direction? Let us begin with heuristic observation. We readily know from (9.13) and (10.5) that

$$\int_{-\infty}^{+\infty} x^k (\lambda^{\sqrt{n}}(x) - \Omega(x))\, dx = \frac{2}{(k+1)(k+2)} \{M_{k+2}(\tau_{\lambda^{\sqrt{n}}}) - M_{k+2}(\tau_\Omega)\}, \quad k = 1, 2, \dots. \tag{11.1}$$

Upon analysis of the right-hand side we prepare the terminology. Recall that $\mathbb{Y}$ denotes the set of all Young diagrams.

Definition 11.1 (Kerov–Olshanski). A function defined on $\mathbb{Y}$ is called a *polynomial function* if it is expressed as a polynomial of coordinates of Young diagrams. The algebra of polynomial functions is denoted by $\mathbb{A}$.

A. Hora and N. Obata: Central Limit Theorem for the Plancherel Measures of the Symmetric Groups. In: A. Hora and N. Obata, Quantum Probability and Spectral Analysis of Graphs, Theoretical and Mathematical Physics, 297–320 (2007)
DOI 10.1007/3-540-48863-4_11

For example, the kth moment of the Rayleigh measure yields a polynomial function. In fact, letting $x_1 < y_1 < \cdots < x_{r-1} < y_{r-1} < x_r$ be the min–max coordinates of $\lambda \in \mathbb{Y}$, we have

$$M_k(\tau_\lambda) = \sum_i x_i^k - \sum_i y_i^k.$$

It is shown that $\mathbb{A}$ is generated by $\{M_k(\tau_\lambda)\,;\,1,2,\ldots\}$. Since there is a polynomial relation between $\{M_k(\mathfrak{m}_\lambda)\}$ and $\{M_k(\tau_\lambda)\}$ by Proposition 9.21, $\lambda \mapsto M_k(\mathfrak{m}_\lambda)$ is a polynomial function too. Some properties of $\mathbb{A}$ will be summarized in Sect. 11.5.

We now anticipate the following fact. For $k = 1,2,\ldots$ let $\tilde{\chi}^\lambda_{(k,1^{n-k})}$ be the normalized irreducible character of $S(n)$ corresponding to the cycle $(k, 1^{n-k})$. We define

$$\Sigma_k(\lambda) = \begin{cases} n^{\downarrow k}\, \tilde{\chi}^\lambda_{(k,1^{n-k})}\,, & |\lambda| = n \geq k, \\ 0, & |\lambda| = n < k, \end{cases} \tag{11.2}$$

where $n^{\downarrow k} = n(n-1)\cdots(n-k+1)$. Then we have

$$\Sigma_k(\lambda) = M_{k+1}(\mathfrak{m}_\lambda) + (\text{lower terms}), \tag{11.3}$$

where the lower terms are understood along an appropriate filtration in $\mathbb{A}$ (see Sect. 11.5 for details).

We consider a family of (11.1) indexed by $k = 1,2,\ldots$. By virtue of (9.18) (and hence the polynomial relations), the right-hand sides of (11.1) are equivalently replaced by

$$n^{-(k+2)/2} M_{k+2}(\mathfrak{m}_\lambda) - M_{k+2}(\mathfrak{m}_\Omega)$$

(up to constants independent of n) since

$$M_{k+2}(\mathfrak{m}_{\lambda^{\sqrt{n}}}) = n^{-(k+2)/2} M_{k+2}(\mathfrak{m}_\lambda)$$

holds. Note that

$$n^{-(k+2)/2} \mathbb{E}^{\mathfrak{P}_n}[M_{k+2}(\mathfrak{m}_\lambda)] \sim M_{k+2}(\mathfrak{m}_\Omega) \quad \text{as } n \to \infty,$$

where $\sim$ means that the ratio of the both sides tends to one. Here $\mathbb{E}^{\mathfrak{P}_n}$ denotes the expectation in the Plancherel measure $\mathfrak{P}_n$. We are hence considering the random variable

$$n^{-(k+2)/2} M_{k+2}(\mathfrak{m}_\lambda) - \mathbb{E}^{\mathfrak{P}_n}[n^{-(k+2)/2} M_{k+2}(\mathfrak{m}_\lambda)]$$

on $(\mathbb{Y}_n, \mathfrak{P}_n)$. Taking (11.3) into account and noting $\mathbb{E}^{\mathfrak{P}_n}[\Sigma_{k+1}(\lambda)] = 0$, we are led to treat the random variable

$$n^{-(k+2)/2} \Sigma_{k+1}(\lambda)$$

on $(\mathbb{Y}_n, \mathfrak{P}_n)$ to describe the deviation $\lambda^{\sqrt{n}}(x) - \Omega(x)$. Rescaled by $\sqrt{n}$ multiple, this agrees with Kerov's central limit theorem for irreducible characters of $S(n)$.

Theorem 11.2 (Kerov's CLT). *Let Σ_k be the random variable on $(\mathbb{Y}_n, \mathfrak{P}_n)$ defined by* (11.2). *Then, for $m \geq 2$ and $x_2, \ldots, x_m \in \mathbb{R}$ we have*

$$\lim_{n\to\infty} \mathfrak{P}_n \left(\left\{ \lambda \in \mathbb{Y}_n \,;\, \begin{matrix} n^{-k/2}\Sigma_k(\lambda) \leq x_k \\ k = 2, 3, \ldots, m \end{matrix} \right\} \right) = \prod_{k=2}^{m} \frac{1}{\sqrt{2\pi k}} \int_{-\infty}^{x_k} \mathrm{e}^{-x^2/2k} dx.$$

In other words, $\Sigma_2, \Sigma_3, \ldots$ are asymptotically independent and Gaussian with respect to the Plancherel measure.

Thus Kerov's CLT zooms in the deviation $\lambda^{\sqrt{n}}(x) - \Omega(x)$ through magnification by $\sqrt{n}$ (just in the diagonal direction of diagrams). The above result is so fundamental in the study of fluctuation of the Young diagrams or equivalently, of the irreducible representations of the symmetric groups that various refinements and extensions have been discussed. We propose a non-commutative extension as one of the promising research directions. Again, the idea of quantum decomposition is crucial. We shall construct an analogous object to an interacting Fock probability space and prove the quantum central limit theorem for adjacency matrices (Theorem 11.13). Then Kerov's central limit theorem follows as a classical reduction so that our approach yields an alternative proof of it.

This chapter deals with the Plancherel measure of the symmetric group. Further development beyond the Plancherel measure will be discussed in the next chapter.

11.2 Use of Quantum Decomposition

We start with some general remarks. Let G be a finite group and $\hat{G}$ the set of equivalence classes of its irreducible representations. The *Plancherel measure* of G is a probability measure on $\hat{G}$ defined by

$$\mathfrak{P}_G(\alpha) = \frac{\dim^2 \alpha}{|G|}, \qquad \alpha \in \hat{G}.$$

For $\alpha \in \hat{G}$ let χ^α be its irreducible character. For each conjugacy class C of G, keeping in mind that χ^α is constant on each conjugacy class, we set

$$\chi^\alpha_C = \chi^\alpha(g), \qquad g \in C.$$

Let A_C denote the adjacency matrix associated with a conjugacy class C, namely, the operator on $\ell^2(G)$ defined by

$$(A_C f)(g) = \sum_{h \in C} f(h^{-1}g), \qquad f \in \ell^2(G).$$

Lemma 11.3. *The adjacency matrix A_C is a self-adjoint operator on $\ell^2(G)$ and its spectral decomposition is given by*

$$A_C = \sum_{\alpha\in\hat{G}} \frac{|C|\chi^\alpha_C}{\dim\alpha} E_\alpha, \tag{11.4}$$

where $\{E_\alpha\,;\,\alpha\in\hat{G}\}$ is a complete system of orthogonal projectors on $\ell^2(G)$.

Lemma 11.4. *For arbitrary conjugacy classes $C_1,\dots,C_p$ of G,*

$$\begin{aligned}\langle\delta_e, A_{C_1}\cdots A_{C_p}\delta_e\rangle &= \frac{1}{|G|}\operatorname{tr}(A_{C_1}\cdots A_{C_p})\\ &= \sum_{\alpha\in\hat{G}} \frac{|C_1|\chi^\alpha_{C_1}}{\dim\alpha}\cdots\frac{|C_p|\chi^\alpha_{C_p}}{\dim\alpha}\mathfrak{P}_G(\alpha).\end{aligned}$$

Proof. Note that $\operatorname{tr} E_\alpha = \dim^2\alpha$ and apply Lemma 11.3. □

Let us go back to the case of $G = S(n)$ and consider the adjacency matrix $A_{(k,1^{n-k})}$, where $k = 2,3,\dots,n$. Since

$$\langle\delta_e, A_{(k,1^{n-k})}\delta_e\rangle = 0, \qquad \langle\delta_e, (A_{(k,1^{n-k})})^2\delta_e\rangle = |C_{(k,1^{n-k})}|, \tag{11.5}$$

the normalization is given by

$$\frac{A_{(k,1^{n-k})}}{\sqrt{|C_{(k,1^{n-k})}|}}, \qquad k = 2,3,\dots,n.$$

For these normalized adjacency matrices, we have the following significant result.

Theorem 11.5 (CLT for the adjacency matrices I). *Let $m = 2,3,\dots$. Then for any $p_2,\dots,p_m\in\{0,1,2,\dots\}$ we have*

$$\begin{aligned}&\lim_{n\to\infty}\left\langle\delta_e, \left(\frac{A_{(2,1^{n-2})}}{\sqrt{|C_{(2,1^{n-2})}|}}\right)^{p_2}\cdots\left(\frac{A_{(m,1^{n-m})}}{\sqrt{|C_{(m,1^{n-m})}|}}\right)^{p_m}\delta_e\right\rangle\\ &= \prod_{k=2}^{m}\frac{1}{\sqrt{2\pi}}\int_{-\infty}^{\infty} x^{p_k} e^{-x^2/2}dx.\end{aligned} \tag{11.6}$$

We first remark the equivalence of Theorems 11.2 and 11.5. In fact, Lemma 11.4 clarifies the connection of joint vacuum spectral distributions of $A_{(k,1^{n-k})}$'s with joint distributions of Σ_k's in the Plancherel measure $\mathfrak{P}_n$. Since the conjugacy class consisting of k-cycles has cardinality

$$|C_{(k,1^{n-k})}| = \frac{n^{\downarrow k}}{k},$$

the assertion of Theorem 11.2 is equivalent to (11.6), both showing that the joint distributions converge to Gaussian ones.

The proof of Theorem 11.5 will be deferred in the next section (directly follows from Theorem 11.12). In fact, we shall first prove the quantum central limit theorem and then obtain Theorem 11.5 as a classical reduction. Our strategy is as follows:

Step 1. Using the length function l on the Cayley graph $(S(n), \mathcal{T})$, we decompose each adjacency matrix into a sum of quantum components:

$$A_{(k,1^{n-k})} = A^+_{(k,1^{n-k})} + A^-_{(k,1^{n-k})} + A^\circ_{(k,1^{n-k})}.$$

Step 2. We introduce an analogue of the Fock space with an orthonormal basis labelled by the Young diagrams in the modified Young graph $\overline{\mathbb{Y}}$ together with creation operator B_k^+ and annihilation operator B_k^-.

Step 3. We show that any matrix element of rescaled $A^+_{(k,1^{n-k})}$'s and $A^-_{(k,1^{n-k})}$'s converges to the same type of matrix element of B_k^+'s and B_k^-'s while equally rescaled $A^\circ_{(k,1^{n-k})}$ vanishes in the limit.

Step 4. We prove that the $*$-algebras generated by $\{B_k^+, B_k^-\}$ are commutative independent with respect to the vacuum state and that the spectral distribution of $B_k^+ + B_k^-$ is Gaussian.

Remark 11.6. The limit operators $B_k^\pm$ are easier to handle than $A^\pm_{(k,1^{n-k})}$ from a combinatorial viewpoint and the spectral structure of $B_k^\pm$ is simpler. In fact, our argument (Sect. 11.3) does not require representation theory of the symmetric group nor theory of symmetric functions. This is an advantage of our approach and supports our principle: taking limits after decomposing operators make the involved combinatorial argument much more transparent.

Remark 11.7. Theorem 11.5 is devoted to the mixed moments of the adjacency matrices of a particular type. We shall deal with a general case in Theorem 11.17.

11.3 Quantum Central Limit Theorem for Adjacency Matrices

For the moment, fixing n, we consider the symmetric group $S(n)$. For $x \in S(n)$ recall that $c(x)$ denotes the number of cycles of x, where cycles of length 1 are counted together. For each $s \in S(n)$ and $\epsilon \in \{+, -, \circ\}$ we define an operator s^ϵ acting on $\ell^2(S(n))$ by

$$s^+\delta_x = \begin{cases} \delta_{sx}, & \text{if } c(sx) < c(x), \\ 0, & \text{otherwise}, \end{cases} \qquad s^-\delta_x = \begin{cases} \delta_{sx}, & \text{if } c(sx) > c(x), \\ 0, & \text{otherwise}, \end{cases}$$

$$s^\circ\delta_x = \begin{cases} \delta_{sx}, & \text{if } c(sx) = c(x), \\ 0, & \text{otherwise}. \end{cases}$$

Clearly,

$$s = s^+ + s^- + s^\circ, \quad (s^+)^* = s^-, \quad (s^\circ)^* = s^\circ,$$

where s is regarded as an operator acting on $\ell^2(S(n))$ through the left regular representation. Let $C \subset S(n)$ be a conjugacy class. Then the corresponding adjacency matrix A_C admits the decomposition

$$A_C = A_C^+ + A_C^- + A_C^\circ,$$

where

$$A_C^\epsilon = \sum_{s \in C} s^\epsilon, \qquad \epsilon \in \{+, -, \circ\}. \tag{11.7}$$

Obviously, A_C^+ and A_C^- are mutually adjoint and A_C° is self-adjoint.

Next we construct an orthonormal set in $\ell^2(S(n))$. Let $\rho \in \overline{\mathbb{Y}}$ satisfying $|\rho| \le n$. Then $\rho \cup (1^{n-|\rho|}) \in \mathbb{Y}_n$ determines a conjugacy class $C_{\rho \cup (1^{n-|\rho|})} \subset S(n)$. We set

$$\xi_{\rho \cup (1^{n-|\rho|})} = \sum_{x \in C_{\rho \cup (1^{n-|\rho|})}} \delta_x, \quad \Phi(\rho \cup (1^{n-|\rho|})) = \frac{\xi_{\rho \cup (1^{n-|\rho|})}}{\sqrt{|C_{\rho \cup (1^{n-|\rho|})}|}}. \tag{11.8}$$

By definition

$$\Phi(\varnothing \cup (1^n)) = \delta_e.$$

Then $\{\Phi(\rho \cup (1^{n-|\rho|}))\,;\, \rho \in \overline{\mathbb{Y}}, |\rho| \le n\}$ becomes an orthonormal system in $\ell^2(S(n))$, the subspace spanned by which is denoted by $\Gamma(S(n))$.

We shall consider the actions of $A^\epsilon_{(j,1^{n-j})}$ on $\Gamma(S(n))$, where $A_{(j,1^{n-j})}$ stands for the adjacency matrix assigned to a conjugacy class $C_{(j,1^{n-j})}$ for $j = 2, 3, \dots, n$. For three Young diagrams $\tau, \rho, \sigma \in \overline{\mathbb{Y}}$ satisfying $|\tau|, |\rho|, |\sigma| \le n$, the intersection number is defined by

$$p^\sigma_{\tau\rho}(n) = |\{z \in C_{\tau \cup (1^{n-|\tau|})}\,;\, z^{-1}x \in C_{\rho \cup (1^{n-|\rho|})}\}|,$$

where $x \in C_{\sigma \cup (1^{n-|\sigma|})}$ is arbitrarily chosen (see (10.15) and (10.34)). Recall the length function $l(x)$ on the Cayley graph of the symmetric group and also $l(\rho)$ on $\overline{\mathbb{Y}}$ defined in (10.16). The latter length function yields a stratification on $\overline{\mathbb{Y}}$ as shown in Fig. 10.3.

Lemma 11.8. *Let $\rho \in \overline{\mathbb{Y}}$ with $|\rho| \le n$ and $j \in \{2, \dots, n\}$. Then we have*

$$A^\pm_{(j,1^{n-j})} \xi_{\rho \cup (1^{n-|\rho|})} = \sum_{i=1}^{j-1} \sum_{\substack{|\sigma| \le n \\ l(\sigma) = l(\rho) \pm i}} p^\sigma_{(j)\rho}(n)\, \xi_{\sigma \cup (1^{n-|\sigma|})},$$

$$A^\circ_{(j,1^{n-j})} \xi_{\rho \cup (1^{n-|\rho|})} = \sum_{\substack{|\sigma| \le n \\ l(\sigma) = l(\rho)}} p^\sigma_{(j)\rho}(n)\, \xi_{\sigma \cup (1^{n-|\sigma|})}.$$

In particular, $\Gamma(S(n))$ is invariant under the actions of $A^\epsilon_{(j,1^{n-j})}$.

Proof. It follows from (11.7) and (11.8) that

$$
\begin{aligned}
A^{+}_{(j,1^{n-j})}\xi_{\rho\cup(1^{n-|\rho|})} &= \sum_{x\in C_{\rho\cup(1^{n-|\rho|})}} \sum_{s\in C_{(j,1^{n-j})}} s^{+}\delta_x \\
&= \sum_{x\in C_{\rho\cup(1^{n-|\rho|})}} \sum_{\substack{s\in C_{(j,1^{n-j})}\\ c(sx)<c(x)}} \delta_{sx} \\
&= \sum_{y\in S(n)} \left|\left\{(s,x)\in C_{(j,1^{n-j})}\times C_{\rho\cup(1^{n-|\rho|})}\,;\, \begin{matrix} c(sx)<c(x)\\ y=sx\end{matrix}\right\}\right| \delta_y \\
&= \sum_{i=1}^{\infty}\sum_{|\sigma|\le n}\sum_{y\in C_{\sigma\cup(1^{n-|\sigma|})}} \\
&\quad\times \left|\left\{(s,x)\in C_{(j,1^{n-j})}\times C_{\rho\cup(1^{n-|\rho|})}\,;\, \begin{matrix} l(\sigma)=l(\rho)+i\\ y=sx\end{matrix}\right\}\right| \delta_y \\
&= \sum_{i=1}^{j-1} \sum_{\substack{|\sigma|\le n\\ l(\sigma)=l(\rho)+i}} p^{\sigma}_{(j)\rho}(n)\,\xi_{\sigma\cup(1^{n-|\sigma|})}\,.
\end{aligned}
$$

The rest is proved similarly. In the case of $\epsilon=-$, the sum over i is actually taken over $\{1,\dots,\min\{j-1,l(\rho)\}\}$. □

We are interested in the normalized adjacency matrices:

$$
\frac{A_{(j,1^{n-j})}}{\sqrt{|C_{(j,1^{n-j})}|}}, \qquad j=2,3,\dots,n. \tag{11.9}
$$

In view of (11.8) and Lemma 11.8 we obtain the following:

Lemma 11.9. *Let $\rho\in\overline{\mathbb{Y}}$ with $|\rho|\le n$ and $j\in\{2,\dots,n\}$. Then we have*

$$
\begin{aligned}
&\frac{A^{\pm}_{(j,1^{n-j})}}{\sqrt{|C_{(j,1^{n-j})}|}}\,\Phi(\rho\cup(1^{n-|\rho|})) \\
&\quad= \sum_{i=1}^{j-1} \sum_{\substack{|\sigma|\le n\\ l(\sigma)=l(\rho)\pm i}} p^{\sigma}_{(j)\rho}(n)\sqrt{\frac{|C_{\sigma\cup(1^{n-|\sigma|})}|}{|C_{(j,1^{n-j})}|\,|C_{\rho\cup(1^{n-|\rho|})}|}}\,\Phi(\sigma\cup(1^{n-|\sigma|})) \qquad (11.10)\\
&\frac{A^{\circ}_{(j,1^{n-j})}}{\sqrt{|C_{(j,1^{n-j})}|}}\,\Phi(\rho\cup(1^{n-|\rho|})) \\
&\quad= \sum_{\substack{|\sigma|\le n\\ l(\sigma)=l(\rho)}} p^{\sigma}_{(j)\rho}(n)\sqrt{\frac{|C_{\sigma\cup(1^{n-|\sigma|})}|}{|C_{(j,1^{n-j})}|\,|C_{\rho\cup(1^{n-|\rho|})}|}}\,\Phi(\sigma\cup(1^{n-|\sigma|})). \qquad (11.11)
\end{aligned}
$$

The asymptotic behaviour of the normalized adjacency operators (11.9) is described by an analogue of the Fock space with creation and annihilation operators. Consider the Hilbert space $\ell^2(\overline{\mathbb{Y}})$ and its canonical orthonormal basis $\{\Psi(\rho)\,;\,\rho\in\overline{\mathbb{Y}}\}$. Let $\Gamma\subset\ell^2(\overline{\mathbb{Y}})$ be a dense subspace spanned by $\{\Psi(\rho)\,;\,\rho\in\overline{\mathbb{Y}}\}$ and for each $j=0,1,2,\dots$ define Γ_j to be a subspace spanned by $\{\Psi(\rho)\,;\,l(\rho)=j\}$, where $l(\rho)$ is the length function on $\overline{\mathbb{Y}}$. Then we obtain a grading structure

$$\Gamma=\bigoplus_{j=0}^{\infty}\Gamma_j.$$

By definition Γ_0 is one-dimensional space spanned by $\Psi(\varnothing)$, which is called the vacuum vector. With each $j=2,3,\dots$, we associate operators $B_j^{\pm}$ defined by the action on $\Psi(\rho)$ as follows:

$$B_j^+\Psi(\rho)=\sqrt{m_j(\rho)+1}\,\Psi(\rho\cup(j)), \tag{11.12}$$

where $\rho\cup(j)\in\overline{\mathbb{Y}}$ is the diagram obtained by adding a j-row to ρ, and $m_j(\rho)$ the number of j-rows of ρ. Similarly,

$$B_j^-\Psi(\rho)=\begin{cases}\sqrt{m_j(\rho)}\,\Psi(\rho\setminus(j)), & \text{if } m_j(\rho)\ge 1,\\ 0, & \text{otherwise,}\end{cases} \tag{11.13}$$

where $\rho\setminus(j)\in\overline{\mathbb{Y}}$ is the diagram obtained by removing a j-row from ρ. It is easy to see that $B_j^{\pm}$ are mutually adjoint. We call

$$\Gamma_{\overline{\mathbb{Y}}}=(\Gamma,\{\Gamma_j\},\{B_j^+\},\{B_j^-\}) \tag{11.14}$$

the Fock space associated with the modified Young graph. B_j^+ and B_j^- are called creation and annihilation operators, respectively, as usual. In fact, we have

$$\begin{aligned}&[B_i^-,B_j^-]=[B_i^+,B_j^+]=[B_i^-,B_j^+]=0\quad\text{for } i\ne j,\\ &[B_j^-,B_j^+]=1,\end{aligned}$$

which coincide with the canonical commutation relation satisfied by the creation and annihilation operators on the Boson Fock space.

Let us recall the Boson Fock space over Hilbert space $\mathcal{H}$. The symmetric tensor product is defined by

$$v_1\hat{\otimes}\cdots\hat{\otimes}v_n=\frac{1}{n!}\sum_{\sigma\in S(n)}v_{\sigma(1)}\otimes\cdots\otimes v_{\sigma(n)},\qquad v_1,\dots,v_n\in\mathcal{H}.$$

Let $\mathcal{H}^{\hat{\otimes}n}$ denote the closed linear span of $\{v_1\hat{\otimes}\cdots\hat{\otimes}v_n\,;\,v_1,\dots,v_n\in\mathcal{H}\}$ and set

$$\Gamma_{\mathrm{Boson}}(\mathcal{H})=\bigoplus_{n=0}^{\infty}\mathcal{H}^{\hat{\otimes}n},$$

where $\mathcal{H}^{\hat{\otimes}0} = \mathbb{C}$ by definition. The inner product of $\mathcal{H}^{\hat{\otimes}n}$ is inherited from that of $\mathcal{H}^{\otimes n}$ as follows:

$$\langle v_1 \otimes \cdots \otimes v_n, v_1 \otimes \cdots \otimes v_n \rangle = \langle v_1, v_1 \rangle \cdots \langle v_n, v_n \rangle, \quad v_1, \ldots, v_n \in \mathcal{H}.$$

Then, the inner product of $\Gamma_{\mathrm{Boson}}(\mathcal{H})$ is defined as

$$\langle (f_n), (f_n) \rangle = \sum_{n=0}^{\infty} \langle f_n, f_n \rangle, \quad f_n \in \mathcal{H}^{\hat{\otimes}n}.$$

The creation operator $A^+(u)$ and the annihilation operator $A^-(u)$ for $u \in \mathcal{H}$ acting on $\Gamma_{\mathrm{Boson}}(\mathcal{H})$ are determined respectively by

$$A^+(u) \underbrace{v\hat{\otimes} \cdots \hat{\otimes} v}_{n} = \sqrt{n+1}\, u \hat{\otimes} \underbrace{v\hat{\otimes} \cdots \hat{\otimes} v}_{n},$$
$$A^-(u) \underbrace{v\hat{\otimes} \cdots \hat{\otimes} v}_{n} = \sqrt{n}\, \langle u, v \rangle \underbrace{v\hat{\otimes} \cdots \hat{\otimes} v}_{n-1},$$

for $v \in \mathcal{H}$. Note that $A^+(u)$ and $A^-(u)$ are mutually adjoint. Now take an orthonormal basis $\{v_2, v_3, \cdots\}$ of $\mathcal{H}$ and set

$$A_j^+ = A^+(v_j), \quad A_j^- = A^-(v_j), \qquad j = 2, 3, \ldots.$$

Then it is straightforward to see that the unitary operator from Γ to $\Gamma_{\mathrm{Boson}}(\mathcal{H})$ defined by

$$\Psi(\rho) \longmapsto \sqrt{\frac{|\rho|\,!}{\prod_{j \geq 2} m_j(\rho)!}} \underbrace{v_2 \hat{\otimes} \cdots \hat{\otimes} v_2}_{m_2(\rho)} \hat{\otimes} \cdots \hat{\otimes} \underbrace{v_j \hat{\otimes} \cdots \hat{\otimes} v_j}_{m_j(\rho)} \hat{\otimes} \cdots$$

intertwines the actions of $B_j^{\pm}$ and $A_j^{\pm}$ for $j = 2, 3, \ldots$.

Proposition 11.10. *The Fock space $\Gamma_{\overline{\mathbb{Y}}}$ defined in (11.14) is isomorphic to the Boson Fock space over an infinite-dimensional separable Hilbert space.*

We note the following:

Theorem 11.11. *For $m \in \mathbb{N}$, non-commutative polynomials $Q_2, \ldots, Q_m$ in two variables, and $\tau, \rho \in \overline{\mathbb{Y}}$ having no j-rows for j such that Q_j is non-trivial (i.e. not a constant), we have*

$$\begin{aligned} &\langle \Psi(\tau), Q_2(B_2^+, B_2^-) \cdots Q_m(B_m^+, B_m^-)\, \Psi(\rho) \rangle_\Gamma \\ &\quad = \langle \Psi(\tau), Q_2(B_2^+, B_2^-) \Psi(\rho) \rangle_\Gamma \cdots \langle \Psi(\tau), Q_m(B_m^+, B_m^-) \Psi(\rho) \rangle_\Gamma . \end{aligned} \tag{11.15}$$

Proof. By linearity it is sufficient to prove (11.15) for monomials $Q_2, \ldots, Q_m$. Using the commutation relation, we rewrite a part of non-normal order (if it exists) as $B_j^- B_j^+ = B_j^+ B_j^- + 1$. Then, again by linearity, only the case where $Q_j(B_j^+, B_j^-) = (B_j^+)^{\alpha_j} (B_j^-)^{\beta_j}$ remains to be checked. By the assumption for ρ and τ we see that, for such Q_j's the both sides of (11.15) vanish unless $\alpha_j = \beta_j = 0$. This completes the proof. □

Theorem 11.12 (CLT for the adjacency matrices II). *Notations being as above, we have*

$$\lim_{n\to\infty}\left\langle \delta_e, \left(\frac{A_{(2,1^{n-2})}}{\sqrt{|C_{(2,1^{n-2})}|}}\right)^{p_2}\cdots\left(\frac{A_{(m,1^{n-m})}}{\sqrt{|C_{(m,1^{n-m})}|}}\right)^{p_m}\delta_e\right\rangle$$
$$=\prod_{j=2}^{m}\langle \Psi(\varnothing),(B_j^+ + B_j^-)^{p_j}\Psi(\varnothing)\rangle, \tag{11.16}$$

for $p_2,\dots,p_m\in\{0,1,2,\dots\}$, $m\geq 2$.

Before going into the proof we prove Theorem 11.5.

Proof of Theorem 11.5. It follows from the definitions (11.12) and (11.13) that the subspace spanned by $\{\Psi(\varnothing),\Psi((j^1)),\Psi((j^2)),\dots\}$ together with $B_j^{\pm}$ agrees with the one-mode Boson Fock space Γ_{Boson}. It is readily known that the spectral distribution of $B_j^+ + B_j^-$ in the vacuum state $\Psi(\varnothing)$ is the Gaussian distribution. Hence, the right-hand sides of (11.16) and (11.6) coincide. □

In fact, Theorem 11.12 follows from Theorem 11.11 and the following:

Theorem 11.13 (QCLT for the adjacency matrices). *Let* $\epsilon_1,\dots,\epsilon_m\in\{+,-\}$ *and* $j_1,\dots,j_m\in\{2,3,\dots\}$, *where* $m=1,2,\dots$. *Then for any* $\tau,\rho\in\overline{\mathbb{Y}}$ *we have*

$$\lim_{n\to\infty}\left\langle \Phi(\tau\cup(1^{n-|\tau|})), \frac{A^{\epsilon_1}_{(j_1,1^{n-j_1})}}{\sqrt{|C_{(j_1,1^{n-j_1})}|}}\cdots\frac{A^{\epsilon_m}_{(j_m,1^{n-j_m})}}{\sqrt{|C_{(j_m,1^{n-j_m})}|}}\Phi(\rho\cup(1^{n-|\rho|}))\right\rangle$$
$$=\langle \Psi(\tau), B^{\epsilon_1}_{j_1}\cdots B^{\epsilon_m}_{j_m}\Psi(\rho)\rangle. \tag{11.17}$$

Moreover, if a conservation component of the form

$$\frac{A^{\circ}_{(j,1^{n-j})}}{\sqrt{|C_{(j,1^{n-j})}|}}$$

appears in the left-hand side of (11.17), *the limit is zero. Therefore, in the sense of stochastic convergence, i.e., in the sense of mixed moments, we have*

$$\left(\frac{A^{+}_{(j,1^{n-j})}}{\sqrt{|C_{(j,1^{n-j})}|}},\frac{A^{-}_{(j,1^{n-j})}}{\sqrt{|C_{(j,1^{n-j})}|}},\frac{A^{\circ}_{(j,1^{n-j})}}{\sqrt{|C_{(j,1^{n-j})}|}}\,;\,j=2,3,\dots\right)$$
$$\longrightarrow\quad (B_j^+,B_j^-,B_j^\circ\,;\,j=2,3,\dots)$$

as $n\to\infty$, *where* $B_j^\circ=0$ *for notational convenience.*

11.4 Proof of QCLT for Adjacency Matrices

We start with estimation of the coefficients of (11.10) and (11.11). Let $j\in\{2,3,\dots\}$ and $\sigma,\rho\in\overline{\mathbb{Y}}$ with $|\sigma|\leq n$, $|\rho|\leq n$. Take $x\in C_{\sigma\cup(1^{n-|\sigma|})}$ in $S(n)$. Note that $x\in C_{\sigma\cup(1^{m-|\sigma|})}$ in $S(m)$ for $m\geq n$. In order to estimate

$$p^{\sigma}_{(j)\rho}(n) = |\{s \in C_{(j,1^{n-j})}\,;\, s^{-1}x \in C_{\rho\cup(1^{n-|\rho|})}\}|,$$

we have to count possible expressions of the form

$$\begin{aligned} s^{-1}x &= (a_j\, a_{j-1}\, \cdots\, a_1)x \\ &= (a_j\, a_{j-1}) \cdots (a_3\, a_2)(a_2\, a_1)x \in C_{\rho\cup(1^{n-|\rho|})}. \end{aligned} \tag{11.18}$$

Let us prepare some notations. We begin with recalling Lemma 10.11 and the succeeding paragraph, especially Fig. 10.4. If a transposition $(a\, b)$ and $y \in C_{\tau\cup(1^{n-|\tau|})}$ satisfies that $(a\, b)y \in C_{\tau'\cup(1^{n-|\tau'|})}$, we have $l(\tau') = l(\tau) + 1$ or $l(\tau) - 1$. According as $+1$ or -1 occurs here, we said that $(a\, b)$ is an up or down action for y, respectively. Note that $y \in C_{\tau\cup(1^{n-|\tau|})}$ is equivalent to $\mathrm{type}(y) = \tau$. Moreover, we have

$$|\tau'| \in \{|\tau| \pm i\,;\, i = 0, 1, 2\}$$

and say that $(a\, b)$ is a $(+i)$ or $(-i)$-size action for y. At every step of multiplying $(a_i\, a_{i-1})$ successively $(j-1)$ times in (11.18), we give one of the three colours R(red), Y(yellow) or G(green) to each $a_1, a_2, \ldots, a_j$. First Y is put on a_1 and G is put on a_2. After performing $(a_2\, a_1)x$, we put R, Y, G on a_1, a_2, a_3 respectively. Next, after performing $(a_3\, a_2)(a_2\, a_1)x$, we put R, R, Y, G on a_1, a_2, a_3, a_4, respectively, and so on. In other words, in the successive multiplying actions in (11.18), each colour indicates the role of a letter in $\mathrm{supp}\, s = \{a_1, \ldots, a_j\}$ at the moment as follows:

R: letter which cannot move any more in the succeeding actions.
Y: letter which moves next once.
G: letter which moves next twice.

An up [resp. down] action increases [resp. decreases] the size of a Young diagram determining a type by 2, 1 or 0. Just before the step $y \mapsto (a\, b)y$ in the multiplying actions in (11.18), a has colour G and b has colour Y. At the moment, this step possibly contributes to the growth of $p^{\sigma}_{(j)\rho}(n)$ with order n if $a \notin \mathrm{supp}\, y$ and with order n^2 if $y = x$, $a \notin \mathrm{supp}\, x$ and $b \notin \mathrm{supp}\, x$. Otherwise, the contribution remains finite (i.e. with order n^0).

Lemma 11.14. *For $j \in \{2, 3, \ldots\}$ and $\sigma, \rho \in \overline{\mathbb{Y}}$ such that neither $\sigma = \rho \cup (j)$ nor $\sigma = \rho \setminus (j)$, we have*

$$p^{\sigma}_{(j)\rho}(n) \sqrt{\frac{|C_{\sigma\cup(1^{n-|\sigma|})}|}{|C_{(j,1^{n-j})}|\,|C_{\rho\cup(1^{n-|\rho|})}|}} = O(n^{-1/2}) \quad \text{as } n \to \infty. \tag{11.19}$$

Proof. Assume σ be neither $\rho \cup (j)$ nor $\rho \setminus (j)$. We divide the $(j-1)$-walks from σ to ρ (see (11.18)) on the modified Young graph $\overline{\mathbb{Y}}$ stratified by the length function into several classes to estimate $p^{\sigma}_{(j)\rho}(n)$. Let u, u_0, u_1 and d denote the number of the steps of up, (± 0)-size up, $(+1)$-size up, and down actions respectively along a $(j-1)$-walk w. Clearly $u + d = j - 1$. The power of the order of the left-hand side of (11.19) is referred to as $(*)$. We shall prove $(*) \le -1/2$ for the following three cases: Steps 1, 3 and 4.

Step 1. Assume that w does not contain (± 2)-size steps. We have $|\rho| \ge |\sigma| + u_1 - d$, and

$$(*) \le u_1 + \frac{1}{2}(|\sigma| - |\rho| - j) \le -\frac{1}{2}(u_0 + 1).$$

Step 2. If a $(+2)$-size action once appears, then neither $(+2)$- nor (-2)-size action appears afterwards. For, after a $(+2)$-size action, we always see a letter of colour R as the left neighbour of a letter of colour Y. A $(+2)$-size action can occur only if a 1-cycle consists of Y. A (-2)-size action can occur only if a 2-row consists of Y and G. Since both cases are impossible, we have no further (± 2)-size actions. On the other hand, if a (-2)-size action once appears, then, except for possible appearance of a $(+2)$-size action at the next step, we do not see afterwards $(+2)$- or (-2)-size actions. For, after the next action of a (-2)-size action, we always see R as the left neighbour of Y.

Step 3. If w contains a (-2)-size step, then (i) it is followed by a $(+2)$-size step or (ii) by a $(+1)$-size step or (iii) the step is final.

(i) We have

$$|\rho| \ge |\sigma| - 2 + 2 + u_1 - (d-1) = |\sigma| + u_1 - d + 1.$$

Since the $(+2)$-size action here is an up step of order n, we get

$$\begin{aligned}(*) &\le u_1 + 1 + \frac{1}{2}(|\sigma| - |\rho| - j)\\ &\le u - u_0 + \frac{1}{2}(-u_1 + d - 1 - j)\\ &= -\frac{1}{2}(u_0 + 1).\end{aligned}$$

Note that $u = u_0 + u_1 + 1$.

(ii) We have

$$|\rho| \ge |\sigma| - 2 + 1 + u_1 - 1 - (d-1) = |\sigma| + u_1 - d - 1.$$

Since the $(+1)$-size action here is an up step of order 1, we get

$$(*) \le u_1 - 1 + \frac{1}{2}(|\sigma| - |\rho| - j) \le -\frac{1}{2}(u_0 + 2).$$

Note that $u = u_0 + u_1$.

(iii) There exists a down without size loss. Indeed, if the last (i.e. $(j-1)$ th) action is a (-2)-size one, there is a 2-cycle consisting of a Y-letter and a G-letter. The 2-cycle comes from a cycle of the form $\cdots$G$*$Y$\cdots$ ($*$ indicates any box) just before. If the '$\cdots$' part actually exists in the left or right, the action of cutting out G$*$ (which turns into YG) does not reduce the size. Otherwise,

G∗Y (= YG∗) comes from $\cdots$G∗∗Y$\cdots$ just before. If one '$\cdots$' part exists at least, the action of cutting out G∗Y does not reduce the size. Tracing back in this way, $s^{-1}x$ is made by removing a j-cycle from x. However, the assumption excludes this situation. We have $|\rho| \geq |\sigma| - 2 + u_1 - (d-1) + 1$ and

$$(*) \leq u_1 + \frac{1}{2}(|\sigma| - |\rho| - j) \leq -\frac{1}{2}(u_0 + 1).$$

Note that $u = u_0 + u_1$.

Step 4. If w contains a $(+2)$-size step, look at the Y-letter in the preceding step. If Y$= a_1$, there exists an up without size gain. In fact, the $(+2)$-size action is the first one and produces a cycle consisting of YR. Then, since the assumption does not allow that letters are added to this cycle one by one successively to the end, there is an up action connecting YR$\cdots$R with a cycle of length ≥ 2, which does not cause increase of the size. Then we have

$$|\rho| \geq |\sigma| + 2 + u_1 - d$$

and

$$(*) \leq u_1 + 2 + \frac{1}{2}(|\sigma| - |\rho| - j) \leq -\frac{1}{2}u_0 \leq -\frac{1}{2}.$$

Note that $u = u_0 + u_1 + 1$. If Y$\neq a_1$, Y came from a non-trivial cycle before, hence the $(+2)$-size step is of order n. Then we have

$$|\rho| \geq |\sigma| + 2 + u_1 - d - 1$$

and

$$(*) \leq u_1 + 1 + \frac{1}{2}(|\sigma| - |\rho| - j) \leq -\frac{1}{2}(u_0 + 1).$$

Again, $u = u_0 + u_1 + 1$. □

Proof of Theorem 11.13. Step 1. If $C_{\rho \cup (1^{n-|\rho|})}$ is a conjugacy class in $S(n)$, then

$$|C_{\rho \cup (1^{n-|\rho|})}| = \frac{n(n-1)\cdots(n-|\rho|+1)}{\prod_{j \geq 2} j^{m_j(\rho)} m_j(\rho)!}.$$

For $\sigma = \rho \cup (j)$ we have $m_j(\sigma) = m_j(\rho) + 1$ and

$$p^{\sigma}_{(j)\rho}(n)\sqrt{\frac{|C_{\sigma \cup (1^{n-|\sigma|})}|}{|C_{(j,1^{n-j})}|\,|C_{\rho \cup (1^{n-|\rho|})}|}} \sim m_j(\sigma)\sqrt{\frac{j \cdot j^{m_j(\rho)} m_j(\rho)!}{j^{m_j(\sigma)} m_j(\sigma)!}}$$
$$= \sqrt{m_j(\rho) + 1},$$

where $\sim$ means that the ratio of both sides tends to one as $n \to \infty$. Similarly, for $\sigma = \rho \setminus (j)$, we have $m_j(\sigma) = m_j(\rho) - 1$ and

$$p^{\sigma}_{(j)\rho}(n)\sqrt{\frac{|C_{\sigma \cup (1^{n-|\sigma|})}|}{|C_{(j,1^{n-j})}|\,|C_{\rho \cup (1^{n-|\rho|})}|}} \sim \frac{1}{j}\sqrt{\frac{j \cdot j^{m_j(\rho)} m_j(\rho)!}{j^{m_j(\sigma)} m_j(\sigma)!}} = \sqrt{m_j(\rho)}.$$

If σ is related to ρ in the other manners, by Lemma 11.14 we obtain

$$\lim_{n\to\infty} p^{\sigma}_{(j)\rho}(n)\sqrt{\frac{|C_{\sigma\cup(1^{n-|\sigma|})}|}{|C_{(j,1^{n-j})}|\,|C_{\rho\cup(1^{n-|\rho|})}|}} = 0.$$

Step 2. We show (11.17) by induction on m. The case of $m = 0$ is trivial. Assume that (11.17) holds for m. It follows from Lemma 11.9 and the argument in Step 1, together with the assumption of induction, that for $\epsilon_1, \dots, \epsilon_m \in \{+, -\}$, $j_1, \dots, j_{m+1} \in \{2, 3, \dots\}$, $\tau, \rho \in \mathbb{Y}$ and $\epsilon_{m+1} = +$,

$$\begin{aligned}
&\lim_{n\to\infty} \Bigg\langle \Phi(\tau \cup (1^{n-|\tau|})), \\
&\qquad \frac{A^{\epsilon_1}_{(j_1,1^{n-j_1})}}{\sqrt{|C_{(j_1,1^{n-j_1})}|}} \cdots \frac{A^{\epsilon_m}_{(j_m,1^{n-j_m})}}{\sqrt{|C_{(j_m,1^{n-j_m})}|}} \frac{A^{+}_{(j_{m+1},1^{n-j_{m+1}})}}{\sqrt{|C_{(j_{m+1},1^{n-j_{m+1}})}|}} \Phi(\rho \cup (1^{n-|\rho|})) \Bigg\rangle \\
&= \lim_{n\to\infty} \sum_{i=1}^{j_{m+1}-1} \sum_{\substack{|\sigma|\le n \\ l(\sigma)=l(\rho)+i}} p^{\sigma}_{(j_{m+1})\rho}(n) \sqrt{\frac{|C_{\sigma\cup(1^{n-|\sigma|})}|}{|C_{(j_{m+1},1^{n-j_{m+1}})}|\,|C_{\rho\cup(1^{n-|\rho|})}|}} \\
&\quad \times \Bigg\langle \Phi(\tau \cup (1^{n-|\tau|})), \frac{A^{\epsilon_1}_{(j_1,1^{n-j_1})}}{\sqrt{|C_{(j_1,1^{n-j_1})}|}} \cdots \frac{A^{\epsilon_m}_{(j_m,1^{n-j_m})}}{\sqrt{|C_{(j_m,1^{n-j_m})}|}} \Phi(\sigma \cup (1^{n-|\sigma|})) \Bigg\rangle \\
&= \sqrt{m_{j_{m+1}}(\rho)+1}\, \langle \Psi(\tau), B^{\epsilon_1}_{j_1} \cdots B^{\epsilon_m}_{j_m} \Psi(\rho \cup (j_{m+1})) \rangle \\
&= \langle \Psi(\tau), B^{\epsilon_1}_{j_1} \cdots B^{\epsilon_m}_{j_m} B^{+}_{j_{m+1}} \Psi(\rho) \rangle.
\end{aligned}$$

A similar computation is applied to the cases of $\epsilon_{m+1} = -$ and $\epsilon_{m+1} = \circ$. Thus the assertion is true also for $m + 1$. □

11.5 Polynomial Functions on Young Diagrams

As we mentioned at the beginning of Sect. 11.1, the algebra of polynomial functions $\mathbb{A}$ provides a nice framework for analysis on Young diagrams. Typical examples of polynomial functions are the moments of the Rayleigh and the transition measures. We assemble some properties relevant to our goal, namely, concentration and fluctuation of Young diagrams and central limit theorem.

For a Young diagram $\lambda \in \mathbb{Y}$, let λ_i and λ'_j be the lengths of the ith row and jth column. (In general, λ' stands for the transposed diagram.) Let $d(\lambda)$ denotes the length of the main diagonal. The Frobenius coordinates of λ is defined by

$$\lambda = (a_1, \dots, a_d; b_1, \dots, b_d), \qquad d = d(\lambda),$$

where

$$a_i = \lambda_i - i, \qquad b_i = \lambda'_i - i, \qquad i = 1, 2, \dots, d.$$

Proposition 11.15. *Let* $\lambda \in \mathbb{Y}$. *Let*

$$(a_1, \ldots, a_d; b_1, \ldots, b_d), \qquad x_1 < y_1 < \cdots < x_{r-1} < y_{r-1} < x_r$$

be its Frobenius coordinates and min–max coordinates, respectively. Then,

$$\frac{(z-y_1)\cdots(z-y_{r-1})}{(z-x_1)\cdots(z-x_r)} = \frac{1}{z}\prod_{i=1}^{d}\frac{(z+b_i)(z-a_i)}{(z-1-a_i)(z+1+b_i)}, \qquad z \in \mathbb{C}. \tag{11.20}$$

Recall that $\{M_k(\tau_\lambda)\,;\, k = 1, 2, \ldots\}$ and $\{M_k(\mathfrak{m}_\lambda)\,; = 1, 2, \ldots\}$ can be expressed each other in polynomials. Therefore, the latter is also a set of generators of $\mathbb{A}$. Moreover, let us see that we can take generators of $\mathbb{A}$ consisting of irreducible characters of the symmetric groups.

For given $\rho \in \mathbb{Y}$, set a function on $\mathbb{Y}$ as

$$\Sigma_\rho(\lambda) = \begin{cases} |\lambda|^{\downarrow|\rho|}\, \tilde{\chi}^{\lambda}_{\rho\cup(1^{|\lambda|-|\rho|})}, & \text{if } |\lambda| \geq |\rho|, \\ 0, & \text{otherwise.} \end{cases} \tag{11.21}$$

When ρ is a 1-row diagram (labelling a conjugacy class of cycles), we write Σ_k for simplicity instead of $\Sigma_{(k)}$. This function was already noted in (11.2). Polynomial relations between $\{M_k(\mathfrak{m}_\lambda)\}$ and $\{\Sigma_k(\lambda)\}$ can be seen through their generating functions. Since we are considering functions on $\mathbb{Y}$, the coefficients of those relations must not contain explicitly the size of a Young diagram, i.e., the variable n.

Proposition 11.16. *Let* $G_{\mathfrak{m}_\lambda}$ *be the Stieltjes transform of the transition measure of* λ, *i.e.,*

$$G_{\mathfrak{m}_\lambda}(z) = \int_{-\infty}^{+\infty} \frac{\mathfrak{m}_\lambda(dx)}{z-x}.$$

Then

$$\begin{aligned} \Sigma_k(\lambda) &= -\frac{1}{k}\,[z^{-1}]\left\{\frac{1}{G_{\mathfrak{m}_\lambda}(z)}\cdots\frac{1}{G_{\mathfrak{m}_\lambda}(z-k+1)}\right\} \\ &= \frac{1}{k}\operatorname*{Res}_{z=\infty}\left\{\frac{1}{G_{\mathfrak{m}_\lambda}(z)}\cdots\frac{1}{G_{\mathfrak{m}_\lambda}(z-k+1)}\right\}. \end{aligned} \tag{11.22}$$

(The conventional notation $[z^{-1}]\{\cdots\}$ *means taking the coefficient of* z^{-1} *of the series.)*

We do not go into the proof but mention some remarks. Note that $G_{\mathfrak{m}_\lambda}(z)$ coincides with (11.20). For the proof the Frobenius formula

$$\begin{aligned} n^{\downarrow k}\tilde{\chi}^{\lambda}_{(k,1^{n-k})} &= -\frac{1}{k}\,[z^{-1}]\left\{z^{\downarrow k}\prod_{i=1}^{n}\frac{z-(\lambda_i-i+n+k)}{z-(\lambda_i-i+n)}\right\} \\ &= \frac{1}{k}\operatorname*{Res}_{z=\infty}\left\{z^{\downarrow k}\prod_{i=1}^{n}\frac{z-(\lambda_i-i+n+k)}{z-(\lambda_i-i+n)}\right\}, \qquad \lambda \in \mathbb{Y}_n, \end{aligned} \tag{11.23}$$

is taken to be a starting point. This expression, however, does not immediately fit well into our aim since the right-hand side contains explicitly the variable n (the size of λ). We need to rewrite this expression in terms of the min–max coordinates.

Let us recall the notions of cumulants of a probability measure. For a probability $\mu \in \mathfrak{P}_{\mathrm{fm}}(\mathbb{R})$ having finite moments of all orders, let $\kappa_n(\mu)$, $R_n(\mu)$ and $B_n(\mu)$ denote the classical, free and Boolean cumulants of degree n, respectively. They are uniquely determined by the moments $\{M_n(\mu)\}$ through the so-called moment-cumulant formulae

$$M_n(\mu) = \sum_{\rho \in \mathbb{Y}_n} |\mathcal{P}_{(\rho)}(n)| \prod_{k=1}^{n} \kappa_k(\mu)^{m_k(\rho)} \tag{11.24}$$

$$= \sum_{\rho \in \mathbb{Y}_n} |\mathcal{P}_{\mathrm{NC}(\rho)}(n)| \prod_{k=1}^{n} R_k(\mu)^{m_k(\rho)} \tag{11.25}$$

$$= \sum_{\rho \in \mathbb{Y}_n} |\mathcal{P}_{\mathrm{I}(\rho)}(n)| \prod_{k=1}^{n} B_k(\mu)^{m_k(\rho)}, \tag{11.26}$$

where $\mathcal{P}_{(\rho)}(n)$, $\mathcal{P}_{\mathrm{NC}(\rho)}(n)$ and $\mathcal{P}_{\mathrm{I}(\rho)}(n)$ denote respectively the sets of partitions, non-crossing partitions and interval partitions with ρ being their block types. In fact, for each cumulant an explicit formula is known by means of the Möbius functions for lattices of the three kind of partitions. Moreover, the cumulants are characterized in analytic way as

$$\log \int_{-\infty}^{\infty} e^{zx} \mu(dx) = \sum_{k=1}^{\infty} \frac{\kappa_k(\mu)}{k!} z^k ,$$

$$K_\mu(z) = \frac{1}{z} + \sum_{k=1}^{\infty} R_k(\mu) z^{k-1}, \tag{11.27}$$

$$\frac{1}{G_\mu(z)} = z - \sum_{k=1}^{\infty} \frac{B_k(\mu)}{z^{k-1}} . \tag{11.28}$$

Here $K_\mu(z)$ is the inverse function of the Stieltjes transform $G_\mu(z)$ on an appropriate domain in $\mathbb{C}$. Moments and cumulants are considered as functions of Young diagrams through their transition measures or Rayleigh measures. Thus $\kappa_k(\mathfrak{m}_\lambda)$, $R_k(\mathfrak{m}_\lambda)$ and $B_k(\mathfrak{m}_\lambda)$ are polynomial functions on $\mathbb{Y}$.

We shall introduce a filtration in the polynomial function algebra $\mathbb{A}$ by means of the *weight degree.* First, we say that

$$M_k(\tau_\lambda) = \sum_i x_i^k - \sum_i y_i^k$$

is homogeneous and has a weight degree k. For an arbitrary polynomial function in $\mathbb{A}$ the weight degree is defined naturally by expressing it in a polynomial of $\{M_k(\tau_\lambda)\}$. We see from (11.24), (11.25) and (11.26) that the kth moments and cumulants of the transition and the Rayleigh measures of Young diagrams are all polynomial functions of weight degree k. Consequently, any choice of $\{M_k\}, \{R_k\}, \{B_k\}, \{\Sigma_k\}$ as a set of generators leads to an equivalent definition of $\mathbb{A}$ with the same filtration.

Since it is shown that Σ_ρ's form a linear basis of $\mathbb{A}$, they satisfy a linearizing formula

$$\Sigma_\rho \Sigma_\sigma = \sum_\tau f_{\rho\sigma}^\tau \Sigma_\tau. \tag{11.29}$$

Ivanov–Kerov [116] and Ivanov–Olshanski [117] describe a method to evaluate

$$g_{\rho\sigma}^\tau = \frac{z_\tau}{z_\rho z_\sigma} f_{\rho\sigma}^\tau,$$

where z_ρ is a conventional notation of (9.1). On the other hand, we have the following linearization formula:

$$A_{\rho\cup(1^{n-|\rho|})} A_{\sigma\cup(1^{n-|\sigma|})} = \sum_{\substack{\tau\in\overline{\mathbb{Y}} \\ |\tau|\le n}} p_{\rho\sigma}^\tau(n) A_{\tau\cup(1^{n-|\tau|})}, \tag{11.30}$$

where

$$\rho, \sigma \in \overline{\mathbb{Y}}, \quad |\rho| \le n, \quad |\sigma| \le n,$$

see also (10.34). Let us see the connection of $g_{\rho\sigma}^\tau$ and $p_{\rho\sigma}^\tau(n)$ through the spectral decomposition of adjacency operators. Since

$$A_{\rho\cup(1^{n-|\rho|})} = \sum_{\lambda\in\mathbb{Y}_n} \frac{1}{z_\rho} \Sigma_\rho(\lambda) E_\lambda, \tag{11.31}$$

by (11.4), comparison of (11.29) and (11.30) yields

$$p_{\rho\sigma}^\tau(n) = \sum_{j=0}^\infty g_{\rho\sigma}^{\tau\cup(1^j)} \binom{n-|\tau|}{j}, \tag{11.32}$$

if $|\rho| + |\sigma|, |\tau| \le n$. Here we note that $|\tau| + j > |\rho| + |\sigma|$ implies $g_{\rho\sigma}^{\tau\cup(1^j)} = 0$.

Combined with some properties of $g_{\rho\sigma}^{\tau\cup(1^j)}$, (11.32) leads us to an alternative proof of Lemma 11.14.

11.6 Kerov's Polynomials

Expanding (11.22), we can express $\Sigma_k(\lambda)$ as a polynomial in $M_j(\mathfrak{m}_\lambda)$'s with weight degrees up to $k+1$ and hence in $R_j(\mathfrak{m}_\lambda)$'s similarly. To be precise, $\Sigma_k(\lambda)$ becomes a polynomial in $R_2(\mathfrak{m}_\lambda), \dots, R_{k+1}(\mathfrak{m}_\lambda)$, that is,

$$\Sigma_k(\lambda) = K_k(R_2(\mathfrak{m}_\lambda), \dots, R_{k+1}(\mathfrak{m}_\lambda)).$$

This polynomial K_k is called *Kerov's polynomial.* By the standard residue calculus we get from (11.27)

$$R_{k+1}(\mathfrak{m}_\lambda) = -\frac{1}{k}\,[z^{-1}]\left(\frac{1}{G_{\mathfrak{m}_\lambda}(z)}\right)^k = \frac{1}{k}\operatorname*{Res}_{z=\infty}\left(\frac{1}{G_{\mathfrak{m}_\lambda}(z)}\right)^k. \qquad (11.33)$$

A nice connection of irreducible characters and free cumulants is suggested by (11.22) and (11.33). Let us compare some coefficients appearing in the expansions of (11.22) and (11.33). For simplicity of notations, we omit the suffix $\mathfrak{m}_\lambda$. Inserting (11.28) into (11.22) and (11.33), we get

$$\begin{aligned}\Sigma_k &= R_{k+1} + (\text{polynomial in } B_j\text{'s with weight degree} \le k),\\ &= R_{k+1} + (\text{polynomial in } R_j\text{'s with weight degree} \le k).\end{aligned} \qquad (11.34)$$

The transposition of a Young diagram $\lambda \mapsto \lambda'$ gives rise to a canonical involution on $\mathbb{A}$ defined by

$$\operatorname{inv}(f)(\lambda) = f(\lambda').$$

Since the transposition of a Young diagram causes parity change of the transition measure and the Rayleigh measure, their moments satisfy

$$\operatorname{inv}(M_k) = (-1)^k M_k.$$

Furthermore, we see from $U_{\lambda'} \simeq U_\lambda \otimes \mathrm{sgn}$ that

$$\operatorname{inv}(\Sigma_k) = (-1)^{k-1}\Sigma_k.$$

We now go back to (11.34). Taking the involution of both sides of (11.34), we see that $\Sigma_k - R_{k+1}$ is a polynomial in $R_2, \dots, R_{k-1}$ with weight degree $\le k-1$ in which the weight degree of each term has the same parity as $k+1$. This is one feature of Kerov's polynomial K_k. It is known that the coefficients of Kerov's polynomial K_k are all integers. Moreover, Kerov conjectured that the coefficients are all non-negative. It would be desirable to characterize the coefficients of Kerov's polynomials by means of some combinatorics.

11.7 Other Extensions of Kerov's Central Limit Theorem

In Sects. 11.2–11.4 we showed a non-commutative extension (quantum version) of Kerov's central limit theorem (Theorem 11.13). In this section we briefly survey other extensions.

We go back to (11.6) in Theorem 11.5, where Kerov's central limit theorem is formulated in terms of adjacency matrices. In fact, (11.6) means that the adjacency matrices

$$A_{(2,1^{n-2})}, \dots, A_{(m,1^{n-m})},$$

are asymptotically independent and Gaussian random variables in the sense of algebraic probability, where we recall that $A_{(k,1^{k-2})}$ corresponds to the cycles of length k. For general adjacency matrices $A_{\rho\cup(1^{n-|\rho|})}$ corresponding to arbitrary conjugacy classes of $S(n)$ the situation becomes more complicated and requires the Hermite polynomials (Definition 1.80). Let $\tilde{H}_n(x)$ be the orthogonal polynomial associated with the standard Gaussian distribution. In other words, $\tilde{H}_n(x)$ is a monic polynomial of degree n which obeys the following recurrence relation:

$$\begin{aligned} &\tilde{H}_0(x) = 1, \qquad \tilde{H}_1(x) = x, \\ &x\tilde{H}_n(x) = \tilde{H}_{n+1}(x) + n\tilde{H}_{n-1}(x), \quad n = 1, 2, \dots . \end{aligned} \tag{11.35}$$

In fact, $\tilde{H}_n(x)$ is a simple modification of the Hermite polynomial, see Theorem 1.81.

Theorem 11.17 (CLT for general adjacency matrices I). *Let $m = 1, 2, \dots$. For $\rho^{(1)}, \dots, \rho^{(m)} \in \overline{\mathbb{Y}}$ and $r_1, \dots, r_m \in \{0, 1, 2, \dots\}$, it holds that*

$$\begin{aligned} &\lim_{n\to\infty} \left\langle \delta_e, \left(\frac{A_{\rho^{(1)}\cup(1^{n-|\rho^{(1)}|})}}{\sqrt{|C_{\rho^{(1)}\cup(1^{n-|\rho^{(1)}|})}|}} \right)^{r_1} \cdots \left(\frac{A_{\rho^{(m)}\cup(1^{n-|\rho^{(m)}|})}}{\sqrt{|C_{\rho^{(m)}\cup(1^{n-|\rho^{(m)}|})}|}} \right)^{r_m} \delta_e \right\rangle \\ &= \prod_{j=2}^{\infty} \frac{1}{\sqrt{2\pi}} \int_{-\infty}^{+\infty} \left(\frac{\tilde{H}_{m_j(\rho^{(1)})}(x)}{\sqrt{m_j(\rho^{(1)})!}} \right)^{r_1} \cdots \left(\frac{\tilde{H}_{m_j(\rho^{(m)})}(x)}{\sqrt{m_j(\rho^{(m)})!}} \right)^{r_m} e^{-x^2/2} dx, \end{aligned} \tag{11.36}$$

where the left-hand side is the inner product of $\ell^2(S(n))$.

The right-hand side of (11.36) is actually a finite product since

$$m_j(\rho^{(1)}) = \cdots = m_j(\rho^{(m)}) = 0$$

for all $j > \max\{|\rho^{(1)}|, \dots, |\rho^{(m)}|\}$. The key observation for the proof is the following:

(H1) The rows of different lengths in Young diagrams behave like statistically independent random variables in the limit.
(H2) The k-fold multiplicity (or interaction) of rows of the same length is described by the Hermite polynomial of degree k in the limit.

We shall outline the proof of Theorem 11.17 from the viewpoint of the quantum central limit theorem developed in Sect. 11.3. In fact, we discuss a generalization of Theorem11.17 as follows.

Theorem 11.18 (CLT for general adjacency matrices II). *Let $m = 1, 2, \dots$ and $\rho^{(1)}, \dots, \rho^{(m)} \in \overline{\mathbb{Y}}$. Then, for any $\tau, \sigma \in \overline{\mathbb{Y}}$ it holds that*

$$\lim_{n\to\infty}\Big\langle \Phi(\tau\cup(1^{n-|\tau|})),$$
$$\frac{A_{\rho^{(1)}\cup(1^{n-|\rho^{(1)}|})}}{\sqrt{|C_{\rho^{(1)}\cup(1^{n-|\rho^{(1)}|})}|}}\cdots\frac{A_{\rho^{(m)}\cup(1^{n-|\rho^{(m)}|})}}{\sqrt{|C_{\rho^{(m)}\cup(1^{n-|\rho^{(m)}|})}|}}\Phi(\sigma\cup(1^{n-|\sigma|}))\Big\rangle$$
$$=\Big\langle \Psi(\tau),\prod_{j=2}^{\infty}\frac{\tilde{H}_{m_j(\rho^{(1)})}(B_j^+ + B_j^-)}{\sqrt{m_j(\rho^{(1)})!}}\cdots\frac{\tilde{H}_{m_j(\rho^{(m)})}(B_j^+ + B_j^-)}{\sqrt{m_j(\rho^{(m)})!}}\Psi(\sigma)\Big\rangle. \quad (11.37)$$

Proof. (Outline) Verifying the property (H1) for the left-hand side of (11.37), we separate in an asymptotic sense the interaction between rows of different lengths. For that purpose it is essentially sufficient to show that for $\rho^{(1)},\rho^{(2)}\in\overline{\mathbb{Y}}$ which do not share any rows of the same length, the following two quantities asymptotically coincide:

$$\Big\langle \Phi(\tau\cup(1^{n-|\tau|})),\frac{A_{\rho^{(1)}\cup\rho^{(2)}\cup(1^{n-|\rho^{(1)}\cup\rho^{(2)}|})}}{\sqrt{|C_{\rho^{(1)}\cup\rho^{(2)}\cup(1^{n-|\rho^{(1)}\cup\rho^{(2)}|})}|}}\Phi(\sigma\cup(1^{n-|\sigma|}))\Big\rangle,$$
$$\Big\langle \Phi(\tau\cup(1^{n-|\tau|})),\frac{A_{\rho^{(1)}\cup(1^{n-|\rho^{(1)}|})}}{\sqrt{|C_{\rho^{(1)}\cup(1^{n-|\rho^{(1)}|})}|}}\frac{A_{\rho^{(2)}\cup(1^{n-|\rho^{(2)}|})}}{\sqrt{|C_{\rho^{(2)}\cup(1^{n-|\rho^{(2)}|})}|}}\Phi(\sigma\cup(1^{n-|\sigma|}))\Big\rangle,$$

where the inner product is taken over $\ell^2(S(n))$. A similar combinatorial argument as in Hora [103] works well for this aim. Note that, under this assumption, we have

$$z_{\rho^{(1)}\cup\rho^{(2)}} = z_{\rho^{(1)}}z_{\rho^{(2)}}$$
$$|C_{\rho^{(1)}\cup\rho^{(2)}\cup(1^{n-|\rho^{(1)}\cup\rho^{(2)}|})}| = |C_{\rho^{(1)}\cup(1^{n-|\rho^{(1)}|})}|\,|C_{\rho^{(2)}\cup(1^{n-|\rho^{(2)}|})}|\,(1+o(1)).$$

We next see how the Hermite polynomials appear as stated in (H2). Essentially, we have to only show

$$\Big\langle \Phi(\tau\cup(1^{n-|\tau|})),\frac{\sqrt{k!}A_{(j^k,1^{n-jk})}}{\sqrt{|C_{(j^k,1^{n-jk})}|}}\Phi(\sigma\cup(1^{n-|\sigma|}))\Big\rangle$$
$$=\Big\langle \Phi(\tau\cup(1^{n-|\tau|})),\tilde{H}_k\Big(\frac{A_{(j,1^{n-j})}}{\sqrt{|C_{(j,1^{n-j})}|}}\Big)\Phi(\sigma\cup(1^{n-|\sigma|}))\Big\rangle+o(1). \quad (11.38)$$

Let us prove (11.38) by induction on k. We start with

$$A_{(j,1^{n-j})}A_{(j^k,1^{n-jk})}=\sum_{\rho}p^{\rho}_{(j)\,(j^k)}(n)A_{\rho\cup(1^{n-|\rho|})}.$$

Applying Lemma 11.14, we come to

$$\left\langle \Phi(\tau \cup (1^{n-|\tau|})), \frac{A_{(j,1^{n-j})}}{\sqrt{|C_{(j,1^{n-j})}|}} \frac{\sqrt{k!}\, A_{(j^k,1^{n-jk})}}{\sqrt{|C_{(j^k,1^{n-jk})}|}} \Phi(\sigma \cup (1^{n-|\sigma|})) \right\rangle$$

$$= \left\langle \Phi(\tau \cup (1^{n-|\tau|})), \frac{\sqrt{(k+1)!}\, A_{(j^{k+1},1^{n-j(k+1)})}}{\sqrt{|C_{(j^{k+1},1^{n-j(k+1)})}|}} \Phi(\sigma \cup (1^{n-|\sigma|})) \right\rangle$$

$$+ k \left\langle \Phi(\tau \cup (1^{n-|\tau|})), \frac{\sqrt{(k-1)!}\, A_{(j^{k-1},1^{n-j(k-1)})}}{\sqrt{|C_{(j^{k-1},1^{n-j(k-1)})}|}} \Phi(\sigma \cup (1^{n-|\sigma|})) \right\rangle$$

$$+ o(1). \tag{11.39}$$

On the other hand, the recurrence formula for the Hermite polynomials yields

$$\left\langle \Phi(\tau \cup (1^{n-|\tau|})), \frac{A_{(j,1^{n-j})}}{\sqrt{|C_{(j,1^{n-j})}|}} \tilde{H}_k\left(\frac{A_{(j,1^{n-j})}}{\sqrt{|C_{(j,1^{n-j})}|}} \right) \Phi(\sigma \cup (1^{n-|\sigma|})) \right\rangle$$

$$= \left\langle \Phi(\tau \cup (1^{n-|\tau|})), \tilde{H}_{k+1}\left(\frac{A_{(j,1^{n-j})}}{\sqrt{|C_{(j,1^{n-j})}|}} \right) \Phi(\sigma \cup (1^{n-|\sigma|})) \right\rangle$$

$$+ k \left\langle \Phi(\tau \cup (1^{n-|\tau|})), \tilde{H}_{k-1}\left(\frac{A_{(j,1^{n-j})}}{\sqrt{|C_{(j,1^{n-j})}|}} \right) \Phi(\sigma \cup (1^{n-|\sigma|})) \right\rangle. \tag{11.40}$$

Combining (11.39) and (11.40), we see that the induction proceeds to obtain (11.38). We now go back to (11.37). The left-hand side is reduced to an expression where the adjacency matrices are of the form $A_{(j,1^{n-j})}$. Then, our assertion follows from the quantum central limit theorem for such adjacency matrices which was already established in Theorem 11.13. □

The right-hand side of (11.37) in Theorem 11.18 is a nice expression. In fact, for any $k = 0, 1, 2, \dots$ and $j = 2, 3, \dots$ we have

$$\tilde{H}_k(B_j^+ + B_j^-) = \sum_{i=0}^{k} \binom{k}{i} (B_j^+)^i (B_j^-)^{k-i}, \tag{11.41}$$

the action of which on the base vector $\Psi(\sigma)$ is easy to handle.

11.8 More Refinements of Fluctuation

Let us recall the heuristic observation in Sect. 11.1. We intended to capture the fluctuation of Young diagrams by rescaling the deviation of scaled Young diagrams from the limit shape $\lambda^{\sqrt{n}} - \Omega$ in the vertical direction. This fluctuation is rigorously described in terms of generalized random variable. In short, there exists a generalized Gaussian random field Δ supported by $[-2, 2]$ such that

$$\lambda^{\sqrt{n}} \sim \Omega + \frac{2}{\sqrt{n}} \Delta \quad \text{as } n \to \infty,$$

with respect to the Plancherel measure. More precisely, we have the following:

Theorem 11.19 (Ivanov–Olshanski). *Given $f_1, \dots, f_p \in C^\infty(\mathbb{R})$, the joint distribution of the family of random variables*

$$\left\langle f_j, \frac{\sqrt{n}}{2}(\lambda^{\sqrt{n}} - \Omega) \right\rangle = \int_{-2}^{+2} \frac{\sqrt{n}}{2}(\lambda^{\sqrt{n}}(x) - \Omega(x)) f_j(x) dx, \quad j = 1, \dots, p, \tag{11.42}$$

in the probability space $(\mathbb{Y}_n, \mathfrak{P}_n)$ converges to the image of the Gaussian measure on the space of distributions supported by $[-2, +2]$ under the map $(\langle f_1, \cdot \rangle, \dots, \langle f_p, \cdot \rangle)$.

The limit of (11.42) is denoted by $\langle f_j, \Delta \rangle$. We do not go into the proof but only mention that the polynomial function algebra $\mathbb{A}$ introduced in Sect. 11.5 plays a central role therein.

A more concrete expression of Δ is known. Let $U_k(x)$ be the Chebyshev polynomial of the second kind (Definition 1.74) and set $\tilde{U}_k(x) = U_k(x/2)$. It is known that $\{\tilde{U}_k(x)\}$ is the orthogonal polynomial with respect to the Wigner semicircle law (Theorem 1.75). Let $\xi_1, \xi_2, \dots$ be a sequence of independent identically distributed random variables, each of which obeys the standard Gaussian distribution. We then have

$$\Delta(x) = \frac{1}{2\pi} \sum_{k=1}^{\infty} \frac{\xi_k \tilde{U}_k(x) \sqrt{4 - x^2}}{\sqrt{k+1}}, \qquad -2 \le x \le 2, \tag{11.43}$$

which is a random Fourier series. Taking $\tilde{U}_j$ as a test function, we have

$$\langle \tilde{U}_j, \Delta \rangle = \frac{\xi_j}{\sqrt{j+1}}. \tag{11.44}$$

Compare (11.1) with (11.42) and (11.44). The moments $M_k(\tau_\lambda)$ of the Rayleigh measure of a Young diagram λ appearing here give a generating set of the polynomial function algebra $\mathbb{A}$, see Sect. 11.5. Ivanov–Olshanski [117] investigated transition rules between the generators given by moments $M_k(\mathfrak{m}_\lambda)$ or $M_k(\tau_\lambda)$ and those of the irreducible characters $\Sigma_k(\lambda)$. By virtue of these transition rules, a central limit theorem for some generators is translated into another central limit theorem for other generators. Again in this framework, Kerov's central limit theorem (for irreducible characters) plays a fundamental role. This method due to Kerov and Ivanov–Olshanski also yields an equivalent form with Theorem 11.17, central limit theorem for arbitrary adjacency matrices $A_{\rho \cup (1^{n-|\rho|})}$, as a central limit theorem for their spectrum $\Sigma_\rho(\lambda)$.

Having obtained a universal Gaussian property of fluctuation, one may be interested in the convergence rate of Kerov's central limit theorem, i.e., an analogue of the celebrated Berry–Esseen theorem. In this direction, we here mention the following result on the uniform norm estimate for distribution functions:

Theorem 11.20 (Fulman). *There exists a universal numerical constant $C > 0$ such that*

$$\left|\mathfrak{P}_n\left(\left\{\lambda \in \mathbb{Y}_n\,;\ \frac{1}{\sqrt{2}\,n}\Sigma_2(\lambda) \le x\right\}\right) - \frac{1}{\sqrt{2\pi}}\int_{-\infty}^{x} \mathrm{e}^{-y^2/2}dy\right| \le C\,n^{-1/4}$$

holds for all $x \in \mathbb{R}$ and $n = 1, 2, \dots$.

Exercises

11.1. Deduce (11.33) from (11.27) by using residue calculus. Give a similar expression for a Boolean cumulant from (11.28).

11.2. Let B^+, B^- be elements in an $*$-algebra which satisfy the commutation relation $B^-B^+ - B^+B^- = 1$. Show that

$$\tilde{H}_k(B^+ + B^-) = \sum_{i=0}^{k} \binom{k}{i} (B^+)^i (B^-)^{k-i}, \qquad k = 0, 1, 2, \dots,$$

where $\tilde{H}_k(x)$ is the modification of the Hermite polynomial defined in (11.35). [Sect. 11.7]

*Notes

The algebra $\mathbb{A}$ of the polynomial functions on the Young diagrams $\mathbb{Y}$ was introduced by Kerov–Olshanski [127].

Kerov's central limit theorem (Theorem 11.2) was proved in Kerov [122] and has been the most fundamental result on fluctuation of the Young diagrams or equivalently of the irreducible representations of the symmetric groups. There are many refinements and extensions, see Fulman [82, 83], Hora [103, 106], Ivanov–Olshanski [117] and Śniady [194].

The non-commutative extension of Kerov's central limit theorem was initiated by Hora [106]. The contents of Sects. 11.2–11.4 are based on Hora [106] though some proofs are refined here. We refer to Bannai–Ito [17] for spectral structure of adjacency matrices of general commutative association schemes.

The theory of various cumulants have been developed considerably within the framework of quantum probability theory. For the free cumulants discussed in Sect. 11.5, see, e.g., Voiculescu–Dykema–Nica [216] and Speicher [197, 198].

For the proof of Proposition 11.15 see Ivanov–Olshanski [117, Proposition 2.6], where the definition of Frobenius coordinates differs from ours by 1/2. Their definition is

$$a_i = \lambda_i - i + \frac{1}{2}, \qquad b_i = \lambda'_i - i + \frac{1}{2},$$

namely, the half of each box along the main diagonal is counted.

For the proof of Proposition 11.16, see Ivanov–Olshanski [117, Proposition 3.2] and Biane [27, Sect. 5]. The Frobenius formula (11.23) is found in, e.g., Macdonald [153, Sect. I.7].

The derivation of Kerov's polynomials in Sect. 11.6 is due to Okounkov, as is mentioned in Biane [27]. The weight degree was introduced by Ivanov–Olshanski [117].

Theorem 11.17 was proved by Hora [103] by means of a purely combinatorial argument without using representation theory of the symmetric group. The properties (H1) and (H2) were the key observation and the Hermite polynomials appeared as matching polynomials of complete graphs. Observing that $\{\Sigma_\rho\,;\,\rho \in \mathbb{Y}\}$ forms a basis of $\mathbb{A}$, Ivanov–Olshanski [117] analyzed asymptotic behaviour of Σ_ρ's by introducing appropriate filtrations in $\mathbb{A}$ and clarified properties (H1) and (H2) more systematically.

Theorem 11.19 is due to Ivanov–Olshanski [117]. A similar result for random matrices is obtained by Johansson [120]. The Berry–Esseen theorem is found in, e.g., Chung [58, Sect. 7.4], Durrett [76, Sect. 2.4].

Theorem 11.20, concerning the Plancherel measure, is due to Fulman [82]. Further generalization to the Jack measures was also achieved in Fulman [83], which will be discussed in the next chapter.

Śniady deeply exploited the moment method for fluctuation of Young diagrams in a wider class of ensembles. In the result of Ivanov–Olshanski mentioned above, the highest terms of Kerov's polynomials were essential. Further analysis on the coefficients of Kerov's polynomials was done by Biane [27] and Śniady [194]. To establish the concentration phenomenon in irreducible decomposition of representations of the symmetric groups, Biane [25,26] introduced the notion of asymptotic factorization property of characters. Roughly speaking, this property means a decay condition of variance (= the second cumulant) of generators of $\mathbb{A}$ stated in Sect. 11.5, which yields concentration (as a weak law of large numbers). Such a situation was already illustrated in the typical Plancherel case in Chap. 10. Śniady [194] treated proper decay conditions of higher cumulants to obtain Gaussian fluctuation of Young diagrams in a wide variety of ensembles including the Plancherel one. In order to analyze mixed moments for elements of $\mathbb{A}$, Śniady brought in the method of genus expansion used in random matrix theory.

12

Deformation of Kerov's Central Limit Theorem

In the previous chapter we studied the adjacency matrix $A_{(k,1^{n-k})}$ corresponding to k-cycles in $S(n)$ and its quantum decomposition:

$$A_{(k,1^{n-k})} = A^+_{(k,1^{n-k})} + A^-_{(k,1^{n-k})} + A^\circ_{(k,1^{n-k})}.$$

In this section, restricting ourselves to the case of $k = 2$, we shall introduce their one-parameter deformation (α-deformation). This deformation is related to the Jack measure on Young diagrams and the Metropolis algorithm on the symmetric group. The associated central limit theorem follows from the quantum central limit theorem (Theorem 11.13), which shows again usefulness of quantum decomposition.

12.1 Jack Symmetric Functions

We start with the Schur functions because the Jack symmetric functions are defined as their one-parameter deformation.

Let $\lambda = (\lambda_1 \geq \lambda_2 \geq \cdots) \in \mathbb{Y}$ be a Young diagram. For $n \geq \mathrm{row}\,(\lambda)$ we consider a polynomial in n variables $x_1, \dots, x_n$ defined by

$$\det(x_i^{\lambda_j+n-j}) = \det \begin{pmatrix} x_1^{\lambda_1+n-1} & x_1^{\lambda_2+n-2} & \cdots & x_1^{\lambda_{n-1}+1} & x_1^{\lambda_n} \\ x_2^{\lambda_1+n-1} & x_2^{\lambda_2+n-2} & \cdots & x_2^{\lambda_{n-1}+1} & x_2^{\lambda_n} \\ \vdots & \vdots & & \vdots & \vdots \\ x_n^{\lambda_1+n-1} & x_n^{\lambda_2+n-2} & \cdots & x_n^{\lambda_{n-1}+1} & x_n^{\lambda_n} \end{pmatrix}. \tag{12.1}$$

Since (12.1) is divisible by $x_i - x_j$ for $i \neq j$, so is by Vandermonde's determinant. Therefore

$$s_\lambda(x_1, \dots, x_n) = \frac{\det(x_i^{\lambda_j+n-j})}{\det(x_i^{n-j})} \tag{12.2}$$

becomes a polynomial which is, as is easily verified, symmetric and homogeneous of degree $|\lambda|$.

A. Hora and N. Obata: Deformation of Kerov's Central Limit Theorem. In: A. Hora and N. Obata, Quantum Probability and Spectral Analysis of Graphs, Theoretical and Mathematical Physics, 321–350 (2007)
DOI 10.1007/3-540-48863-4_12

Definition 12.1. The polynomial $s_\lambda(x_1, \dots, x_n)$ defined by (12.2) is called the *Schur polynomial* in the variables $x_1, \dots, x_n$ corresponding to the Young diagram λ.

Example 12.2 (Schur polynomials in three variables).

$$
\begin{aligned}
s_{(1)}(x_1, x_2, x_3) &= x_1 + x_2 + x_3 \\
s_{(2)}(x_1, x_2, x_3) &= x_1^2 + x_2^2 + x_3^2 + x_1x_2 + x_2x_3 + x_3x_1 \\
s_{(1^2)}(x_1, x_2, x_3) &= x_1x_2 + x_2x_3 + x_3x_1 \\
s_{(3)}(x_1, x_2, x_3) &= x_1^3 + x_2^3 + x_3^3 \\
&\qquad + x_1^2(x_2 + x_3) + x_2^2(x_3 + x_1) + x_3^2(x_1 + x_2) + x_1x_2x_3 \\
s_{(12)}(x_1, x_2, x_3) &= x_1^2(x_2 + x_3) + x_2^2(x_3 + x_1) + x_3^2(x_1 + x_2) + 2x_1x_2x_3 \\
s_{(1^3)}(x_1, x_2, x_3) &= x_1x_2x_3.
\end{aligned}
$$

Remark 12.3. The Schur polynomials give irreducible characters of $U(n)$. Actually (12.2) is known as Weyl's character formula for $U(n)$.

Let $\mathbf{\Lambda}_n^k$ denote the linear space of polynomials in n variables which are symmetric and homogeneous of degree k (including the zero polynomial). For $\lambda = (\lambda_1 \geq \lambda_2 \geq \cdots) \in \mathbb{Y}$ and $n \geq \mathrm{row}\,(\lambda)$ we define

$$
m_\lambda(x_1, \dots, x_n) = \sum_{(\alpha_1, \dots, \alpha_n)} x_1^{\alpha_1} \cdots x_n^{\alpha_n},
$$

where $(\alpha_1, \dots, \alpha_n)$ runs over all distinct permutations of $(\lambda_1, \dots, \lambda_n)$. Obviously, m_λ is symmetric and homogeneous of degree $|\lambda|$.

Lemma 12.4. *For $n \geq 1$ and $k \geq 0$, $\{m_\lambda\,;\, \mathrm{row}\,(\lambda) \leq n,\ |\lambda| = k\}$ forms a linear basis of $\mathbf{\Lambda}_n^k$. In particular, if $n \geq k$, so does $\{m_\lambda\,;\, |\lambda| = k\}$.*

It is sometimes convenient to consider a formal series in infinitely many variables since the number of variables of a symmetric function is not very essential. Let k be fixed and take $m \geq n$. There is a natural projection $\mathbf{\Lambda}_m^k \to \mathbf{\Lambda}_n^k$ defined by $f(x_1, \dots, x_n, \dots, x_m) \mapsto f(x_1, \dots, x_n, 0, \dots, 0)$. Equipped with these projections, $\{\mathbf{\Lambda}_n^k\}_n$ becomes a projective system of linear spaces. Define

$$
\mathbf{\Lambda}^k = \operatorname*{proj\,lim}_{n \to \infty} \mathbf{\Lambda}_n^k .
$$

It is shown that the canonical projection $\mathbf{\Lambda}^k \to \mathbf{\Lambda}_n^k$ is a linear isomorphism whenever $n \geq k$. Since

$$
m_\lambda(x_1, x_2, \dots, x_n, 0) = m_\lambda(x_1, x_2, \dots, x_n), \qquad n \geq \mathrm{row}\,(\lambda),
$$

there would be no confusion to use the same symbol m_λ for the inverse image of the canonical projection $\mathbf{\Lambda}^k \to \mathbf{\Lambda}_n^k$. Moreover, the inverse image is identified with the formal sum of monomials:

$$m_\lambda(x_1, x_2, \dots) = \sum_{(\alpha_1,\alpha_2,\dots)} x_1^{\alpha_1} x_2^{\alpha_2} \cdots, \tag{12.3}$$

where $(\alpha_1, \alpha_2, \dots)$ runs over all distinct permutations of $(\lambda_1, \lambda_2, \dots)$. This $m_\lambda(x_1, x_2, \dots)$ is called the *monomial symmetric function* corresponding to the Young diagram λ. The term 'function' is used because m_λ is no longer a polynomial but a formal sum of monomials.

It follows from Lemma 12.4 that $\{m_\lambda\,;\, |\lambda| = k\}$ forms a linear basis of $\mathbf{\Lambda}^k$. We set

$$\mathbf{\Lambda} = \bigoplus_{k=0}^{\infty} \mathbf{\Lambda}^k,$$

which is the linear space spanned by $\{m_\lambda\,;\, \lambda \in \mathbb{Y}\}$. An element of $\mathbf{\Lambda}$ is called a *symmetric function*. Equipped with an obvious multiplication, $\mathbf{\Lambda}$ becomes an algebra, which is called the *algebra of symmetric functions*.

For an integer $k \geq 0$ the kth *power sum* is defined by

$$p_k(x_1, x_2, \dots) = m_{(k)}(x_1, x_2, \dots) = x_1^k + x_2^k + \cdots, \qquad k \geq 1,$$
$$p_0(x_1, x_2, \dots) = 1.$$

Obviously, $p_k \in \mathbf{\Lambda}^k$. For $\lambda = (\lambda_1 \geq \lambda_2 \geq \cdots) \in \mathbb{Y}$ we set

$$p_\lambda = p_{\lambda_1} p_{\lambda_2} \cdots.$$

It is obvious that p_λ is symmetric and homogeneous of degree $|\lambda|$. Moreover, $\{p_\lambda\,;\, \lambda \in \mathbb{Y}\}$ forms a linear basis of $\mathbf{\Lambda}$.

Now let us go back to the Schur polynomials $s_\lambda(x_1, \dots, x_n)$. It is readily known that $s_\lambda(x_1, \dots, x_n) \in \mathbf{\Lambda}_n^k$, where $k = |\lambda|$. We see by (12.2) that

$$s_\lambda(x_1, \dots, x_n, 0) = s_\lambda(x_1, \dots, x_n), \qquad n \geq \operatorname{row}(\lambda),$$

so that, just as in the case of m_λ, we use the same symbol s_λ for the inverse image under the canonical projection $\mathbf{\Lambda}^k \to \mathbf{\Lambda}_n^k$. It is known that $\{s_\lambda\,;\, |\lambda| = k\}$ forms a linear basis of $\mathbf{\Lambda}^k$. We call $s_\lambda \in \mathbf{\Lambda}^k$, $|\lambda| = k$, a *Schur function*.

Let us introduce an inner product on $\mathbf{\Lambda}$. For $\lambda \in \mathbb{Y}$ we set

$$z_\lambda = \prod_{j=1}^{\infty} j^{m_j(\lambda)} m_j(\lambda)!,$$

where $m_j(\lambda)$ is the number of j-rows in λ, see also (9.1). Taking into account that $\{p_\lambda\,;\, \lambda \in \mathbb{Y}\}$ forms a linear basis of $\mathbf{\Lambda}$, we define an inner product of $\mathbf{\Lambda}$ by

$$\langle p_\lambda, p_\mu \rangle = \delta_{\lambda\mu} z_\lambda, \qquad \lambda, \mu \in \mathbb{Y}. \tag{12.4}$$

The Schur functions are obtained by means of orthogonalization of the monomial symmetric functions. We need the *natural partial order* (also called *dominance partial order*) on $\mathbb{Y}_n$. For $\lambda, \mu \in \mathbb{Y}_n$ we write $\lambda \geq \mu$ if

$$\lambda_1 + \cdots + \lambda_i \geq \mu_1 + \cdots + \mu_i \quad \text{for all } i = 1, \ldots, n.$$

We write $\lambda > \mu$ if $\lambda \geq \mu$ and $\lambda \neq \mu$.

Proposition 12.5. *The Schur functions $\{s_\lambda\}$ are characterized by the following two properties:*

(i) $\langle s_\lambda, s_\mu \rangle = 0$ *for* $\lambda \neq \mu$*;*
(ii) s_λ *is expressible in the form*

$$s_\lambda = m_\lambda + \sum_{\mu < \lambda} v_{\lambda\mu} m_\mu, \tag{12.5}$$

with some constants $v_{\lambda\mu}$.

Thus, we see from (12.5) that the transition matrix from the monomial symmetric functions to the Schur functions is strictly upper unitriangular.

Now we introduce a one-parameter deformation of the inner product (12.4). With each $\alpha > 0$ we associate an inner product on $\mathbf{\Lambda}$ defined by

$$\langle p_\lambda, p_\mu \rangle^{(\alpha)} = \delta_{\lambda,\mu} \alpha^{\mathrm{row}\,(\lambda)} z_\lambda, \qquad \lambda, \mu \in \mathbb{Y}.$$

This will be referred to as the *α-inner product.* The Jack symmetric functions are introduced as a natural extension of Proposition 12.5.

Definition 12.6. Let $\alpha > 0$ be a fixed parameter. The *Jack symmetric functions* $\{P_\lambda^{(\alpha)}\}$ are characterized by the following properties:

(i) $\langle P_\lambda^{(\alpha)}, P_\mu^{(\alpha)} \rangle^{(\alpha)} = 0$ for $\lambda \neq \mu$;
(ii) $P_\lambda^{(\alpha)}$ is expressible in the form

$$P_\lambda^{(\alpha)} = m_\lambda + \sum_{\mu < \lambda} v_{\lambda\mu}^{(\alpha)} m_\mu$$

with some constants $v_{\lambda\mu}^{(\alpha)}$.

By definition, the Jack symmetric function $P_\lambda^{(\alpha)}$ is a homogeneous symmetric function of degree $|\lambda|$. The relation to the Schur functions is apparent. The Jack symmetric functions with parameter $\alpha = 1$ are the Schur functions, i.e.,

$$P_\lambda^{(1)} = s_\lambda, \qquad \lambda \in \mathbb{Y}. \tag{12.6}$$

Example 12.7. A few Jack symmetric functions:

$$\begin{aligned}
P_{(1)}^{(\alpha)} &= m_{(1)} \\
P_{(2)}^{(\alpha)} &= m_{(2)} + \frac{2}{\alpha+1} m_{(1^2)} \\
P_{(1^2)}^{(\alpha)} &= m_{(1^2)} \\
P_{(3)}^{(\alpha)} &= m_{(3)} + \frac{3}{2\alpha+1} m_{(12)} + \frac{6}{(2\alpha+1)(\alpha+1)} m_{(1^3)} \\
P_{(12)}^{(\alpha)} &= m_{(12)} + \frac{6}{\alpha+2} m_{(1^3)} \\
P_{(1^3)}^{(\alpha)} &= m_{(1^3)} .
\end{aligned}$$

12.2 Jack Graphs

Recall that $\{s_\lambda\,;\,\lambda \in \mathbb{Y}\}$ forms a linear basis of the algebra $\mathbf{\Lambda}$ of symmetric functions. Hence $s_{(1)} s_\lambda$ is expressible as a (finite) linear combination of those bases. Explicitly we have the following.

Theorem 12.8 (Pieri's formula). *For any $\lambda \in \mathbb{Y}$ it holds that*

$$s_{(1)} s_\lambda = \sum_{\Lambda:\lambda \nearrow \Lambda} s_\Lambda . \tag{12.7}$$

Thus, the Young graph (Fig. 9.4) is reconstructed from Pieri's formula. In fact, Λ is adjacent to λ in the Young graph if and only if s_Λ appears with a non-zero (in fact, the value is 1) coefficient on the right-hand side of (12.7). This viewpoint is the key to a Jack analogue of the Young graph.

First we need to extend Pieri's formula for Jack symmetric functions. Let $\lambda \in \mathbb{Y}$ and consider (i,j)-box $b \in \lambda$. The *arm length* and the *leg length* of the box b are defined by

$$a_\lambda(b) = \lambda_i - j, \qquad l_\lambda(b) = \lambda'_j - i,$$

respectively, see Fig. 12.1(a). A hook is thus decomposed into an arm and a leg. Denoting by $h_\lambda(b)$ the hook length, we have

$$h_\lambda(b) = a_\lambda(b) + l_\lambda(b) + 1.$$

Next consider $\lambda, \Lambda \in \mathbb{Y}$ with $\lambda \nearrow \Lambda$. Then $b = \Lambda \setminus \lambda$ is a corner of Λ. Let $\mathrm{ver}\,(\Lambda \setminus \lambda)$ denote the boxes in λ which lie strictly above the corner b. Similarly, $\mathrm{hor}\,(\Lambda \setminus \lambda)$ denotes the boxes strictly on the left of b, see Fig. 12.1(b).

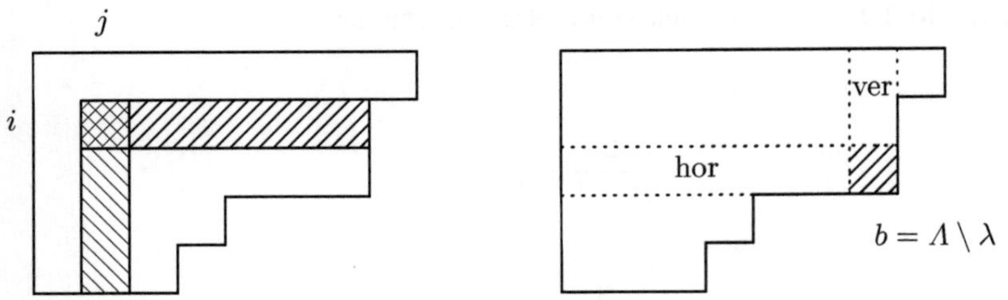

Fig. 12.1. (a) Arm and leg, (b) ver and hor

Theorem 12.9 (Pieri's formula for Jack symmetric functions). *Let* $\alpha > 0$. *For* $\lambda \in \mathbb{Y}$ *it holds that*

$$P^{(\alpha)}_{(1)} P^{(\alpha)}_{\lambda} = \sum_{\Lambda:\lambda\nearrow\Lambda} \kappa^{(\alpha)}(\lambda, \Lambda)\, P^{(\alpha)}_{\Lambda}, \tag{12.8}$$

where

$$\kappa^{(\alpha)}(\lambda, \Lambda) = \prod_{b\in \mathrm{ver}\,(\Lambda\setminus\lambda)} \frac{(\alpha a_\lambda(b) + l_\lambda(b) + \alpha)(\alpha a_\lambda(b) + l_\lambda(b) + 2)}{(\alpha a_\lambda(b) + l_\lambda(b) + 1)(\alpha a_\lambda(b) + l_\lambda(b) + \alpha + 1)}. \tag{12.9}$$

The *Jack graph* with parameter $\alpha > 0$, denoted by $\mathbb{J}^{(\alpha)}$, is the Young graph $\mathbb{Y}$ equipped with the weight function $\kappa^{(\alpha)}$, where each edge $\lambda \nearrow \Lambda$ carries a weight $\kappa^{(\alpha)}(\lambda, \Lambda)$. Strictly speaking, the Jack graph is not a graph in general but a *network*.

For $\alpha = 1$, Theorem 12.9 is reduced to Pieri's formula for Schur functions (Theorem 12.8). In fact, we have

$$P^{(1)}_{\lambda} = s_\lambda, \qquad \kappa^{(1)}(\lambda, \Lambda) = 1,$$

which follow from (12.6) and (12.9), respectively. Hence the Jack graph $\mathbb{J}^{(1)}$ is nothing else the Young graph. Many notions associated with the Young graph can be extended to the Jack graph.

Remark 12.10. The limiting case of $\alpha \to \infty$ is also interesting. In view of (12.9), we have

$$\kappa^{(\infty)}(\lambda, \Lambda) = m_k(\Lambda), \tag{12.10}$$

where k is determined in such a way that the box $\Lambda \setminus \lambda$ is contained in a k-row of Λ. On the other hand, the monomial symmetric functions satisfy Pieri's formula with coefficients (12.10). Actually, $P^{(\alpha)}_{\lambda}$ is known to be well defined in this limiting case as

$$P^{(\infty)}_{\lambda} = m_\lambda.$$

The Jack graph $\mathbb{J}^{(\infty)}$ with edge multiplicities (12.10) is called the *Kingman graph.*

As before, let $\mathfrak{T}$ denote the set of infinite paths in the Young graph $\mathbb{Y}$ starting at $\varnothing$. This symbol is also used for the Jack graph without confusion for the graph structure of the Jack graph agrees with $\mathbb{Y}$. With each finite path

$$u = (\varnothing = \lambda^{(0)} \nearrow \lambda^{(1)} \nearrow \cdots \nearrow \lambda^{(n)} = \lambda)$$

we associate the *weight* defined by

$$w_u = \prod_{i=1}^{n} \kappa^{(\alpha)}(\lambda^{(i-1)}, \lambda^{(i)}). \tag{12.11}$$

For the Young graph (i.e., $\alpha = 1$) we have $w_u = 1$ for all finite paths u. The *combinatorial dimension function* of the Jack graph $\mathbb{J}^{(\alpha)}$ is defined by

$$d^{(\alpha)}(\lambda) = \sum_{u} w_u, \qquad \lambda \in \mathbb{Y}, \tag{12.12}$$

where u runs over the finite paths $(\varnothing \nearrow \cdots \nearrow \lambda)$. In the case of the Young graph we have

$$d^{(1)}(\lambda) = f^{\lambda} = \dim \lambda$$

and f^{λ} is given by the hook formula (Theorem 9.7), i.e.,

$$d^{(1)}(\lambda) = \frac{n!}{\prod_{b\in\lambda} h_\lambda(b)}, \qquad \lambda \in \mathbb{Y}_n.$$

This formula is extended as follows.

Theorem 12.11 (Stanley). *The combinatorial dimension function of the Jack graph* $\mathbb{J}^{(\alpha)}$ *satisfies*

$$d^{(\alpha)}(\lambda) = \frac{\alpha^n n!}{\prod_{b\in\lambda}(\alpha a_\lambda(b) + l_\lambda(b) + \alpha)}, \qquad \lambda \in \mathbb{Y}_n. \tag{12.13}$$

12.3 Deformed Young Diagrams

We are now in a position to introduce an *α-deformed Young diagram* or an *anisotropic Young diagram* and related concepts. A Young diagram is disposed on the plane keeping a symmetric balance, see Fig. 9.7. We now tip the balance by changing the unit length of the row axis. Given $\alpha > 0$, the *α-deformed Young diagram* is obtained by magnifying the row axis with a magnification of α, hence consists of boxes (rectangles) of size $\sqrt{2}\,\alpha \times \sqrt{2}$. For example, Fig. 12.2 illustrates the case of $\alpha = 2$.

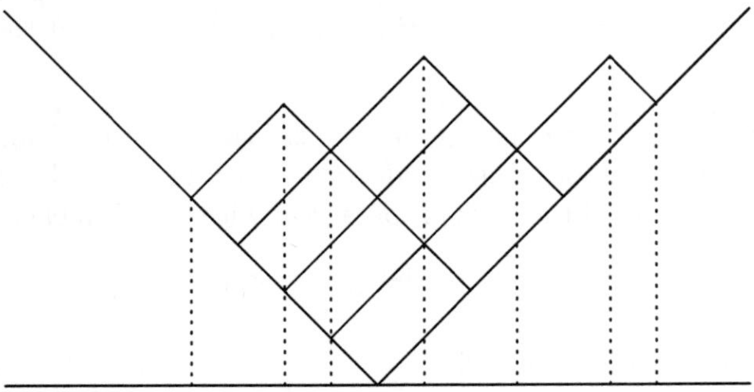

Fig. 12.2. Min–max coordinates of the α-deformed Young diagram ($\alpha = 2$)

We maintain the min–max coordinates for the α-deformation of a Young diagram $\lambda \in \mathbb{Y}$, which are denoted as

$$x_1^{(\alpha)} < y_1^{(\alpha)} < \cdots < y_{r-1}^{(\alpha)} < x_r^{(\alpha)}. \tag{12.14}$$

Accordingly, the *α-deformed transition measure* is defined by

$$\mathfrak{m}_\lambda^{(\alpha)} = \sum_{i=1}^{r} \mu_i \delta_{x_i^{(\alpha)}}, \tag{12.15}$$

where

$$\mu_i = \frac{(x_i^{(\alpha)} - y_1^{(\alpha)}) \cdots (x_i^{(\alpha)} - y_{r-1}^{(\alpha)})}{(x_i^{(\alpha)} - x_1^{(\alpha)}) \cdots (x_i^{(\alpha)} - x_{i-1}^{(\alpha)})(x_i^{(\alpha)} - x_{i+1}^{(\alpha)}) \cdots (x_i^{(\alpha)} - x_r^{(\alpha)})}. \tag{12.16}$$

Obviously, $\mu_i > 0$. As in (9.16), we have

$$\frac{(z - y_1^{(\alpha)}) \cdots (z - y_{r-1}^{(\alpha)})}{(z - x_1^{(\alpha)}) \cdots (z - x_r^{(\alpha)})} = \int_{-\infty}^{+\infty} \frac{\mathfrak{m}_\lambda^{(\alpha)}(dx)}{z - x}, \tag{12.17}$$

from which by taking $z \to \infty$ (after multiplying z) we see that $\mathfrak{m}_\lambda^{(\alpha)}$ is a probability measure.

Lemma 12.12. *For $\alpha > 0$ consider the α-deformation of a Young diagram $\lambda \in \mathbb{Y}$ with min–max coordinates given as in* (12.14)*. Then for the α-deformed transition measure we have*

$$\mathfrak{m}_\lambda^{(\alpha)}(\{x_s^{(\alpha)}\}) = \prod_{b \in \mathrm{ver}\,(\Lambda^{(s)} \setminus \lambda)} \frac{\alpha a_\lambda(b) + l_\lambda(b) + \alpha}{\alpha a_\lambda(b) + l_\lambda(b) + \alpha + 1}$$
$$\times \prod_{b \in \mathrm{hor}\,(\Lambda^{(s)} \setminus \lambda)} \frac{\alpha a_\lambda(b) + l_\lambda(b) + 1}{\alpha a_\lambda(b) + l_\lambda(b) + \alpha + 1},$$

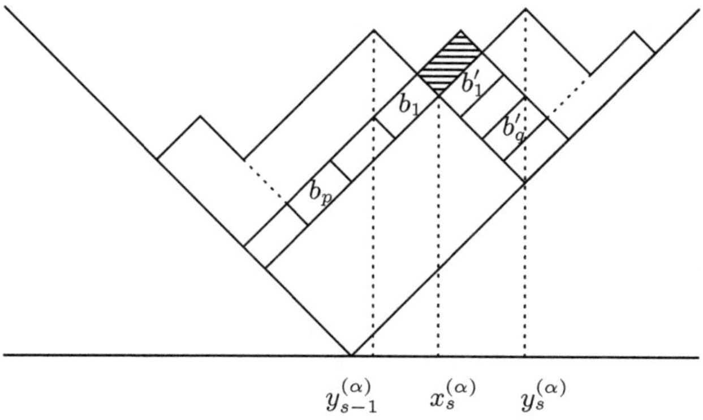

Fig. 12.3. $\Lambda^{(s)}$: $b_1, \dots, b_p$ and $b'_1, \dots, b'_q$

for $s = 1, \dots, r$, *where* $\Lambda^{(s)}$ *denotes the* α*-deformed Young diagram obtained by putting a box (rectangle) at the* s*th valley* $x_s^{(\alpha)}$ *of* λ.

Proof. The argument is similar to the proof of Lemma 9.26. Consider boxes $b_1, \dots, b_p$ in the first block of $\mathrm{hor}\,(\Lambda^{(s)} \setminus \lambda)$ and $b'_1, \dots, b'_q$ in the first block of $\mathrm{ver}\,(\Lambda^{(s)} \setminus \lambda)$, see Fig. 12.3. We then have

$$\prod_{b \in \{b_1, \dots, b_p\}} \frac{\alpha a_\lambda(b) + l_\lambda(b) + 1}{\alpha a_\lambda(b) + l_\lambda(b) + \alpha + 1} = \frac{\alpha a_\lambda(b_1) + l_\lambda(b_1) + 1}{\alpha a_\lambda(b_p) + l_\lambda(b_p) + \alpha + 1}$$

$$= \frac{x_s^{(\alpha)} - y_{s-1}^{(\alpha)}}{y_{s-1}^{(\alpha)} - x_{s-1}^{(\alpha)} + x_s^{(\alpha)} - y_{s-1}^{(\alpha)}} = \frac{x_s^{(\alpha)} - y_{s-1}^{(\alpha)}}{x_s^{(\alpha)} - x_{s-1}^{(\alpha)}}.$$

Therefore, taking all blocks in the horizontal zone into account, we get

$$\prod_{b \in \mathrm{hor}\,(\Lambda^{(s)} \setminus \lambda)} \frac{\alpha a_\lambda(b) + l_\lambda(b) + 1}{\alpha a_\lambda(b) + l_\lambda(b) + \alpha + 1} = \frac{x_s^{(\alpha)} - y_{s-1}^{(\alpha)}}{x_s^{(\alpha)} - x_{s-1}^{(\alpha)}} \cdots \frac{x_s^{(\alpha)} - y_1^{(\alpha)}}{x_s^{(\alpha)} - x_1^{(\alpha)}}. \tag{12.18}$$

Similarly, for the vertical zone we get

$$\prod_{b \in \{b'_1, \dots, b'_q\}} \frac{\alpha a_\lambda(b) + l_\lambda(b) + \alpha}{\alpha a_\lambda(b) + l_\lambda(b) + \alpha + 1}$$

$$= \frac{\alpha a_\lambda(b'_1) + l_\lambda(b'_1) + \alpha}{\alpha a_\lambda(b'_q) + l_\lambda(b'_q) + \alpha + 1} = \frac{y_s^{(\alpha)} - x_s^{(\alpha)}}{x_{s+1}^{(\alpha)} - x_s^{(\alpha)}};$$

hence,

$$\prod_{b \in \mathrm{ver}\,(\Lambda^{(s)} \setminus \lambda)} \frac{\alpha a_\lambda(b) + l_\lambda(b) + \alpha}{\alpha a_\lambda(b) + l_\lambda(b) + \alpha + 1} = \frac{y_s^{(\alpha)} - x_s^{(\alpha)}}{x_{s+1}^{(\alpha)} - x_s^{(\alpha)}} \cdots \frac{y_{r-1}^{(\alpha)} - x_s^{(\alpha)}}{x_r^{(\alpha)} - x_s^{(\alpha)}}. \tag{12.19}$$

The desired result follows by combining (12.18), (12.19) and (12.16). □

Remark 12.13. Let (i,j) be the indices of the shaded box b in Fig. 12.3. Then we have

$$x_s^{(\alpha)} = \alpha(j-1) - (i-1),$$

which is called the *α-content* of box b. If $\alpha = 1$, the α-content coincides with the usual content $j - i$.

In Sect. 9.5 we mentioned a relationship between the hook lengths and the min–max coordinates. During the proof of Lemma 9.26 we established

$$\frac{\dim \Lambda^{(s)}}{(n+1)\dim\lambda} = \frac{\prod_{b\in\lambda} h_\lambda(b)}{\prod_{b\in\Lambda^{(s)}} h_{\Lambda^{(s)}}(b)}, \tag{12.20}$$

where $\Lambda^{(s)} \in \mathbb{Y}_{n+1}$ is the Young diagram obtained by putting a box at the sth valley of $\lambda \in \mathbb{Y}_n$. Calculating the right-hand side of (12.20) in terms of the min–max coordinates, we obtained

$$\mathfrak{m}_\lambda(\{x_s\}) = \frac{\prod_{b\in\lambda} h_\lambda(b)}{\prod_{b\in\Lambda^{(s)}} h_{\Lambda^{(s)}}(b)}. \tag{12.21}$$

The formula in Lemma 12.12 is an α-deformation of (12.21).

12.4 Jack Measures

In Sect. 9.6 we introduced the Plancherel measure, which is the most fundamental probability measure on the path space $\mathfrak{T}$ associated with the Young graph $\mathbb{Y}$. Having discussed the α-deformation of the Young graph, i.e., the Jack graph, we shall now introduce the α-deformed Plancherel measure, i.e., the Jack measure. For the definition we employ the same idea as in Sect. 9.6. Going back to formula (12.21), we remark that $\mathfrak{m}_\lambda$ is a probability measure. Then we have

$$\frac{1}{\prod_{b\in\lambda} h_\lambda(b)} = \sum_s \frac{1}{\prod_{b\in\Lambda^{(s)}} h_{\Lambda^{(s)}}(b)} = \sum_{\Lambda:\lambda\nearrow\Lambda} \frac{1}{\prod_{b\in\Lambda} h_\Lambda(b)},$$

which means that $1/\prod_{b\in\lambda} h_\lambda(b)$ is a harmonic function on the Young graph. This harmonic function played a crucial role in the definition of the Plancherel measure. So we start with the following.

Definition 12.14. A function φ on $\mathbb{Y}$ is called *harmonic* on the Jack graph $\mathbb{J}^{(\alpha)}$ if

$$\varphi(\lambda) = \sum_{\Lambda:\lambda\nearrow\Lambda} \kappa^{(\alpha)}(\lambda,\Lambda)\varphi(\Lambda), \qquad \lambda\in\mathbb{Y}. \tag{12.22}$$

Proposition 12.15. *The function φ defined by*

$$\varphi(\lambda) = \frac{1}{\prod_{b\in\lambda}(\alpha a_\lambda(b) + l_\lambda(b) + 1)}, \qquad \lambda\in\mathbb{Y}, \tag{12.23}$$

is a positive, normalized harmonic function on the Jack graph $\mathbb{J}^{(\alpha)}$.

Proof. Obviously, $\varphi(\lambda) > 0$ for all $\lambda \in \mathbb{Y}$ and $\varphi(\varnothing) = 1$ (normalized). Take λ and Λ such that $\lambda \nearrow \Lambda$. Factorizing $\varphi(\Lambda)$ into three parts:

$$\begin{aligned}\varphi(\Lambda) &= \prod_{b\in\Lambda} \frac{1}{\alpha a_\Lambda(b) + l_\Lambda(b) + 1} \\ &= \prod_{b\in \mathrm{ver}\,(\Lambda\setminus\lambda)} \frac{1}{\alpha a_\lambda(b) + l_\lambda(b) + 2} \prod_{b\in \mathrm{hor}\,(\Lambda\setminus\lambda)} \frac{1}{\alpha a_\lambda(b) + l_\lambda(b) + \alpha + 1} \\ &\quad \times \prod_{\mathrm{rest}} \frac{1}{\alpha a_\lambda(b) + l_\lambda(b) + 1},\end{aligned}$$

we have

$$\begin{aligned}\kappa^{(\alpha)}(\lambda, \Lambda)\varphi(\Lambda) = \varphi(\lambda) &\prod_{b\in \mathrm{ver}\,(\Lambda\setminus\lambda)} \frac{\alpha a_\lambda(b) + l_\lambda(b) + \alpha}{\alpha a_\lambda(b) + l_\lambda(b) + \alpha + 1} \\ \times &\prod_{b\in \mathrm{hor}\,(\Lambda\setminus\lambda)} \frac{\alpha a_\lambda(b) + l_\lambda(b) + 1}{\alpha a_\lambda(b) + l_\lambda(b) + \alpha + 1}.\end{aligned}$$

Letting $\Lambda^{(1)}, \dots, \Lambda^{(r)}$ denote the upper level diagrams of λ (i.e. $\lambda \nearrow \Lambda^{(s)}$) and using Lemma 12.12, we have

$$\sum_{\Lambda:\lambda\nearrow\Lambda} \kappa^{(\alpha)}(\lambda, \Lambda)\varphi(\Lambda) = \varphi(\lambda) \sum_{s=1}^{r} \mathfrak{m}_\lambda^{(\alpha)}(\{x_s^{(\alpha)}\}) = \varphi(\lambda),$$

which proves that φ is harmonic on $\mathbb{J}^{(\alpha)}$. □

As before, $\mathfrak{T}$ stands for the set of all infinite paths in $\mathbb{Y}$ starting at $\varnothing$. Given a finite path

$$u = (\varnothing \nearrow \lambda^{(1)} \nearrow \cdots \nearrow \lambda^{(n)} = \lambda),$$

we define a cylindrical subset of $\mathfrak{T}$ by

$$C_u = \{t \in \mathfrak{T};\, t(j) = \lambda^{(j)},\, j = 0, 1, \dots, n\}.$$

Let us set

$$\mathfrak{J}^{(\alpha)}(C_u) = w_u \varphi(\lambda), \tag{12.24}$$

where w_u is the weight of u defined in (12.11) and φ the harmonic function defined in (12.23). Note that (12.24) determines consistently a finitely additive probability on cylindrical subsets of $\mathfrak{T}$. In fact, it is obvious that

$$\mathfrak{J}^{(\alpha)}(\mathfrak{T}) = \varphi(\varnothing) = 1.$$

By virtue of harmonicity of φ, we see that

$$\begin{aligned}\sum_{\Lambda:\lambda\nearrow\Lambda} \mathfrak{J}^{(\alpha)}(C_{(u\nearrow\Lambda)}) &= \sum_{\Lambda:\lambda\nearrow\Lambda} w_{(u\nearrow\Lambda)}\varphi(\Lambda) \\ &= \sum_{\Lambda:\lambda\nearrow\Lambda} w_u\kappa^{(\alpha)}(\lambda,\Lambda)\varphi(\Lambda) \\ &= w_u\varphi(\lambda) = \mathfrak{J}^{(\alpha)}(C_u).\end{aligned}$$

Thus $\mathfrak{J}^{(\alpha)}$ is uniquely extended to a probability on $\mathfrak{T}$ by the Hopf extension theorem, which we call the *Jack measure*. It is a central measure in the sense that $\mathfrak{J}^{(\alpha)}(C_u)/w_u$ depends only on the terminal vertex of the path u. The distribution on the nth stratum induced by the Jack measure, denoted by $\mathfrak{J}_n^{(\alpha)}$, is called the *Jack measure on* $\mathbb{Y}_n$. By definition, for $\lambda \in \mathbb{Y}_n$ we have

$$\begin{aligned}\mathfrak{J}_n^{(\alpha)}(\lambda) &= \mathfrak{J}^{(\alpha)}(\{t\in\mathfrak{T}\,;\, t(n)=\lambda\}) \\ &= \sum_u \mathfrak{J}^{(\alpha)}(C_u) = \sum_u w_u\varphi(\lambda) = d^{(\alpha)}(\lambda)\varphi(\lambda),\end{aligned}$$

where the sum is taken over all finite paths $u = (\varnothing \nearrow \cdots \nearrow \lambda)$. Applying Stanley's formula (Theorem 12.11), we obtain

Proposition 12.16. *For the Jack measure $\mathfrak{J}_n^{(\alpha)}$ on $\mathbb{Y}_n$, we have*

$$\mathfrak{J}_n^{(\alpha)}(\lambda) = \frac{\alpha^n\, n!}{\prod_{b\in\lambda}(\alpha a_\lambda(b)+l_\lambda(b)+\alpha)(\alpha a_\lambda(b)+l_\lambda(b)+1)}, \qquad \lambda\in\mathbb{Y}_n.$$

Our goal is to establish an α-deformed version of Kerov's central limit theorem (Theorem 11.2). Recall that Kerov's central limit theorem deals with the family of random variables $\{\Sigma_\rho(\lambda)\,;\,\rho\in\mathbb{Y}\}$ with respect to the Plancherel measure (in fact, ρ is restricted to cycles); for the definition of $\Sigma_\rho(\lambda)$ see (11.21). The α-deformed Plancherel measure has been just introduced, that is the Jack measure. So, in this last part of the present section, we clarify the random variables to be discussed in the α-deformed version of Kerov's central limit theorem.

As a consequence from the Frobenius character formula, the irreducible characters of $S(n)$ appear as the transition matrix between the Schur functions and the power sums that are homogeneous of degree n, namely,

$$s_\lambda = \sum_{\rho\in\mathbb{Y}_n} \frac{1}{z_\rho}\chi_\rho^\lambda p_\rho, \qquad \lambda\in\mathbb{Y}_n. \tag{12.25}$$

Now we define

$$J_\lambda^{(\alpha)} = P_\lambda^{(\alpha)} \prod_{b\in\lambda}(\alpha a_\lambda(b)+l_\lambda(b)+1), \tag{12.26}$$

which is a constant multiple of the Jack symmetric function $P_\lambda^{(\alpha)}$ (Definition 12.6). Then there exists a base transition matrix

$$\Theta(\alpha) = \big(\theta_\rho^\lambda(\alpha)\big), \tag{12.27}$$

where the (ρ,λ)-entry is $\theta_\rho^\lambda(\alpha)$, such that

$$J_\lambda^{(\alpha)} = \sum_{\rho\in\mathbb{Y}_n} \theta_\rho^\lambda(\alpha)\, p_\rho, \qquad \lambda\in\mathbb{Y}_n. \tag{12.28}$$

Since $s_\lambda = P_\lambda^{(1)}$, putting $\alpha=1$ in (12.28), we obtain

$$s_\lambda = P_\lambda^{(1)} = \sum_{\rho\in\mathbb{Y}_n} \frac{1}{\prod_{b\in\lambda} h_\lambda(b)}\, \theta_\rho^\lambda(1)\, p_\rho. \tag{12.29}$$

Comparing (12.25) and (12.29), we have

$$\theta_\rho^\lambda(1) = \frac{\prod_{b\in\lambda} h_\lambda(b)}{z_\rho}\, \chi_\rho^\lambda = \frac{n!}{z_\rho \dim\lambda}\, \chi_\rho^\lambda, \qquad \lambda,\rho\in\mathbb{Y}_n.$$

For $\rho,\lambda\in\mathbb{Y}$ satisfying $|\rho|\le|\lambda|=n$, we have

$$\theta_{(\rho,1^{n-|\rho|})}^\lambda(1) = \frac{n!}{z_{(\rho,1^{n-|\rho|})}}\, \tilde{\chi}_{(\rho,1^{n-|\rho|})}^\lambda = \frac{1}{\binom{n-|\rho|+m_1(\rho)}{m_1(\rho)} z_\rho}\, \Sigma_\rho(\lambda),$$

where $\Sigma_\rho(\lambda)$ is defined in (11.21). In particular,

Proposition 12.17. *Let $\rho\in\overline{\mathbb{Y}}$, i.e., $m_1(\rho)=0$, and $n\ge|\rho|$. Then it holds that*

$$\theta_{(\rho,1^{n-|\rho|})}^\lambda(1) = \frac{1}{z_\rho}\, \Sigma_\rho(\lambda), \qquad \lambda\in\mathbb{Y}_n. \tag{12.30}$$

The above relation (12.30) suggests that $\{\theta_{(\rho,1^{n-|\rho|})}^\lambda(\alpha)\,;\ \rho\in\overline{\mathbb{Y}}\}$ are the random variables to be discussed for the α-deformed version of Kerov's central limit theorem. In fact, dealing with particular random variables $\theta_{(2,1^{n-2})}^\lambda(\alpha)$, we shall achieve our goal in Sect. 12.6, where the limit distribution of the random variable $\theta_{(2,1^{n-2})}^\lambda(\alpha)$ as $n\to\infty$ will be described. Here we only mention the following.

Proposition 12.18. *It holds that*

$$\theta_{(2,1^{n-2})}^\lambda(\alpha) = \alpha n(\lambda') - n(\lambda), \qquad \lambda\in\mathbb{Y}_n, \tag{12.31}$$

where

$$n(\lambda) = \sum_{j=1}^{\mathrm{col}\,(\lambda)} \binom{\lambda'_j}{2} = \sum_{i=1}^{n} (i-1)\lambda_i. \tag{12.32}$$

Recall that the number $n(\lambda)$ also appeared in (9.5).

12.5 Deformed Adjacency Matrices

Our strategy of investigating the limit distribution of the random variables $\theta^{\lambda}_{(2,1^{n-2})}(\alpha)$ is to use an appropriate algebraic realization with quantum decomposition. To this end we shall introduce the α-deformed adjacency matrix.

We start with the review of the case of $\alpha = 1$. Let $\{\delta_x \,;\, x \in S(n)\}$ be the canonical basis of $\ell^2(S(n))$. For a Young diagram $\rho \in \mathbb{Y}_n$ let C_ρ denote the corresponding conjugacy class of $S(n)$ and set

$$\xi_\rho = \sum_{x\in C_\rho} \delta_x, \qquad \Phi(\rho) = \frac{1}{\sqrt{|C_\rho|}}\,\xi_\rho = \frac{1}{\sqrt{|C_\rho|}} \sum_{x\in C_\rho} \delta_x.$$

Then $\{\Phi(\rho)\,;\, \rho \in \mathbb{Y}_n\}$ becomes an orthonormal set of $\ell^2(S(n))$. Let $\Gamma(S(n)) \subset \ell^2(S(n))$ denote the subspace spanned by $\{\Phi(\rho)\,;\, \rho \in \mathbb{Y}_n\}$. The adjacency matrix A_ρ, defined by

$$(A_\rho)_{xy} = \begin{cases} 1, & xy^{-1} \in C_\rho\,, \\ 0, & \text{otherwise}, \end{cases}$$

acts on $\ell^2(S(n))$ in a usual manner. Then

$$\xi_\rho = A_\rho \delta_e.$$

On the other hand, A_ρ being identified with an element of the group $*$-algebra $\mathbb{C}[S(n)]$, we write

$$A_\rho = \sum_{x\in C_\rho} x.$$

The centre $\mathcal{Z}(\mathbb{C}[S(n)])$ is a linear span of $\{A_\rho\,;\, \rho \in \mathbb{Y}_n\}$. We define the intersection numbers $p^{\mu}_{\lambda\rho}$ by

$$A_\lambda A_\rho = \sum_{\mu\in\mathbb{Y}_n} p^{\mu}_{\lambda\rho} A_\mu, \qquad \lambda,\rho \in \mathbb{Y}_n. \tag{12.33}$$

Moreover, the correspondence

$$A_\rho \mapsto A_\rho\delta_e = \xi_\rho, \qquad \rho \in \mathbb{Y}_n,$$

extends uniquely a linear isomorphism from $\mathcal{Z}(\mathbb{C}[S(n)])$ onto $\Gamma(S(n))$. Then, in view of (12.33) we have

$$A_\lambda \xi_\rho = A_\lambda A_\rho \delta_e = \sum_{\mu\in\mathbb{Y}_n} p^{\mu}_{\lambda\rho} A_\mu \delta_e = \sum_{\mu\in\mathbb{Y}_n} p^{\mu}_{\lambda\rho} \xi_\mu,$$

so that $\mathcal{Z}(\mathbb{C}[S(n)]) \cong \Gamma(S(n))$ as $\mathcal{Z}(\mathbb{C}[S(n)])$-module.

By general theory (see Lemma 11.3) there exists a complete system $\{E_\lambda\,;\,\lambda\in\mathbb{Y}_n\}$ of orthogonal projections on $\ell^2(S(n))$ such that

$$A_\rho=\sum_{\lambda\in\mathbb{Y}_n}|C_\rho|\,\frac{\chi^\lambda_\rho}{\dim\lambda}\,E_\lambda,\qquad \rho\in\mathbb{Y}_n. \tag{12.34}$$

Taking into account that $E_\lambda\in\mathcal{Z}(\mathbb{C}[S(n)])$, let $\eta_\lambda\in\Gamma(S(n))$ be the corresponding element under the isomorphism $\mathcal{Z}(\mathbb{C}[S(n)])\cong\Gamma(S(n))$, namely, $\eta_\lambda=E_\lambda\delta_e$. It then follows from (12.34) that

$$\xi_\rho=\sum_{\lambda\in\mathbb{Y}_n}|C_\rho|\,\frac{\chi^\lambda_\rho}{\dim\lambda}\,\eta_\lambda.$$

Then, using the orthogonality relation (see Exercise 12.4)

$$\sum_{g\in S(n)}\chi^\lambda(g)\chi^\mu(g^{-1})=n!\,\delta_{\lambda\mu}, \tag{12.35}$$

we have

$$\frac{\eta_\lambda}{\dim\lambda}=\sum_{\rho\in\mathbb{Y}_n}\frac{\chi^\lambda_\rho}{n!}\,\xi_\rho. \tag{12.36}$$

Recall that $\mathbf{\Lambda}^n$ denotes the space of homogeneous symmetric functions of degree n. The correspondence

$$I:\sqrt{z_\rho}\,\Phi(\rho)\mapsto p_\rho,\qquad \rho\in\mathbb{Y}_n, \tag{12.37}$$

extends to a unitary operator, denoted by I again, from $\Gamma(S(n))$ onto $\mathbf{\Lambda}^n$. We see from (12.36) that

$$\frac{\sqrt{n!}}{\dim\lambda}\,\eta_\lambda=\sum_{\rho\in\mathbb{Y}_n}\frac{\chi^\lambda_\rho}{\sqrt{n!}}\,\xi_\rho=\sum_{\rho\in\mathbb{Y}_n}\frac{\chi^\lambda_\rho}{\sqrt{n!}}\,\frac{\sqrt{|C_\rho|}}{\sqrt{z_\rho}}\,\sqrt{z_\rho}\,\Phi(\rho),$$

which becomes

$$\frac{\sqrt{n!}}{\dim\lambda}\,\eta_\lambda=\sum_{\rho\in\mathbb{Y}_n}\frac{\chi^\lambda_\rho}{z_\rho}\,\sqrt{z_\rho}\,\Phi(\rho),$$

where $z_\rho|C_\rho|=n!$ is used. Then, applying I defined in (12.37) and (12.25), we obtain

$$I\biggl(\frac{\sqrt{n!}}{\dim\lambda}\eta_\lambda\biggr)=\sum_{\rho\in\mathbb{Y}_n}\frac{\chi^\lambda_\rho}{z_\rho}\,p_\rho=s_\lambda. \tag{12.38}$$

Theorem 12.19. *For $\rho,\lambda\in\mathbb{Y}_n$ it holds that*

$$IA_\rho I^{-1}s_\lambda=|C_\rho|\,\frac{\chi^\lambda_\rho}{\dim\lambda}\,s_\lambda. \tag{12.39}$$

Proof. We see from (12.38) that

$$A_\rho I^{-1} s_\lambda = \frac{\sqrt{n!}}{\dim\lambda} A_\rho \eta_\lambda = \frac{\sqrt{n!}}{\dim\lambda} A_\rho E_\lambda \delta_e. \tag{12.40}$$

Here we note from (12.34) that

$$A_\rho E_\lambda = |C_\rho| \frac{\chi_\rho^\lambda}{\dim\lambda} E_\lambda.$$

Then (12.40) becomes

$$\begin{aligned} A_\rho I^{-1} s_\lambda &= \frac{\sqrt{n!}}{\dim\lambda} |C_\rho| \frac{\chi_\rho^\lambda}{\dim\lambda} E_\lambda \delta_e \\ &= |C_\rho| \frac{\chi_\rho^\lambda}{\dim\lambda} \frac{\sqrt{n!}}{\dim\lambda} \eta_\lambda = |C_\rho| \frac{\chi_\rho^\lambda}{\dim\lambda} I^{-1} s_\lambda. \end{aligned}$$

Thus (12.39) follows. □

It follows from Theorem 12.19 that the Schur functions $\{s_\lambda\,;\,\lambda \in \mathbb{Y}_n\}$ form a basis of eigenvectors of $IA_\rho I^{-1}$ and (12.39) describes its spectral decomposition. The goal of this section is to develop the α-deformed version of Theorem 12.19, though we shall restrict our consideration to the case of $\rho = (2, 1^{n-2}) \in \mathbb{Y}_n$.

In Sect. 11.3 the quantum decomposition of $A_{(j,1^{n-j})}$ was studied. It follows from Lemma 11.8 that $\Gamma(S(n))$ is also invariant under the quantum components $A^\epsilon_{(j,1^{n-j})}$ and the explicit actions are known. In particular, the quantum decomposition of $A_{(2,1^{n-2})}$ is of the form

$$A_{(2,1^{n-2})} = A^+_{(2,1^{n-2})} + A^-_{(2,1^{n-2})}.$$

We introduce a particular diagonal operator on $\Gamma(S(n))$. For $x \in S(n)$ let shape (x) denote the Young diagram of size n indicating the cycle type of x. Define an operator N_n on $\ell^2(S(n))$ by

$$N_n \delta_x = n(\text{shape}\,(x)')\delta_x, \qquad x \in S(n),$$

where $n(\rho)$ is defined as in (12.32). Then we have

$$N_n \xi_\rho = n(\rho')\xi_\rho, \qquad \rho \in \mathbb{Y}_n, \tag{12.41}$$

which means that N_n is a diagonal operator on $\Gamma(S(n))$.

Remark 12.20. By definition shape (x) includes trivial cycles, i.e., one-box rows, while type (x) defined in Sect. 10.3 does not.

Given $\alpha > 0$, we consider the *α-deformed adjacency matrix* defined by

$$A_n^{(\alpha)} = A^+_{(2,1^{n-2})} + \frac{1}{\alpha} A^-_{(2,1^{n-2})} + \frac{\alpha-1}{\alpha} N_n. \tag{12.42}$$

The above α-deformation is somewhat abrupt though, as we shall see in Sect. 12.7, there is an idea of Metropolis algorithm behind. It is readily known that $\Gamma(S(n))$ is invariant under each component on the right-hand side of (12.42). In what follows, these operators restricted to $\Gamma(S(n))$ are denoted by the same symbols. We shall establish an α-deformed analogue of Theorem 12.19 for $A_n^{(\alpha)}$ in terms of Jack symmetric functions.

Lemma 12.21. *Let $\{p^\mu_{\lambda\rho}\}$ be the intersection numbers of $\{A_\lambda\}$ defined in* (12.33)*. It holds that*

$$|C_\mu|\, p^\mu_{\lambda\rho} = |C_\rho|\, p^\rho_{\lambda\mu}, \qquad \rho, \mu, \lambda \in \mathbb{Y}_n.$$

Proof. Consider

$$s = \left|\{(x,y,z) \in S(n)^3\,;\, xy^{-1} \in C_\lambda, yz^{-1} \in C_\rho, xz^{-1} \in C_\mu\}\right|.$$

We have

$$\begin{aligned} s &= \sum_{\substack{x,z\in S(n)\\ xz^{-1}\in C_\mu}} \left|\left\{y \in S(n)\,;\, \begin{matrix} xy^{-1} \in C_\lambda \\ yz^{-1} \in C_\rho \end{matrix}\right\}\right| = \sum_{\substack{x,z\in S(n)\\ xz^{-1}\in C_\mu}} p^\mu_{\lambda\rho} \\ &= |S(n)||C_\mu|\, p^\mu_{\lambda\rho}. \end{aligned} \tag{12.43}$$

In a similar manner,

$$s = \sum_{\substack{y,z\in S(n)\\ yz^{-1}\in C_\rho}} \left|\left\{x \in S(n)\,;\, \begin{matrix} xy^{-1} \in C_\lambda \\ xz^{-1} \in C_\mu \end{matrix}\right\}\right| = |S(n)||C_\rho|\, p^\rho_{\lambda\mu}. \tag{12.44}$$

The assertion then follows from (12.43) and (12.44). □

Proposition 12.22. *For $\rho \in \mathbb{Y}_n$ we have*

$$\begin{aligned} (IA_n^{(\alpha)}I^{-1})p_\rho &= \sum_{\substack{\sigma\in\mathbb{Y}_n\\ l(\sigma)=l(\rho)+1}} p^\rho_{(2,1^{n-2})\sigma}\, p_\sigma \\ &\quad + \frac{1}{\alpha} \sum_{\substack{\sigma\in\mathbb{Y}_n\\ l(\sigma)=l(\rho)-1}} p^\rho_{(2,1^{n-2})\sigma}\, p_\sigma + \frac{\alpha-1}{\alpha} n(\rho')p_\rho\,. \end{aligned} \tag{12.45}$$

Proof. In view of Lemma 11.8 we see that

$$A^+_{(2,1^{n-2})} I^{-1} p_\rho = \sqrt{z_\rho}\, A^+_{(2,1^{n-2})} \Phi(\rho)$$
$$= \sqrt{z_\rho} \sum_{\substack{\sigma \in \mathbb{Y}_n \\ l(\sigma)=l(\rho)+1}} p^\sigma_{(2,1^{n-2})\rho} \sqrt{\frac{|C_\sigma|}{|C_\rho|}}\, \Phi(\sigma). \tag{12.46}$$

Similarly,

$$A^-_{(2,1^{n-2})} I^{-1} p_\rho = \sqrt{z_\rho} \sum_{\substack{\sigma \in \mathbb{Y}_n \\ l(\sigma)=l(\rho)-1}} p^\sigma_{(2,1^{n-2})\rho} \sqrt{\frac{|C_\sigma|}{|C_\rho|}}\, \Phi(\sigma). \tag{12.47}$$

In view of (12.41) we have

$$N_n I^{-1} p_\rho = \sqrt{z_\rho}\, n(\rho') \Phi(\rho). \tag{12.48}$$

Then (12.45) follows by combining (12.46)–(12.48) with Lemma 12.21. □

We now consider the differential operator

$$D(\alpha) = \frac{\alpha}{2} \sum_{i=1}^m x_i^2 \frac{\partial^2}{\partial x_i^2} + \sum_{i \neq j} \frac{x_i^2}{x_i - x_j} \frac{\partial}{\partial x_i} \tag{12.49}$$

acting on $\mathbf{\Lambda}^n_m$, the space of homogeneous symmetric polynomials of degree n in m variables with $m > n$.

Proposition 12.23. *As operators on* $\mathbf{\Lambda}^n_m$ *it holds that*

$$I A^{(\alpha)}_n I^{-1} = \frac{1}{\alpha}\big(D(\alpha) - (m-1)n\big). \tag{12.50}$$

Proof. We prove (12.50) by observing explicit actions on the power sums $p_\rho = p_\rho(x_1, \ldots, x_m)$ for $\rho \in \mathbb{Y}_n$ being fixed.

We compute the three terms on the right-hand side of (12.45). Note that $p^\rho_{(2,1^{n-2})\sigma} > 0$ is possible only if ρ and σ are two shapes communicating each other by multiplying a transposition to their corresponding elements in $S(n)$.

1st term. Note first that $l(\sigma) = l(\rho) + 1$ if and only if two rows in ρ are lumped to make σ. Take the two rows, say, ρ_r and ρ_s, then

$$p_\sigma = p_\rho \frac{p_{\rho_r + \rho_s}}{p_{\rho_r} p_{\rho_s}}.$$

We can make a transposition to lump the two rows by picking up one letter from each row ($\rho_r \rho_s$ possibilities). Hence we have

$$\sum_{\substack{\sigma \in \mathbb{Y}_n \\ l(\sigma)=l(\rho)+1}} p^\rho_{(2,1^{n-2})\sigma} p_\sigma = \frac{1}{2} \sum_{r \neq s} \frac{\rho_r \rho_s p_{\rho_r+\rho_s}}{p_{\rho_r} p_{\rho_s}}\, p_\rho. \tag{12.51}$$

2nd term. Note that $l(\sigma) = l(\rho) - 1$ if and only if one row in ρ is divided into two to make σ. If the row of ρ_k is divided into j and $\rho_k - j$, then

$$p_\sigma = p_\rho \frac{p_j p_{\rho_k - j}}{p_{\rho_k}}.$$

Here j runs over $\{1, \dots, \rho_k - 1\}$ and, for each j, ρ_k transpositions are taken. Notifying double counting, we have

$$\sum_{\substack{\sigma \in \mathbb{Y}_n \\ l(\sigma) = l(\rho) - 1}} P^\rho_{(2,1^{n-2})\sigma} p_\sigma = \sum_k \sum_{j=1}^{\rho_k - 1} \frac{\rho_k}{2} \frac{p_j p_{\rho_k - j}}{p_{\rho_k}} p_\rho. \tag{12.52}$$

3rd term. We have

$$n(\rho') = \binom{\rho_1}{2} + \binom{\rho_2}{2} + \cdots = \sum_k \frac{1}{2} \rho_k (\rho_k - 1). \tag{12.53}$$

Summing up (12.51)–(12.53), we get

$$\begin{aligned} I A_n^{(\alpha)} I^{-1} p_\rho = \frac{1}{2} p_\rho \Bigg\{ &\sum_{r \neq s} \frac{\rho_r \rho_s p_{\rho_r + \rho_s}}{p_{\rho_r} p_{\rho_s}} \\ &+ \frac{1}{\alpha} \sum_k \sum_{j=1}^{\rho_k - 1} \frac{\rho_k p_j p_{\rho_k - j}}{p_{\rho_k}} + \frac{\alpha - 1}{\alpha} \sum_k \rho_k (\rho_k - 1) \Bigg\}. \end{aligned} \tag{12.54}$$

We next consider the action of $D(\alpha)$. First we see that

$$\sum_{i=1}^m x_i^2 \frac{\partial^2}{\partial x_i^2} p_\rho = p_\rho \sum_{k=1}^m \rho_k (\rho_k - 1) + p_\rho \sum_{k=1}^m \sum_{j \neq k} \frac{\rho_j \rho_k p_{\rho_j + \rho_k}}{p_{\rho_j} p_{\rho_k}}. \tag{12.55}$$

Second we have

$$\begin{aligned} \sum_{i \neq j} \frac{x_i^2}{x_i - x_j} \frac{\partial}{\partial x_i} p_\rho &= \sum_{k=1}^m \sum_{i<j} \frac{\rho_k}{p_{\rho_k}} p_\rho \frac{x_i^{\rho_k + 1} - x_j^{\rho_k + 1}}{x_i - x_j} \\ &= \sum_{k=1}^m \frac{\rho_k p_\rho}{p_{\rho_k}} \frac{1}{2} \Bigg\{ \sum_{i,j} (x_i^{\rho_k} + x_i^{\rho_k - 1} x_j + \cdots \\ &\qquad \cdots + x_i x_j^{\rho_k - 1} + x_j^{\rho_k}) - \sum_i (\rho_k + 1) x_i^{\rho_k} \Bigg\} \\ &= \frac{1}{2} p_\rho \sum_{k=1}^m \rho_k (2m - \rho_k - 1) + \frac{1}{2} p_\rho \sum_{k=1}^m \frac{\rho_k}{p_{\rho_k}} \sum_{j=1}^{\rho_k - 1} p_j p_{\rho_k - j}. \end{aligned} \tag{12.56}$$

Combining (12.55) and (12.56), we obtain

$$D(\alpha)p_\rho = \frac{\alpha}{2}\, p_\rho \left\{ \sum_{k=1}^{m} \rho_k(\rho_k - 1) + \frac{1}{\alpha} \sum_{k=1}^{m} \rho_k(2m - \rho_k - 1) \right.$$
$$\left. + \sum_{r \neq s} \frac{\rho_r \rho_s p_{\rho_r + \rho_s}}{p_{\rho_r} p_{\rho_s}} + \frac{1}{\alpha} \sum_{k=1}^{m} \frac{\rho_k}{p_{\rho_k}} \sum_{j=1}^{\rho_k - 1} p_j p_{\rho_k - j} \right\}. \tag{12.57}$$

The desired identity (12.50) follows by comparing (12.54) and (12.57). □

Here we need the following.

Theorem 12.24 (Macdonald). *The Jack symmetric function $P_\lambda^{(\alpha)}$, considered as a polynomial in $\mathbf{\Lambda}_m^n$, is an eigenfunction of $D(\alpha)$ with eigenvalue $\theta^\lambda_{(2,1^{n-2})}(\alpha) + (m-1)n$.*

Combining Proposition 12.23 and Theorem 12.24, we obtain the α-deformed analogue of Theorem 12.19, that is, a diagonalization of $A_n^{(\alpha)}$.

Theorem 12.25. *For $\alpha > 0$ we have*

$$IA_n^{(\alpha)} I^{-1} P_\lambda^{(\alpha)} = \frac{1}{\alpha}\, \theta^\lambda_{(2,1^{n-2})}(\alpha) P_\lambda^{(\alpha)}, \qquad \lambda \in \mathbb{Y}_n. \tag{12.58}$$

12.6 Central Limit Theorem for the Jack Measures

Throughout $\alpha > 0$ is fixed. Recall that $\Gamma(S(n))$ is a Hilbert space equipped with the orthonormal basis $\{\Phi(\rho)\,;\, \rho \in \mathbb{Y}_n\}$. We need to investigate the matrix elements of $A_n^{(\alpha)k}$ with respect to this basis. Recall the transition matrix $\Theta(\alpha) = \big(\theta^\lambda_\rho(\alpha)\big)$ introduced in (12.27), which is characterized by

$$J_\lambda^{(\alpha)} = \sum_{\rho \in \mathbb{Y}_n} \theta^\lambda_\rho(\alpha)\, p_\rho, \qquad \lambda \in \mathbb{Y}_n. \tag{12.59}$$

Proposition 12.26. *For any $\sigma, \rho \in \mathbb{Y}_n$ and $k = 1, 2, \ldots$ we have*

$$\langle \Phi(\sigma), A_n^{(\alpha)k} \Phi(\rho) \rangle = \sqrt{\frac{z_\sigma}{z_\rho}} \sum_{\lambda \in \mathbb{Y}_n} \theta^\lambda_\sigma(\alpha) \left(\frac{1}{\alpha} \theta^\lambda_{(2,1^{n-2})}(\alpha) \right)^k (\Theta(\alpha)^{-1})_{\lambda\rho}. \tag{12.60}$$

Proof. Let $\Delta(\alpha)$ be the diagonal matrix acting on $\Gamma(S(n))$ with diagonal elements

$$\Delta(\alpha)_{\lambda\lambda} = \frac{1}{\alpha}\, \theta^\lambda_{(2,1^{n-2})}(\alpha).$$

Recall that $J_\lambda^{(\alpha)}$ is a constant multiple of $P_\lambda^{(\alpha)}$ by definition (12.26). Then, in terms of the matrix notation, (12.58) in Theorem 12.25 yields

$$\big[(IA_n^{(\alpha)}I^{-1})J_\lambda^{(\alpha)}\big]_{\lambda\in\mathbb{Y}_n} = \big[J_\lambda^{(\alpha)}\big]_{\lambda\in\mathbb{Y}_n}\Delta(\alpha).$$

Applying (12.59), we have

$$\big[(IA_n^{(\alpha)}I^{-1})p_\rho\big]_{\rho\in\mathbb{Y}_n} = \big[p_\rho\big]_{\rho\in\mathbb{Y}_n}\Theta(\alpha)\Delta(\alpha)\Theta(\alpha)^{-1},$$

and hence,

$$\big[(IA_n^{(\alpha)}I^{-1})^k p_\rho\big]_{\rho\in\mathbb{Y}_n} = \big[p_\rho\big]_{\rho\in\mathbb{Y}_n}\Theta(\alpha)\Delta(\alpha)^k\Theta(\alpha)^{-1}. \tag{12.61}$$

Taking the inner product of p_σ and the ρth column of (12.61), we obtain

$$\begin{aligned}\langle\Phi(\sigma), A_n^{(\alpha)k}\Phi(\rho)\rangle &= \sqrt{\frac{z_\sigma}{z_\rho}}\left(\Theta(\alpha)\Delta(\alpha)^k\Theta(\alpha)^{-1}\right)_{\sigma\rho} \\ &= \sqrt{\frac{z_\sigma}{z_\rho}}\sum_{\lambda\in\mathbb{Y}_n}\theta_\sigma^\lambda(\alpha)\left(\frac{1}{\alpha}\theta_{(2,1^{n-2})}^\lambda(\alpha)\right)^k(\Theta(\alpha)^{-1})_{\lambda\rho},\end{aligned}$$

which completes the proof. □

We need some notations. For $\lambda\in\mathbb{Y}$ set

$$\begin{aligned}c_\lambda(\alpha) &= \prod_{b\in\lambda}(\alpha a_\lambda(b)+l_\lambda(b)+1),\\ c'_\lambda(\alpha) &= \prod_{b\in\lambda}(\alpha a_\lambda(b)+l_\lambda(b)+\alpha).\end{aligned}$$

Then, for example,

$$J_\lambda^{(\alpha)} = c_\lambda(\alpha)P_\lambda^{(\alpha)},\quad d^{(\alpha)}(\lambda) = \frac{\alpha^n n!}{c'_\lambda(\alpha)},\quad \mathfrak{J}_n^{(\alpha)}(\lambda) = \frac{\alpha^n n!}{c_\lambda(\alpha)c'_\lambda(\alpha)}.$$

Lemma 12.27. *We have*

$$\langle P_\lambda^{(\alpha)}, P_\lambda^{(\alpha)}\rangle^{(\alpha)} = \frac{c'_\lambda(\alpha)}{c_\lambda(\alpha)},\qquad \lambda\in\mathbb{Y}.$$

Accepting this without proof, we shall show the following.

Lemma 12.28. *We have*

$$\langle p_{(1^n)}, P_\lambda^{(\alpha)}\rangle^{(\alpha)} = d^{(\alpha)}(\lambda)\frac{c'_\lambda(\alpha)}{c_\lambda(\alpha)},\qquad \lambda\in\mathbb{Y}_n.$$

Proof. Let $p_{(1)}^\perp$ be the adjoint operator of $p_{(1)}$ with respect to the α-inner product, i.e.,

$$\langle p_{(1)}f, g\rangle^{(\alpha)} = \langle f, p_{(1)}^\perp g\rangle^{(\alpha)},\qquad f,g\in\mathbf{\Lambda}.$$

Obviously, $p_{(1)}^{\perp} g \in \mathbf{\Lambda}^n$ for $g \in \mathbf{\Lambda}^{n+1}$. Applying Pieri's formula for Jack symmetric functions (Theorem 12.9) to Lemma 12.27, we obtain

$$\begin{aligned}\langle P_\lambda^{(\alpha)}, p_{(1)}^{\perp} P_\Lambda^{(\alpha)} \rangle^{(\alpha)} &= \sum_{M:\lambda\nearrow M} \kappa^{(\alpha)}(\lambda, M) \langle P_M^{(\alpha)}, P_\Lambda^{(\alpha)} \rangle^{(\alpha)} \\ &= \kappa^{(\alpha)}(\lambda, \Lambda) \frac{c'_\Lambda(\alpha)}{c_\Lambda(\alpha)}, \qquad \lambda \in \mathbb{Y}_n, \quad \Lambda \in \mathbb{Y}_{n+1}.\end{aligned}$$

Since the coefficient of $P_\lambda^{(\alpha)}$ in $p_{(1)}^{\perp} P_\Lambda^{(\alpha)}$, which is homogeneous of degree n, is given by

$$\frac{\langle P_\lambda^{(\alpha)}, p_{(1)}^{\perp} P_\Lambda^{(\alpha)} \rangle^{(\alpha)}}{\langle P_\lambda^{(\alpha)}, P_\lambda^{(\alpha)} \rangle^{(\alpha)}},$$

we have

$$p_{(1)}^{\perp} P_\Lambda^{(\alpha)} = \sum_{\lambda:\lambda\nearrow\Lambda} \kappa^{(\alpha)}(\lambda, \Lambda) \frac{c'_\Lambda(\alpha)}{c_\Lambda(\alpha)} \frac{c_\lambda(\alpha)}{c'_\lambda(\alpha)} P_\lambda^{(\alpha)}. \tag{12.62}$$

By repeated application we also have

$$(p_{(1)}^{\perp})^n P_\Lambda^{(\alpha)} = \sum_u w_u \frac{c'_\Lambda(\alpha)}{c_\Lambda(\alpha)} \frac{c_{(1)}(\alpha)}{c'_{(1)}(\alpha)} P_{(1)}^{(\alpha)}, \tag{12.63}$$

where u runs over the paths $(1) \nearrow \cdots \nearrow \Lambda$. Note that (12.63) is valid for $n = 0$ and $\Lambda = (1)$. Consequently,

$$\begin{aligned}\langle p_{(1^n)}, P_\lambda^{(\alpha)} \rangle^{(\alpha)} &= \langle p_{(1)}^n, P_\lambda^{(\alpha)} \rangle^{(\alpha)} \\ &= \langle p_{(1)}, (p_{(1)}^{\perp})^{n-1} P_\lambda^{(\alpha)} \rangle^{(\alpha)} = \frac{c'_\lambda(\alpha)}{c_\lambda(\alpha)} d^{(\alpha)}(\lambda),\end{aligned}$$

which completes the proof. □

Proposition 12.29.

$$(\Theta(\alpha)^{-1})_{\lambda(1^n)} = \mathfrak{J}_n^{(\alpha)}(\lambda), \qquad \lambda \in \mathbb{Y}_n. \tag{12.64}$$

Proof. We see from (12.59) that

$$p_{(1^n)} = \sum_{\lambda \in \mathbb{Y}_n} J_\lambda^{(\alpha)} (\Theta(\alpha)^{-1})_{\lambda\,(1^n)}.$$

Taking the α-inner product of both sides with $J_\lambda^{(\alpha)}$ and applying Lemmas 12.27 and 12.28, we obtain

$$(\Theta(\alpha)^{-1})_{\lambda\,(1^n)} = \frac{\langle p_{(1^n)}, P_\lambda^{(\alpha)} \rangle^{(\alpha)}}{c_\lambda(\alpha) \langle P_\lambda^{(\alpha)}, P_\lambda^{(\alpha)} \rangle^{(\alpha)}} = \frac{d_\lambda^{(\alpha)}}{c_\lambda(\alpha)} = \mathfrak{J}_n^{(\alpha)}(\lambda),$$

as desired. □

Proposition 12.30. *For any* $\lambda \in \mathbb{Y}_n$ *we have*

$$\theta^{\lambda}_{(1^n)}(\alpha) = 1. \tag{12.65}$$

Proof. By Lemma 12.28,

$$\begin{aligned}\theta^{\lambda}_{(1^n)}(\alpha) &= \frac{\langle p_{(1^n)}, J^{(\alpha)}_{\lambda} \rangle^{(\alpha)}}{\langle p_{(1^n)}, p_{(1^n)} \rangle^{(\alpha)}} \\ &= \frac{c_\lambda(\alpha) \langle p_{(1^n)}, P^{(\alpha)}_{\lambda} \rangle^{(\alpha)}}{\alpha^n n!} = \frac{d^{(\alpha)}(\lambda) c'_\lambda(\alpha)}{\alpha^n n!} = 1,\end{aligned}$$

which completes the proof. □

We are now ready to discuss the central limit theorem for

$$A^{(\alpha)}_n = A^{+}_{(2,1^{n-2})} + \frac{1}{\alpha} A^{-}_{(2,1^{n-2})} + \frac{\alpha - 1}{\alpha} N_n.$$

We first recall the action of N_n. For $\rho \in \overline{\mathbb{Y}}$ and $n \geq |\rho|$ we have

$$(\rho \cup (1^{n-|\rho|}))' = (\rho'_1 + n - |\rho| \geq \rho'_2 \geq \rho'_3 \geq \cdots).$$

Therefore,

$$N_n \Phi(\rho \cup (1^{n-|\rho|})) = (\rho'_2 + 2\rho'_3 + \cdots) \Phi(\rho \cup (1^{n-|\rho|}). \tag{12.66}$$

We next need normalization of $A^{(\alpha)}_n$.

Lemma 12.31. *The variance of* $A^{(\alpha)}_n$ *with respect to the vacuum state is given by*

$$V_n = \langle \delta_e, A^{(\alpha)2}_n \delta_e \rangle = \frac{1}{\alpha} \binom{n}{2}. \tag{12.67}$$

Proof. By straightforward calculation.

$$\begin{aligned}\langle \delta_e, A^{(\alpha)2}_n \delta_e \rangle &= \left\langle \delta_e, \frac{1}{\alpha} A^{-}_{(2,1^{n-2})} A^{+}_{(2,1^{n-2})} \delta_e \right\rangle + \left\langle \delta_e, \left(\frac{\alpha - 1}{\alpha} N_n \right)^2 \delta_e \right\rangle \\ &= \frac{1}{\alpha} \langle \delta_e, A^{-}_{(2,1^{n-2})} A^{+}_{(2,1^{n-2})} \delta_e \rangle \\ &= \frac{1}{\alpha} \langle \delta_e, A^{2}_{(2,1^{n-2})} \delta_e \rangle \\ &= \frac{1}{\alpha} \binom{n}{2},\end{aligned}$$

which shows the assertion. □

Thus, the normalized adjacency matrix is given by

$$\frac{A_n^{(\alpha)}}{\sqrt{V_n}} = \frac{1}{\sqrt{V_n}}\left(A^+_{(2,1^{n-2})} + \frac{1}{\alpha}\,A^-_{(2,1^{n-2})} + \frac{\alpha-1}{\alpha}\,N_n\right).$$

The limit is described by the Fock space associated with the modified Young graph $\overline{\mathbb{Y}}$, see Sect. 11.3. Recall that Γ denotes the dense subspace of $\ell^2(\overline{\mathbb{Y}})$ spanned by the orthonormal basis $\{\Psi(\rho)\,;\,\rho\in\overline{\mathbb{Y}}\}$. The operators $B_2^\pm$ are defined by

$$B_2^+\Psi(\rho) = \sqrt{m_2(\rho)+1}\,\Psi(\rho\cup(2)),$$

$$B_2^-\Psi(\rho) = \begin{cases}\sqrt{m_2(\rho)}\,\Psi(\rho\setminus(2)), & \text{if } m_2(\rho)\ge 1,\\ 0, & \text{otherwise.}\end{cases}$$

Theorem 12.32 (QCLT for the α-deformed adjacency matrices). *Let $\alpha>0$ be fixed and set*

$$A_n^{(\alpha)+} = A^+_{(2,1^{n-2})}, \quad A_n^{(\alpha)-} = \frac{1}{\alpha}\,A^-_{(2,1^{n-2})}, \quad A_n^{(\alpha)\circ} = \frac{\alpha-1}{\alpha}\,N_n.$$

Let $\epsilon_1,\dots,\epsilon_m\in\{+,-,\circ\}$, $m=1,2,\dots$. Then for any $\tau,\rho\in\overline{\mathbb{Y}}$ we have

$$\lim_{n\to\infty}\left\langle \Phi(\tau\cup(1^{n-|\tau|})), \frac{A_n^{(\alpha)\epsilon_1}}{\sqrt{V_n}}\cdots\frac{A_n^{(\alpha)\epsilon_m}}{\sqrt{V_n}}\,\Phi(\rho\cup(1^{n-|\rho|}))\right\rangle$$
$$= \langle \Psi(\tau), \tilde{B}^{\epsilon_1}\cdots\tilde{B}^{\epsilon_m}\Psi(\rho)\rangle,$$

where

$$\tilde{B}^+ = \sqrt{\alpha}\,B_2^+, \quad \tilde{B}^- = \frac{1}{\sqrt{\alpha}}\,B_2^-, \quad \tilde{B}^\circ = 0.$$

Proof. We see from (12.66) that the action of the diagonal part $A_n^{(\alpha)\circ}/\sqrt{V_n}$ becomes negligible in the limit as $n\to\infty$. Then the assertion is a simple consequence of the quantum central limit theorem (Theorem 11.13) for the quantum components of $A_n^{(\alpha)}$. □

Let us come back to the random variable $\theta^\lambda_{(2,1^{n-2})}(\alpha)$ defined on $(\mathbb{Y}_n, \mathfrak{J}_n^{(\alpha)})$, where $\mathfrak{J}_n^{(\alpha)}$ is the Jack measure. Setting $\sigma=\rho=(1^n)$ in Proposition 12.26, we obtain

$$\langle\delta_e, A_n^{(\alpha)k}\delta_e\rangle = \sum_{\lambda\in\mathbb{Y}_n}\left(\frac{1}{\alpha}\,\theta^\lambda_{(2,1^{n-2})}(\alpha)\right)^k \mathfrak{J}_n^{(\alpha)}(\lambda), \tag{12.68}$$

where Propositions 12.29 and 12.30 are taken into account. In particular,

$$\frac{1}{\sqrt{V_n}}\,\frac{1}{\alpha}\,\theta^\lambda_{(2,1^{n-2})}(\alpha) = \left\{\alpha\binom{n}{2}\right\}^{-1/2}\theta^\lambda_{(2,1^{n-2})}(\alpha)$$

is the normalization.

Theorem 12.33 (CLT for the Jack measure). *Notations being as above,*

$$\lim_{n\to\infty}\sum_{\lambda\in\mathbb{Y}_n}\left[\left\{\alpha\binom{n}{2}\right\}^{-1/2}\theta^{\lambda}_{(2,1^{n-2})}(\alpha)\right]^k \mathfrak{J}^{(\alpha)}_n(\lambda)$$
$$=\frac{1}{\sqrt{2\pi}}\int_{-\infty}^{+\infty}x^k e^{-x^2/2}dx, \qquad k=0,1,2,\dots. \tag{12.69}$$

Proof. By (12.68),

$$\sum_{\lambda\in\mathbb{Y}_n}\left[\left\{\alpha\binom{n}{2}\right\}^{-1/2}\theta^{\lambda}_{(2,1^{n-2})}(\alpha)\right]^k \mathfrak{J}^{(\alpha)}_n(\lambda)=\left\langle\delta_e,\left(\frac{A^{(\alpha)}_n}{\sqrt{V_n}}\right)^k\delta_e\right\rangle.$$

Then, applying Theorem 12.32, we have

$$\lim_{n\to\infty}\left\langle\delta_e,\left(\frac{A^{(\alpha)}_n}{\sqrt{V_n}}\right)^k\delta_e\right\rangle=\langle\Psi(\varnothing),(\tilde{B}^+ +\tilde{B}^- +\tilde{B}^\circ)^k\Psi(\varnothing)\rangle$$
$$=\left\langle\Psi(\varnothing),\left(\sqrt{\alpha}\,B_2^+ +\frac{1}{\sqrt{\alpha}}\,B_2^-\right)^k\Psi(\varnothing)\right\rangle.$$

Since $\sqrt{\alpha}\,B_2^+$ and $B_2^-/\sqrt{\alpha}$ satisfy the canonical commutation relation, their sum obeys the standard Gaussian distribution in the vacuum state, that is, (12.69) follows. □

12.7 The Metropolis Algorithm and Hanlon's Theorem

Consider a simple random walk on $S(n)$ with transition matrix

$$S(x,y)=\begin{cases}\binom{n}{2}^{-1}, & \text{if } yx^{-1} \text{ is a transposition,}\\ 0, & \text{otherwise.}\end{cases} \tag{12.70}$$

It is noteworthy that $S(x,y)$ is directly related to the adjacency matrix $A_{(2,1^{n-2})}$. In fact, we have

$$S(x,y)=\binom{n}{2}^{-1}\langle\delta_y,A_{(2,1^{n-2})}\delta_x\rangle. \tag{12.71}$$

This simple random walk is clearly symmetric (with respect to the uniform probability on $S(n)$) and its distribution converges to the uniform probability as time goes by.

Let $Q(x,y)$, $x,y\in X$, be the transition matrix of a symmetric Markov chain on a finite set X. Consider an arbitrary distribution π on X such that

$\pi(x) > 0$ for all $x \in X$. The *Metropolis algorithm* is a famous procedure to modify the original symmetric random walk according to π as follows. If $x \neq y$ and $\pi(x) \leq \pi(y)$, then the walker moves from x to y at probability $Q_\pi(x, y) = Q(x, y)$, which is the same as the original walk. If $\pi(x) > \pi(y)$, then the walker takes a new probability $Q_\pi(x, y) = Q(x, y)\pi(y)/\pi(x)$, which is smaller than the original $Q(x, y)$. The walker adjusts the probability $Q_\pi(x, x)$ of staying at x in order to keep stochasticity of Q_π. In this way, the transition matrix Q_π produces a new Markov chain called the *Metropolis chain.* Note that Q_π is symmetric with respect to π, i.e.,

$$\pi(x)Q_\pi(x, y) = \pi(y)Q_\pi(y, x), \qquad x, y \in X.$$

If the original chain is irreducible and aperiodic, so is the Metropolis chain. Then, its distribution converges to the unique invariant probability π as time goes by.

Let $\alpha \geq 1$ be a constant number. Consider a distribution π on $S(n)$ defined in such a way that

$$\pi(x) \propto \alpha^{-c(x)}, \qquad x \in S(n),$$

where $c(x)$ denotes the number of cycles in x. Obviously, the unit e carries the smallest mass. Note also that

$$\sum_{x\in S(n)} \alpha^{-c(x)} = \frac{1}{\alpha}\left(\frac{1}{\alpha}+1\right)\cdots\left(\frac{1}{\alpha}+n-1\right). \tag{12.72}$$

We take the simple random walk S defined by (12.70) as the original chain Q. Applying the Metropolis algorithm to S and the distribution π above, we obtain a Metropolis chain on $X = S(n)$, which is denoted by $S^{(\alpha)}$.

Lemma 12.34. *The transition matrix of the Metropolis chain $S^{(\alpha)}$ is given as follows:*

(i) *for $x \neq y$ such that yx^{-1} is a transposition,*

$$S^{(\alpha)}(x, y) = \begin{cases} \binom{n}{2}^{-1}, & \text{if } c(y) = c(x) - 1, \\ \left\{\alpha\binom{n}{2}\right\}^{-1}, & \text{if } c(y) = c(x) + 1; \end{cases}$$

(ii) *for $x \neq y$ such that yx^{-1} is not a transposition,*

$$S^{(\alpha)}(x, y) = 0;$$

(iii) *for $x = y$,*

$$S^{(\alpha)}(x, x) = \frac{\alpha-1}{\alpha}\binom{n}{2}^{-1} n(\mathrm{shape}\,(x)').$$

Proof. (i) and (ii) are immediate by construction. Fix $x \in S(n)$ and set $\rho = \text{shape}(x) \in \mathbb{Y}_n$. Consider $y \in S(n)$ such that yx^{-1} is a transposition. Then $c(y) = c(x) - 1$ or $c(y) = c(x) + 1$ occurs. The latter case corresponds to dividing a row of x into two by picking up two letters in the row and multiplying their transposition by x. The number of such y's is

$$\sum_k \binom{\rho_k}{2} = n(\rho').$$

Hence

$$\sum_{y \in S(n)} S^{(\alpha)}(x,y) = S^{(\alpha)}(x,x) + \left\{\binom{n}{2} - n(\rho')\right\}\binom{n}{2}^{-1} + n(\rho')\left\{\alpha\binom{n}{2}\right\}^{-1}$$

$$= S^{(\alpha)}(x,x) + 1 - \left(1 - \frac{1}{\alpha}\right)\binom{n}{2}^{-1} n(\rho').$$

Since the left-hand side is equal to 1, we obtain $S^{(\alpha)}(x,x)$ as in (iii). □

Lemma 12.35. *For $\alpha \geq 1$ let $A_n^{(\alpha)}$ be the α-deformed adjacency matrix defined in* (12.42). *Then,*

$$S^{(\alpha)}(x,y) = \binom{n}{2}^{-1} \langle \delta_y, A_n^{(\alpha)} \delta_x \rangle, \qquad x, y \in S(n). \tag{12.73}$$

Lemma 12.36. *Let G be a finite group and $\Gamma(G) \subset \ell^2(G)$ be the subspace of functions taking a constant value on each conjugacy class of G. Let A be a linear operator on $\ell^2(G)$ with matrix representation (A_{xy}) with respect to the basis $\{\delta_x\,;\, x \in G\}$. Then the following are equivalent:*

(i) *for any conjugacy class C of G, the function $x \mapsto \sum_{y \in C} A_{xy}$ is constant on each conjugacy class of G;*
(ii) *$\Gamma(G)$ is invariant under A.*

The proofs of Lemmas 12.35 and 12.36 are straightforward.

Recall that $\Gamma(S(n))$ is kept invariant under the actions of $A_n^{(\alpha)}$ and its adjoint. Hence, applying Lemmas 12.35 and 12.36, we see that the function

$$x \mapsto \sum_{y \in C_\sigma} S^{(\alpha)}(x,y), \qquad x \in S(n),$$

is a constant on each conjugacy class of $S(n)$. Thus, for $\rho \in \mathbb{Y}_n$, taking $x \in C_\rho$, we may define

$$T^{(\alpha)}(\rho,\sigma) = \sum_{y \in C_\sigma} S^{(\alpha)}(x,y), \qquad \rho, \sigma \in \mathbb{Y}_n. \tag{12.74}$$

Obviously, $T^{(\alpha)} = (T^{(\alpha)}(\rho,\sigma))$ is a stochastic matrix of size $|\mathbb{Y}_n|$ so that we obtain a Markov chain on $\mathbb{Y}_n$, which is called the *lumped chain* of $S^{(\alpha)}$.

Theorem 12.37 (Hanlon). *Let $\alpha \geq 1$ and $k = 1, 2, \dots$. Then, for $\rho, \sigma \in \mathbb{Y}_n$ we have*

$$(T^{(\alpha)k})(\rho,\sigma) = \sum_{\lambda\in\mathbb{Y}_n} \theta^\lambda_\sigma(\alpha)\left[\left\{\alpha\binom{n}{2}\right\}^{-1}\theta^\lambda_{(2,1^{n-2})}(\alpha)\right]^k (\Theta(\alpha)^{-1})_{\lambda\rho}. \quad (12.75)$$

Proof. Inserting (12.73) into (12.74), we see that for $\rho, \sigma \in \mathbb{Y}_n$,

$$\begin{aligned} T^{(\alpha)}(\rho,\sigma) &= \frac{1}{|C_\rho|}\binom{n}{2}^{-1}\sum_{x\in C_\rho}\sum_{y\in C_\sigma}\langle\delta_y, A_n^{(\alpha)}\delta_x\rangle \\ &= \frac{1}{|C_\rho|}\binom{n}{2}^{-1}\langle\xi_\sigma, A_n^{(\alpha)}\xi_\rho\rangle \\ &= \sqrt{\frac{z_\rho}{z_\sigma}}\binom{n}{2}^{-1}\langle\Phi(\sigma), A_n^{(\alpha)}\Phi(\rho)\rangle. \end{aligned}$$

Hence we have

$$(T^{(\alpha)k})(\rho,\sigma) = \sqrt{\frac{z_\rho}{z_\sigma}}\left\langle\Phi(\sigma), \left[\binom{n}{2}^{-1}A_n^{(\alpha)}\right]^k\Phi(\rho)\right\rangle. \quad (12.76)$$

The assertion follows by combining (12.76) and Proposition 12.26. □

Corollary 12.38. *The probability that the lumped chain $T^{(\alpha)}$ starting at the origin (1^n) stays at σ at the time k is given by*

$$\sum_{\lambda\in\mathbb{Y}_n} \theta^\lambda_\sigma(\alpha)\left[\left\{\alpha\binom{n}{2}\right\}^{-1}\theta^\lambda_{(2,1^{n-2})}(\alpha)\right]^k \mathfrak{J}_n^{(\alpha)}(\lambda).$$

Proof. The probability in question is equal to $(T^{(\alpha)k})((1^n),\sigma)$. By Theorem 12.37 we have

$$(T^{(\alpha)k})((1^n),\sigma) = \sum_{\lambda\in\mathbb{Y}_n} \theta^\lambda_\sigma(\alpha)\left[\left\{\alpha\binom{n}{2}\right\}^{-1}\theta^\lambda_{(2,1^{n-2})}(\alpha)\right]^k (\Theta(\alpha)^{-1})_{\lambda(1^n)}.$$

By Proposition 12.29 we have $(\Theta(\alpha)^{-1})_{\lambda(1^n)} = \mathfrak{J}_n^{(\alpha)}(\lambda)$, so that the desired result follows immediately. □

Corollary 12.39. *The probability that the lumped chain $T^{(\alpha)}$ starting at the origin (1^n) returns to the origin at the time k is given by*

$$\sum_{\lambda\in\mathbb{Y}_n} \left[\left\{\alpha\binom{n}{2}\right\}^{-1}\theta^\lambda_{(2,1^{n-2})}(\alpha)\right]^k \mathfrak{J}_n^{(\alpha)}(\lambda).$$

Proof. We only need to set $\sigma = (1^n)$ in Corollary 12.38 and apply Proposition 12.30. □

Exercises

12.1. Let $m \geq n \geq 1$ be integers. Show that the natural projection $\mathbf{\Lambda}_m^k \to \mathbf{\Lambda}_n^k$ sends $m_\lambda(x_1, \dots, x_m)$ to $m_\lambda(x_1, \dots, x_n)$ or to 0 based on whether $\text{row}(\lambda) \leq n$ or $\text{row}(\lambda) > n$.

12.2. Show the following identities for the Schur functions.

$$\begin{aligned} s_{(1)}s_{(1)} &= s_{(2)} + s_{(1^2)}, \\ s_{(1)}s_{(2)} &= s_{(3)} + s_{(12)}, \\ s_{(1)}s_{(1^2)} &= s_{(12)} + s_{(1^3)}. \end{aligned}$$

12.3. Show Pieri's formula for the monomial m_λ.

12.4 (Orthogonality relation). Let G be a finite group and $\hat{G}$ the set of equivalence classes of irreducible representations of G. Let χ^λ be the character of $\lambda \in \hat{G}$. Show the orthogonality relation:

$$\frac{1}{|G|} \sum_{g \in G} \chi^\lambda(g)\chi^\mu(g^{-1}) = \delta_{\lambda\mu}, \qquad \lambda, \mu \in \hat{G}.$$

[(12.35)]

12.5. Prove the identity (12.72).

Notes

The facts on Jack symmetric functions used in this chapter without proofs are found in the basic references Macdonald [153] and Stanley [200].

Proposition 12.5 is found in Macdonald [153, Sect. VI.1], and Pieri's formula for Jack symmetric functions (Theorem 12.9) in Macdonald [153, Sect. VI.6].

Many results on the Plancherel measure on the path space of the Young graph have been extended to the Jack graph, see Kerov [124, 126] for the general theory along this line. Theorem 12.11 is due to Stanley [200]. The notion of α-deformation of a Young diagram and related topics were developed by Kerov [125].

For the deformation of irreducible characters of the symmetric group see Macdonald [153, Sect. VI.10]. Proposition 12.18 is quoted from Macdonald [153, Sect. VI.10, Example 1].

The extension of Jack symmetric functions is due to Macdonald [153]. See also Kerov [126], where such an extension is called *generalized Hall–Littlewood polynomials*. For further structure of the Jack graph and relevant potential theory, see Borodin–Olshanshi [34] and Kerov–Okounkov–Olshanski [128].

Lemma 12.21 is a special case of a general equality for a (symmetric) association scheme, see Bannai–Ito [17, Sect. 2.2].

The proof of Theorem 12.24 is found in Macdonald [153, Sect. VI.4] and Stanley [200]. The proof of Lemma 12.27 is found in Macdonald [153, Sect. VI.10]. The method of the proof of Lemma 12.28 is due to Fulman [83, Lemma 3.3]. For the adjoint operator $f^{\perp}$ see Macdonald [153, Sect. I.5].

The central limit theorem for the Jack measure (Theorem 12.33) in connection with the Metropolis algorithm was initiated by Fulman [83], where an error estimate of distribution functions is also obtained. In this sense Theorem 12.33 is a weaker version of Fulman's central limit theorem. We wish to emphasize, nevertheless, that the idea of quantum decomposition naturally fits with the Metropolis algorithm and creates an alternative approach to the asymptotic analysis of the Jack graphs. The proof in Sect. 12.6 is new and is first presented in this book.

The main result reconstructed in Sect. 12.7 is due to Hanlon [91]. See also Diaconis–Hanlon [71] and references cited therein for the relevant Metropolis algorithm.

References

1. L. Accardi, A. Bach: Quantum central limit theorems for strongly mixing random variables. Z. Wahr. Verw. Gebiete **68** (1985), 393–402.
2. L. Accardi, A. Ben Ghorbal, N. Obata: Monotone independence, comb graphs and Bose–Einstein condensation. Infin. Dimens. Anal. Quantum Probab. Relat. Topics **7** (2004), 419–435.
3. L. Accardi, M. Bożejko: Interacting Fock spaces and Gaussianization of probability measures. Infin. Dimens. Anal. Quantum Probab. Relat. Topics **1** (1998), 663–670.
4. L. Accardi, Y. Hashimoto, N. Obata: Notions of independence related to the free group. Infin. Dimens. Anal. Quantum Probab. Relat. Topics **1** (1998), 201–220.
5. L. Accardi, Y. Hashimoto, N. Obata: Singleton independence. Banach Center Publ. **43** (1998), 9–24.
6. L. Accardi, Y. Hashimoto, N. Obata: A role of singletons in quantum central limit theorems. J. Korean Math. Soc. **35** (1998), 675–690.
7. L. Accardi, Y.-G. Lu: Quantum central limit theorems for weakly dependent maps (I). Acta Math. Hungar. **63** (1994), 183–212.
8. L. Accardi, Y.-G. Lu: Quantum central limit theorems for weakly dependent maps (II). Acta Math. Hungar. **63** (1994), 249–282.
9. L. Accardi, Y.-G. Lu, I. Volovich: *Quantum Theory and Its Stochastic Limit.* Berlin: Springer-Verlag, 2002.
10. L. Accardi, M. Nahni: Interacting Fock spaces and orthogonal polynomials in several variables. In: *Non-Commutativity, Infinite-Dimensionality and Probability at the Crossroads*, N. Obata, T. Matsui, A. Hora (ed). River Edge, NJ: World Scientific, 2002, pp. 192–205 (QP–PQ: Quantum Probab. White Noise Anal., Vol. 16).
11. N. I. Akhiezer: *The Classical Moment Problem and Some Related Questions in Analysis.* New York: Hafner 1965.
12. M. Akiyama, H. Yoshida: The distributions for linear combinations of a free family of projections and their orthogonal polynomials. Infin. Dimens. Anal. Quantum Probab. Relat. Topics **2** (1999), 627–643.
13. S. T. Ali, J.-P. Antoine, J.-P. Gazeau: *Coherent States, Wavelets and Their Generalizations*, Graduate Texts in Contemporary Physics. New York, Springer-Verlag, 2000.

14. K. Aomoto, Y. Kato: Green functions and spectra on free products of cyclic groups. Ann. Inst. Fourier **38** (1988), 59–85.
15. D. Avitzour: Free products of C^*-algebras. Trans. Am. Math. Soc. **271** (1982), 423–435.
16. J. Baik, P. Deift, K. Johansson: On the distribution of the length of the longest increasing subsequence of random permutations. J. Am. Math. Soc. **12** (1999), 1119–1178.
17. E. Bannai, T. Ito: *Algebraic Combinatorics I. Association Schemes.* Menlo Park, CA: The Benjamin/Cummings 1984.
18. R. Balakrishnan, K. Ranganathan: *A Textbook of Graph Theory* New York: Springer-Verlag, 2000.
19. G. Baldi, R. Burioni, D. Cassi: Localized states on comb lattices. Phys. Rev. E **70** (2004), 031111 (6 pages).
20. A. Ben Ghorbal, M. Schürmann: Non-commutative notions of stochastic independence. Math. Proc. Cambridge Philos. Soc. **133** (2002), 531–561.
21. A. Ben Ghorbal, M. Schürmann: Quantum stochastic calculus on Boolean Fock space. Infin. Dimens. Anal. Quantum Probab. Relat. Topics **7** (2004), 631–650.
22. D. Bertacchi, F. Zucca: Uniform asymptotic estimates of transition probabilities on combs. J. Aust. Math. Soc. **75** (2003), 325–353.
23. P. Biane: Permutation model for semi-circular systems and quantum random walks. Pacific J. Math. **171** (1995), 373–387.
24. P. Biane: On the free convolution with a semi-circular distribution. Indiana Univ. Math. J. **46** (1997), 705–718.
25. P. Biane: Representations of symmetric groups and free probability. Adv. Math. **138** (1998), 126–181.
26. P. Biane: Approximate factorization and concentration for characters of symmetric groups. Int. Math. Res. Notices **2001** (2001), 179–192.
27. P. Biane: Characters of symmetric groups and free cumulants. In: *Asymptotic Combinatorics with Applications to Mathematical Physics*, A. M. Vershik (ed). Berlin: Springer, 2003, pp. 185–200 (Lecture Notes in Math., Vol. 1815).
28. P. Biane, R. Speicher: Free diffusions, free entropy and free Fisher information. Ann. Inst. H. Poincaré Probab. Stat. **37** (2001), 581–606.
29. N. Biggs: Some odd graph theory. Ann. New York Acad. Sci. **319** (1979), 71–81.
30. N. Biggs: *Algebraic Graph Theory*, 2nd ed. Cambridge Mathematical Library. Cambridge: Cambridge Univ. Press, 1993.
31. W. R. Bloom, H. Heyer: *Harmonic Analysis of Probability Measures on Hypergroups*, de Gruyter Studies in Math. 20. Berlin: Walter de Gruyter, 1995.
32. B. Bollobás: *Modern Graph Theory*, Graduate Texts in Mathematics, Vol. 184. New York: Springer-Verlag, 1998.
33. A. Borodin, A. Okounkov, G. Olshanski: Asymptotics of Plancherel measures for symmetric groups. J. Am. Math. Soc. **13** (2000), 481–515.
34. A. Borodin, G. Olshanski: Harmonic functions on multiplicative graphs and interpolation polynomials. Electron. J. Combin. **7** (2000), Research Paper 28, 39 pp. (electronic).
35. M. Bożejko: On $\Lambda(p)$ sets with minimal constant in discrete noncommutative groups. Proc. Am. Math. Soc. **51** (1975), 407–412.
36. M. Bożejko: Positive definite functions on the free group and the noncommutative Riesz product. Boll. Un. Mat. Ital. A (6) **5** (1986), 13–21.
37. M. Bożejko: Uniformly bounded representations of free groups. J. Reine Angew. Math. **377** (1987), 170–186.

38. M. Bożejko: *Positive and Negative Definite Kernels on Discrete Groups.* University of Heidelberg, 1987 (Unpublished Lecture Notes).
39. M. Bożejko: Positive-definite kernels, length functions on groups and a noncommutative von Neumann inequality. Studia Math. **95** (1989), 107–118.
40. M. Bożejko, W. Bryc: On a class of free Levy laws related to a regression problem. J. Funct. Anal. **236** (2006), 59–77.
41. M. Bożejko, A. D. Krystek, Ł. J. Wojakowski: Remarks on the r and Δ convolutions. Math. Z. **253** (2006), 177–196.
42. M. Bożejko, B. Kümmerer, R. Speicher: q-Gaussian processes: Non-commutative and classical aspects. Comm. Math. Phys. **185** (1997), 129–154.
43. M. Bożejko, M. Leinert, R. Speicher: Convolution and limit theorems for conditionally free random variables. Pacific J. Math. **175** (1996), 357–388.
44. M. Bożejko, R. Speicher: ψ-independent and symmetrized white noises. In: *Quantum Probability and Related Topics VI*, L. Accardi (ed). River Edge, NJ: World Scientific 1991, pp. 219–236.
45. M. Bożejko, R. Speicher: An example of a generalized Brownian motion. Comm. Math. Phys. **137** (1991), 519–531.
46. M. Bożejko, R. Speicher: Completely positive maps on Coxeter groups, deformed commutation relations, and operator spaces. Math. Ann. **300** (1994), 97–120.
47. M. Bożejko, R. Speicher: Interpolations between bosonic and fermionic relations given by generalized Brownian motions. Math. Z. **222** (1996), 135–159.
48. M. Bożejko, J. Wysoczański: New examples of convolution and non-commutative central limit theorems. Banach Center Publ. **175** (1998), 95–103.
49. M. Bożejko, J. Wysoczański: Remarks on t-transformations of measures and convolutions. Ann. Inst. H. Poincaré Probab. Stat. **37** (2001), 737–761.
50. A. E. Brouwer, A. M. Cohen, A. Neumaier: *Distance-Regular Graphs.* Berlin: Springer-Verlag, 1989.
51. R. Burioni, D. Cassi, I. Meccoli, M. Rasetti, S. Regina, P. Sodano, A. Vezzani: Bose–Einstein condensation in inhomogeneous Josephson arrays. Europhys. Lett. **52** (2000), 251–256.
52. R. Burioni, D. Cassi, M. Rasetti, P. Sodano, A. Vezzani: Bose–Einstein condensation on inhomogeneous complex networks. J. Phys. B: At. Mol. Opt. Phys. **34** (2001), 4697–4710.
53. R. Burioni, D. Cassi, A. Vezzani: Topology, hidden spectra and Bose-Einstein condensation on low-dimensional complex networks. J. Phys. A: Math. Gen. **35** (2002), 1245–1252.
54. T. Cabanal-Duvillard, V. Ionescu: Un théorème central limite pour des variables aléatoires non-commutatives. C. R. Acad. Sci. Paris **325** Série I (1997), 1117–1120.
55. D. I. Cartwright, W. Młotkowski: Harmonic analysis for groups acting on triangle buildings. J. Aust. Math. Soc. Ser. **A 56** (1994), 345–383.
56. T. S. Chihara: *An Introduction to Orthogonal Polynomials.* Gordon and Breach, New York 1978.
57. I. Chiswell: Abstract length functions in groups. Math. Proc. Camb. Phil. Soc. **80** (1976), 451–463.
58. K. L. Chung: *A Course in Probability Theory*, 3rd ed. New York: Academic Press, 2001.
59. A. M. Cockroft, R. L. Hudson: Quantum mechanical Wiener processes. J. Multivariate Anal. **7** (1977), 107–124.

60. J. M. Cohen, A. R. Trenholme: Orthogonal polynomials with a constant recursion formula and an application to harmonic analysis. J. Funct. Anal. **59** (1984), 175–184.
61. D. D. Cushen, R. L. Hudson: A quantum mechanical central limit theorem. J. Appl. Probab. **8** (1971), 454–469.
62. D. M. Cvetković, M. Doob, H. Sachs: *Spectra of Graphs.* New York: Academic Press, 1979.
63. D. M. Cvetković, P. Rowlinson, S. Simić: *Eigenspaces of Graphs.* Cambridge, U.K.: Cambridge Univ. Press, 1997.
64. P. K. Das: Eigenvectors of backwardshift on a deformed Hilbert space. Int. J. Theor. Phys. **37** (1998), 2363–2369.
65. P. K. Das: Erratum: Eigenvectors of backwardshift on a deformed Hilbert space. Int. J. Theor. Phys. **38** (1999), 2063–2064.
66. P. K. Das: Coherent states and squeezed states in interacting Fock space. Int. J. Theor. Phys. **41** (2002), 1099–1106.
67. M. de Giosa, Y. G. Lu: The free creation and annihilation operators as the central limit of quantum Bernoulli process. Random Oper. Stochastic Equations **5** (1997), 227–236.
68. M. de Giosa, Y. G. Lu: From quantum Bernoulli process to creation and annihilation operators on interacting q-Fock space. Japan. J. Math. **24** (1998), 149–167
69. P. Deift: *Orthogonal Polynomials and Random Matrices: A Riemann–Hilbert Approach*, Courant Lect. Notes Vol. 3. Providence, RI: Am. Math. Soc., 1998).
70. P. Diaconis: *Group Representations in Probability and Statistics*, IMS Lecture Notes–Monograph Series Vol. 11. Hayward; CA: Institute of Mathematical Statistics, 1988.
71. P. Diaconis, P. Hanlon: Eigen-analysis for some examples of the Metropolis algorithm. Contemporary Math. **138** (1992), 99–117.
72. R. Diestel: *Graph Theory*, 3rd edn. Graduate Texts in Mathematics Vol. 173. Berlin: Springer-Verlag, 2005.
73. W. F. Donoghue, Jr.: *Monotone Matrix Functions and Analytic Continuation* Berlin: Springer-Verlag, 1974.
74. S. N. Dorogovtsev, J. F. F. Mendes: *Evolution of Networks.* New York: Oxford Univ. Press, 2003.
75. N. Dunford, J. T. Schwartz: *Linear Operators Part I: General Theory.* New York: Wiley, 1988.
76. R. Durrett: *Probability: Theory and Examples.* Belmont, CA: Duxbury Press, 1991.
77. W. Feller: *An Introduction to Probability Theory and Its Applications, Vol. I*, 2nd edn. New York: Wiley, 1957.
78. G. Fendler: Central limit theorems for Coxeter systems and Artin systems of extra large type. Infin. Dimens. Anal. Quantum Probab. Relat. Topics **6** (2003), 537–548.
79. U. Franz: Monotone independence is associative. Infin. Dimens. Anal. Quantum Probab. Relat. Topics **4** (2001), 401–407.
80. U. Franz: What is stochastic independence. In: *Non-Commutativity, Infinite-Dimensionality and Probability at the Crossroads*, N. Obata, T. Matsui, A. Hora (ed). River Edge, NJ: World Scientific 2002, pp. 254–274. (QP–PQ: Quantum Probab. White Noise Anal., Vol. 16).

81. U. Franz: Unification of boolean, monotone, anti-monotone, and tensor independence and Lévy processes. Math. Z. **243** (2003), 779–816.
82. J. Fulman: Stein's method and Plancherel measure of the symmetric group. Trans. Am. Math. Soc. **357** (2005), 555–570.
83. J. Fulman: Stein's method, Jack measure, and the Metropolis algorithm. J. Combin. Theory Ser. A **108** (2004), 275–296.
84. W. Fulton, J. Harris: *Representation Theory: A First Course*, Graduate Texts in Mathematics, Vol. 129 New York: Springer-Verlag, 1991.
85. E. Gutkin: Green's functions of free products of operators, with applications to graph spectra and to random walks. Nagoya Math. J. **149** (1998), 93–116.
86. N. Giri, W. von Waldenfels: An algebraic version of the central limit theorem. Z. Wahr. Verw. Gebiete **42** (1978), 129–134.
87. C. Godsil, G. Royle: *Algebraic Graph Theory*, Graduate Texts in Mathematics, Vol. 207. New York: Springer-Verlag, 2001.
88. M. Guţă, H. Maassen: Generalized Brownian motion and second quantization. J. Funct. Anal. **191** (2002), 241–275.
89. U. Haagerup: An example of a nonnuclear C*-algebra which has the metric approximation property. Invent. Math. **50** (1979), 279–293.
90. J. M. Hammersley: A few seedlings of research. In: *Proc. 6th Berkeley Symp. Math. Stat. Prob.*, Vol. 1 University of California Press, 1972, pp. 345–394.
91. P. Hanlon: A Markov chain on the symmetric group and Jack symmetric function. Discrete Math. **99** (1992), 123–140.
92. Y. Hashimoto: Deformations of the semicircle law derived from random walks on free groups. Probab. Math. Statist. **18** (1998), 399–410.
93. Y. Hashimoto: Samples of algebraic central limit theorems based on $\mathbb{Z}/2\mathbb{Z}$. In: *Infinite Dimensional Harmonic Analysis*, H. Heyer, T. Hirai, N. Obata (ed). Tübingen: Gräbner, 1999, pp. 115–126.
94. Y. Hashimoto: Quantum decomposition in discrete groups and interacting Fock spaces. Infin. Dimens. Anal. Quantum Probab. Relat. Topics **4** (2001), 277–287.
95. Y. Hashimoto: Creation-annihilation processes on cellar complecies. In: *Non-Commutativity, Infinite-Dimensionality and Probability at the Crossroads*, N. Obata, T. Matsui, A. Hora (ed). River Edge, NJ: World Scientific, 2002, pp. 275–287 (QP–PQ: Quantum Probab. White Noise Anal. Vol. 16).
96. Y. Hashimoto, A. Hora, N. Obata: Central limit theorems for large graphs: Method of quantum decomposition. J. Math. Phys. **44** (2003), 71–88.
97. Y. Hashimoto, N. Obata, N. Tabei: A quantum aspect of asymptotic spectral analysis of large Hamming graphs. In: *Quantum Information III*, T. Hida, K. Saitô (ed). River Edge, NJ: World Scientific, 2001, pp. 45–57.
98. G. C. Hegerfeldt: A quantum characterization of Gaussianness. In *Quantum Probability and Related Topics*, L. Accardi et al. (ed). River Edge, NJ: World Scientific, 1992, pp. 165–173.
99. F. Hiai, D. Petz: *The Semicircle Law, Free Random Variables and Entropy* Providence, RI: Amer. Math. Soc., 2000.
100. P. Hilton, J. Pedersen: Catalan numbers, their generalization, and their uses. Math. Intelligencer **13** (1991), 64–75.
101. O. Hiwatashi, T. Kuroda, M. Nagisa, H. Yoshida: The free analogue of noncentral chi-square distributions and symmetric quadratic forms in free random variables. Math. Z. **230** (1999), 63–77.

102. A. Hora: Central limit theorems and asymptotic spectral analysis on large graphs. Infin. Dimens. Anal. Quantum Probab. Relat. Topics **1** (1998), 221–246.
103. A. Hora: Central limit theorem for the adjacency operators on the infinite symmetric group. Comm. Math. Phys. **195** (1998), 405–416.
104. A. Hora: Gibbs state on a distance-regular graph and its application to a scaling limit of the spectral distributions of discrete Laplacians. Probab. Theory Relat. Fields **118** (2000), 115–130.
105. A. Hora: Scaling limit for Gibbs states for Johnson graphs and resulting Meixner classes. Infin. Dimens. Anal. Quantum Probab. Relat. Topics **6** (2003), 139–143.
106. A. Hora: A noncommutative version of Kerov's Gaussian limit for the Plancherel measure of the symmetric group. In: *Asymptotic Combinatorics with Applications to Mathematical Physics.* A. M. Vershik (ed). Berlin: Springer, 2003, pp. 77–88. (Lecture Notes in Math., Vol. 1815).
107. A. Hora: Asymptotic spectral analysis on the Johnson graphs in infinite degree and zero temperature limit. Interdiscip. Inform. Sci. **10** (2004), 1–10.
108. A. Hora: Jucys–Murphy element and walks on modified Young graph. Banach Center Publ. **73** (2006), 223–235.
109. A. Hora: The limit shape of Young diagrams for Weyl groups of type B. Oberwolfach Reports **2**, no. 2 (2005).
110. A. Hora, N. Obata: Quantum decomposition and quantum central limit theorem. In: *Fundamental Problems in Quantum Mechanics*, L. Accardi, S. Tasaki (ed). River Edge, NJ: World Scientific, 2003, pp. 284–305.
111. A. Hora, N. Obata: An interacting Fock space with periodic Jacobi parameter obtained from regular graphs in large scale limit. In: *Quantum Information V*, T. Hida, K. Saitô (ed). River Edge, NJ: World Scientific, 2006, pp. 121–144.
112. A. Hora, N. Obata: Asymptotic spectral analysis of growing regular graphs. Trans. Am. Math. Soc. in press.
113. R. L. Hudson, K. R. Parthasarathy: Quantum Itô's formula and stochastic evolutions. Comm. Math. Phys. **93** (1984), 301–323.
114. J. E. Humphreys: *Reflection Groups and Coxeter Groups*, Cambridge Studies in Advanced Mathematics, Vol. 29. Cambridge, U.K.: Cambridge Univ. Press, 1990.
115. D. Igarashi, N. Obata: Asymptotic spectral analysis of growing graphs: Odd graphs and spidernets. Banach Center Publ. **73** (2006), 245–265.
116. V. Ivanov, S. Kerov: The algebra of conjugacy classes in symmetric groups, and partial permutations. J. Math. Sci. (New York) **107** (2001), 4212–4230.
117. V. Ivanov, G. Olshanski: Kerov's central limit theorem for the Plancherel measure on Young diagrams. In: *Symmetric Functions 2001: Surveys of Developments and Perspectives.* S. Fomin (ed). Dordrecht: Kluwer Academic Publishers, 2002, pp. 93–151. (NATO Sci. Ser. II, Math. Phys. Chem. **74**).
118. G. D. James: *The Representation Theory of the Symmetric Groups.* Berlin: Springer, 1978. (Lecture Notes in Math., Vol. 682).
119. U. C. Ji, N. Obata: Quantum white noise calculus. In: *Non-Commutativity, Infinite-Dimensionality and Probability at the Crossroads*, N. Obata, T. Matsui, A. Hora (ed). River Edge, NJ: World Scientific, 2002, pp. 143–191 (QP–PQ: Quantum Probab. White Noise Anal., Vol. 16).
120. K. Johansson: On fluctuations of eigenvalues of random Hermitian matrices. Duke Math. J. **91** (1998), 151–204.

121. K. Johansson: Discrete orthogonal polynomial ensembles and the Plancherel measure. Ann. Math. **153**(2) (2001), 259–296.
122. S. Kerov: Gaussian limit for the Plancherel measure of the symmetric group. C. R. Acad. Sci. Paris Sér. I Math. **316** (1993), 303–308.
123. S. V. Kerov: Transition probabilities for continual Young diagrams and the Markov moment problem. Funct. Anal. Appl. **27** (1993), 104–117.
124. S. Kerov: The boundary of Young lattice and random Young tableaux. In: *Formal Power Series and Algebraic Combinatorics*, Providence, RI: Amer. Math. Soc., 1996, pp. 133–158 (DIMACS Ser. Discrete Math. Theoret. Comput. Sci., Vol. 24).
125. S. V. Kerov: Anisotropic Young diagrams and symmetric Jack functions. Funct. Anal. Appl. **34** (2000), 41–51.
126. S. V. Kerov: *Asymptotic Representation Theory of the Symmetric Group and Its Applications in Analysis*, Providence, Rhode Island: Amer. Math. Soc., 2003 (Translations of Mathematical Monographs, Vol. 219).
127. S. Kerov, G. Olshanski: Polynomial functions on the set of Young diagrams. C. R. Acad. Sci. Paris Sér. I Math. **319** (1994), 121–126.
128. S. Kerov, A. Okounkov, G. Olshanski: The boundary of the Young graph with Jack edge multiplicities. Int. Math. Res. Notices **1998** (1998), 173–199.
129. S. Kerov, G. Olshanski, A. Vershik: Harmonic analysis on the infinite symmetric group. A deformation of the regular representation. C. R. Acad. Sci. Paris Sér. I Math. **316** (1993), 773–778.
130. S. Kerov, G. Olshanski, A. Vershik: Harmonic analysis on the infinite symmetric group. Invent. Math. **158** (2004), 551–642.
131. H. Kesten: Symmetric random walk on groups. Trans. Amer. Math. Soc. **92** (1959), 336–354.
132. H. Kesten (ed): *Probability on Discrete Structures.* Berlin: Springer, 2004. (Encyclopaedia of Mathematical Sciences, Vol. 110).
133. J. R. Klauder: The action option and a Feynman quantization of spinor fields in terms of ordinary c-numbers. Ann. Phys. **11** (1960), 123–168.
134. J. R. Klauder, B.-S. Skagerstam: *Coherent States.* River Edge, NJ: World Scientific, 1985.
135. B. Krawczyk, R. Speicher: Combinatorics of free cumulants. J. Combin. Theory Ser. **A90** (2000), 267–292.
136. A. Krystek, H. Yoshida: The combinatorics of the r-free convolution. Infin. Dimens. Anal. Quantum Probab. Relat. Topics **6** (2003), 619–627.
137. A. Krystek, H. Yoshida: Generalized t-transformations of probability measures and deformed convolutions. Probab. Math. Statist. **24** (2004), 97–119.
138. F. Lehner: Cumulants, lattice paths, and orthogonal polynomials. Discrete Math. **270** (2003), 177–191.
139. F. Lehner: Cumulants in noncommutative probability theory II. Generalized Gaussian random variables. Probab. Theory Relat. Fields **127** (2003), 407–422.
140. F. Lehner: Cumulants in noncommutative probability theory I. Noncommutative exchangeability systems. Math. Z. **248** (2004), 67–100.
141. F. Lehner: Cumulants in noncommutative probability theory III. Creation and annihilation operators on Fock spaces. Infin. Dimens. Anal. Quantum Probab. Relat. Topics **8** (2005), 407–437.
142. R. Lenczewski: On sums of q-independent $SU_q(2)$ quantum variables. Comm. Math. Phys. **154** (1993), 127–134.

143. R. Lenczewski: Unification of independence in quantum probability. Infin. Dimens. Anal. Quantum Probab. Relat. Topics **1** (1998), 383–405.
144. R. Lenczewski: Filtered random variables, bialgebras, and convolutions. J. Math. Phys. **42** (2001), 5876–5903.
145. R. Lenczewski: Reduction of free independence to tensor independence. Infin. Dimens. Anal. Quantum Probab. Relat. Topics **7** (2004), 337–360.
146. R. Lenczewski: On noncommutative independence. In: *Quantum Probability and Infinite Dimensional Analysis.* M. Schürmann, U. Franz (ed). River Edge, NJ: World Scientific, 2005, pp. 320–336 (QP–PQ: Quantum Probab. White Noise Anal., Vol. 18).
147. V. Liebscher: On a central limit theorem for monotone noise. Infin. Dimens. Anal. Quantum Probab. Relat. Topics **2** (1999), 155–167.
148. B. F. Logan, L. A. Shepp: A variational problem for random Young tableaux. Adv. Math. **26** (1977), 206–222.
149. Y. G. Lu: An interacting free Fock space and the arcsine law. Probab. Math. Stat. **17** (1997), 149–166.
150. Y. G. Lu: On the interacting free Fock space and the deformed Wigner law. Nagoya Math. J. **145** (1997), 1–28.
151. R. Lyndon: Length functions in groups. Math. Scand. **12** (1963), 209–234.
152. H. Maassen: Addition of freely independent random variables. J. Funct. Anal. **106** (1992), 409–438.
153. I. G. Macdonald: *Symmetric Functions and Hall Polynomials*, 2nd edn, Oxford Mathematical Monographs (Oxford: Univ. Press, Oxford, 1995).
154. H. D. MacPherson: Infinite distance transitive graphs of finite valency. Combinatorica **2** (1982), 63–69.
155. V. A. Marchenko, L. A. Pastur: Distribution of eigenvalues for some sets of random matrices. USSR Sb. **1** (1967), 457–483.
156. T. Matsui: BEC of free bosons on networks. Infin. Dimens. Anal. Quantum Probab. Relat. Topics **9** (2006), 1–26.
157. B. D. McKay: The expected eigenvalue distribution of a large regular graph. Linear Alg. Appl. **40** (1981), 203–216.
158. P.-A. Meyer: *Quantum Probability for Probabilists.* Berlin: Springer-Verlag, 1993 (Lecture Notes in Math., Vol. 1538).
159. W. Młotkowski: Operator-valued version of conditionally free product. Studia Math. **153** (2004), 13–30.
160. W. Młotkowski: Λ-free probability. Infin. Dimens. Anal. Quantum Probab. Relat. Topics **7** (2004), 27–41.
161. W. Młotkowski: Limit theorems in Λ-Boolean probability. Infin. Dimens. Anal. Quantum Probab. Relat. Topics **7** (2004), 449–459.
162. W. Młotkowski, R. Szwarc: Nonnegative linearization for polynomials orthogonal with respect to discrete measures. Constr. Approx. **17** (2001), 413–429.
163. N. Muraki: A new example of noncommutative 'de Moivre–Laplace theorem.' In: *Probability Theory and Mathematical Statistics.* S. Watanabe, M. Fukushima, Yu. V. Prohorov, A. N. Shiryarev (ed). River Edge, NJ: World Scientific, 1996, pp. 353–362.
164. N. Muraki: Noncommutative Brownian motion in monotone Fock space. Comm. Math. Phys. **183** (1997), 557–570.
165. N. Muraki: Monotonic independence, monotonic central limit theorem and monotonic law of small numbers. Infin. Dimens. Anal. Quantum Probab. Relat. Topics **4** (2001), 39–58.

166. N. Muraki: The five independences as quasi-universal products. Infin. Dimens. Anal. Quantum Probab. Relat. Topics **5** (2002), 113–134.
167. N. Muraki: Monotonic convolution and monotonic Lévy-Hinčin formula. Unpublished manuscript, 2000.
168. N. Muraki: The five independences as natural products. Infin. Dimens. Anal. Quantum Probab. Relat. Topics **6** (2003), 337–371.
169. N. Obata: Quantum probabilistic approach to spectral analysis of star graphs. Interdiscip. Inform. Sci. **10** (2004), 41–52.
170. N. Obata: Notions of independence in quantum probability and spectral analysis of graphs. Sugaku Expositions in press.
171. N. Obata: Positive Q-matrices of graphs. Studia Math. **179** (2007), 81–97.
172. A. Okounkov: Random matrices and random permutations. Int. Math. Res. Notices **2000** (2000), 1043–1095.
173. A. Okounkov, A. Vershik: A new approach to representation theory of symmetric groups. Selecta Math. (N.S.) **2** (1996), 581–605.
174. F. Oravecz: Fermi convolution. Infin. Dimens. Anal. Quantum Probab. Relat. Topics **5** (2002), 235–242.
175. A. Papoulis: *Probability, Random Variable, mad Stochastic Processes*, 2nd edn. New York: McGraw-Hill, 1984.
176. K. R. Parthasarathy: *An Introduction to Quantum Stochastic Calculus*. Cambridge, MA: Birkhäuser, 1992.
177. G. K. Pedersen: *C^*-Algebras and Their Automorphism Groups*, London–New York: Academic Press, 1979. (London Math. Soc. Monographs, Vol. 14).
178. N. Privault: Quantum stochastic calculus for the uniform measure and Boolean convolution. In: *Séminaire de Probabilités XXXV*. Berlin: Springer, 2001, pp. 28–47 (Lecture Notes in Math., Vol. 1755).
179. D. P. Proskurin: On monotone independence families of operators. Methods Funct. Anal. Topology **10** (2004), 64–68.
180. D. P. Proskurin, A. M. Iksanov: The interpolation between the classical and monotone independence given by the twisted CCR. Opuscula Math. **23** (2003), 63–69.
181. M. Reed, B. Simon: *Methods of Modern Mathematical Physics I: Functional Analysis*. New York: Academic Press, 1980.
182. B. E. Sagan: *The Symmetric Group: Representations, Combinatorial Algorithms, and Symmetric Functions*, 2nd edn. Berlin: Springer, 2001 (Graduate Texts in Mathematics, Vol. 203).
183. N. Saitoh, H. Yoshida: A q-deformed Poisson distribution based on orthogonal polynomials. J. Phys. A: Math. Gen. **33** (2000), 1435–1444.
184. N. Saitoh, H. Yoshida: q-deformed Poisson random variables on q-Fock space. J. Math. Phys. **41** (2000), 5767–5772.
185. N. Saitoh, H. Yoshida: The infinite divisibility and orthogonal polynomials with a constant recursion formula in free probability theory. Probab. Math. Statist. **21** (2001), 159–170.
186. W. Schoutens: *Stochastic Processes and Orthogonal Polynomials*. Berlin: Springer-Verlag, 2000 (Lecture Notes in Stat., Vol. 146).
187. M. Schürmann: *White Noise on Bialgebras*. Berlin: Springer-Verlag, 1993 (Lecture Notes in Math., Vol. 1544).
188. M. Schürmann: Direct sums of tensor products and non-commutative independence. J. Funct. Anal. **133** (1995), 1–9.

189. J.-P. Serre: *Trees.* Berlin: Springer-Verlag, 2003 (Springer Monographs in Mathematics).
190. J. A. Shohat, J. D. Tamarkin: *The Problem of Moments.* Providence, RI: Amer. Math. Soc., 1943.
191. R. Simion, D. Ullman: On the structure of the lattice of noncrossing partitions. Discr. Math. **98** (1991), 193–206.
192. B. Simon: *Representations of Finite and Compact Groups*, Graduate Studies in Mathematics, Vol. 10. Providence, RI: Amer. Math. Soc., 1996.
193. B. Simon: *Orthogonal Polynomials on the Unit Circle*, Colloq. Publ. Vol. 54. Providence, RI: Amer. Math. Soc., 2005.
194. P. Śniady: Asymptotics of characters of symmetric groups, genus expansion and free probability. Discrete Math. **306** (2006), 624–665.
195. R. Speicher: A new example of 'independence' and 'white noise.' Probab. Theory Relat. Fields **84** (1990), 141–159.
196. R. Speicher: On universal products. In: *Free Probability Theory*, D. Voiculescu (ed). Providence, RI: Amer. Math. Soc., 1997, pp. 257–266 (Fields Inst. Commun. Vol. 12).
197. R. Speicher: *Combinatorial Theory of the Free Product with Amalgamation and Operator-Valued Free Probability*, Providence, RI: Amer. Math. Soc., 1998 (Mem. Amer. Math. Soc., Vol. 132).
198. R. Speicher: Free calculus. In: *Quantum Probability Communications, Vol. XII*, J. M. Lindsay, S. Attal (ed). River Edge, NJ: World Scientific 2003, pp. 209–235.
199. R. Speicher, R. Woroudi: Boolean convolution. In: *Free Probability Theory*, D. Voiculescu (ed). Amer. Math. Soc., Providence: RI, 1997, pp. 267–279 (Fields Inst. Commun. Vol. 12).
200. R. P. Stanley: Some combinatorial properties of Jack symmetric functions. Adv. Math. **77** (1989), 76–115.
201. G. Stoica: Limit laws for normed and weighted Boolean convolutions. J. Math. Anal. Appl. **309** (2005), 369–374.
202. G. Szegö: *Orthogonal Polynomials*, 4th edn. Providence, RI: Amer. Math. Soc., 1975.
203. R. Szwarc: Structure of geodesics in the Cayley graph of infinite Coxeter groups. Colloq. Math. **95** (2003), 79–90.
204. M. Takesaki: *Theory of Operator Algebra I.* Berlin: Springer-Verlag, 2002 (Encyclopaedia of Mathematical Sciences, Vol. 124).
205. I. Terada, K. Harada: *Group Theory.* Iwanami Shoten, Tokyo, 1997 (in Japanese).
206. H. van Leeuwen, H. Maassen: A q deformation of the Gauss distribution. J. Math. Phys. **36** (1995), 4743–4756.
207. H. Urakawa: Heat kernel and Green kernel comparison theorems for infinite graphs. J. Funct. Anal. **146** (1997), 206–235.
208. H. Urakawa: The Cheeger constant, the heat kernel, and the Green kernel of an infinite graph. Monatsh. Math. **138** (2003), 225–237.
209. A. M. Vershik: Asymptotic combinatorics and algebraic analysis. In: *Proc. Int. Congress Mathematicians (Zürich, 1994), Vol. 2*, (Birkhäuser, Basel, 1995), pp. 1384–1394.
210. A. M. Vershik: Two lectures on the asymptotic representation theory and statistics of Young diagrams. In: *Asymptotic Combinatorics with Applications to*

Mathematical Physics. A. M. Vershik (ed). Berlin: Springer, 2003, pp. 161–182 (Lecture Notes in Math. Vol. 1815).
211. A. M. Vershik, S. V. Kerov: Asymptotics of the Plancherel measure of the symmetric group and the limiting form of Young tables. Soviet Math. Dokl. **18** (1977), 527–531.
212. A. M. Vershik, S. V. Kerov: Asymptotic theory of characters of the symmetric group. Funct. Anal. Appl. **15** (1981), 246–255.
213. D. Voiculescu: Symmetries of some reduced free C^*-algebras. In: *Operator Algebras and their Connections with Topology and Ergodic Theory.* H. Araki et al. (ed). Berlin: Springer-Verlag, 1985, pp. 556–588 (Lecture Notes in Math., Vol. 1132).
214. D. Voiculescu: Addition of non-commuting random variables. J. Funct. Anal. **66** (1986), 323–346.
215. D. Voiculescu: Free noncommutative random variables, random matrices and the II_1 factors of free groups. In: *Quantum Probability and Related Fields, Vol. VI.* L. Accardi et al. (ed). River Edge, NJ: World Scientific, 1991, pp. 473–487.
216. D. V. Voiculescu, K. J. Dykema, A. Nica: *Free Random Variables*, CRM Monograph Series Vol. 1 Providence, RI: Amer. Math. Soc., 1992.
217. M. Voit: Central limit theorems for random walks on $\mathbf{N}_0$ that are associated with orthogonal polynomials. J. Multivariate Anal. **34** (1990), 290–322.
218. M. Voit: A product formula for orthogonal polynomials associated with infinite distance-transitive graphs. J. Approx. Theory **120** (2003), 337–354.
219. J. von Neumann: *Mathematischen Grundlagen der Quantenmechanik.* Berlin: Springer-Verlag, 1932.
220. W. von Waldenfels: An approach to the theory of pressure broadening of spectral lines. In: *Probability and Information Theory II.* Berlin: Springer, 1973, pp. 19–69 (Lecture Notes in Math., Vol. 296).
221. W. von Waldenfels: Interval partitions and pair interactions. In: *Séminaire de Probabilités IX.* P.-A. Meyer (ed). Berlin: Springer-Verlag, 1975, pp. 565–588 (Lecture Notes in Math., Vol. 465).
222. H. S. Wall: *Analytic Theory of Continued Fractions.* Amer. Math. Soc. Chelsea Pub., 1948.
223. E. P. Wigner: Characteristic vectors of bordered matrices with infinite dimensions. Ann. Math. **62**(2) (1955), 548–564.
224. E. P. Wigner: Characteristic vectors of bordered matrices with infinite dimensions II. Ann. Math. **65**(2) (1957), 203–207.
225. E. P. Wigner: On the distribution of the roots of certain symmetric matrices. Ann. Math. **67**(2) (1958), 325–327.
226. W. Woess: *Random Walks on Infinite Graphs and Groups.* Cambridge, U.K.: Cambridge Univ. Press, 2000.
227. J. Wysoczański: Monotonic independence on the weakly monotone Fock space and related Poisson type theorem. Infin. Dimens. Anal. Quantum Probab. Relat. Topics **8** (2005), 259–275.
228. H. Yoshida: Remarks on the s-free convolution. In: *Non-Commutativity, Infinite-Dimensionality and Probability at the Crossroads*, N. Obata, T. Matsui, A. Hora (ed). River Edge, NJ: World Scientific, 2002, pp. 412–433 (QP–PQ: Quantum Probab. White Noise Anal., Vol. 16).
229. D. V. Znoĭko: Free products of nets and free symmetrizers of graphs. Math. Sb. (N.S.) **98**(140) (1975), 463–480 (English translation).

Index

Theoretical and Mathematical Physics

Quantum Probability and Spectral Analysis of Graphs
By A. Hora and N. Obata

From Nucleons to Nucleus
Concepts of Microscopic Nuclear Theory
By J. Suhonen

Concepts and Results in Chaotic Dynamics: A Short Course
By P. Collet and J.-P. Eckmann

The Theory of Quark and Gluon Interactions
4th Edition
By F. J. Ynduráin

From Microphysics to Macrophysics
Methods and Applications of Statistical Physics
Volume I, Study Edition
By R. Balian

From Microphysics to Macrophysics
Methods and Applications of Statistical Physics
Volume II, Study Edition
By R. Balian

Titles published before 2006 in *Texts and Monographs in Physics*

The Statistical Mechanics of Financial Markets
3rd Edition
By J. Voit

Magnetic Monopoles
By Y. Shnir

Coherent Dynamics of Complex Quantum Systems
By V. M. Akulin

Geometric Optics on Phase Space
By K. B. Wolf

General Relativity
By N. Straumann

Quantum Entropy and Its Use
By M. Ohya and D. Petz

Statistical Methods in Quantum Optics 1
By H. J. Carmichael

Operator Algebras and Quantum Statistical Mechanics 1
By O. Bratteli and D. W. Robinson

Operator Algebras and Quantum Statistical Mechanics 2
By O. Bratteli and D. W. Robinson

Aspects of Ergodic, Qualitative and Statistical Theory of Motion
By G. Gallavotti, F. Bonetto and G. Gentile

The Frenkel-Kontorova Model
Concepts, Methods, and Applications
By O. M. Braun and Y. S. Kivshar

The Atomic Nucleus as a Relativistic System
By L. N. Savushkin and H. Toki

The Geometric Phase in Quantum Systems
Foundations, Mathematical Concepts, and Applications in Molecular and Condensed Matter Physics
By A. Bohm, A. Mostafazadeh, H. Koizumi, Q. Niu and J. Zwanziger

Relativistic Quantum Mechanics
2nd Edition
By H. M. Pilkuhn

Physics of Neutrinos
and Applications to Astrophysics
By M. Fukugita and T. Yanagida

High-Energy Particle Diffraction
By E. Barone and V. Predazzi

Foundations of Fluid Dynamics
By G. Gallavotti

Many-Body Problems and Quantum Field Theory An Introduction
2nd Edition
By Ph. A. Martin, F. Rothen, S. Goldfarb and S. Leach

Statistical Physics of Fluids
Basic Concepts and Applications
By V. I. Kalikmanov

Statistical Mechanics A Short Treatise
By G. Gallavotti

Quantum Non-linear Sigma Models
From Quantum Field Theory
to Supersymmetry, Conformal Field Theory,
Black Holes and Strings
By S. V. Ketov

Perturbative Quantum Electrodynamics and Axiomatic Field Theory
By O. Steinmann

The Nuclear Many-Body Problem
By P. Ring and P. Schuck

Magnetism and Superconductivity
By L.-P. Lévy

Information Theory and Quantum Physics
Physical Foundations for Understanding the Conscious Process
By H. S. Green

Quantum Field Theory in Strongly Correlated Electronic Systems
By N. Nagaosa

Quantum Field Theory in Condensed Matter Physics
By N. Nagaosa

Conformal Invariance and Critical Phenomena
By M. Henkel

Statistical Mechanics of Lattice Systems
Volume 1: Closed-Form and Exact Solutions
2nd Edition
By D. A. Lavis and G. M. Bell

Statistical Mechanics of Lattice Systems
Volume 2: Exact, Series and Renormalization Group Methods
By D. A. Lavis and G. M. Bell

Fields, Symmetries, and Quarks
2nd Edition
By U. Mosel

Renormalization An Introduction
By M. Salmhofer

Multi-Hamiltonian Theory of Dynamical Systems
By M. Błaszak

Quantum Groups and Their Representations
By A. Klimyk and K. Schmüdgen

Quantum The Quantum Theory of Particles, Fields, and Cosmology
By E. Elbaz

Effective Lagrangians for the Standard Model
By A. Dobado, A. Gómez-Nicola, A. L. Maroto and J. R. Peláez

Scattering Theory of Classical and Quantum *N*-Particle Systems
By. J. Derezinski and C. Gérard

Quantum Relativity A Synthesis of the Ideas of Einstein and Heisenberg
By D. R. Finkelstein

The Mechanics and Thermodynamics of Continuous Media
By M. Šilhavý

Local Quantum Physics Fields, Particles, Algebras
2nd Edition
By R. Haag

Relativistic Quantum Mechanics and Introduction to Field Theory
By F. J. Ynduráin

Supersymmetric Methods in Quantum and Statistical Physics
By G. Junker

Path Integral Approach to Quantum Physics An Introduction
2nd printing
By G. Roepstorff

Finite Quantum Electrodynamics
The Causal Approach
2nd edition
By G. Scharf

From Electrostatics to Optics
A Concise Electrodynamics Course
By G. Scharf

Geometry of the Standard Model of Elementary Particles
By A. Derdzinski

Quantum Mechanics II
By A. Galindo and P. Pascual

Generalized Coherent States and Their Applications
By A. Perelomov

The Elements of Mechanics
By G. Gallavotti

Essential Relativity Special, General, and Cosmological Revised
2nd edition
By W. Rindler

MIX
Papier aus verantwortungsvollen Quellen
Paper from responsible sources
FSC® C105338

If you have any concerns about our products,
you can contact us on
ProductSafety@springernature.com

In case Publisher is established outside the EU,
the EU authorized representative is:
Springer Nature Customer Service Center GmbH
Europaplatz 3, 69115 Heidelberg, Germany

Printed by Libri Plureos GmbH
in Hamburg, Germany